Advances in the Astonishing World of Phytochemicals: State-of-the-Art for Antioxidants

Advances in the Astonishing World of Phytochemicals: State-of-the-Art for Antioxidants

Editors

Antonella D'Anneo
Marianna Lauricella

Basel • Beijing • Wuhan • Barcelona • Belgrade • Novi Sad • Cluj • Manchester

Editors
Antonella D'Anneo
University of Palermo
Palermo
Italy

Marianna Lauricella
University of Palermo
Palermo
Italy

Editorial Office
MDPI
St. Alban-Anlage 66
4052 Basel, Switzerland

This is a reprint of articles from the Special Issue published online in the open access journal *Antioxidants* (ISSN 2076-3921) (available at: https://www.mdpi.com/journal/antioxidants/special_issues/Advances_Phytochemicals_State_Art_Antioxidants).

For citation purposes, cite each article independently as indicated on the article page online and as indicated below:

Lastname, A.A.; Lastname, B.B. Article Title. *Journal Name* **Year**, *Volume Number*, Page Range.

ISBN 978-3-0365-8604-5 (Hbk)
ISBN 978-3-0365-8605-2 (PDF)
doi.org/10.3390/books978-3-0365-8605-2

Contents

About the Editors

Antonella D'Anneo

Antonella D'Anneo is an Associate Professor in Biochemistry at the Department of Biological, Chemical and Pharmaceutical Sciences and Technologies (STEBICEF), University of Palermo. She is also a Lecturer in Biochemistry for the Bachelor's degree in Biological Sciences, University of Palermo. She is a member of the Italian Society of Biochemistry and Molecular Biology (SIB). She is also the Associate Editor in Chief of *Cancer Management and Research* and Associate Editor for *Frontiers in Pharmacology*. She began her studies in the field of tumor biology and cell death mechanisms (apoptosis, necrosis, necroptosis, autophagy and anoikis), activated by natural and synthetic compounds, and biochemical pathways that can be activated in tumor cells. To complement her graduate training in biochemistry, she trained, as a visiting scholar, at Rangos Research Center (University of Pittsburgh, USA), taking part in projects concerning gene therapy approaches in order to determine the tolerization of diabetic patients against the cells of islet donors. More recently, her studies have focused on the identification of the nutraceutical properties (anti-tumor, anti-inflammatory, anti-obesity and antioxidant) of phytochemicals in order to disclose their biological activity and potential medicinal use. She is the author of 70 scientific publications in ISI-indexed journals.

Marianna Lauricella

Marianna Lauricella is an Associate Professor in Biochemistry at the Department of Biomedicine, Neuroscience and Advanced Diagnostics, School of Medicine, University of Palermo. She is a Lecturer in Biochemistry on the master's degree Course in Medicine, University of Palermo. She is a member of the Italian Society of Biochemistry and Molecular Biology (SIB). She is an Associate Editor of *BMC Cancer* and an Editorial Board Member of *The International Journal of Molecular Sciences*. Her research activity mainly concerns the study of the mechanisms governing the control of survival and death in cultured tumor cells. In particular, the aim of her research is to identify natural and synthetic compounds that can selectively induce cell death processes (apoptosis, autophagy, necroptosis). She also has expertise in the study of the role exerted by oxidative stress in inducing obesity and the anti-obesity effects of natural compounds in in vitro models of obesity. She is the author of over 75 scientific publications in ISI-indexed journals.

 *antioxidants*

Editorial

Advances in the Astonishing World of Phytochemicals: State-of-the-Art for Antioxidants

Marianna Lauricella [1],* and Antonella D'Anneo [2],*

[1] Department of Biomedicine, Neurosciences and Advanced Diagnostics (BIND), Institute of Biochemistry, University of Palermo, 90127 Palermo, Italy
[2] Laboratory of Biochemistry, Department of Biological, Chemical and Pharmaceutical Sciences and Technologies (STEBICEF), University of Palermo, 90127 Palermo, Italy
* Correspondence: marianna.lauricella@unipa.it (M.L.); antonella.danneo@unipa.it (A.D.)

Citation: Lauricella, M.; D'Anneo, A. Advances in the Astonishing World of Phytochemicals: State-of-the-Art for Antioxidants. *Antioxidants* **2023**, *12*, 1581. https://doi.org/10.3390/antiox12081581

Received: 19 July 2023
Revised: 28 July 2023
Accepted: 4 August 2023
Published: 8 August 2023

In recent years, research on phytochemicals has underscored pleiotropic actions and medicinal and health-promoting properties which certainly deserve serious attention. Natural-derived molecules, such as phytohormones, glycosides, terpenoids, alkaloids, and phenolic compounds, offer a protective or preventative shield against many several pathological conditions such as aging, cardiovascular diseases, diabetes, obesity, cancer, asthma, and neurodegenerative disorders [1,2]. On the other hand, the multi-faceted potentials of phytocompounds isolated from different parts of plants or fruits stimulate the interest of the pharmaceutical, nutraceutical, and cosmetic industries. A main goal of these companies is to identify new and innovative phytomolecules to use as they are as natural reservoirs in plants or to appropriately modify them with the insertion of pharmacophore groups to design enhanced derivatives [3].

In the light of these considerations, we put together this Special Issue, titled "Advances in the Astonishing World of Phytochemicals: State-of-the-Art for Antioxidants", containing seventeen papers (fourteen research articles, one review, one comment, and a reply).

The scientific evidence reported in the SI analyzed both distribution and pleiotropic beneficial effects (antidiabetic, antitumor, antiflogistic, antibacterial, etc.) of some bioactive compounds with antioxidant properties. However, it also has to be considered that the relative abundance as well as the distribution of phytochemicals in plants or fruits is consistently affected by different parameters, such as environmental edaphic conditions, ripeness degree of fruits, and right season harvest [4].

In their contribution, Ali et al., for example, demonstrated that Australian fruits and spices such as mountain pepper berries (*Tasmannia lanceolata*), rosella (*Hibiscus sabdariffa*), lemon aspen (*Acronychia acidula*), and strawberry gum (*Eucalyptus olida*) represent a rich reservoir of bioactive phenolic metabolites (phenolic acids, flavonoids, isoflavonoids, tannins, stilbenes, lignans, and limonoids). Among these, the analysis provided evidence that both *Eucalyptus olida* and *Tasmannia lanceolata* possess the highest antioxidant and antidiabetic potential [5], a property that could be exploited in the development of specific biopharmaceuticals.

The effect of environmental conditions on the content and quality of phytochemicals was recently reported in studies performed on Amaranth, a leafy vegetable capable of growing under several salinity and drought-stress-induced conditions [6]. Salt stress has been demonstrated to enrich the amount of bioactive compounds with antioxidant properties. Indeed, the application of salt eustress conditions (25–100 mM NaCl) was able to boost the profile of microelements, macro-elements, phytochemicals, and phenolic acids in *Amaranthus gangeticus*, contributing to providing excellent quality in the end product for its antioxidant properties [7].

Studies performed on Romanian *Armoracia rusticana* L., a horseradish plant widely appreciated for its medicinal and aromatic properties, offered a complete profile of the low-molecular-weight metabolites of the plant grown in Romania. Nine categories of secondary metabolites (glucosilates, fatty acids, isothiocyanates, amino acids, phenolic acids, flavonoids, terpenoids, coumarins, and miscellaneous) were identified, and the development of phyto-engineered carrier systems capable of merging the biological properties of horseradish and kaolinite was proposed [8]. As a whole, the conclusion is that these systems could represent a possible controlled drug release system to apply to cancer-specific targeting.

In another study, Rani et al. explored the biological potential of dichloromethane and methanol root and shoot extracts of *Dryopteris juxtapostia*, a species belonging to the *Dryopteris* genus growing in the states of the north temperate zone. The study demonstrated that both extracts exerted radical-scavenging and anti-inflammatory and antitumoral effects in vitro as well as hepatoprotective actions in rats. *D. juxtapostia* root dichloromethane extracts exhibited the highest biological potential compared to other extracts, thus demonstrating the importance of using dichloromethane to obtain extracts enriched in phenolic components [9].

In addition, Vieira et al. demonstrated the anti-inflammatory effects of roots and flowers extracts of *E. purpurea*, a plant whose extracts are traditionally used to treat cold and flu. The study compared the effects of dichloromethanolic and ethanolic root and flowers extracts with alkylamide-rich extracts obtained by using the accelerated solvent extractor system, a green and innovative extraction technique. The authors concluded that all the extracts were capable of reducing the IL-6 levels as well as the intracellular levels of ROS/RNS in lipopolysaccharide-stimulated human-monocyte-derived macrophages. However, the alkylamide fractions possessed the strongest anti-inflammatory effects, thus evidencing these compounds as the main active extract components [10].

A fruit particularly rich in phytocompounds is tomato (*Lycopersicon esculentum* Mill.), a food largely consumed for its nutritive and nutraceutical properties [11]. Noteworthy, the different phytonutrient composition and antioxidant properties of the tomato are related to the different ripening times. On these bases, the study of Gambino et al. compared the different phytonutrients composition and properties of golden tomato (GT), a food product harvested at an incomplete ripening stage with respect to red tomato (RT), harvested at full maturation [12]. The authors demonstrated that GT contains a higher level of naringenin and chlorogenic acid, two polyphenols with antilipemic effects [13,14], than RT. Regarding biological activities, GT displays a better reducing power compared to RT [15]. Interestingly, GT oral supplementation in high-fed rats reduced body-weight gain and LDL cholesterol levels, as well as lowered oxidative stress markers both in the blood and liver, thus suggesting a potential of "GT" oral supplementation.

The biological properties of *Urtica dioica* (UD), *Matricaria chamomilla* (MC), and *Murraya koenigii* (MK), traditionally used in Ayurvedic medicine as nerve relaxants and cognition enhancers [16], were evaluated in the study of Shabir et al. [17]. Considering the effects of these plants on the nervous system, the authors investigated the ability of aqueous and ethanolic extracts of UD, MC, and MK to ameliorate the toxic effects of rotenone, a neurotoxic natural pesticide, in wild-type *Drosophila melanogaster*. The study evidenced the ability of plant extracts to exert neuroprotective effects on *Drosophila melanogaster* by alleviating rotenone-induced oxidative stress, enhancing locomotion, and restoring acetylcholine levels, thus suggesting a potential use of these extracts to treat neurological diseases. Of course, the right recovery of phytochemicals also depends either on the type of extraction techniques or solvents applied in the extraction procedure. This aspect was clearly addressed by Boyadzhieva et al., demonstrating a good recovery efficiency of phytochemicals from different parts (leaves, flowers, and stems) of *Gnaphalium viscosum* (Kunth, such as the antioxidants kaempferol, kaempferol-3-O-β-d-glucoside, and chlorogenic acid). Interestingly, for the first time, this study also demonstrated the presence in this species of leontopodic acids A and B, two highly potent antioxidants derived from glucaric acid [18].

In a study performed in yarrow (*Achillea millefolium L.*), a flowering plant commonly used in folk medicine to alleviate symptoms related to gastrointestinal discomfort [19],

Villalva at al. used a supercritical antisolvent fractionation process [20] to obtain two different fractions containing polar phenolic compounds and monoterpenes and sesquiterpenes, respectively. Both the fractions explained the antibacterial effects observed against *Helicobacter pylori* strains. Furthermore, the extracted fractions exerted antioxidant and anti-inflammatory effects in *Helicobacter pylori*-infected human gastric AGS cells. From this study, we can conclude that yarrow extracts can be useful against *Helicobacter pylori* infection. The Villalva's data have been criticized by Franski and Beszterda-Buszczak [21]. Although they do not question the quality of the paper, these authors raised questions about the correctness of some compounds identified by Villalva et al. However, Villalva clarified all the doubts of Franski and Beszterda-Buszczak in a reply paper [22].

Nowadays, there is great interest in the bio-waste products of agriculture for the presence of bioactive healthy compounds [23,24]. Thinning young apples (TAPs) are usually discarded to guarantee the output and to increase the quality of harvested apples. However, it has been shown that TAPs contain more than 10-fold polyphenols with respect to harvested apples [25]. In their contribution, Ferrario et al. characterized the profile of polyphenols in TAP using a dual LC-HRMS metabolomic approach to identify a total of 68 polyphenols. According to this investigation, TAP fractions exert both antioxidant and anti-inflammatory effects by up-regulating the nuclear-factor-erythroid-2-related factor (Nrf2) signaling pathways and inhibiting NF-kB activation in cell models [26]. These results evidenced TAP as a source of bioactive molecules endowed with antioxidant properties.

The presence of bioactive compounds has also been Identified in marine environments. For example, seaweeds, such as red (*Rhodophyta*), green (*Chlorophyta*), and brown algae (*Phaeophyta*), which are not included in the diet of the Western world, are widely spread in Asian and Chinese nutrition for their high-quality profile in bioactive molecules as phenolic compounds, vitamins, pigments, and essential minerals. The use of a green pressured liquid extraction technique allowed Perez-Vazquez et al., under specific experimental conditions of temperature, type of used solvent, extraction time, and pressure, to recover a high yield of active biomolecules to exploit on both a pharmaceutical and food industrial scale [27].

Notably, a recent study of Liberti et al. demonstrated the antioxidant and anti-inflammatory properties of sulfated exopolysaccharides (s-EPSs) and phycoerythrin (PE), two molecules naturally produced by the red marine microalga *Porphyridium cruentum*. In particular, s-EPSs were able to prevent GSH depletion and lipid peroxidation on a cell-based system but not in vitro, while PE showed high ROS scavenging capacity both in vitro and on a cell-based system. Interestingly, both the compounds were capable of inhibiting the pro-inflammatory enzyme COX-2 and promoting a fast scratch closure [28]. Altogether, the data obtained support the use of these compounds isolated by *P. cruentum* as anti-inflammatory components of medical patches.

The identification of phytomolecules with potential tailored applications represents a significant goal in the phytochemistry field. Particularly significant is the research discussed by Notaro et al. exploring the biochemical action of methyl gallate (MG), a gallotannin widely used in traditional Chinese phytotherapy to alleviate several cancer symptoms [29]. The findings reported by the authors shed light on the antitumor potential of MG. This phytocompound preferentially targeted HCT116 colon cancer cells, with respect to differentiated Caco-2 cells, an enterocyte-like cell model. In colon cancer cells, MG induced an oxidative injury sustained by ROS generation and endoplasmic reticulum stress as well as an upregulation in intracellular calcium content. In the first phase of treatment, oxidative events were accompanied by an autophagic process, that, for longer times of incubation, culminated in the apoptotic cell demise with DNA fragmentation and p53 and H2Ax activation. A particular role in the MG-induced mechanism was played by the oncosuppressor p53 protein. The conclusion of this research revealed the existence of an intertwined relationship between oxidative stress and p53 as a causative event in apoptotic cell death. Such a study paves the way to future investigations of MG alone or in combination treatment as a preventative or adjuvant phytocompound to apply in colon cancer treatment.

However, beyond these effects, bioactive compounds present in plants have also been demonstrated to play a protective role against oxidative injury, an aspect recently studied by Lv et al. in Caco-2 cells. The use of proanthocyanidins purified from kiwi leaves (*Actinidia chinensis*) counteracted both H_2O_2-induced oxidative damage as well as malondialdehyde increase [30]. Such an effect was accompanied with an upregulation of antioxidant systems (GSH-px, CAT, T-SOD) and the corresponding mRNA targets of Nrf2, the master regulator of the cellular stress response [31]. The conclusion of this interesting study is that the characterization of the antioxidant properties of kiwi leaves proanthocyanidins emphasizes their possible functional application either for a policy of circular economy or for sustainable industrial use.

The whole antioxidant activity of a sample cannot be ascribed only to a single bioactive component, but in many cases the overall potential is the result of the combinatorial effect of more components, acting in a synergistic, antagonistic, or additive manner. The comparative analysis of 10 phenolic acids (protocatechuic, gentisic, gallic, vanillic, syringic, p-coumaric, caffeic, ferulic, sinapic, and rosmarinic acid) used alone and in different combination mixtures provided evidence of the high antioxidant activity of gallic acid by a ferric reducing antioxidant power (FRAP) technique and a good oxygen radical absorbance capacity of rosmarinic acid by ORAC assays [32]. A relevant aspect of this study relied on the observation that hydroxybenzoic acid mixtures containing gentisic acid showed a clear synergistic action. These data strongly sustain the idea that the biological activity of a mixture, in some cases, cannot be ascribed to a single compound, but it has to be searched in the combination of compounds present and their ability to interact with each other.

We would like to share our gratitude to all authors who submitted their outstanding research to this Special Issue. Their manuscripts highlighted the role of natural-derived compounds with antioxidant potential action to apply as preventative or adjuvant molecules in the treatment of some chronic human diseases. Additionally, the identification of extraction techniques and solvents that can maximize the extraction of bioactive compounds is of great importance.

Author Contributions: Conceptualization, M.L. and A.D.; writing—original draft preparation, A.D. and M.L.; review and editing A.D and M.L. All authors have read and agreed to the published version of the manuscript.

Acknowledgments: We thank all the authors who contributed to the research topic of this Special Issue and reviewers for the perceptive suggestions and comments. A particular acknowledgment is for all Editorial staff from *Antioxidants* journal for their huge support in the preparation of this Special Issue.

Conflicts of Interest: The authors declare no conflict of interest.

References

1. Akula, R.; Ravishankar, G.A. Influence of Abiotic Stress Signals on Secondary Metabolites in Plants. *Plant Signal. Behav.* **2011**, *6*, 1720–1731. [CrossRef] [PubMed]
2. Forni, C.; Facchiano, F.; Bartoli, M.; Pieretti, S.; Facchiano, A.; D'Arcangelo, D.; Norelli, S.; Valle, G.; Nisini, R.; Beninati, S.; et al. Beneficial Role of Phytochemicals on Oxidative Stress and Age-Diseases. *BioMed Res. Int.* **2019**, *2019*, 8748253. [CrossRef] [PubMed]
3. Hidalgo, D.; Sanchez, R.; Lalaleo, L.; Bonfill, M.; Corchete, P.; Palazon, J. Biotechnological Production of Pharmaceuticals and Biopharmaceuticals in Plant Cell and Organ Cultures. *CMC* **2018**, *25*, 3577–3596. [CrossRef] [PubMed]
4. Manach, C.; Scalbert, A.; Morand, C.; Rémésy, C.; Jiménez, L. Polyphenols: Food Sources and Bioavailability. *Am. J. Clin. Nutr.* **2004**, *79*, 727–747. [CrossRef] [PubMed]
5. Ali, A.; Cottrell, J.J.; Dunshea, F.R. Antioxidant, Alpha-Glucosidase Inhibition Activities, In Silico Molecular Docking and Pharmacokinetics Study of Phenolic Compounds from Native Australian Fruits and Spices. *Antioxidants* **2023**, *12*, 254. [CrossRef]
6. Sarker, U.; Oba, S. Augmentation of Leaf Color Parameters, Pigments, Vitamins, Phenolic Acids, Flavonoids and Antioxidant Activity in Selected Amaranthus Tricolor under Salinity Stress. *Sci. Rep.* **2018**, *8*, 12349. [CrossRef]
7. Sarker, U.; Ercisli, S. Salt Eustress Induction in Red Amaranth (*Amaranthus gangeticus*) Augments Nutritional, Phenolic Acids and Antiradical Potential of Leaves. *Antioxidants* **2022**, *11*, 2434. [CrossRef]

8. Segneanu, A.-E.; Vlase, G.; Chirigiu, L.; Herea, D.D.; Pricop, M.-A.; Saracin, P.-A.; Tanasie, Ș.E. Romanian Wild-Growing *Armoracia rusticana* L.—Untargeted Low-Molecular Metabolomic Approach to a Potential Antitumoral Phyto-Carrier System Based on Kaolinite. *Antioxidants* **2023**, *12*, 1268. [CrossRef]
9. Rani, A.; Uzair, M.; Ali, S.; Qamar, M.; Ahmad, N.; Abbas, M.W.; Esatbeyoglu, T. Dryopteris Juxtapostia Root and Shoot: Determination of Phytochemicals; Antioxidant, Anti-Inflammatory, and Hepatoprotective Effects; and Toxicity Assessment. *Antioxidants* **2022**, *11*, 1670. [CrossRef]
10. Vieira, S.F.; Gonçalves, S.M.; Gonçalves, V.M.F.; Llaguno, C.P.; Macías, F.; Tiritan, M.E.; Cunha, C.; Carvalho, A.; Reis, R.L.; Ferreira, H.; et al. *Echinacea purpurea* Fractions Represent Promising Plant-Based Anti-Inflammatory Formulations. *Antioxidants* **2023**, *12*, 425. [CrossRef]
11. Cruz-Carrión, Á.; Ruiz De Azua, M.J.; Bravo, F.I.; Aragonès, G.; Muguerza, B.; Suárez, M.; Arola-Arnal, A. Tomatoes Consumed In-Season Prevent Oxidative Stress in Fischer 344 Rats: Impact of Geographical Origin. *Food Funct.* **2021**, *12*, 8340–8350. [CrossRef]
12. Gambino, G.; Giglia, G.; Allegra, M.; Di Liberto, V.; Zummo, F.P.; Rappa, F.; Restivo, I.; Vetrano, F.; Saiano, F.; Palazzolo, E.; et al. "Golden" Tomato Consumption Ameliorates Metabolic Syndrome: A Focus on the Redox Balance in the High-Fat-Diet-Fed Rat. *Antioxidants* **2023**, *12*, 1121. [CrossRef]
13. Lee, M.-K.; Moon, S.-S.; Lee, S.-E.; Bok, S.-H.; Jeong, T.-S.; Park, Y.B.; Choi, M.-S. Naringenin 7-O-Cetyl Ether as Inhibitor of HMG-CoA Reductase and Modulator of Plasma and Hepatic Lipids in High Cholesterol-Fed Rats. *Bioorganic Med. Chem.* **2003**, *11*, 393–398. [CrossRef] [PubMed]
14. Cho, A.-S.; Jeon, S.-M.; Kim, M.-J.; Yeo, J.; Seo, K.-I.; Choi, M.-S.; Lee, M.-K. Chlorogenic Acid Exhibits Anti-Obesity Property and Improves Lipid Metabolism in High-Fat Diet-Induced-Obese Mice. *Food Chem. Toxicol.* **2010**, *48*, 937–943. [CrossRef] [PubMed]
15. Di Majo, D.; La Neve, L.; La Guardia, M.; Casuccio, A.; Giammanco, M. The Influence of Two Different PH Levels on the Antioxidant Properties of Flavonols, Flavan-3-Ols, Phenolic Acids and Aldehyde Compounds Analysed in Synthetic Wine and in a Phosphate Buffer. *J. Food Compos. Anal.* **2011**, *24*, 265–269. [CrossRef]
16. Anand, U.; Tudu, C.K.; Nandy, S.; Sunita, K.; Tripathi, V.; Loake, G.J.; Dey, A.; Proćków, J. Ethnodermatological Use of Medicinal Plants in India: From Ayurvedic Formulations to Clinical Perspectives—A Review. *J. Ethnopharmacol.* **2022**, *284*, 114744. [CrossRef]
17. Shabir, S.; Yousuf, S.; Singh, S.K.; Vamanu, E.; Singh, M.P. Ethnopharmacological Effects of *Urtica dioica, Matricaria chamomilla*, and *Murraya koenigii* on Rotenone-Exposed *D. melanogaster*: An Attenuation of Cellular, Biochemical, and Organismal Markers. *Antioxidants* **2022**, *11*, 1623. [CrossRef]
18. Boyadzhieva, S.; Coelho, J.A.P.; Errico, M.; Reynel-Avilla, H.E.; Yankov, D.S.; Bonilla-Petriciolet, A.; Stateva, R.P. Assessment of *Gnaphalium viscosum* (Kunth) Valorization Prospects: Sustainable Recovery of Antioxidants by Different Techniques. *Antioxidants* **2022**, *11*, 2495. [CrossRef]
19. Garcia-Oliveira, P.; Barral, M.; Carpena, M.; Gullón, P.; Fraga-Corral, M.; Otero, P.; Prieto, M.A.; Simal-Gandara, J. Traditional Plants from Asteraceae Family as Potential Candidates for Functional Food Industry. *Food Funct.* **2021**, *12*, 2850–2873. [CrossRef]
20. Villalva, M.; Silvan, J.M.; Alarcón-Cavero, T.; Villanueva-Bermejo, D.; Jaime, L.; Santoyo, S.; Martinez-Rodriguez, A.J. Antioxidant, Anti-Inflammatory, and Antibacterial Properties of an *Achillea millefolium* L. Extract and Its Fractions Obtained by Supercritical Anti-Solvent Fractionation against *Helicobacter pylori*. *Antioxidants* **2022**, *11*, 1849. [CrossRef]
21. Frański, R.; Beszterda-Buszczak, M. Comment on Villalva et al. Antioxidant, Anti-Inflammatory, and Antibacterial Properties of an *Achillea millefolium* L. Extract and Its Fractions Obtained by Supercritical Anti-Solvent Fractionation against *Helicobacter pylori*. *Antioxidants* **2022**, *11*, 1849. *Antioxidants* **2023**, *12*, 1226. [CrossRef] [PubMed]
22. Villalva, M.; Silvan, J.M.; Alarcón-Cavero, T.; Villanueva-Bermejo, D.; Jaime, L.; Santoyo, S.; Martinez-Rodriguez, A.J. Reply to Frański, R.; Beszterda-Buszczak, M. Comment on "Villalva et al. Antioxidant, Anti-Inflammatory, and Antibacterial Properties of an *Achillea millefolium* L. Extract and Its Fractions Obtained by Supercritical Anti-Solvent Fractionation against *Helicobacter pylori*. *Antioxidants* **2022**, *11*, 1849.". *Antioxidants* **2023**, *12*, 1384. [CrossRef]
23. Pratelli, G.; Carlisi, D.; D'Anneo, A.; Maggio, A.; Emanuele, S.; Palumbo Piccionello, A.; Giuliano, M.; De Blasio, A.; Calvaruso, G.; Lauricella, M. Bio-Waste Products of *Mangifera indica* L. Reduce Adipogenesis and Exert Antioxidant Effects on 3T3-L1 Cells. *Antioxidants* **2022**, *11*, 363. [CrossRef] [PubMed]
24. Osorio, L.L.D.R.; Flórez-López, E.; Grande-Tovar, C.D. The Potential of Selected Agri-Food Loss and Waste to Contribute to a Circular Economy: Applications in the Food, Cosmetic and Pharmaceutical Industries. *Molecules* **2021**, *26*, 515. [CrossRef] [PubMed]
25. Zheng, H.-Z.; Kim, Y.-I.; Chung, S.-K. A Profile of Physicochemical and Antioxidant Changes during Fruit Growth for the Utilisation of Unripe Apples. *Food Chem.* **2012**, *131*, 106–110. [CrossRef]
26. Ferrario, G.; Baron, G.; Gado, F.; Della Vedova, L.; Bombardelli, E.; Carini, M.; D'Amato, A.; Aldini, G.; Altomare, A. Polyphenols from Thinned Young Apples: HPLC-HRMS Profile and Evaluation of Their Anti-Oxidant and Anti-Inflammatory Activities by Proteomic Studies. *Antioxidants* **2022**, *11*, 1577. [CrossRef]
27. Perez-Vazquez, A.; Carpena, M.; Barciela, P.; Cassani, L.; Simal-Gandara, J.; Prieto, M.A. Pressurized Liquid Extraction for the Recovery of Bioactive Compounds from Seaweeds for Food Industry Application: A Review. *Antioxidants* **2023**, *12*, 612. [CrossRef]
28. Liberti, D.; Imbimbo, P.; Giustino, E.; D'Elia, L.; Silva, M.; Barreira, L.; Monti, D.M. Shedding Light on the Hidden Benefit of Porphyridium Cruentum Culture. *Antioxidants* **2023**, *12*, 337. [CrossRef]

29. Notaro, A.; Lauricella, M.; Di Liberto, D.; Emanuele, S.; Giuliano, M.; Attanzio, A.; Tesoriere, L.; Carlisi, D.; Allegra, M.; De Blasio, A.; et al. A Deadly Liaison between Oxidative Injury and P53 Drives Methyl-Gallate-Induced Autophagy and Apoptosis in HCT116 Colon Cancer Cells. *Antioxidants* **2023**, *12*, 1292. [CrossRef]
30. Lv, J.-M.; Gouda, M.; Ye, X.-Q.; Shao, Z.-P.; Chen, J.-C. Evaluation of Proanthocyanidins from Kiwi Leaves (*Actinidia chinensis*) against Caco-2 Cells Oxidative Stress through Nrf2-ARE Signaling Pathway. *Antioxidants* **2022**, *11*, 1367. [CrossRef]
31. Hiebert, P. The Nrf2 Transcription Factor: A Multifaceted Regulator of the Extracellular Matrix. *Matrix Biol. Plus* **2021**, *10*, 100057. [CrossRef] [PubMed]
32. Skroza, D.; Šimat, V.; Vrdoljak, L.; Jolić, N.; Skelin, A.; Čagalj, M.; Frleta, R.; Generalić Mekinić, I. Investigation of Antioxidant Synergisms and Antagonisms among Phenolic Acids in the Model Matrices Using FRAP and ORAC Methods. *Antioxidants* **2022**, *11*, 1784. [CrossRef] [PubMed]

 *antioxidants*

Article

Antioxidant, Alpha-Glucosidase Inhibition Activities, In Silico Molecular Docking and Pharmacokinetics Study of Phenolic Compounds from Native Australian Fruits and Spices

Akhtar Ali [1], Jeremy J. Cottrell [1] and Frank R. Dunshea [1,2,*]

[1]   School of Agriculture and Food, The University of Melbourne, Parkville, VIC 3010, Australia
[2]   Faculty of Biological Sciences, The University of Leeds, Leeds LS2 9JT, UK
*   Correspondence: fdunshea@unimelb.edu.au

Abstract: Native Australian fruits and spices are enriched with beneficial phytochemicals, especially phenolic compounds, which are not fully elucidated. Therefore, this study aimed to analyze native Australian mountain-pepper berries (*Tasmannia lanceolata*), rosella (*Hibiscus sabdariffa*), lemon aspen (*Acronychia acidula*), and strawberry gum (*Eucalyptus olida*) for phenolic and non-phenolic metabolites and their antioxidant and alpha-glucosidase inhibition activities. Liquid chromatography–mass spectrometry–electrospray ionization coupled with quadrupole time of flight (LC-ESI-QTOF-MS/MS) was applied to elucidate the composition, identities, and quantities of bioactive phenolic metabolites in Australian native commercial fruits and spices. This study identified 143 phenolic compounds, including 31 phenolic acids, 70 flavonoids, 10 isoflavonoids, 7 tannins, 3 stilbenes, 7 lignans, 10 other compounds, and 5 limonoids. Strawberry gum was found to have the highest total phenolic content (TPC—36.57 ± 1.34 milligram gallic acid equivalent per gram (mg GAE/g), whereas lemon aspen contained the least TPC (4.40 ± 0.38 mg GAE/g). Moreover, strawberry gum and mountain pepper berries were found to have the highest antioxidant and anti-diabetic potential. In silico molecular docking and pharmacokinetics screening were also conducted to predict the potential of the most abundant phenolic compounds in these selected plants. A positive correlation was observed between phenolic contents and biological activities. This study will encourage further research to identify the nutraceutical and phytopharmaceutical potential of these native Australian fruits.

Keywords: mountain pepper; rosella; strawberry gum; lemon aspen; flavonoids; anthocyanins; bioavailability; LC-MS/MS

Citation: Ali, A.; Cottrell, J.J.; Dunshea, F.R. Antioxidant, Alpha-Glucosidase Inhibition Activities, In Silico Molecular Docking and Pharmacokinetics Study of Phenolic Compounds from Native Australian Fruits and Spices. *Antioxidants* **2023**, *12*, 254. https://doi.org/10.3390/antiox12020254

Academic Editors: Antonella D'Anneo and Marianna Lauricella

Received: 22 December 2022
Revised: 18 January 2023
Accepted: 19 January 2023
Published: 23 January 2023

1. Introduction

Diabetes mellitus is one of the leading causes of death around the globe [1] and is characterized by high blood glucose levels. Alpha-glucosidase (α-glucosidase) is the main enzyme with a significant role in hydroxylation, digestion, and absorption of sugars in the human body. Therefore, inhibiting α-glucosidase is an effective strategy for treating and minimizing type 2 diabetes. There is increasing interest in using natural sources to treat diabetes. Various nutraceuticals and bioactive compounds have been investigated to control/inhibit the complications of diabetes. Using phenolic metabolites is a therapeutic approach to suppressing the prevalence of pre- or post-diabetic conditions [1]. Therefore, detailed characterization and identification of phenolic metabolites is required to understand the potent role of polyphenols in food and human health.

Fruits, vegetables, herbs, spices, and medicinal plants contain large amounts of phytochemicals, including polyphenols. When they encounter living tissues, they exhibit a beneficial effect on human health [1]. Flavonoids are the largest subclass of polyphenols, with more than ten thousand compounds being reported [2]. According to nutritionists, foods high in polyphenols may reduce or remove the risk for certain malignancies, degenerative diseases, cardiovascular ailments, and chronic inflammation in humans [3].

Excessive production of reactive oxygen species (ROS) and reactive nitrogen species (RNS) leads to oxidative stress in the body, the leading cause of the above-mentioned pathological conditions [4]. Tasmannia lanceolata, commonly known as the mountain pepper berry, is used in food flavoring and traditional medicine to treat venereal diseases, skin disorders, and stomachaches [5]. Lemon aspen is a pale/yellow, 2–2.5 cm in diameter fruit that is endemic to Queensland, Australia. It is traditionally used in curries and on meats as a seasoning, though syrups, sauces, and infused vinegar are also made from lemon aspen [6]. Strawberry gum and rosella are also used in traditional medicine.

Considering that bioactive phenolic compounds have strong antioxidant and antidiabetic potential, it was hypothesized that selected native Australian plants could have considerable bioactive potential. In this context, we comprehensively characterized native Australian mountain pepper berries, rosella, lemon aspen, and strawberry gum for their phenolic compounds' antioxidant and α-glucosidase inhibition potential. Previously, these native Australian plants had not been studied for radical scavenging and α-glucosidase inhibition activities. Therefore, total monomeric anthocyanin content (TMAC), total phenolic content (TPC), total condensed tannins (TCT), and total flavonoid content (TFC) were measured. Furthermore, antioxidant activities, through the ferrous ion chelating assay (FICA), ferric reducing antioxidant power (FRAP) test, 2,2′-diphenyl-1-picrylhydrazyl (DPPH) reducing power assay (RPA), 2,2′-azinobis-(3-ethylbenzothiazoline-6-sulfonic acid (ABTS) assay, phosphomolybdate assay (PMA), and hydroxyl-radical scavenging assay ($^{\bullet}$OH-RSA), were also measured in this study. The anti-diabetic potential of these selected native Australian plants was also measured through the α-glucosidase inhibition activity. Moreover, LC-ESI-qTOF-MS/MS was used to characterize and screen polyphenols from these native Australian plants. Furthermore, the binding affinities of most abundant phenolic metabolites in selected native Australian plants for the active sites of a-glucosidase (5NN8) were predicted using the in silico molecular docking. Nowadays, in silico molecular docking is widely used in drug discovery, which enables us to understand the behavior of drug molecules in the binding sites of the α-glucosidase protein and explains the basic biochemical processes [7–9]. Moreover, oral bioavailability, drug-likeness, absorption, distribution, metabolism, excretion, and toxicity of abundant phenolic compounds were computed to predict their suitability as therapeutic agents. This research explores the use of native Australian plants in the medicinal, pharmaceutical, food, and feed industries.

2. Materials and Methods

2.1. Chemicals and Reagents

Analytical, HPLC, and LCMS-grade chemicals were used as described [1,10].

2.2. Preparation and Extraction of Phenolic Compounds

Mountain pepper berries (whole dried), rosella (freeze-dried powder), and strawberry gum (finely ground) were purchased from the Australian superfoods Company (www.australiansuperco.com.au) accessed on 21 September 2021. Lemon aspen (freeze-dried powder) was purchased from Australian Creative Native Foods (www.creativenativefoods.com.au) accessed on 21 September 2021. The bioactive phenolic compounds from the selected native Australian fruits were extracted in triplicate by following the method of Ali et al. [1]. The extracts were stored at -20 °C, and all the analyses were conducted within a week.

2.3. Measurement of Phenolic Contents and Biological Activities

The TPC of selected native Australian plants was examined by following the method of Ali et al. [1]. A 25 μL sample extract or standard, 25 μL Folin-Ciocalteu reagent (25% in Milli-Q water) and 200 μL of H_2O were mixed in a 96-well plate and incubated for 5 min at room temperature. Then, 25 μL of 10% sodium carbonate was mixed and again incubated for 60 min at room temperature in the dark. Gallic acid monohydrate ($\geq$99%) in analytical grade ethanol (0–200 μg/mL) was used to generate standard curve at

765 nm. Then, the method of Sharifi-Rad et al. [11] was used to quantify the TFC of native Australian fruits and spices. The TCT and TMAC of selected plants were determined using the procedures of Ali et al. [1,10]. The DPPH and ABTS activities were measured using the methods of Chou et al. [12] and Zahid et al. [13]. The PMA, RPA and FRAP potential of these selected plants were quantified by adopting the methods of Ali et al. [10]. The FICA and the •OH-RSA potential of selected plants were quantified by adopting the methods of Bashmil et al. [14] and Ali et al. [10]. α-Glucosidase inhibition activity was determined by following our previously published method [1], and acarbose (Aca) was used as a reference drug ($\geq$95%).

2.4. LC-MS/MS Analysis

LC-ESI-Q-TOF-MS/MS was used to analyze the untargeted phenolic metabolites from native Australian mountain pepper berries, rosella, strawberry gum, and lemon aspen by following the methods of Ali at al. [1,15]. The heatmap hierarchical clustering was conducted by using MetaboAnalyst 5.0 (www.metaboanalyst.ca) accessed on 7 November 2022.

2.5. Molecular Docking and Pharmacokinetic Properties of Abundant Phenolic Compounds

The pharmacokinetic properties of the most abundant phenolic compounds tentatively identified in the plants were predicted by following the methods of Ali et al. [16] and Daina et al. [17]. Oral bioavailability, absorption, distribution, metabolism, excretion, and toxicity of the abundant phenolic compounds were predicted. Moreover, in silico molecular docking was also conducted to predict the α-glucosidase potential of the selected phenolic compounds from native Australian fruits and spices, as described by Ali et al. [1]. Grid box dimensions were x = -12.95, y = -36.99, and y = 87.77 while docking ligands with a length lower than 20 Å.

2.6. Statistical Analysis

Minitab (version 18.0, Minitab, LLC, State College, PA, USA) and XLSTAT-2019.1.3 software were used for and analysis of variance (ANOVA), Pearson correlation, and a biplot analysis. The results of phenolic contents and their biological activities are represented as mean $\pm$ standard deviation.

3. Results and Discussion

3.1. Measurement of Total Polyphenols (TPC, TFC, TMAC, TCT)

Phytochemicals, especially plants' secondary metabolites, are vital for human health [18]. Phenolic acids and flavonoids are critical secondary bioactive metabolites with various health benefits. They are considered multi-functional metabolites, as metal chelators, hydrogen atom donors, free radical scavengers, and reducing agents [18].

In this study, we investigated Australian mountain pepper berries, rosella, lemon aspen, and strawberry gum for phenolic and non-phenolic compounds. TPC, TFC, TCT, and TMAC quantified in these native Australian plants are given in Table 1.

Total phenols represent phenolic acids, flavonoids, stilbenes, lignans, coumarins and derivatives, tyrosols, and other small molecules. In this context, strawberry gum was found to have the highest TPC (36.57 ± 1.34 mg GAE/g) of the selected Australian native plants. The TPC of strawberry gum was comparable to the previously quantified TPC of Australian-grown thyme (43.16 ± 1.54 mg GAE/g), basil (39.91 ± 1.39 mg GAE/g), and allspice (40.49 ± 1.92 mg GAE/g) [10,18]. Previously, the levels of phenolic compounds in Australian native lemon myrtle and Tasmanian pepper berry were found to be in the range of 16.9 to 31.4 mg GAE/g [19]. Moreover, the TPC of mountain pepper berries was comparable to the TPC reported by Cáceres-Vélez et al. [20] and Vélez et al. [21]. The concentrations of phenolic contents in mountain pepper berry, rosella, and lemon aspen are 2 to 3-fold higher than in Australian-grown cherries [22]. Previously, Lukmanto et al. [23] also measured the TPC of 8.63 mg GAE/g, which is comparatively higher than our results. The TPC of strawberry gum is also comparable to that of villous amomum fruit (46.02 ± 1.12 mg GAE/g) and that of citron fruit (46.22 ± 1.01 mg GAE/g) reported by Liu et al. [24].

Table 1. Quantification of phenolic contents in Australian native fruits and spices.

Variables	TPC mg GAE/g	TFC mg QE/g	TCT mg CE/g	TMAC mg C3GE/g
Rosella	5.65 ± 0.48 [b]	1.33 ± 0.10 [c]	1.26 ± 1.13 [d]	0.08 ± 0.02
Mountain pepper berries	6.10 ± 0.34 [cd]	1.73 ± 0.15 [b]	2.37 ± 0.10 [b]	0.17 ± 0.03
Lemon aspen	4.40 ± 0.38 [c]	0.79 ± 0.04 [d]	1.80 ± 0.35 [c]	0.00 ± 0.00
Strawberry gum	36.57 ± 1.34 [a]	15.69 ± 2.69 [a]	8.05 ± 0.52 [a]	0.00 ± 0.00

Total phenolic content (TPC), total flavonoid content (TFC), total condensed tannins (TCT), total monomeric anthocyanin content (TMAC), cyanidin 3-glucoside equivalent (C3GE), gallic acid equivalent (GAE), quercetin equivalent (QE), catechin equivalent (CE). Values are presented as mean ± standard deviation (n = 3) per gram of dry weight. Values within the same column with different superscripts ([a–d]) are significantly different.

The highest TFC (15.69 ± 2.69 mg QE/g) was quantified in strawberry gum, and the lowest TFC (0.79 ± 0.04 mg QE/g) was quantified in lemon aspen. The highest TCT (8.05 ± 0.52 mg CE/g) was also measured in strawberry gum, and the lowest TCT (1.26 ± 1.13 mg CE/g) was measured in rosella. The TMAC was only measured in mountain pepper berry (0.17 ± 0.03 mg/g) and rosella (0.08 ± 0.02 mg/g). Previously, we measured higher amounts of total anthocyanins in the Davidson plum and quandong peach than berries [1]. Flavonoids are the most abundant class of phenolic compounds in fruits, herbs, and medicinal plants, and they have gained much interest due to their health properties. Previously, a limited number of studies have been conducted to investigate the total flavonoid contents in these plants. There are significant differences in total phenolics, and flavonoids found in each study conducted on these plants due to the aforementioned factors.

3.2. Biological Activities of Native Australian Fruits and Spices

Phenolic compounds include diverse antioxidant constituents present in plants that have various health effects. According to several studies, certain plants' antioxidant properties vary due to their diverse bioactive components and mostly depend on the extraction technique and method used to quantify them. Numerous studies have been carried out to estimate the antioxidant activities of plants from different geographical locations [25–30], but the information on native Australian plants is limited. Therefore, we conducted various antioxidant activity tests to understand the targeted antioxidant potential of native Australian fruits. Various antioxidant activity tests should help in understanding the potential of these native Australian herbs and medicinal plants.

In this study, seven in vitro antioxidant assays (DPPH, ABTS, FRAP, RPA, •OH-RSA, FICA, and PMA) were conducted, and α-glucosidase inhibition activity was tested, to measure the antioxidant and anti-diabetic potential of native Australian native mountain pepper berries, rosella, strawberry gum, and lemon aspen (Table S1, Figure 1).

DPPH and ABTS are the widely used in vitro antioxidant assays for total antioxidant potential measurement of plant extracts based on scavenging the free radicals in the biological system. The highest DPPH (49.70 ± 3.21 mg AAE/g) was measured in strawberry gum, and the lowest DPPH (24.94 ± 0.70 mg AAE/g) was quantified in sea lemon aspen. ABTS$^+$ radical cation inhibition is based on the characteristic wavelength of 734 nm [31]. The ABTS values of strawberry gum (87.65 ± 3.17 mg AAE/g) and mountain pepper berries (85.60 ± 2.32 mg AAE/g) were higher than those of rosella (59.27 ± 1.50 mg AAE/g) and lemon aspen (46.18 ± 0.38 mg AAE/g). Some other studies also reported higher ABTS values for rosemary, oregano, and mint [18]. This indicates that strawberry gum has a higher antioxidant potential than mountain pepper berries, rosella, and lemon aspen. The Fe^{+3}–TPTZ complex, which reduces the antioxidant compounds' ability to form an Fe^{+2}–TPTZ complex in the biological system, was evaluated through the FRAP assay [10]. The FRAP values of strawberry gum (26.57 ± 3.10 mg AAE/g) and rosella (14.30 ± 1.92 mg AAE/g) were higher than those of the other selected fruits and spices ($p < 0.05$). Previously, the highest FRAP values were found in rosemary (17.21 ± 0.54 mg AAE/g) and oregano (10.72 ± 1.44 mg AAE/g). Fenugreek was found to have the lowest value of FRAP (1.48 ± 1.21 mg AAE/g). Furthermore, Wojdyło et al. [32] also reported higher FRAP for rosemary than the other plants selected in our study.

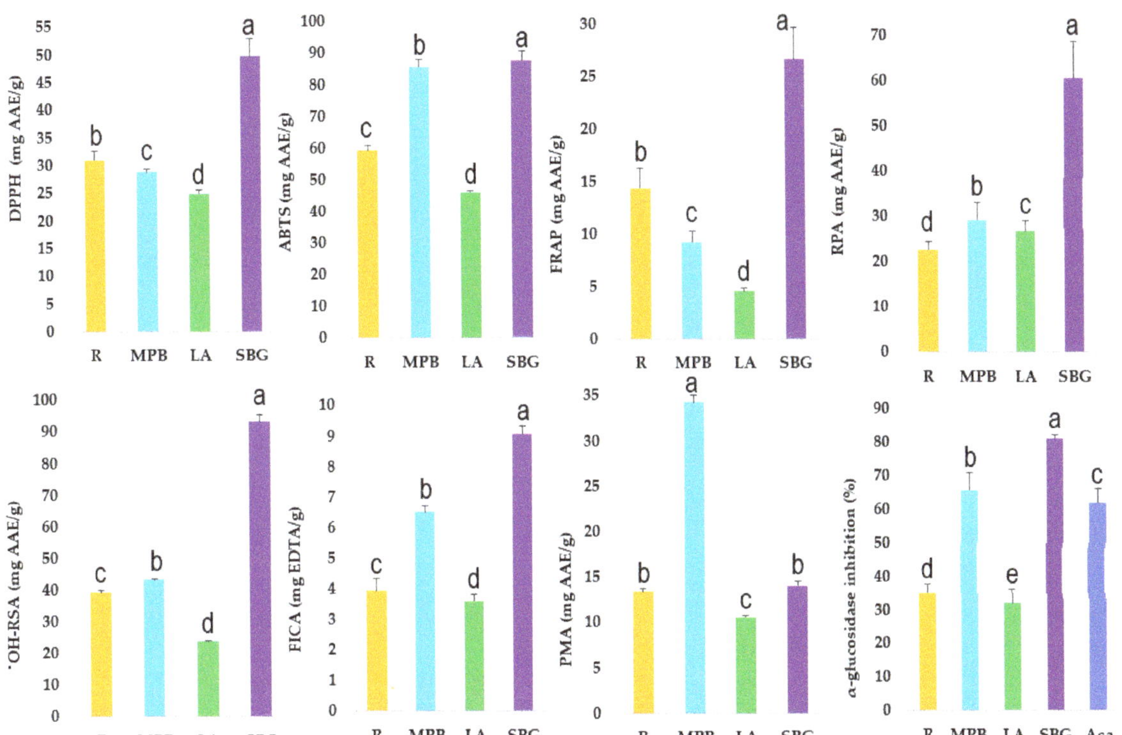

Figure 1. Biological activities (2,2′-diphenyl-1-picrylhydrazyl (DPPH), 2,2′-azinobis-(3-ethylbenzothiazoline-6-sulfonic acid (ABTS), ferric reducing antioxidant power (FRAP), reducing power assay (RPA), hydroxyl-radical scavenging assay (•OH-RSA), ferrous ion chelating assay (FICA), phosphomolybdate assay (PMA)) and α-glucosidase inhibition activity of native Australian rosella (R), mountain pepper berries (MPB), lemon aspen (LA), and strawberry gum (SBG); acarbose (Aca). The vales with letters (a–e) are significantly different from each other.

Excessive amounts of different reactive oxygen species (ROS), such as hydrogen peroxide (H_2O_2), hydroxy radicals (•OH), and super-oxide radicals (O_2•), cause various pathologies. The •OH radicals cause lipid peroxidation and DNA damage due to high oxidative stress, and the daily consumption of antioxidant-rich fruits is crucial in order to protect the human body from these pathologies [18]. The highest •OH-RSA value (93.29 ± 2.20 mg AAE/g) was achieved by strawberry gum. This is vital because •OH scavenging inhibits lipid peroxidation by inhibiting the transition of oxidized metal ions [33,34]. The metal chelating ability of native Australian fruits and spices was estimated by using the ferrous ion chelating assay (FICA), and the highest FICA result (9.05 ± 0.27 mg EDTA/g) was achieved by strawberry gum. The principal antioxidant ingredients are flavonoids, according to a significant association between the antioxidant properties and flavonoids. These fruits and spices can contain different reducing agents which can bind with free radicals to terminate or stabilize the chain reactions in the biological systems [35]. Thus, high reduction power for a fruit extract indicates high antioxidant capacity. Free radicals can be produced by metabolic processes within bodily tissues and brought from outside sources such as food, medications, and pollution. Natural antioxidants are increasingly being used as food additives to neutralize free radicals. This is due to their scavenging abilities and the fact that they are all-natural, non-synthetic items that are well-liked by consumers. Furthermore, α-glucosidase inhibition activity of strawberry gum (81.01 ± 4.6 %) and mountain pepper berries (65.78 ± 5.01 %) was quantified higher than acarbose (standard), rosella, and lemon aspen. Previously, Syabana et al. [36] quantified the IC_{50} of *Cosmos caudatus* (61.33 ± 1.21 µg/mL), *Etlingera elaitor* (53.13 ± 2.87 µg/mL), *Pluchea indica* (12.17 ± 0.18 µg/mL), and *Syzygium polyanthum* (11.76 ± 0.32 µg/mL). The inhibitory activity of strawberry gum (12.01 ± 1.2 µg/mL) was higher than the activity of *Cosmos caudatus* and *Etlingera elaitor* and comparable to the activity of *Pluchea indica* and *Syzygium polyanthum* (Table S1). The inhibitory activity levels of rosella (79.09 ± 7.52 µg/mL) and lemon aspen (83.07 ± 9.03 µg/mL) were lower than those of these plants (Table S1). Moreover, the inhibitory activity of strawberry gum was comparable to that of the methanolic extract of *Satureja cuneifolia* (10.66 µg/mL) reported by Taslimi et al. [37].

Fruits, herbs, spices, and medicinal plants are used as antioxidant sources in the human diet because they inhibit or deactivate the free radicals in the body [38]. Generally, phenolic molecules are regarded as the active antioxidant components in fruits, herbs, and medicinal plants, thereby having potent health benefits. They act as metal chelators, anti-radicals, hydrogen-ion donors, and reducing agents in the biological system [10]. It has been reported that there are many methods to measure a plant extract's total antioxidant potential due to the diverse nature of antioxidant compounds, mainly phenolic constituents [1,18]. The plant's bioactive compounds, mainly polyphenols, depend on the type of cultivar, area, and climatic conditions. There are several techniques to assess the antioxidant capacities of bioactive phenolic metabolites, each with its advantages and disadvantages [11,14,19]. Generally, no approach measures the exact antioxidant capacity of bioactive phenolic compounds because of the complexity of phenolic compounds and the variety of processes of reactions in the human body [39]. These results demonstrate that further research is required to identify and quantify the individual phenolic compounds in these selected native Australian plants. Thus, LC-MS/MS was used to elucidate plant extracts' structure, composition, and bioactive metabolites. The proper quantification and identification of individual phenolic compounds via the process of LC-MS/MS in these plants might help make the essential role of these bioactive metabolites in antioxidant activities understandable.

3.3. Correlation Analysis

Correlation analysis was executed between the phenolic contents (TPC and TFC) of the Australian native herbs and their antioxidant activities generated by the eight different assays (Table 2).

Table 2. Pearson correlation between phenolic contents and biological activities.

Variables	TPC	TFC	TCT	TMAC	DPPH	ABTS	FRAP	PMA	FICA	•OH-RSA	RPA
TFC	**1.00**										
TCT	**0.99**	**0.99**									
TMAC	−0.48	−0.47	−0.44								
DPPH	**0.98**	**0.98**	**0.95**	−0.38							
ABTS	0.62	0.63	0.66	0.38	0.66						
FRAP	**0.92**	**0.92**	0.87	−0.31	**0.98**	0.64					
PMA	−0.21	−0.20	−0.13	0.90	−0.18	0.62	−0.19				
FICA	0.88	0.88	**0.91**	−0.04	0.87	**0.91**	0.80	0.28			
•OH-RSA	**0.97**	**0.97**	**0.96**	−0.27	**0.99**	0.76	**0.96**	−0.04	**0.93**		
RPA	**0.99**	**0.99**	**1.00**	−0.46	**0.95**	0.65	0.86	−0.14	**0.91**	**0.95**	
* α-glu	0.79	0.80	0.84	0.11	0.79	**0.95**	0.71	0.43	**0.99**	0.87	0.83

* = α-glucosidase inhibition activity (%), values in bold are different from 0 with a significance level alpha = 0.1.

It is observed that a positive correlation ($p \leq 0.1$) of TPC was observed with the TFC ($r = 1.00$), TCT ($r = 0.99$), FRAP ($r = 0.92$), $^\bullet$OH-RSA ($r = 0.97$), and RPA ($r = 0.99$); the TFC had a significant positive correlation with TCT ($r = 0.99$), DPPH ($r = 0.98$), FRAP ($r = 0.92$), $^\bullet$OH-RSA ($r = 0.97$), and RPA ($r = 0.99$). This appears to show a direct association between the phenolic compounds in the strawberry gum and the antioxidant processes of peroxyl inhibition, ferric chelation, and free radical scavenging. There was a strong correlation of flavonoids with hydroxyl inhibition, but there were lesser ones with the free radical scavenging, the phosphomolybdate assay outcome, and ferric ion chelation activity. This also indicates the diversity of phenolic and non-phenolic metabolites present in the extracts of native Australian plants. This may be connected to the fact that the flavonoid's ability to operate as an antioxidant often depends on where the hydroxyl group is located on the B-ring and whether it can provide a free radical, either a hydrogen or an electron [1]. Additionally, the experimental conditions, the mechanisms of the antioxidant reactions, and the synergistic or antagonistic effects of various compounds present in the reaction mixture can all impact the associations between antioxidant activity and phenolic compounds [15,18].

A biplot (Figure 2) exhibits that the higher TPC, TFC, and TCT in strawberry gum significantly contributed to all antioxidant activities except that shown by the phospho-molybdate assay. Furthermore, it is depicted that mountain pepper berries have higher concentrations of total monomeric anthocyanin than other plants. Interestingly, rosella and lemon aspen are negatively correlated with all biological activities, which indicates that these have low concentrations of phenolic compounds and flavonoids. Previously, it has been demonstrated that a greater number of OH groups in a flavonoid is favorable for biological activities. Furthermore, each ring's structural arrangement its number of hydroxyl groups, a catechol group in the B ring, and several double bonds in the C ring determine the antioxidant capacity of phenolic metabolites in extracts [40]. Numerous investigations of herbs and medicinal plants have shown a significant, positive association between phenolic content and antioxidant activity [41]. Previously, we reported a positive correlation between phenolic contents of herbs and spices and their antioxidant activities [10,18]. Additionally, two other studies showed that phenolic contents in native Australian fruits and other plants also had positive relationships with their biological activities [1,15].

3.4. LC-MS Analysis

Nutritionists and food scientists have concentrated on exploring the thorough characterization of fruits, spices, and medicinal plants in response to the growing interest in and understanding the antioxidant potential and associated health benefits of phenolic chemicals. The untargeted characterization and screening of phenolic compounds from Australian native fruits and medicinal plants (mountain pepper berries, strawberry gum, rosella, and lemon aspen) were achieved using LC-ESI-QTOF-MS/MS. In this context, a total of 143 phenolic and non-phenolic metabolites, including 31 phenolic acids, 70 flavonoids, 10 isoflavonoids, 7 tannins, 3 stilbenes, 7 lignans, 10 other compounds, and 5 limonoids, were tentatively characterized through the analysis of their MS/MS spectra (Table 3, Figures S1 and S2).

3.4.1. Phenolic Acids

Phenolic acids are diverse plant metabolites from the secondary class produced via the phenylpropanoid pathway by shikimic acid [42]. They are broadly utilized in beauty, health, pharmacology, and medicinal industries due to their anti-aging, antioxidant, antimicrobial, anti-cancer, cardio-protective, antitumor, and anti-inflammatory properties [43].

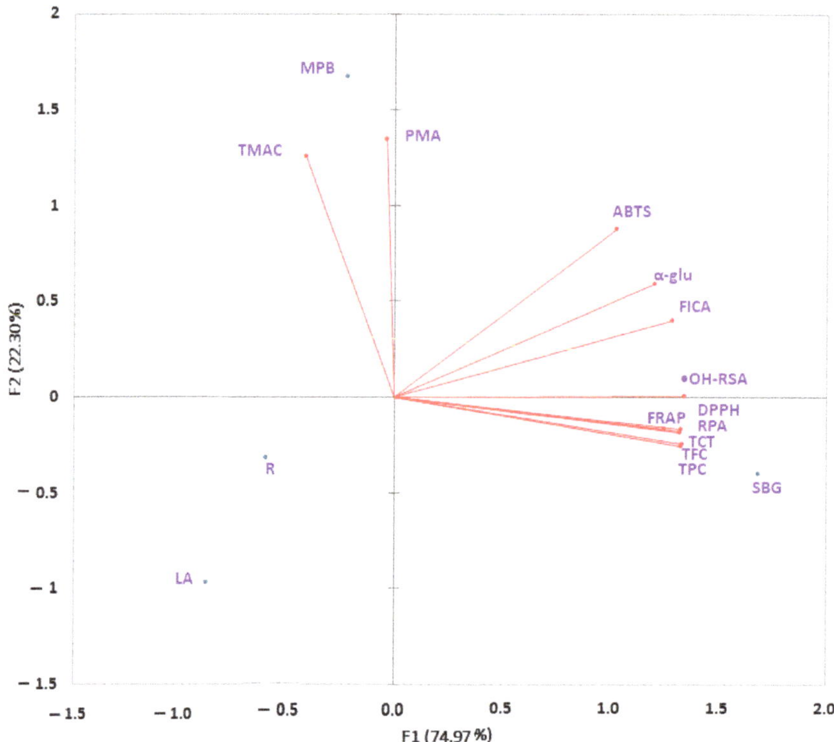

Figure 2. Biplot analysis of phenolic contents and biological activities in native Australian lemon aspen (**LA**), rosella (R), strawberry gum (**SBG**), and mountain pepper berries (**MPB**).

Benzoic acid and Its Derivatives

Benzoic acid is the simplest aromatic carboxylic acid and has a range of derivatives. Eight hydroxybenzoic acids in total were tentatively identified in these native Australian plants. Compounds **1**, **3**, and **8** produced fragment ions at m/z 169, 153, and 137 after the loss of the glycosyl moiety $[M-H-162]^-$ from the precursor ions, respectively. Compounds **1**, **3**, and **8** were tentatively identified as gallic acid 4-*O*-glucoside, protocatechuic acid 4-*O*-glucoside, and 4-hydroxybenzoic acid 4-*O*-glucoside, respectively. Compounds **2**, **4**, and **5** generated product ions at m/z 125, 109, and 93 after the loss of CO_2 (44 Da) from the parent ions, respectively [10]. Compounds **2**, **4**, and **5** were identified through pure standards as gallic acid, protocatechuic acid, and *p*-hydroxybenzoic acid. Compound **2** (gallic acid—$C_7H_6O_5$) and compound **4** (protocatechuic acid—$C_7H_6O_4$) were identified in mountain pepper berries and strawberry gum. Gallic acid 4-*O*-glucoside was only identified in mountain pepper berries, whereas benzoic acid (compound **4**) was identified in all four fruits. The presence of the unique phenolic acid protocatechuic acid has been reported in many therapeutic plants [44]. It has said to possess several health benefits, including anti-inflammatory, antioxidant, anti-cancer, anti-ulcer, anti-diabetic, hepato-protective, and neuro-protective activities [44].

Table 3. LC-MS/MS characterization of phenolic metabolites in Australian native fruits and spices.

No.	Proposed Compounds	Molecular Formula	RT (min)	ESI +/−	Theoretical (m/z)	Observed (m/z)	Mass Error (ppm)	MS/MS	Samples
	Phenolic acids								
	Hydroxybenzoic acids and derivatives								
1	Gallic acid 4-O-glucoside	$C_{13}H_{16}O_{10}$	5.212	[M−H]$^-$	331.0671	331.0676	1.5	169, 125	MPB
2	Gallic acid	$C_7H_6O_5$	6.288	[M−H]$^-$	169.0142	169.0140	−1.2	125	MPB, SBG, R
3	Protocatechuic acid 4-O-glucoside	$C_{13}H_{16}O_9$	8.861	[M−H]$^-$	315.0721	315.0699	−7.0	153, 109	R, MPB
4	Protocatechuic acid	$C_7H_6O_4$	9.341	[M−H]$^-$	153.0193	153.0179	−9.1	109	MPB, SBG, R
5	p-Hydroxybenzoic acid	$C_7H_6O_3$	12.359	[M−H]$^-$	137.0244	137.0244	0.0	93	R, MPB
6	3-O-Methylgallic acid	$C_8H_8O_5$	14.708	[M−H]$^-$	183.0299	183.0296	−1.6	168, 124, 78	R, SBG
7	Benzoic acid	$C_7H_6O_2$	26.507	[M−H]$^-$	121.0295	121.0300	4.1	77	R, LA, SBG, MBP
8	4-Hydroxybenzoic acid 4-O-glucoside	$C_{13}H_{16}O_8$	35.326	[M−H]$^-$	299.0772	299.0773	0.3	255, 137	SBG, R
	Cinnamic acids and derivatives								
9	3-Feruloyquinic acid lactone	$C_{17}H_{18}O_8$	3.926	[M−H]$^-$	349.0929	349.0921	−2.3	193, 191, 178	SBG
10	p-Coumaroyl malic acid	$C_{13}H_{12}O_7$	4.495	[M−H]$^-$	279.0510	279.0503	−2.5	163	LA, SBG
11	Ferulic acid 4-O-glucuronide	$C_{16}H_{18}O_{10}$	6.585	[M−H]$^-$	369.0827	369.0828	0.3	193	R, MPB, SBG, LA
12	1-O-Caffeoyl-β GREEK-D-glucose	$C_{15}H_{18}O_9$	13.291	[M−H]$^-$	341.0873	341.0870	−0.9	179, 135	SBG
13	Feruloyl tartaric acid	$C_{14}H_{14}O_9$	14.091	[M−H]$^-$	325.0565	325.0562	−0.9	193, 149, 105	SBG
14	Rosmarinic acid	$C_{18}H_{16}O_8$	14.395	[M−H]$^-$	359.0772	359.0753	−5.3	197, 179, 135	MPB
15	*trans* p-Coumaric acid 4-glucoside	$C_{15}H_{18}O_8$	15.213	[M−H]$^-$	325.0929	325.0926	−0.9	163, 119	SBG, LA, MPB
16	3-Caffeoylquinic acid	$C_{16}H_{18}O_9$	15.691	[M−H]$^-$	353.0878	353.0875	−0.8	191, 179, 161, 135	R, SBG, LA, MPB
17	Caffeic acid	$C_9H_8O_4$	16.614	[M−H]$^-$	179.0350	179.0343	−3.9	135	SBG, MPB, R, LA
18	Cinnamoyl glucose	$C_{15}H_{18}O_7$	16.808	[M−H]$^-$	309.0979	309.0979	0.0	147	R
19	1-O-Sinapoyl-β GREEK-D-glucose	$C_{17}H_{22}O_{10}$	16.945	[M−H]$^-$	385.1140	385.1139	−0.3	223	LA, MPB
20	Cinnamic acid	$C_9H_8O_2$	17.423	[M−H]$^-$	147.0451	147.0458	4.8	103	SBG, MPB
21	3-p-Coumaroylquinic acid	$C_{16}H_{18}O_8$	18.518	[M−H]$^-$	337.0929	337.0918	−3.3	191, 119	MPB, LA, R
22	3-Feruloylquinic acid	$C_{17}H_{20}O_9$	19.200	[M−H]$^-$	367.1034	367.1029	−1.4	191	R, LA
23	3-Sinapoylquinic acid	$C_{18}H_{22}O_{10}$	21.997	[M−H]$^-$	397.1140	397.1141	0.3	223, 191	LA, R
24	p-Coumaroyl tartaric acid	$C_{13}H_{12}O_8$	22.056	[M−H]$^-$	295.0459	295.0464	1.7	115	LA
25	p-Coumaric acid	$C_9H_8O_3$	22.309	[M−H]$^-$	163.0400	163.0400	0.0	119	R, LA, MPB
26	1,2-Disinapoylgentiobiose	$C_{34}H_{42}O_{19}$	23.851	[M−H]$^-$	753.2247	753.2247	0.0	223, 207	SBG, LA
27	Sinapic acid	$C_{11}H_{12}O_5$	24.062	[M−H]$^-$	223.0612	223.0618	2.7	193, 179, 149, 134	LA
28	Feruloyl glucose	$C_{16}H_{20}O_9$	26.100	[M−H]$^-$	355.1034	355.1038	1.1	223, 207	LA, MPB
29	1,2-Diferuloylgentiobiose	$C_{32}H_{38}O_{17}$	26.359	[M−H]$^-$	693.2036	693.2042	0.9	193, 134	LA, MPB, SBG
30	1,5-Dicaffeoylquinic acid	$C_{25}H_{24}O_{12}$	26.770	[M−H]$^-$	515.1195	515.1197	0.4	191, 179, 135	R, SBG
31	p-Coumaroyl glycolic acid	$C_{11}H_{10}O_5$	60.679	[M+H]$^+$	223.0601	223.0605	1.8	205, 147, 119	R, MPB
	Flavonoids								
	Anthocyanins								
32	Delphinidin 3-O-sambubioside	$C_{26}H_{29}O_{16}$	11.988	[M]$^+$	597.1456	597.1471	2.5	303	R, MBP

Table 3. *Cont.*

No.	Proposed Compounds	Molecular Formula	RT (min)	ESI +/−	Theoretical (*m/z*)	Observed (*m/z*)	Mass Error (ppm)	MS/MS	Samples
33	Cyanidin 3-sambubioside	$C_{26}H_{29}O_{15}$	13.177	[M]$^+$	581.1506	581.1526	3.4	287	R
34	Cyanidin	$C_{15}H_{11}O_6$	13.926	[M]$^+$	287.0556	287.0522	−11.8	231, 139, 69	MPB, R
35	Cyanidin 3-rutinoside	$C_{27}H_{31}O_{15}$	13.621	[M]$^+$	595.1663	595.1660	−0.5	287	MPB
36	Cyanidin 3-*O*-glucoside	$C_{21}H_{21}O_{11}$	14.461	[M]$^+$	449.1084	449.0994	20.4	287	MPB, R
37	Peonidin 3-*O*-(6″-p-coumaroyl-glucoside)	$C_{31}H_{29}O_{13}$	16.378	[M]$^+$	609.1608	609.1617	1.5	301	MPB
38	Delphinidin 3-rutinoside	$C_{27}H_{31}O_{16}$	20.460	[M]$^+$	611.1612	611.1623	1.8	449, 303	R
39	Delphinidin 3-galatoside	$C_{21}H_{21}O_{12}$	20.460	[M]$^+$	465.1033	465.1033	0.0	303	R
40	Delphinidin 3-*O*-(6″-p-coumaroyl-glucoside)	$C_{30}H_{27}O_{14}$	20.528	[M]$^+$	611.1401	611.1430	4.7	303	MPB, R
41	Delphinidin	$C_{15}H_{11}O_7$	20.528	[M]$^+$	303.0505	303.0495	−3.3	303	MPB, R
	Flavanols								
42	Theaflavin 3-*O*-gallate	$C_{36}H_{28}O_{16}$	4.172	[M+H]$^+$	717.1450	717.1418	−4.5	699, 565, 139	SBG
43	Prodelphinidin trimer GC-GC-C	$C_{45}H_{38}O_{20}$	6.333	[M−H]$^-$	897.1883	897.1906	2.6	879, 305, 289, 125	SBG, R
44	(−)-Epigallocatechin	$C_{15}H_{14}O_7$	12.207	[M−H]$^-$	305.0667	305.0650	−5.6	289, 245, 179	R
45	4′,4″-Dimethylepigallocatechin gallate	$C_{24}H_{22}O_{11}$	13.524	[M−H]$^-$	485.1089	485.1092	0.6	441, 319, 183, 139	SBG
46	(−)-Epicatechin	$C_{15}H_{14}O_6$	15.19	[M−H]$^-$	289.0717	289.0711	−2.1	245, 205	SBG, R, MPB,
47	Cinnamtannin A2	$C_{60}H_{50}O_{24}$	17.559	[M−H]$^-$	1153.2619	1153.2602	−1.5	1135, 577, 289, 125	SBG
48	Catechin 3′-glucoside	$C_{21}H_{24}O_{11}$	20.08	[M−H]$^-$	451.1246	451.1253	1.6	289, 245	LA, MPB
	Flavanones								
49	Naringin 6′-malonate	$C_{30}H_{34}O_{17}$	3.858	[M−H]$^-$	665.1723	665.1701	−3.3	579	SBG
50	6″-Acetylliquiritin	$C_{23}H_{24}O_{10}$	6.288	[M−H]$^-$	459.1297	459.1313	3.5	441, 255	SBG, R
51	Narirutin 4′-*O*-glucoside	$C_{33}H_{42}O_{19}$	20.803	[M−H]$^-$	741.2247	741.2269	3.0	579, 271	LA, SBG
52	Hesperetin 5-glucoside	$C_{22}H_{24}O_{11}$	23.183	[M−H]$^-$	463.1246	463.1252	1.3	301	LA
53	Hesperetin 3′-*O*-glucuronide	$C_{22}H_{22}O_{12}$	23.249	[M−H]$^-$	477.1038	477.1052	2.9	301	SBG, LA
54	6-Geranylnaringenin	$C_{25}H_{28}O_5$	23.288	[M−H]$^-$	407.1864	407.1864	0.0	287, 271	SBG
55	Naringenin-7-*O*-glucoside	$C_{21}H_{22}O_{10}$	25.833	[M−H]$^-$	433.1135	433.1154	4.4	301, 271, 151, 119	MPB, SBG
56	Eriodictyol-7-*O*-glucoside	$C_{21}H_{22}O_{11}$	27.312	[M−H]$^-$	449.1084	449.1046	−8.5	287, 151	SBG
57	Hesperetin	$C_{16}H_{14}O_6$	27.569	[M−H]$^-$	301.0717	301.0719	0.7	265, 221, 177, 137	LA, SBG, MPB
58	Hesperetin 5′,7-*O*-diglucuronide	$C_{28}H_{30}O_{18}$	37.143	[M−H]$^-$	653.1359	653.1341	−2.8	301	LA
59	Eriodictyol	$C_{15}H_{12}O_6$	37.692	[M−H]$^-$	287.0556	287.0572	5.6	151, 135	SBG
60	Naringenin	$C_{15}H_{12}O_5$	43.436	[M−H]$^-$	271.0607	271.0623	5.9	227, 151, 119, 107	SBG
61	8-Prenylnaringenin	$C_{20}H_{20}O_5$	49.532	[M−H]$^-$	339.1238	339.1230	−2.4	221, 147	MPB
62	5,7-Dihydroxyflavanone	$C_{15}H_{12}O_6$	52.05	[M−H]$^-$	255.0658	255.0671	5.1	213, 151	SBG
63	Hesperidin	$C_{28}H_{34}O_{15}$	53.498	[M+H]$^+$	611.1971	611.1966	−0.8	303	SBG

Table 3. *Cont.*

No.	Proposed Compounds	Molecular Formula	RT (min)	ESI +/−	Theoretical (m/z)	Observed (m/z)	Mass Error (ppm)	MS/MS	Samples
64	3′,4′,5′-Trimethoxyflavone	$C_{18}H_{16}O_5$	56.326	[M−H]⁻	311.0920	311.0890	−9.6	296, 267	SBG
	Flavones and isoflavones								
65	3′-O-Methylmaysin	$C_{28}H_{30}O_{14}$	3.942	[M−H]⁻	589.1563	589.1571	1.4	589	R, LA
66	Tetramethylscutellarein	$C_{19}H_{18}O_6$	5.213	[M−H]⁻	341.103	341.1030	0.0	341	R, MPB
67	Velutin	$C_{17}H_{14}O_6$	6.265	[M−H]⁻	313.0717	313.0713	−1.3	313	SBG
68	Diosmin	$C_{28}H_{32}O_{15}$	16.891	[M−H]⁻	607.1668	607.1669	0.2	301	MPB, LA
69	Azaleatin 3-arabinoside	$C_{21}H_{20}O_{11}$	21.170	[M−H]⁻	447.0928	447.0903	−5.6	299, 269	MBP
70	Syringetin-3-O-glucoside	$C_{23}H_{24}O_{13}$	24.105	[M−H]⁻	507.1144	507.1160	3.2	345	LA
71	Luteolin	$C_{15}H_{10}O_6$	28.520	[M−H]⁻	285.0404	285.0423	6.7	267, 175, 133, 107	LA, MBP, R
72	Biochanin A 7-O-glucoside	$C_{22}H_{22}O_{10}$	31.319	[M−H]⁻	445.1135	445.1156	4.7	283	SBG
73	Apigenin 6-C-glucoside	$C_{21}H_{20}O_{10}$	32.237	[M−H]⁻	431.0983	431.0967	-3.7	271	MPB
74	Apigenin	$C_{15}H_{10}O_5$	38.457	[M−H]⁻	269.0450	269.0467	6.3	225, 149	MPB
75	Chrysoeriol 7-O-glucoside	$C_{22}H_{22}O_{11}$	40.140	[M−H]⁻	461.1089	461.1068	−4.6	299	MPB
76	Diosmetin	$C_{16}H_{12}O_6$	40.251	[M−H]⁻	299.0561	299.0567	2.0	284, 265, 133	LA, MPB
77	Wogonin	$C_{16}H_{12}O_5$	51.656	[M−H]⁻	283.0607	283.0587	−7.1	268	MPB
78	Glycitein	$C_{16}H_{12}O_5$	52.368	[M−H]⁻	283.0607	283.0617	3.5	268	MPB
79	Chrysin	$C_{15}H_{10}O_4$	52.451	[M−H]⁻	253.0501	253.0515	5.5	253	SBG
	Flavonols and dihydroflavonols								
80	6-Hydroxykaempferol 3,6-diglucoside 7-glucuronide	$C_{33}H_{38}O_{23}$	14.082	[M−H]⁻	801.1726	801.1826	12.5	447, 285	MBP
81	Limocitrin	$C_{17}H_{14}O_8$	17.162	[M−H]⁻	345.0616	345.0604	−3.5	330, 315, 301, 181	LA
82	Myricetin 3-O-glucoside	$C_{21}H_{20}O_{13}$	19.041	[M−H]⁻	479.0831	479.0816	−3.1	317	R
83	Quercetin 3-(2-galloylglucoside)	$C_{28}H_{24}O_{16}$	20.371	[M−H]⁻	615.0986	615.0936	−8.1	301, 169	SBG
84	* Rutin	$C_{27}H_{30}O_{16}$	20.530	[M−H]⁻	609.1461	609.1443	−3.0	301, 300, 271, 255	MPB, R
85	Myricetin 3-O-rhamnoside (myricitrin)	$C_{21}H_{20}O_{12}$	21.328	[M−H]⁻	463.0882	463.0849	−7.1	317	SBG, R
86	Kaempferol 3-rutinoside	$C_{27}H_{30}O_{15}$	24.911	[M−H]⁻	593.1507	593.1511	0.7	285, 151	MPB
87	Kaempferol 3-O-arabinoside	$C_{20}H_{18}O_{10}$	25.902	[M−H]⁻	417.0822	417.0793	−7.0	285	MPB
88	Quercetin 3-rhamnoside (quercitrin)	$C_{21}H_{20}O_{11}$	26.142	[M−H]⁻	447.0928	447.0941	2.9	301	SBG
89	Dihydroquercetin 3-O-rhamnoside	$C_{21}H_{22}O_{11}$	26.305	[M−H]⁻	449.1089	449.1095	1.3	303	SBG, MPB
90	Quercetin 3-O-arabinoside	$C_{20}H_{18}O_{11}$	25.269	[M−H]⁻	433.0776	433.0769	−1.6	301, 271, 151	MPB
91	Isorhamnetin 3-O-glucoside 7-O-rhamnoside	$C_{28}H_{32}O_{16}$	26.574	[M−H]⁻	623.1617	623.1607	−1.6	315	R, LA
92	* Quercetin-3-O-glucoside	$C_{21}H_{20}O_{12}$	27.258	[M−H]⁻	463.0882	463.0842	−8.6	301, 271, 255, 151	SBG, MBP, R
93	* Isorhamnetin	$C_{16}H_{12}O_7$	29.384	[M−H]⁻	315.0510	315.0491	−6.0	300, 271, 151	LA, MPB, SBG
94	Kaempferol 3-O-glucoside	$C_{21}H_{20}O_{11}$	30.214	[M−H]⁻	447.0928	447.0927	−0.2	285, 255, 147	MPB, R, LA
95	* Myricetin	$C_{15}H_{10}O_8$	30.613	[M−H]⁻	317.0298	317.0314	3.3	179, 151	SBG, MPB, R

Table 3. *Cont.*

No.	Proposed Compounds	Molecular Formula	RT (min)	ESI +/−	Theoretical (*m/z*)	Observed (*m/z*)	Mass Error (ppm)	MS/MS	Samples
96	* Taxifolin	$C_{15}H_{12}O_7$	31.176	[M−H]⁻	303.0510	303.0505	−1.7	217, 125	LA, MPB
97	Isorhamnetin 3-*O*-glucuronide	$C_{22}H_{20}O_{13}$	31.344	[M−H]⁻	491.0831	491.0819	−2.4	315	R
98	* Quercetin	$C_{15}H_{10}O_7$	39.148	[M−H]⁻	301.0353	301.0352	−0.3	271, 179, 151, 121	SBG, LA, MPB, R
	Chalcones								
99	Xanthohumol	$C_{21}H_{22}O_5$	10.842	[M−H]⁻	353.1389	353.1399	2.8	295, 233	SBG
100	Phloretin	$C_{15}H_{14}O_5$	28.333	[M−H]⁻	273.0768	273.0780	4.3	167, 119	SBG
101	Phloretin-2′-*O*-glucoside	$C_{21}H_{24}O_{10}$	28.33	[M−H]⁻	435.1296	435.1303	1.5	273, 167	SBG
	Isoflavonoids								
102	Dihydroformononetin	$C_{16}H_{14}O_4$	3.991	[M−H]⁻	269.0819	269.0816	−1.1	253, 239, 223	SBG, MPB
103	Equol 7-*O*-glucuronide	$C_{21}H_{22}O_9$	6.310	[M−H]⁻	417.1191	417.1201	2.4	241	SBG
104	6″-*O*-Malonyldaidzin	$C_{24}H_{22}O_{12}$	14.474	[M−H]⁻	501.1038	501.1027	−2.2	415	MPB, LA
105	6″-*O*-Acetyldaidzin	$C_{23}H_{22}O_{10}$	16.980	[M−H]⁻	457.1140	457.1138	−0.4	415	SBG, R
106	Daidzin 4′-*O*-glucuronide	$C_{27}H_{28}O_{15}$	21.029	[M−H]⁻	591.1355	591.1353	−0.3	415, 253	LA, R, SBG
107	3′,4′,7-Trihydroxyisoflavanone	$C_{15}H_{12}O_5$	27.308	[M−H]⁻	271.0612	271.0611	−0.4	239, 135, 121	SBG
108	3′-*O*-Methylviolanone	$C_{18}H_{18}O_6$	27.762	[M−H]⁻	329.1030	329.1029	−0.3	285, 163	MPB, LA
109	Daidzein 7-*O*-glucuronide	$C_{21}H_{18}O_{10}$	33.742	[M−H]⁻	429.0827	429.0807	−4.7	253	R
110	3′-Hydroxymelanettin	$C_{16}H_{12}O_6$	40.251	[M−H]⁻	299.0561	299.0567	2.0	284	LA, MPB
111	2′-Hydroxyformononetin	$C_{16}H_{12}O_5$	52.124	[M−H]⁻	283.0612	283.0606	−2.1	268	SBG, MPB
	Tannins								
112	Gallagic acid	$C_{28}H_{12}O_{16}$	3.075	[M−H]⁻	603.0052	603.0041	−1.8	587, 559, 549, 446	R, MPB
113	2-*O*-Galloylpunicalin	$C_{41}H_{26}O_{26}$	6.333	[M−H]⁻	933.0639	933.0645	0.6	781, 169, 125	SBG
114	Glucosyringic acid	$C_{15}H_{20}O_{10}$	7.546	[M−H]⁻	359.0978	359.0914	−17.8	315, 197, 153, 125	MPB
115	Procyanidin trimer C1	$C_{45}H_{38}O_{18}$	16.230	[M−H]⁻	865.1985	865.2012	3.1	739, 713, 695	SBG, R
116	Procyanidin B2	$C_{30}H_{26}O_{12}$	19.039	[M−H]⁻	577.1351	577.1323	−4.9	451, 425, 289, 245	MPB, LA, SBG, R
117	Punicafolin	$C_{41}H_{30}O_{26}$	19.102	[M−H]⁻	937.0952	937.0966	1.5	169, 125	SBG
118	Ellagic acid	$C_{14}H_6O_8$	55.906	[M−H]⁻	300.9990	300.9988	−0.7	284, 257	LA, MBP, R
	Stilbenes								
119	Piceatannol	$C_{14}H_{12}O_4$	5.594	[M−H]⁻	243.0663	243.0653	−4.1	225, 201	SBG, MPB
120	Polydatin	$C_{20}H_{22}O_8$	21.854	[M−H]⁻	389.1242	389.1245	0.8	227	LA, MPB, SBG
121	Piceatannol 3-*O*-glucoside	$C_{20}H_{22}O_9$	30.064	[M−H]⁻	405.1191	405.1188	−0.7	243	MPB
	Lignans								
122	Sesamin	$C_{20}H_{18}O_6$	4.879	[M−H]⁻	353.1030	353.1015	−4.2	338, 163	MPB
123	2-Hydroxyenterolactone	$C_{18}H_{18}O_5$	6.371	[M−H]⁻	313.1081	313.1091	3.2	255, 163	LA, MPB
124	Silibinin	$C_{25}H_{22}O_{10}$	16.794	[M−H]⁻	481.1140	481.1151	2.3	301, 179, 165, 151	MPB
125	7-Oxomatairesinol	$C_{20}H_{20}O_7$	21.997	[M−H]⁻	371.1136	371.1138	0.5	358, 343, 328	LA
126	Arctigenin	$C_{21}H_{24}O_6$	23.288	[M−H]⁻	371.1500	371.1497	−0.8	356, 312, 295	SBG
127	Enterolactone	$C_{18}H_{18}O_4$	47.739	[M−H]⁻	299.1288	299.1299	4.3	281, 187, 165	MPB, SBG

Table 3. *Cont.*

No.	Proposed Compounds	Molecular Formula	RT (min)	ESI +/−	Theoretical (m/z)	Observed (m/z)	Mass Error (ppm)	MS/MS	Samples
128	2-Hydroxyenterodiol	$C_{18}H_{22}O_5$	53.013	$[M-H]^-$	317.1394	317.1395	0.3	299, 287, 269, 257	MPB
	Other compounds								
129	Pyrogallol	$C_6H_6O_3$	7.009	$[M-H]^-$	125.0244	125.0242	−1.6	107, 97, 79	MPB
130	[6]-Gingerol	$C_{17}H_{26}O_4$	12.249	$[M-H]^-$	293.1758	293.1768	3.4	137	SBG, MPB
131	Quinic acid	$C_7H_{12}O_6$	4.189	$[M-H]^-$	191.0561	191.0578	9.0	85	MPB, SBG
132	1,2,4,6-Tetragalloyl-β GREEK-D-glucopyranose	$C_{34}H_{28}O_{22}$	19.144	$[M-H]^-$	787.0999	787.0953	−5.9	169, 125	SBG
133	Umbelliferone	$C_9H_6O_3$	19.396	$[M-H]^-$	161.0244	161.0246	1.2	133	MPB, R
134	2-Hydroxybenzaldehyde	$C_7H_6O_2$	20.620	$[M-H]^-$	121.0269	121.0276	5.8	92, 77	MPB
135	p-Coumaraldehyde	$C_9H_8O_2$	29.139	$[M-H]^-$	147.0451	147.0463	8.0	119	MPB
136	Xanthotoxol	$C_{11}H_6O_4$	47.927	$[M-H]^-$	201.0193	201.0191	−1.0	171	MPB, LA, SBG
137	Carnosic acid	$C_{20}H_{28}O_4$	55.899	$[M-H]^-$	331.1915	331.1927	3.6	287	SBG, R
138	Mellein	$C_{10}H_{10}O_3$	62.141	$[M+H]^+$	179.0703	179.0694	−5.0	135	LA
	Limonoids								
139	Limonin	$C_{26}H_{30}O_8$	19.039	$[M-H]^-$	469.1868	469.1859	−1.9	229	LA
140	Obacunoic acid	$C_{26}H_{32}O_8$	25.201	$[M-H]^-$	471.2024	471.2027	0.6	471	LA, MPB, SBG
141	Nomilin	$C_{28}H_{34}O_9$	51.280	$[M+H]^+$	515.2276	515.2280	0.8	515	MPB, SBG
142	Obacunone	$C_{26}H_{30}O_7$	19.253	$[M-H]^-$	455.2065	455.2065	0.0	407, 163	SBG, R, MPB, LA
143	Citrusin	$C_{28}H_{34}O_{11}$	55.330	$[M+H]^+$	547.2174	547.2162	−2.2	547	LA

Mountain pepper berries (MPB), rosella (R), lemon aspen (LA), and strawberry gum (SBG). Compounds with asterisk (*) were identified with pure standards.

Cinnamic Acids and Derivatives

The most prevalent phenolic acid class is hydroxycinnamic in fruits, herbs, and medicinal plants. Sixteen hydroxycinnamic acids were identified, and their fragmentation patterns were verified using MS/MS. The removal of CO_2 and the hexosyl moiety from the parent ions is the primary way that phenolic acids exhibit the fragmentation pattern [10]. Rosmarinic acid, caffeic acid, sinapic acid, *p*-coumaric acid, 3-caffeoylquinic acid (chlorogenic acid), and cinnamic acid were confirmed through pure standards. The quinic acid derivatives **9** (*m/z* 349.0921), **20** (*m/z* 337.0934) **21** (*m/z* 367.1025), **22** (*m/z* 397.1141), and **29** (*m/z* 515.1197) are known as -feruloyquinic acid lactone, 3-*p*-coumaroylquinic acid, 3-feruloylquinic acid, 3-sinapoylquinic acid, and 1,5-dicaffeoylquinic acid, respectively. Compound 21 (3-feruloylquinic acid) was tentatively identified in rosella and lemon aspen at *m/z* 367.1029, which generated characteristic fragment ions of ferulic acid and quinic acid at *m/z* 193 and 191 in negative-ion mode. Compound **29** (1,5-dicaffeoylquinic acid) was detected in rosella and strawberry gum in negative mode. This was confirmed through MS/MS, where it produced fragment ions at *m/z* 353, 191, and 179 after the breakdown of precursor ions into 5-caffeoylquinic acid (*m/z* 353), quinic acid (*m/z* 191), and caffeic acid (*m/z* 179) units, respectively [10]. Previously, 1,5-dicaffeoylquinic acid was also identified in cumin [10]. Compound **10** at ESI$^-$ *m/z* 279.0503 was identified in lemon aspen and strawberry gum, which generated a product ion at *m/z* 163 (coumaric acid) after the loss of $C_4H_4O_4$ (116 Da) from the precursor ion in the MS/MS scan. Therefore, compound **10** was putatively identified as *p*-coumaroyl malic acid. Rosmarinic acid produced a characteristic fragment at *m/z* 197 after the removal of a hexose moiety (162 Da), which further broke down into a caffeic acid unit (*m/z* 179) through the loss of H_2O, and a caffeic acid (*m/z* 179) fragment at *m/z* 135 represents the loss of a CO_2 [M−H−44]$^-$ unit [45,46]. Rosmarinic acid (compound **14**) is one of herbs' and medicinal plants' most plentiful phenolic acids. Compound **24** at ESI$^-$ *m/z* 163.0400 was identified in mountain pepper berries, rosella, and lemon aspen, which generated a product ion at *m/z* 119 after the loss of CO_2 [M−H−44]$^-$ from the precursor ion (*m/z* 163.0400). Compound **24** was tentatively identified as *p*-coumaric acid. Compound **31** at ESI$^+$ *m/z* 223.0605 generated product ions at *m/z* 205 and 147, and 119, after the loss of a unit of H_2O (18 Da), the glycolic acid moiety (76 Da), and both from the precursor ion, respectively. In contrast, a product ion at *m/z* 119 is a specific fragment ion of *p*-coumaric acid. As a result, compound **31** was tentatively identified as *p*-coumaroyl glycolic acid in rosella and mountain pepper berries. Previously, Kadam et al. [47] also reported *p*-coumaroyl glycolic in *Lepidium sativum* seedcake.

3.4.2. Flavonoids

Flavonoids are widely used in nutraceutical, pharmaceutical, and cosmetic industries due to their anti-carcinogenic, antimicrobial, anti-inflammatory, anti-mutagenic, and antioxidant properties. In this study, we tentatively identified seventy flavonoids (Table 2).

Anthocyanins

Anthocyanins are water-soluble, colored plant pigments. The main positions of their hydroxyls are 3, 5, and 7 in ring A and 3′ and 5′ in ring B [48]. The screening, identification, and characterization of anthocyanins in native Australian rosella and mountain pepper berries were conducted. This work identified nine anthocyanins using their MS/MS spectra (Table 2). The native Australian quandong peach and Davidson plum were used as reference plants to understand anthocyanins' structural and spectral characteristics further; these fruits are abundant in anthocyanins [1]. The removal of sugar units from anthocyanins (162 Da for hexoses, 150 Da for xyloses, 132 Da for pentoses, and 308 Da for the rutinoside moiety from the basic aglycone of corresponding anthocyanins) results in the formation of MS/MS product ions (303 Da for delphinidin, 331 Da for malvidin, 301 Da for peonidin, 317 Da for petunidin, and 287 Da for cyanidin) [1]. Compounds **33**, 35, and **36** at ESI$^+$ *m/z* 581.1526, 595.1660, and 449.0994 generated a characteristic fragment ion at *m/z* 287 (cyanidin). Thus, compounds **33**, **35**, and **36** were tentatively identified as cyanidin 3-sambubioside, cyanidin 3-rutinoside, and cyanidin-3-O-glucoside, respectively. Compound **36** (cyanidin 3-*O*-glucoside) was identified in mountain pepper berries and rosella. Cyanidin-3-*O*-glucoside was quantified in grapes from 2.7 to 51.7 μg/mL by Oh et al. [49]. Compounds **32**, **38**, **39**, **40**, and **41** produced a distinctive fragment of delphinidin at m/z 303 in positive-ion mode (Table 2). Compounds **33**, **38**, and **39** were only identified in rosella. Due to their positively charged oxygen atom, anthocyanins have higher antioxidant activity than other flavonoids [50].

Flavanols

We identified monomeric flavanols in our samples, including epicatechin, epigallo-catechin, and derivatives [48]. In this study, seven flavanols, including polymerized and derivative substances, were tentatively identified in mountain pepper berries, rosella, strawberry gum, and lemon aspen. Flavanols are also called catechins, having no double bond between C2 and C3, and there is no carbonyl group in ring C (C4) [51]. Compound **44** at ESI$^-$ m/z 305.0650 generated product ions at m/z 289, 169, and 125 from the ion precursor. Compound **44** was putatively identified as (−)-epigallocatechin ($C_{15}H_{14}O_7$). They have been reported abundantly for their potent antioxidant and cardio-protective effects in tea and cocoa. Compound **46** at ESI$^-$ m/z 289.0711 was identified in strawberry gum, rosella, and mountain pepper berries, which produced product ions at m/z 245, 205, and 179 after CO_2 loss [M−H−44]$^-$, flavonoid A ring [M−H−84]$^-$ loss, and flavonoid B ring [M−H−110]$^-$ loss from the precursor ion, respectively. Compound **46** was tentatively identified as epicatechin ($C_{15}H_{14}O_6$) [52]. These compounds are the building blocks of proanthocyanidins (condensed tannins). The most prevalent flavonoids: flavanols, and flavan-3-ols have a variety of chemical and biological properties.

Flavanones

Flavanones do not have double bond between C2 and C3, but they have a carbonyl ring at C4 in ring C [48]. Sixteen compounds were identified as flavanones. Compounds **49** (naringin 6′-malonate), **54** (6-geranylnaringenin), **56** (eriodictyol-7-*O*-glucoside), **59** (eriodictyol), **60** (naringenin), **62** (5,7-dihydroxyflavanone), **63** (Hesperidin), and **64** (3′,4′,5′-trimethoxyflavone) were only identified in strawberry gum; and compounds **52** (hesperetin 5-glucoside) and **58** (hesperetin 5′,7-*O*-diglucuronide) were only identified in lemon aspen. Compound **61** (8-Prenylnaringenin) was only identified in mountain pepper berries. Compounds **51**, **52**, **55**, and **56** generated product ions at m/z 579, 301, 271, and 287 after the loss of a glycosyl moiety from their precursor ions, respectively. Therefore, compounds **51**, **52**, **55**, and **56** were tentatively identified as narirutin 4′-*O*-glucoside, hesperetin 5-glucoside, naringenin-7-*O*-glucoside, and eriodictyol-7-*O*-glucoside, respectively.

Flavones and Isoflavones

Flavones are characterized by a non-saturated C3 chain and have a double bond between C2 and C3 [48]. Sixteen compounds were characterized as flavones and flavanones in mountain pepper berries, rosella, strawberry gum, and lemon aspen. Compounds **67** (velutin), **72** (biochanin A 7-*O*-glucoside), and **79** (chrysin) were only identified in strawberry gum; and compounds **69** (azaleatin 3-arabinoside), **73** (Apigenin 6-C-glucoside), **75** (apigenin), **75** (chrysoeriol 7-*O*-glucoside), **77** (wogonin), and **78** (glycitein) were only identified in mountain pepper berries. Compounds **68** (diosmin), **71** (luteolin), **74** (apigenin), **76** (diosmetin), and **79** (chrysin) were identified via the MS/MS spectra of pure standards. Compounds **70** (syringetin-3-*O*-glucoside), **72** (biochanin A 7-*O*-glucoside), **73** (apigenin 6-C-glucoside), and **75** (chrysoeriol 7-*O*-glucoside) generated product ions at m/z 299, 345, 271, and 299, respectively, after the loss of glycosyl moiety from their parent ions.

Flavonols, Dihydroflavonols, and Chalcones

Flavonols have a double bond between C2 and C3, and there is a carbonyl in ring C (C4) and a OH group at C3 [51]. These compounds have strong absorption at 340–380 nm. Eighteen compounds were identified as flavonols and dihydroflavonols. Compound **81** (limocitrin) was only identified in lemon aspen, and compounds **82** (myricetin 3-*O*-glucoside) and **97** (isorhamnetin 3-*O*-glucuronide) were only identified in rosella. Compounds **82** (m/z 479.0816), **92** (m/z 463.0842), and **94** (Kaempferol 3-*O*-glucoside) generated product ions at m/z **317** (myricetin), 301 (quercetin), and 285 (kaempferol) after the loss of a hexose moiety (162 Da) from the precursor ions, respectively. Compounds **82**, **92**, and **94** were putatively identified as myricetin 3-*O*-glucoside, quercetin-3-*O*-glucoside, and kaempferol 3-*O*-glucoside, respectively. Moreover, compounds **85**, **86**, **87**, **88**, **89**, **90**, **and** **97** produced fragment ions at m/z 317 (myricetin), 285 (kaempferol), 301 (quercetin), 303 (dihydroquercetin), and 315 (isorhamnetin) after the loss of sugar moieties, including rhamnoside (146 Da), rutinoside (308 Da), arabinoside (132 Da), and glucuronide (176 Da), from their precursor ions. Compounds **83** (quercetin 3-(2-galloylglucoside) and **88** (quercitrin) were only identified in lemon aspen while compounds **80** (6-hydroxykaempferol 3,6-diglucoside 7-glucuronide), **86** (kaempferol 3-rutinoside), **87** (kaempferol 3-*O*-arabinoside), and **90** (quercetin 3-*O*-arabinoside) were only identified in mountain pepper berries. Compounds **84** (rutin), **92** (quercetin-3-*O*-glucoside), **93** (Isorhamnetin), **95** (myricetin), **96** (taxifolin), and **98** (quercetin) were identified through the MS/MS spectra of pure standards [53]. The resulting ions at m/z 300 and 271, which correspond to the loss of CH_3 and CO_2 from the precursor [1,18], were used to identify isorhamnetin (compound **93** at ESI^- m/z 315.0504), which was identified in mountain pepper berries, lemon aspen, and strawberry gum. In addition to repairing iron-induced DNA oxidation, myricetin 3-*O*-rhamnoside (compound **85**) also inhibits the activity of digestive, lipid, fecal, and colonic bacterial enzymes and functions as an anti-allergenic, anti-obesity, and anti-cancer compound [54]. Flavonols are also frequently found in Australian native fruits and medicinal plants. According to a comparison of the flavonoid literature, the aglycone derivatives of kaempferol, myricetin, and quercetin are the most often found flavonols in these plants. These aglycone derivatives are renowned for having highly effective anti-diabetic properties. These aglycone compounds are eight times more potent than the diabetic medication acarbose, according to some research [55]. In many earlier investigations, quercetin and kaempferol were connected to rutinoside, galactosides, and glucosides; previously, these flavonoid-3-*O*-glycosides were not described in selected native Australian plants. Three phenolic compounds, **99**, **100**, and **101**, were only identified in strawberry gum.

3.4.3. Isoflavonoids

Isoflavonoids differ from flavonoids, as the isoflavonoid skeleton was biogenetically engineered from the 2-phenylchroman skeleton. In isoflavonoids, ring A (phenyl ring) is fused with the C-ring (six-membered heterocyclic ring) and another phenyl B-ring at C3, whereas the B-ring is substituted at C2 position in flavonoids [15]. Ten phenolic compounds were identified as isoflavonoids. Compounds **103** (equol 7-*O*-glucuronide) and **107** (3′,4′,7-trihydroxyisoflavanone) were only identified in strawberry gum; and **109** (daidzein 7-*O*-glucuronide) and **110** (3′-hydroxymelanettin) were only identified in rosella and lemon aspen. Compounds **103**, **106** and **109** generated product ions at m/z 241 (equol), 415 (daidzin), and 253 (daidzein) after the loss of [M−H−176] from their precursor ions, respectively. Compounds **103**, **106**, and **109** were tentatively identified as equol 7-*O*-glucuronide, daidzin 4′-*O*-glucuronide, and daidzein 7-*O*-glucuronide, respectively. As per our knowledge, no previous research has been conducted in such a comprehensive way to identify these isoflavonoids in the selected Australian native plants.

3.4.4. Tannins

Proanthocyanidins (condensed tannins) are condensed flavanols. Seven compounds were identified as tannins (proanthocyanidins, hydrolyzable and complex tannins) [56]. Compound **115** at m/z 865.2004 produced fragment ions at m/z 739, 713, and 695 in negative-ion mode. The daughter ion at m/z 739 formed after the loss of ring "A" because of the fission of the heterocyclic ring $[M-H-126]^-$ from the precursor ion, RDA (152 Da), and a water unit (18 Da) from the latter product ion (m/z 713). Compound **115** was putatively identified as the procyanidin trimer C1 in strawberry gum and rosella. Compound **116** at ESI$^-$ m/z 577.1353 was tentatively identified in mountain pepper berries, lemon aspen, and strawberry gum, which generated fragment ions at m/z 451, 425, and 289; C4, C5 and O-C2 showed cleavage of one pyran ring, which led to phloroglucinol molecule loss (A-ring) from the precursor ion [52], which resulted in product ions at m/z 451 $[M-H-126]^-$ and 425 $[M-H-152]^-$. Compound **116** was putatively identified as the procyanidin B2. Previously, procyanidin B2 and procyanidin trimmer C1 were recognized in nutmeg and cinnamon [10]. They have been reported to have anti-cancer, antioxidant, cardio-protective, and anti-inflammatory activities [56,57]. Compounds **113** (2-*O*-galloylpunicalin) and **117** (punicafolin) were only identified in strawberry gum, and compound **114** (glucosyringic acid) was only identified in mountain pepper berries.

3.4.5. Lignans and Stilbenes

Stilbenes are natural phytochemicals that contain a 1,2-diphenylethylene (a basic skeleton of stilbenoids), and lignans are a group of diphenol derivatives with dibenzylbutane skeleton structures [15]. Due to their diverse structural makeup and established advantages for human health, lignans and stilbenes are among the most studied secondary plant metabolites [15]. Ten metabolites that fit into these classes were putatively discovered in this investigation. A total of three stilbenes (piceatannol, polydatin, and piceatannol 3-*O*-glucoside) and seven lignans were tentatively identified in these selected Australian native fruits and medicinal plants. Compound **119** (piceatannol) resulted in a deprotonated precursor ion at m/z 243 that formed a fragment ion at m/z 225 following the removal of a water unit $[M-H-H_2O]$, and a second product ion at m/z 201 due to the neutral loss of C_2H_2O (42 Da) from the precursor ion. Previously, piceatannol was found in fenugreek and dill leaves [18] and has been reported to have strong anti-mutagenic, antioxidant, anti-inflammatory, and anti-cancer properties. Compound **127** at ESI$^+$ m/z 299.1279 was putatively identified in mountain pepper berries and strawberry gum, which generated product ions at m/z 281, 187, and 165 after the loss of $[M-H-H_2O]$, $[M-H-C_6H_8O_2]$, and $[M-H-C_9H_8O_2]$, respectively, from the precursor ion. Compound **127** was characterized as enterolactone. Enterolactone has been acknowledged for its antioxidant [58] and anti-cancer activities [59]. Compounds **122** (sesamin), **124** (silibinin), and **128** (2-hydroxyenterodiol) were only identified in mountain pepper berries.

3.4.6. Other Compounds

Ten compounds were identified as other compounds. Compounds **137** (carnosic acid) and **138** (mellein) generated product ions at m/z 287 and 135, respectively, after the loss of CO_2 (44 Da). Compound **129** at ESI$^-$ m/z 125.0242 was identified in mountain pepper berries and generated product ions at m/z 107, 97, and 79 after the loss of H_2O (18 Da) and CO (28 Da), and the removal of H_2O after the loss of CO (18 Da). Compound **133** at ESI$^-$ m/z 161.0242 was identified in mountain pepper berries and rosella, which produced fragment ion s at m/z 133, 117, and 105 through the removal of $[M-H-CO]$, $[M-H-CO_2]$, and $[M-H-C_2H_2]$ from the precursor ion and the former product ion, respectively. Compound **133** was tentatively characterized as umbelliferone. Compound **134** (2-hydroxybenzaldehyde) was tentatively identified only in mountain pepper berries, which produced fragment ions at m/z 92 and 77 after the loss of CO (28 Da) and CO_2 (44 Da), respectively, from the precursor ion. Compound **132** (1,2,4,6-tetragalloyl-β GREEK-D-glucopyranose) was only identified in strawberry gum; and compounds **129** (pyrogallol), **134** (2-hydroxybenzaldehyde), and **135** (*p*-coumaraldehyde) were only identified in mountain pepper berries. A total of five limonoids were putatively detected in these native Australian fruits and spices. Compounds **139** (limonin) and **143** (citrusin) were tentatively identified only in lemon aspen.

The screening and profiling of the phenolic compounds give an overall idea of antioxidation compounds in selected Australian native plants. Strawberry gum is an excellent source of phenolic compounds, especially flavonoids used in the food, feed, cosmetics, and pharmaceutical industries because several of them have already been shown to possess high antioxidation capabilities.

3.5. Quantification/Semi-Quantification of Targeted Phenolic Compounds

A total of 26 compounds were quantified in Australian native mountain pepper berries, strawberry gum, rosella, and lemon aspen, which are given in Table S2. Flavonoids are the most abundant class in these selected Australian native plants. Strawberry gum was found to have the highest concentration of flavonoids, and quercitrin had the highest concentration among them (1274.04 $\pm$ 43.78 µg/g). Myricetin 3-*O*-rhamnoside (394.71 $\pm$ 16.21 µg/g), 3′,4′,5′-trimethoxyflavone (615.15 $\pm$ 21.63 µg/g), quercetin 3-*O*-arabinoside (371.54 $\pm$ 14.26 µg/g), quercetin 3-(2-galloylglucoside) (309.15 $\pm$ 20.38 µg/g), chrysin (35.52 $\pm$ 2.77 µg/g), and naringenin (24.72 $\pm$ 1.83 µg/g) were only quantified in strawberry gum. Chlorogenic acid (3-caffeoylquinic acid) is the most abundant phenolic acid in mountain pepper berries (134.05 $\pm$ 12.67 µg/g), and the lowest concentration of chlorogenic acid was quantified in strawberry gum. Previously, Konczak et al. [60] also quantified the higher concentration of chlorogenic acid in Tasmanian pepper berries. Protocatechuic acid was quantified in strawberry gum (63.56 $\pm$ 4.67 µg/g) and mountain pepper berries (44.57 $\pm$ 5.82 µg/g); *p*-hydroxybenzoic acid was quantified in rosella (11.74 $\pm$ 1.56 µg/g) and mountain pepper berries (21.91 $\pm$ 3.41 µg/g). The highest concentration of caffeic acid was found in mountain pepper berries (23.49 $\pm$ 1.92 µg/g), and the lowest concentration of caffeic acid was found in strawberry gum (15.51 $\pm$ 2.09 µg/g). Gallic acid (19.24 $\pm$ 3.12 µg/g) and *p*-coumaric acid (10.56 $\pm$ 1.35 µg/g) were found in mountain pepper berries. Gallic acid was also found in strawberry gum (23.54 $\pm$ 3.19 µg/g) and rosella (17.21 $\pm$ 2.17 µg/g). The highest concentration (39.52 $\pm$ 3.65 µg/g) of procyanidin B2 was found in strawberry gum, and the lowest concentration (11.32 $\pm$ 1.48 µg/g) was measured in lemon aspen. Rutin (56.61 $\pm$ 5.48 µg/g) was only found in mountain pepper berries. Previously, Konczak et al. [60] also found rutin in mountain pepper berries. The highest concentration of quercetin was found in mountain pepper berries (71.46 $\pm$ 4.52 µg/g), and the lowest concentration was measured in lemon aspen (18.31 $\pm$ 2.34 µg/g). The highest concentrations of isorhamnetin (26.83 $\pm$ 2.86 µg/g) and myricetin (23.67 $\pm$ 3.71 µg/g) were found in mountain pepper berries, and the lowest concentrations of isorhamnetin (12.52 $\pm$ 1.08 µg/g) and myricetin (13.16 $\pm$ 0.89 µg/g) were found in strawberry gum. A total of six anthocyanin compounds were also found in mountain pepper berries and rosella. Delphinidin 3-*O*-sambubioside (196.61 $\pm$ 17.91 µg/g) and cyanidin 3-rutinoside (142.98 $\pm$ 13.01 µg/g) were found in rosella; and delphinidin 3-*O*-sambubioside (59.67 $\pm$ 5.24 µg/g) and cyanidin 3-rutinoside (82.91 $\pm$ 7.25 µg/g) were found in mountain pepper berries. Cyanidin-3-sambubioside (72.21 $\pm$ 8.63 µg/g) and delphinidin 3-rutinoside (17.23 $\pm$ 1.61 µg/g) were only found in rosella.

Furthermore, hierarchical heatmap clustering (Figure 3) was conducted by using MetaboAnalyst 5.0 (www.metaboanalyst.ca) accessed on 7 November 2022.

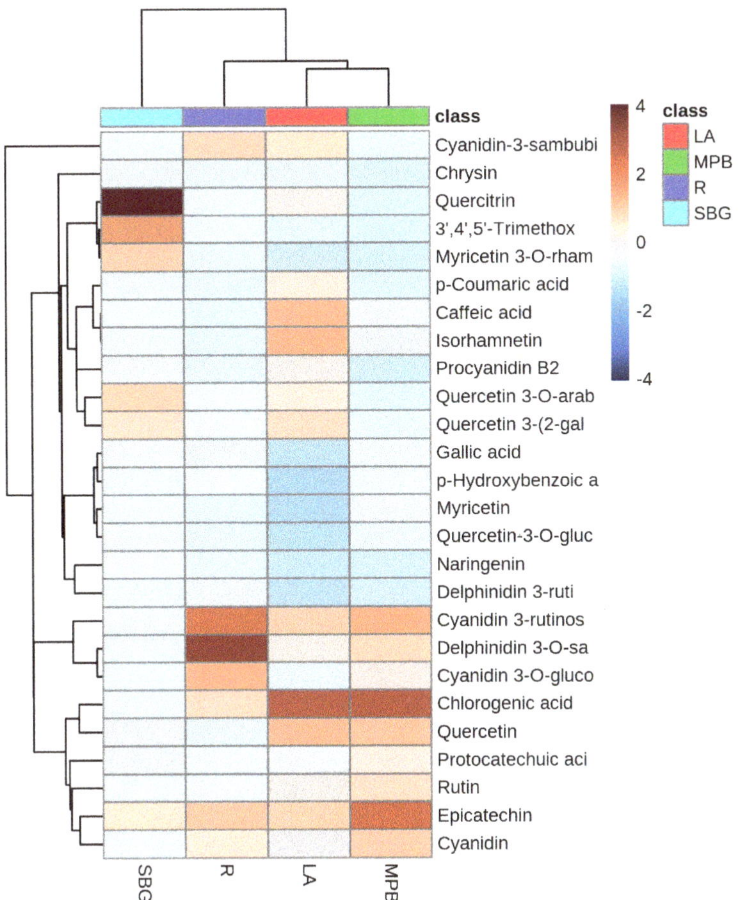

Figure 3. Heatmap hierarchical clustering of quantified phenolic compounds in mountain pepper berries (**MPB**), lemon aspen (**LS**), rosella (**R**), and strawberry gum (**SBG**).

It depicted in the heatmap that quercitrin, 3′,4′,5′-trimethoxyflavone, myricetin 3-*O*-rhamnoside, quercetin 3-*O*-arabnoside, quercetin 3-(2-galloylglucoside), and epicatechin had higher concentrations than other quantified phenolic compounds in strawberry gum; and delphinidin 3-*O*-sambubioside, cyanidin 3-rutinoside, cyanidin 3-glucoside, cyanidin-3-sambubioside, cyanidin, epicatechin, and chlorogenic acid had higher concentrations in rosella. The highest concentrations of chlorogenic acid were found in mountain pepper berries and lemon aspen. Mountain pepper berries had higher concentrations of chlorogenic acid, epicatechin, cyanidin 3-rutinoside, quercetin, cyanidin, delphinidin 3-*O*-sambubiode, rutin, and protocatechuic acid.

3.6. Molecular Docking

In silico molecular docking was conducted to predict the roles of abundant phenolic compounds in α-glucosidase inhibition activity. The estimated binding geometry 2D and 3D structures of myricitrin and chlorogenic acid in α-glucosidase protein (5NN8) are given in Figure 4A,B; and the calculated binding energy, glide energy, and binding geometry 2D of selected phenolic compounds are given in Table S3 and Figure S3.

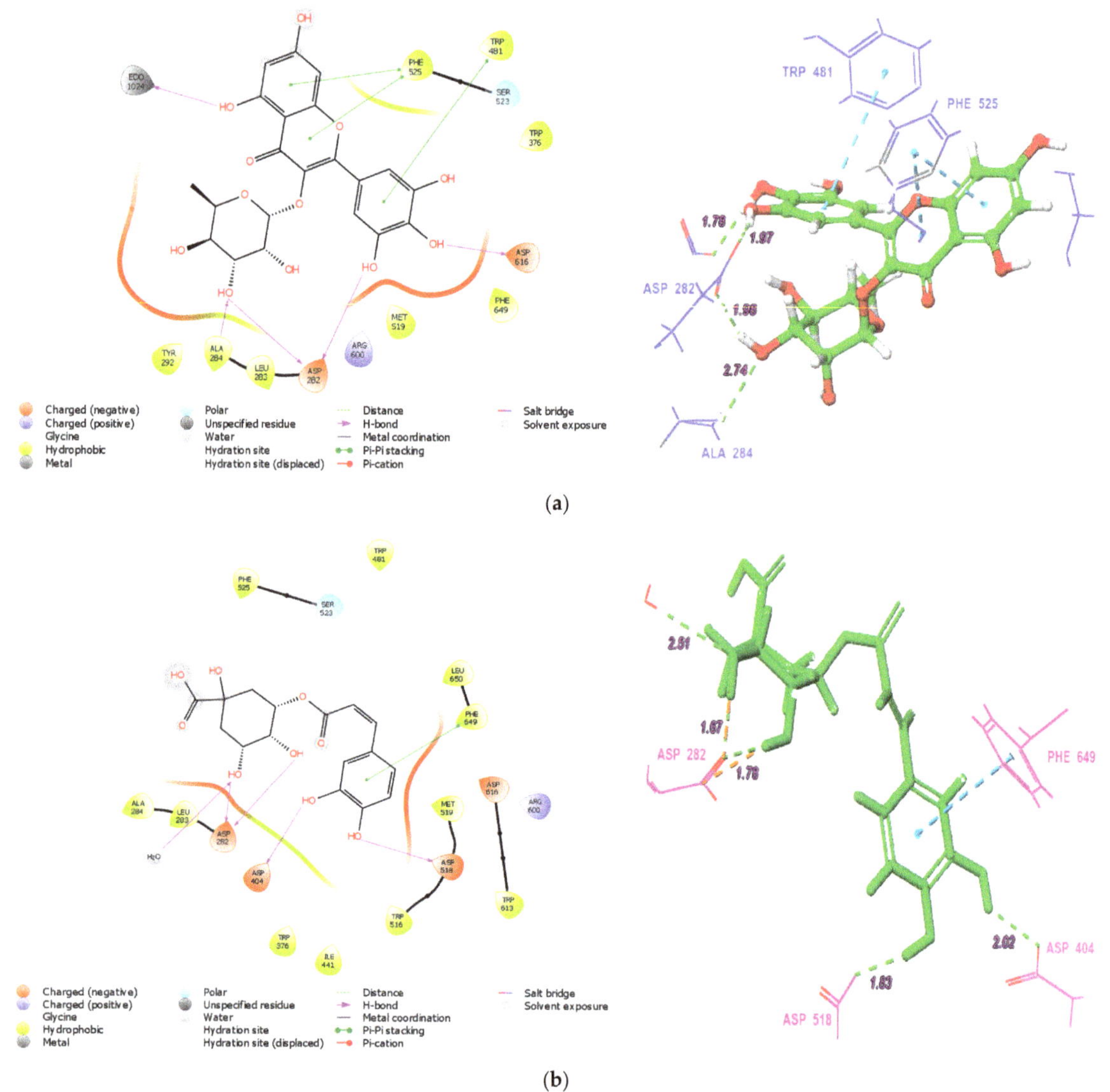

Figure 4. (**a**) The estimated binding geometry (2D (**left**) and 3D (**right**)) of myricitrin in 5NN8. The active side residues are named with three letters. (**b**) The estimated binding geometry (2D (**left**) and 3D (**right**)) of chlorogenic acid in 5NN8. The active side residues are named with three letters.

All compounds were properly docked in 5NN8. Myricitrin (Figure 4A) and quercitrin made two hydrogen bonds with ASP 282 (negatively charged) and one each with ASP 616 (negatively charged), ALA 284 (hydrophobic), and EDO 1024. They had double Pi–Pi stacking hydrophobically with PHE 525 and a single Pi–Pi bond with TRP 481. Chlorogenic acid (Figure 4B) made two hydrogen bonds with ASP 282, one hydrogen bond each with ASP 518 and ASP 404, and one with a water molecule; and it had π–π staking with the hydrophobic PHE 649. Quercetin 3-(2-galloylglucoside) made hydrogen bonds with ASP 282 (negatively charged), ASP 616 (negatively charged), ASP 518 (negatively charged), ASP 404 (negatively charged), ALA 284 (hydrophobic), ARG 600 (positively charged), PHE 525 (hydrophobic) and water molecules; and Pi–Pi bonds with TRP 481,

PHE 525, and PHE 649. Moreover, delphinidin 3-rutinoside, delphinidin 3-sambubioside, and rutin made four hydrogen bonds (two with ASP 616, one with ASP 404, and one with ASP 518), six hydrogen bonds (two ASP 282 and one each with ASP 518, ASP 403, EDO 1024, and a water molecule), and six hydrogen bonds (ASP 282, ASP 404, ASP 518, ASP 616, ASN 524 and SER 523); they also had π–π stacking in one (TRP 481), two (TRP 481 and PHE 649), and four (two with PHE 649, one with TRP 481, and one with TRP 376) bonds, respectively. Acarbose made twelve hydrogen bonds (two each with ASP 518, ASP 404, ASP 282, and SER 523; three OH groups from water molecules, which further made hydrogen bonds with ASP 645 and ARG 281). Naringin made hydrogen bonds with ASP 282, PHE 525, LEU 678, EDO 1024, and ARG 281 and one Pi–Pi stacking interaction with TRP 481. Furthermore, diosmin made three hydrogen bonds with the negatively charged ASP 282 and one with the negatively charged ASP 616 (Figure S3). The binding energies of quercetin 3-(2-galloylglucoside), delphinidin 3-rutinoside, cyanidin 3-O-rutinoside, delphinidin 3-sambubioside, rutin, acarbose, cyanidin 3-rhamnoside 5-glucoside, delphinidin, procyanidin B2, myricitrin, 3-feruloylquinic acid, taxifolin, diosmin, quercitrin, chlorogenic acid, naringin, 3-*p*-coumaroylquinic acid, myricetin, quercetin, isorhamnetin, quinic acid, luteolin, (-)-epicatechin, hesperetin, and gallic acid in 5NN8 were calculated as -11.09, -11.08, -10.90, -10.38, -10.14, -9.65, -9.46, -8.48, -8.05, -7.59, -7.32, -7.13, -6.84, -6.72, -6.62, -6.40, -6.35, -6.28, -5.95, -5.68, -6.65, -5.52, -5.36, -5.28, and -5.15 kcal/mol, respectively (Table S3). From the given results, it is predicted that quercetin 3-(2-galloylglucoside) identified in strawberry gum has higher α-glucosidase-inhibiting activity than acarbose. Overall, flavonoids are predicted to have higher binding affinities than the other selected phenolic compounds. Interestingly, 3-feruloylquinic acid has a higher binding affinity than taxifolin, diosmin, quercitrin, naringin, myricetin, quercetin, isorhamnetin, and luteolin chlorogenic acid; and 3-*p*-coumaroylquinic acid has a higher binding affinity than myricetin, quercetin, isorhamnetin, luteolin, (-)-epicatechin, hesperetin, diosmetin, and naringenin (Table S3). In silico molecular docking is a prediction of possible interactions between target proteins (5NN8) and potential inhibitors. Therefore, it is critical to assess the inhibitory activities of individual purified phenolic compounds to establish the precise roles of individual bioactive compounds in the inhibition of α-glucosidase. Moreover, the insights into inhibitory mechanisms of bioactive polyphenolic compounds against α-glucosidase and other proteins involved in diabetic conditions can be revealed through advanced molecular dynamics techniques and free-energy calculations, and through inverse molecular docking [61].

3.7. Pharmacokinetics Study of Selected Phenolic Compounds

Using computational methods to test the potential drug metabolites helps reduce the number of experimental studies and improve the success rate in pharmacokinetics studies. Absorption, distribution, metabolism, excretion, and toxicological (ADMET) screening were also conducted to validate this study for drug discovery. The interaction of inhibitors with a target receptor cannot guarantee the suitability of phenolic metabolites as drugs for the target pathology. Therefore, ADMET screening of compounds is critical in drug discovery. Unfavorable characteristics of ADMET in the biological system are the main reasons for the failure of drug molecules during clinical experiments [7]. This study evaluated the most abundant phenolic compounds identified in selected plants for ADMET properties.

3.7.1. Absorption and Distribution

The absorption of the phenolic compounds was predicted through the BIOLED-Egg method and using the pkCSM platform. The results of absorption are given in Figure 5 and Table S4 and Table S5.

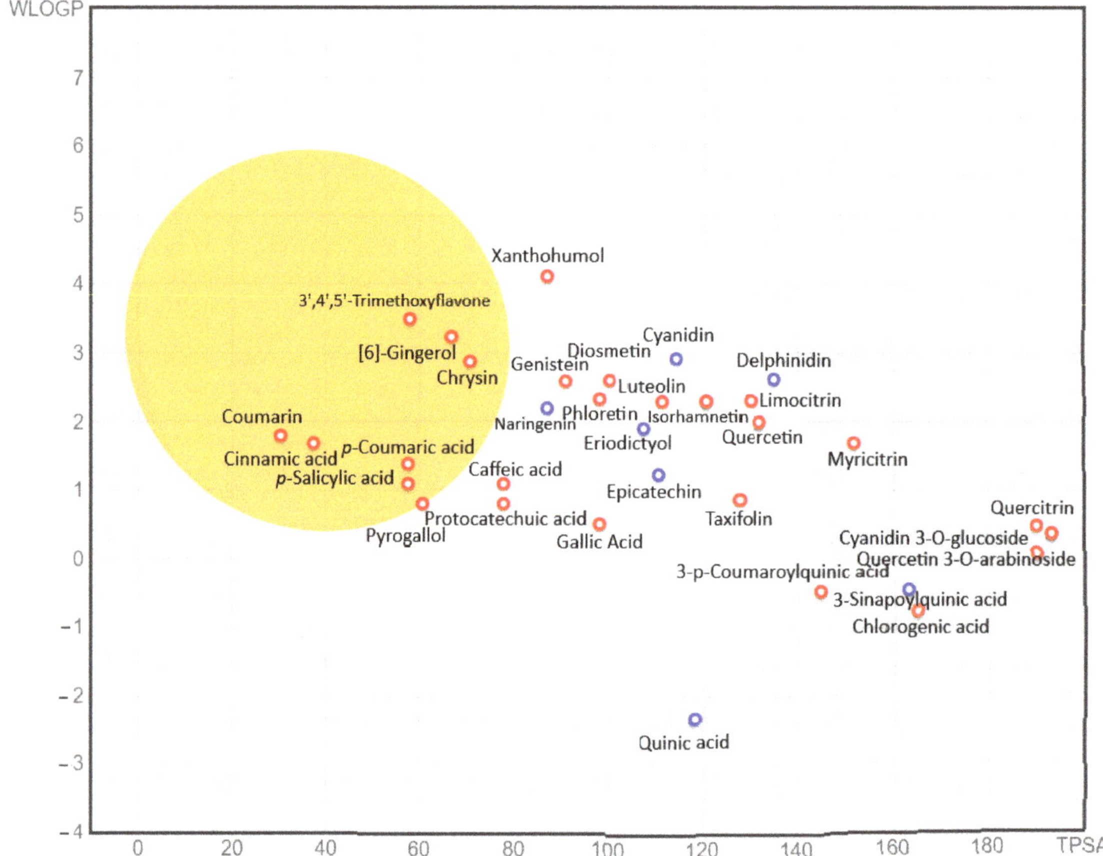

Figure 5. Evaluation of abundant phenolic compounds through the BOILED-Egg method. The blue dots indicate molecules predicted to be expelled from the CNS by P-glycoprotein, and the red dots indicate molecules predicted not to be expelled from the CNS by P-glycoprotein. The egg-yolk area predicts the phenolic metabolites that will passively penetrate the blood–brain barrier. In contrast, the egg-white area predicts which phenolic compounds will be absorbed through the gastrointestinal tract.

Figure 5 predicts that cinnamic acid, coumarin, *p*-coumaric acid, *p*-hydroxybenzoic acid, chrysin, [6]-gingerol, and 3′,4′,5′-trimethoxyflavone pass through the blood–brain barrier; and gallic acid, protocatechuic acid, caffeic acid, pyrogallol, cyanidin, taxifolin, epicatechin, delphinidin, naringenin, genistein, phloretin, quercetin, diosmetin, isorhamnetin, limocitrin, and eriodictyol should be absorbed through the gastrointestinal tract. Moreover, the results predict that the cinnamic acid found in mountain pepper berries and strawberry gum will more readily cross the blood–brain barrier than other phenolic compounds (Table S4). 3′,4′,5′-Trimethoxyflavone (98.1%), coumarin (97.3%), cinnamic acid (94.8%), chrysin (93.8%), *p*-coumaric acid (93.5%), genistein (93.4%), [6]-gingerol (92.4%) naringenin (91.3%), xanthohumol (89.9%), cyanidin (87.3%), *p*-hydroxybenzoic acid (84%), pyrogallol (83.6%), and luteolin (81.1%) are predicted to have the highest human intestinal absorption. Coumarin is the only compound which is predicted to pass through the skin. It is worth noting that anthocyanin aglycones with sugar moieties are predicted to have no human intestinal absorption (Table S4). Therefore, we can predict that anthocyanins with sugar moieties may play a role in gut modulation after the breakdown through colonic fermentation into their basic aglycones, or they will play a role as prebiotic polyphenols. Additionally, cinnamic (1.72), coumarin (1.65), 3′,4′,5′-trimethoxyflavone (1.39), *p*-coumaric acid (1.21), *p*-hydroxybenzoic acid (1.15), pyrogallol (1.12), naringenin (1.03), chrysin (0.95), [6]-gingerol (0.94), and taxifolin (0.92), are predicted to have the highest Caco-2 cell permeability. If the Caco2 permeability value is higher than 0.90, a compound is considered to have high Caco-2 permeability. Furthermore, the compounds which have Caco-2 permeability, gastrointestinal absorption, a good bioavailability score, and obey Lipinski's rule of five while not being able to pass through the BBB, not acting as P-gp substrates, and having poor skin permeability should be successful drugs [62].

Most of the flavonoids that are not absorbed in the gastrointestinal tract can be metabolized by gut microbiota into small phenolic metabolites, where they tend to be absorbed in the colon [48]. Flavonoids are bound to albumin and transported to the liver through the portal vein after absorption. However, the bioavailability of flavonoids is low due to the limited absorption, extensive metabolism, and rapid excretion [63].

3.7.2. Drug-Likeness

The bioavailability radars of selected compounds were obtained by following the method of Daina et al. [17] to predict the drug-likeness to assess the oral bioavailability of compounds (Figure 6).

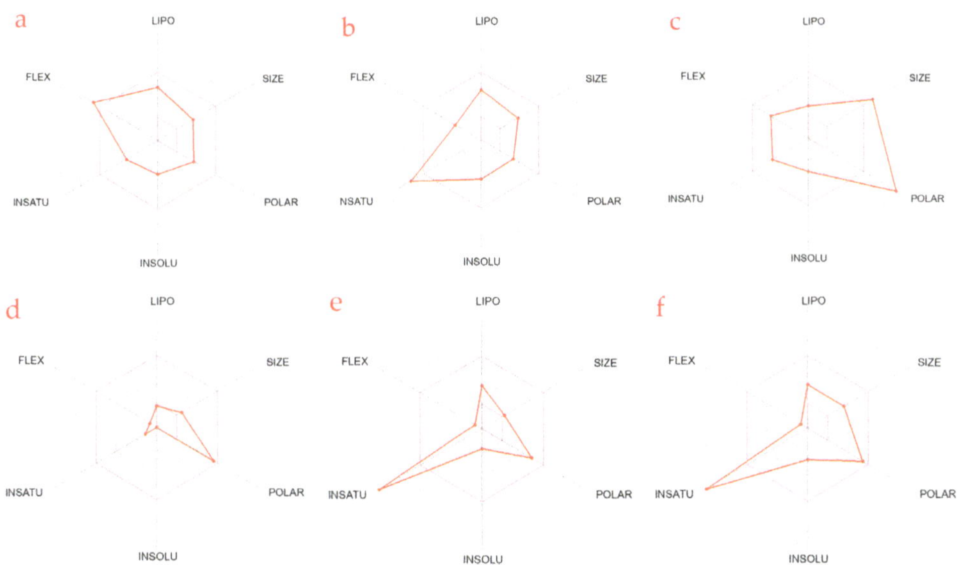

Figure 6. The pink area of the bioavailability radar represents the optimal range for each property. Radars of [6]-gingerol (**a**), 3′,4′,5′-trimethoxyflavone (**b**), naringin (**c**), quinic acid (**d**), gallic acid (**e**), and cyanidin (**f**) were obtained.

Figure 6 and Table S6 depict that no compound predicted oral bioavailability except quinic acid. To predict the oral bioavailability of selected compounds, six physiochemical properties (size, polarity, lipophilicity, flexibility, saturation, and solubility) were considered and analyzed through the bioavailability radar.

3.7.3. Metabolism, Excretion, and Toxicity

Cytochrome P450 (CYP) plays a vital role in the metabolism of bioactive compounds (drugs) [63]. The predicted metabolism and excretion of the phenolic compounds are given in Table S7. Metabolism was predicted through the CYP model for substrate or inhibitor

(CYP1A2, CYP2D6, CYP3A4, CYP2C9, and CYP2C19). Bioactive compounds that inhibit the CYP pathway may cause elevated concentrations of other bioactive compounds, resulting in higher toxicity of that compound and vice versa. Bioactive compounds with higher total clearance are predicted to have higher bioavailability and metabolism in the liver (Table S5). Virtual toxicological screening of the bioactive compounds is provided in Table S8. The predicted results indicate that all bioactive compounds do not inhibit the hERG 1 channel, and no compound predicted AIME toxicity, hepatotoxicity, skin sensitization, *Tetrahymena pyriformis* toxicity, or Minnow toxicity except 3′,4′,5′-trimethoxyflavone, which predicted toxicity in Minnow.

4. Conclusions

In this study, native Australian fruits and spices were comprehensively analyzed for polyphenols, and a total of 143 metabolites were identified. Twenty-six of these compounds were quantified. Strawberry gum had higher total phenolic content, antioxidant capacity, and α-glucosidase inhibition activity than rosella, lemon aspen, and mountain pepper berries. Furthermore, in silico molecular docking predicted that flavonoids have a significant role in the inhibition of α-glucosidase. Additionally, simulated pharmacokinetics predicted that all screening phenolic compounds from native Australian fruits and spices are safe and do not have any toxicity, and small phenolic metabolites such as phenolic acids have higher absorption in Caco-2 cells and the gastrointestinal tract than other phenolic compounds. This study demonstrates that strawberry gum has a significant medicinal and pharmaceutical potential that could be utilized in food, feed, cosmetic, and pharmaceutical industries with the further proved in vivo data.

Supplementary Materials: The following supporting information can be downloaded at: https://www.mdpi.com/article/10.3390/antiox12020254/s1. Figure S1: Base peak chromatograms (BPC) of mountain pepper berries, rosella, lemon aspen, and strawberry gum in positive (black) and negative (blue) modes of ionization. Figure S2: MS/MS spectra of some selected compounds. Figure S3: Two-dimensional binding geometry of some selected compounds. Table S1: Antioxidant activities of native Australian fruits and spices. Table S2: Quantification/semi-quantification of phenolic metabolites in Australian native fruits and spices (µg/g). Table S3: The calculated binding energies of selected compounds. Table S4: Predicted absorption and distribution of selected compounds. Table S5: Pharmacokinetic properties of selected compounds. Table S6: Radar bioavailability properties of selected compounds. Table S7: Metabolism and excretion of selected compounds. Table S8: Predicted toxicity of abundant phenolic compounds.

Author Contributions: Conceptualization, methodology, formal analysis, investigation, software, validation, data curation, visualization, writing—original draft preparation, A.A.; writing—review and editing, supervision, resources, project administration, funding acquisition, F.R.D. and J.J.C. All authors have read and agreed to the published version of the manuscript.

Funding: This research received no external funding.

Institutional Review Board Statement: Not applicable.

Informed Consent Statement: Not applicable.

Data Availability Statement: The supporting data are available in the Supplementary Materials.

Acknowledgments: We owe incredible thanks to William Nikolas, Asif Noor, and Swati Varshney from the Mass Spectrometry Proteomics Facility, Bio21 Molecular Institute, VIC, Australia, for providing their support in training and learning.

Conflicts of Interest: Authors declare no conflict of interest.

References

1. Ali, A.; Cottrell, J.J.; Dunshea, F.R. Identification and characterization of anthocyanins and non-anthocyanin phenolics from australian native fruits and their antioxidant, antidiabetic, and anti-alzheimer potential. *Food Res. Int.* **2022**, *162*, 111951. [CrossRef]

2. Ali, A.; Zahid, H.F.; Cottrell, J.J.; Dunshea, F.R. A comparative study for nutritional and phytochemical profiling of coffea arabica (c. Arabica) from different origins and their antioxidant potential and molecular docking. *Molecules* **2022**, *27*, 5126.

3. Tsao, R. Chemistry and biochemistry of dietary polyphenols. *Nutrients* **2010**, *2*, 1231–1246.

4. Kiloni, S.M.; Akhtar, A.; Cáceres-Vélez, P.R.; Dunshea, F.; Jusuf, P. P06-05 zebrafish embryo acute toxicity and antioxidant characterization of native australian plants: Towards safe and effective glaucoma treatments. *Toxicol. Lett.* **2022**, *368*, S115.

5. Cock, I.E. Medicinal and aromatic plants–australia. *Ethnopharmacol. Encycl. Life Support Syst. EOLSS* **2011**, *1*, 1–173.

6. Richmond, R.; Bowyer, M.; Vuong, Q. Australian native fruits: Potential uses as functional food ingredients. *J. Funct. Foods* **2019**, *62*, 103547.

7. Attique, S.A.; Hassan, M.; Usman, M.; Atif, R.M.; Mahboob, S.; Al-Ghanim, K.A.; Bilal, M.; Nawaz, M.Z. A molecular docking approach to evaluate the pharmacological properties of natural and synthetic treatment candidates for use against hypertension. *Int. J. Environ. Res. Public Health* **2019**, *16*, 923. [CrossRef] [PubMed]

8. Bahun, M.; Jukić, M.; Oblak, D.; Kranjc, L.; Bajc, G.; Butala, M.; Bozovičar, K.; Bratkovič, T.; Podlipnik, Č.; Poklar Ulrih, N. Inhibition of the sars-cov-2 3cl(pro) main protease by plant polyphenols. *Food Chem.* **2022**, *373*, 131594. [PubMed]

9. Jukič, M.; Janežič, D.; Bren, U. Potential novel thioether-amide or guanidine-linker class of sars-cov-2 virus rna-dependent rna polymerase inhibitors identified by high-throughput virtual screening coupled to free-energy calculations. *Int. J. Mol. Sci.* **2021**, *22*, 11143. [CrossRef] [PubMed]

10. Ali, A.; Wu, H.; Ponnampalam, E.N.; Cottrell, J.J.; Dunshea, F.R.; Suleria, H.A.R. Comprehensive profiling of most widely used spices for their phenolic compounds through lc-esi-qtof-ms2 and their antioxidant potential. *Antioxidants* **2021**, *10*, 721.

11. Sharifi-Rad, J.; Song, S.; Ali, A.; Subbiah, V.; Taheri, Y.; Suleria, H.A.R. Lc-esi-qtof-ms/ms characterization of phenolic compounds from pyracantha coccinea m. Roem. And their antioxidant capacity. *Cell. Mol. Biol.* **2021**, *67*, 201–211. [CrossRef] [PubMed]

12. Chou, O.; Ali, A.; Subbiah, V.; Barrow, C.J.; Dunshea, F.R.; Suleria, H.A.R. Lc-esi-qtof-ms/ms characterisation of phenolics in herbal tea infusion and their antioxidant potential. *Fermentation* **2021**, *7*, 73.

13. Zahid, H.F.; Ali, A.; Ranadheera, C.S.; Fang, Z.; Dunshea, F.R.; Ajlouni, S. In vitro bioaccessibility of phenolic compounds and alpha-glucosidase inhibition activity in yoghurts enriched with mango peel powder. *Food Biosci.* **2022**, *50*, 102011.

14. Bashmil, Y.M.; Ali, A.; BK, A.; Dunshea, F.R.; Suleria, H.A.R. Screening and characterization of phenolic compounds from australian grown bananas and their antioxidant capacity. *Antioxidants* **2021**, *10*, 1521.

15. Ali, A.; Cottrell, J.J.; Dunshea, F.R. Lc-ms/ms characterization of phenolic metabolites and their antioxidant activities from australian native plants. *Metabolites* **2022**, *12*, 1016. [PubMed]

16. Ali, A.; Kiloni, S.M.; Cáceres-Vélez, P.R.; Jusuf, P.R.; Cottrell, J.J.; Dunshea, F.R. Phytochemicals, antioxidant activities, and toxicological screening of native australian fruits using zebrafish embryonic model. *Foods* **2022**, *11*, 4038.

17. Daina, A.; Michielin, O.; Zoete, V. Swissadme: A free web tool to evaluate pharmacokinetics, drug-likeness and medicinal chemistry friendliness of small molecules. *Sci. Rep.* **2017**, *7*, 42717.

18. Ali, A.; Bashmil, Y.M.; Cottrell, J.J.; Suleria, H.A.R.; Dunshea, F.R. Lc-ms/ms-qtof screening and identification of phenolic compounds from australian grown herbs and their antioxidant potential. *Antioxidants* **2021**, *10*, 1770. [PubMed]

19. Konczak, I.; Zabaras, D.; Dunstan, M.; Aguas, P. Antioxidant capacity and phenolic compounds in commercially grown native australian herbs and spices. *Food Chem.* **2010**, *122*, 260–266. [CrossRef]

20. Cáceres-Vélez, P.R.; Ali, A.; Fournier-Level, A.; Dunshea, F.R.; Jusuf, P.R. Phytochemical and safety evaluations of finger lime, mountain pepper, and tamarind in zebrafish embryos. *Antioxidants* **2022**, *11*, 1280. [CrossRef] [PubMed]

21. Vélez, P.R.C.; Ali, A.; Fournier-Level, A.; Dunshea, F.; Jusuf, P.R. P06-04 antioxidant activity and embryotoxicity of citrus australasica, tasmannia lanceolata and diploglottis australis extracts in zebrafish. *Toxicol. Lett.* **2022**, *368*, S114. [CrossRef]

22. Hu, T.; Subbiah, V.; Wu, H.; Bk, A.; Rauf, A.; Alhumaydhi, F.A.; Suleria, H.A.R. Determination and characterization of phenolic compounds from australia-grown sweet cherries (prunus avium l.) and their potential antioxidant properties. *ACS Omega* **2021**, *6*, 34687–34699. [CrossRef]

23. Lukmanto, S.; Roesdiyono, N.; Ju, Y.-H.; Indraswati, N.; Soetaredjo, F.E.; Ismadji, S. Supercritical co2 extraction of phenolic compounds in roselle (hibiscus sabdariffa l.). *Chem. Eng. Commun.* **2013**, *200*, 1187–1196. [CrossRef]

24. Liu, H.; Qiu, N.; Ding, H.; Yao, R. Polyphenols contents and antioxidant capacity of 68 chinese herbals suitable for medical or food uses. *Food Res. Int.* **2008**, *41*, 363–370.

25. Yang, W.-J.; Li, D.-P.; Li, J.-K.; Li, M.-H.; Chen, Y.-L.; Zhang, P.-Z. Synergistic antioxidant activities of eight traditional chinese herb pairs. *Biol. Pharm. Bull.* **2009**, *32*, 1021–1026. [CrossRef] [PubMed]

26. Yoo, K.M.; Lee, C.H.; Lee, H.; Moon, B.; Lee, C.Y. Relative antioxidant and cytoprotective activities of common herbs. *Food Chem.* **2008**, *106*, 929–936. [CrossRef]

27. Tsai, M.-L.; Lin, C.-C.; Lin, W.-C.; Yang, C.-H. Antimicrobial, antioxidant, and anti-inflammatory activities of essential oils from five selected herbs. *Biosci. Biotechnol. Biochem.* **2011**, *75*, 1977–1983.

28. Lin, C.-C.; Yang, C.-H.; Wu, P.-S.; Kwan, C.-C.; Chen, Y.-S. Antimicrobial, anti-tyrosinase and antioxidant activities of aqueous aromatic extracts from forty-eight selected herbs. *J. Med. Plants Res.* **2011**, *5*, 6203–6209.

29. Garg, D.; Muley, A.; Khare, N.; Marar, T. Comparative analysis of phytochemical profile and antioxidant activity of some indian culinary herbs. *Res. J. Pharm. Biol. Chem. Sci.* **2012**, *3*, 845–854.

30. Chen, I.-C.; Chang, H.-C.; Yang, H.-W.; Chen, G.-L. Evaluation of total antioxidant activity of several popular vegetables and chinese herbs: A fast approach with abts/h2o2/hrp system in microplates. *J. Food Drug Anal.* **2004**, *12*, 29–33. [CrossRef]

31. Köksal, E.; Bursal, E.; Gülçin, İ.; Korkmaz, M.; Çağlayan, C.; Gören, A.C.; Alwasel, S.H. Antioxidant activity and polyphenol content of turkish thyme (thymus vulgaris) monitored by liquid chromatography and tandem mass spectrometry. *Int. J. Food Prop.* **2017**, *20*, 514–525. [CrossRef]

32. Wojdyło, A.; Oszmiański, J.; Czemerys, R. Antioxidant activity and phenolic compounds in 32 selected herbs. *Food Chem.* **2007**, *105*, 940–949.

33. Gülçin, İ. Antioxidant activity of caffeic acid (3, 4-dihydroxycinnamic acid). *Toxicology* **2006**, *217*, 213–220. [CrossRef] [PubMed]

34. Gülçin, İ.; Huyut, Z.; Elmastaş, M.; Aboul-Enein, H.Y. Radical scavenging and antioxidant activity of tannic acid. *Arab. J. Chem.* **2010**, *3*, 43–53.

35. Gülçın, İ.; Oktay, M.; Kıreçcı, E.; Küfrevıoğlu, Ö.İ. Screening of antioxidant and antimicrobial activities of anise (pimpinella anisum l.) seed extracts. *Food Chem.* **2003**, *83*, 371–382.

36. Syabana, M.A.; Yuliana, N.D.; Batubara, I.; Fardiaz, D. Antidiabetic activity screening and nmr profile of vegetable and spices commonly consumed in indonesia. *Food Sci. Technol.* **2020**, *41*, 254–264. [CrossRef]

37. Taslimi, P.; Köksal, E.; Gören, A.C.; Bursal, E.; Aras, A.; Kılıç, Ö.; Alwasel, S.; Gülçin, İ. Anti-alzheimer, antidiabetic and antioxidant potential of satureja cuneifolia and analysis of its phenolic contents by lc-ms/ms. *Arab. J. Chem.* **2020**, *13*, 4528–4537.

38. Cottrell, J.J.; Le, H.H.; Artaiz, O.; Iqbal, Y.; Suleria, H.A.; Ali, A.; Celi, P.; Dunshea, F.R. Recent advances in the use of phytochemicals to manage gastrointestinal oxidative stress in poultry and pigs. *Anim. Prod. Sci.* **2022**, *62*, 1140–1146.

39. Granato, D.; Shahidi, F.; Wrolstad, R.; Kilmartin, P.; Melton, L.D.; Hidalgo, F.J.; Miyashita, K.; van Camp, J.; Alasalvar, C.; Ismail, A.B. Antioxidant activity, total phenolics and flavonoids contents: Should we ban in vitro screening methods? *Food Chem.* **2018**, *264*, 471–475. [CrossRef]

40. Freeman, B.L.; Eggett, D.L.; Parker, T.L. Synergistic and antagonistic interactions of phenolic compounds found in navel oranges. *J. Food Sci.* **2010**, *75*, C570–C576. [CrossRef]

41. Shan, B.; Cai, Y.Z.; Sun, M.; Corke, H. Antioxidant capacity of 26 spice extracts and characterization of their phenolic constituents. *J. Agric. Food Chem.* **2005**, *53*, 7749–7759.

42. Mandal, S.M.; Chakraborty, D.; Dey, S. Phenolic acids act as signaling molecules in plant-microbe symbioses. *Plant Signal. Behav.* **2010**, *5*, 359–368.

43. Goleniowski, M.; Bonfill, M.; Cusido, R.; Palazón, J. Phenolic acids. *Nat. Prod.* **2013**, *2013*, 1951–1953.

44. Kakkar, S.; Bais, S. A review on protocatechuic acid and its pharmacological potential. *ISRN Pharmacol.* **2014**, *2014*, 952943. [PubMed]

45. Hossain, M.B.; Rai, D.K.; Brunton, N.P.; Martin-Diana, A.B.; Barry-Ryan, C. Characterization of phenolic composition in lamiaceae spices by lc-esi-ms/ms. *J. Agric. Food Chem.* **2010**, *58*, 10576–10581. [PubMed]

46. Dong, J.; Zhu, Y.; Gao, X.; Chang, Y.; Wang, M.; Zhang, P. Qualitative and quantitative analysis of the major constituents in chinese medicinal preparation dan-lou tablet by ultra high performance liquid chromatography/diode-array detector/quadrupole time-of-flight tandem mass spectrometry. *J. Pharm. Biomed. Anal.* **2013**, *80*, 50–62. [CrossRef]

47. Kadam, D.; Palamthodi, S.; Lele, S.S. Lc-esi-q-tof-ms/ms profiling and antioxidant activity of phenolics from l. Sativum seedcake. *J. Food Sci. Technol.* **2018**, *55*, 1154–1163. [PubMed]

48. Aherne, S.A.; O'Brien, N.M. Dietary flavonols: Chemistry, food content, and metabolism. *Nutrition* **2002**, *18*, 75–81.

49. Oh, Y.S.; Lee, J.H.; Yoon, S.H.; Oh, C.H.; Choi, D.S.; Choe, E.; Jung, M.Y. Characterization and quantification of anthocyanins in grape juices obtained from the grapes cultivated in korea by hplc/dad, hplc/ms, and hplc/ms/ms. *J. Food Sci.* **2008**, *73*, C378–C389.

50. Liu, Y.; Tikunov, Y.; Schouten, R.E.; Marcelis, L.F.M.; Visser, R.G.F.; Bovy, A. Anthocyanin biosynthesis and degradation mechanisms in solanaceous vegetables: A review. *Front. Chem.* **2018**, *6*, 52. [CrossRef]

51. Murkovic, M. Phenolic compounds: Occurrence, classes, and analysis. In *Encyclopedia of Food and Health*; Caballero, B., Finglas, P.M., Toldrá, F., Eds.; Academic Press: Oxford, UK, 2016; pp. 346–351.

52. Rockenbach, I.I.; Jungfer, E.; Ritter, C.; Santiago-Schübel, B.; Thiele, B.; Fett, R.; Galensa, R. Characterization of flavan-3-ols in seeds of grape pomace by ce, hplc-dad-msn and lc-esi-fticr-ms. *Food Res. Int.* **2012**, *48*, 848–855.

53. Zahid, H.F.; Ali, A.; Ranadheera, C.S.; Fang, Z.; Ajlouni, S. Identification of phenolics profile in freeze-dried apple peel and their bioactivities during in vitro digestion and colonic fermentation. *Int. J. Mol. Sci.* **2023**, *24*, 1514.

54. Thuan, N.H.; Pandey, R.P.; Thuy, T.T.; Park, J.W.; Sohng, J.K. Improvement of regio-specific production of myricetin-3-o-α-l-rhamnoside in engineered escherichia coli. *Appl. Biochem. Biotechnol.* **2013**, *171*, 1956–1967. [CrossRef]

55. Habtemariam, S. A-glucosidase inhibitory activity of kaempferol-3-o-rutinoside. *Nat. Prod. Commun.* **2011**, *6*, 201–203. [CrossRef]

56. Chang, Z.; Zhang, Q.; Liang, W.; Zhou, K.; Jian, P.; She, G.; Zhang, L. A comprehensive review of the structure elucidation of tannins from terminalia linn. *Evid. Based Complement. Altern. Med.* **2019**, *2019*, 8623909. [CrossRef] [PubMed]

57. Hou, K.; Wang, Z. Application of nanotechnology to enhance adsorption and bioavailability of procyanidins: A review. *Food Rev. Int.* **2021**, *38*, 738–752. [CrossRef]

58. Kitts, D.D.; Yuan, Y.V.; Wijewickreme, A.N.; Thompson, L.U. Antioxidant activity of the flaxseed lignan secoisolariciresinol diglycoside and its mammalian lignan metabolites enterodiol and enterolactone. *Mol. Cell. Biochem.* **1999**, *202*, 91–100.

59. Liu, Z.; Fei, Y.J.; Cao, X.H.; Xu, D.; Tang, W.J.; Yang, K.; Xu, W.X.; Tang, J.H. Lignans intake and enterolactone concentration and prognosis of breast cancer: A systematic review and meta-analysis. *J. Cancer* **2021**, *12*, 2787–2796. [PubMed]

60. Konczak, I.; Zabaras, D.; Dunstan, M.; Aguas, P. Antioxidant capacity and hydrophilic phytochemicals in commercially grown native australian fruits. *Food Chem.* **2010**, *123*, 1048–1054.
61. Pantiora, P.; Furlan, V.; Matiadis, D.; Mavroidi, B.; Perperopoulou, F.; Papageorgiou, A.C.; Sagnou, M.; Bren, U.; Pelecanou, M.; Labrou, N.E. Monocarbonyl curcumin analogues as potent inhibitors against human glutathione transferase p1-1. *Antioxidants* **2023**, *12*, 63. [CrossRef]
62. Khalfaoui, A.; Noumi, E.; Belaabed, S.; Aouadi, K.; Lamjed, B.; Adnan, M.; Defant, A.; Kadri, A.; Snoussi, M.; Khan, M.A.; et al. Lc-esi/ms-phytochemical profiling with antioxidant, antibacterial, antifungal, antiviral and in silico pharmacological properties of algerian asphodelus tenuifolius (cav.) organic extracts. *Antioxidants* **2021**, *10*, 628. [CrossRef] [PubMed]
63. Khan, J.; Deb, P.K.; Priya, S.; Medina, K.D.; Devi, R.; Walode, S.G.; Rudrapal, M. Dietary flavonoids: Cardioprotective potential with antioxidant effects and their pharmacokinetic, toxicological and therapeutic concerns. *Molecules* **2021**, *26*, 4021. [CrossRef] [PubMed]

Article

Salt Eustress Induction in Red Amaranth (*Amaranthus gangeticus*) Augments Nutritional, Phenolic Acids and Antiradical Potential of Leaves

Umakanta Sarker [1,*] and Sezai Ercisli [2]

[1] Department of Genetics and Plant Breeding, Faculty of Agriculture, Bangabandhu Sheikh Mujibur Rahman Agricultural University, Gazipur 1706, Bangladesh
[2] Department of Horticulture, Faculty of Agriculture, Ataturk University, Erzurum 25240, Turkey
*   Correspondence: umakanta@bsmrau.edu.bd

Abstract: Earlier researchers have highlighted the utilization of salt eustress for boosting the nutritional and phenolic acid (PA) profiles and antiradical potential (ARP) of vegetables, which eventually boost food values for nourishing human diets. Amaranth is a rapidly grown, diversely acclimated C_4 leafy vegetable with climate resilience and salinity resistance. The application of salinity eustress in amaranth has a great scope to augment the nutritional and PA profiles and ARP. Therefore, the *A. gangeticus* genotype was evaluated in response to salt eustress for nutrients, PA profile, and ARP. Antioxidant potential and high-yielding genotype (LS1) were grown under four salt eustresses (control, 25 mM, 50 mM, 100 mM NaCl) in a randomized completely block design (RCBD) in four replicates. Salt stress remarkably augmented microelements, proximate, macro-elements, phytochemicals, PA profiles, and ARP of *A. gangeticus* leaves in this order: control < low sodium chloride stress (LSCS) < moderate sodium chloride stress (MSCS) < severe sodium chloride stress (SSCS). A large quantity of 16 PAs, including seven cinnamic acids (CAs) and nine benzoic acids (BAs) were detected in *A. gangeticus* genotypes. All the microelements, proximate, macro-elements, phytochemicals, PA profiles, and ARP of *A. gangeticus* under MSCS, and SSCS levels were much higher in comparison with the control. It can be utilized as preferential food for our daily diets as these antiradical compounds have strong antioxidants. Salt-treated *A. gangeticus* contributed to excellent quality in the end product in terms of microelements, proximate, macro-elements, phytochemicals, PA profiles, and ARP. *A. gangeticus* can be cultivated as an encouraging substitute crop in salt-affected areas of the world.

Keywords: *A. gangeticus*; protein and dietary fiber; minerals; phytochemicals; HPLC-UV DPPH; ABTS+; PA profiles; NaCl

Citation: Sarker, U.; Ercisli, S. Salt Eustress Induction in Red Amaranth (*Amaranthus gangeticus*) Augments Nutritional, Phenolic Acids and Antiradical Potential of Leaves. *Antioxidants* **2022**, *11*, 2434. https://doi.org/10.3390/antiox11122434

Academic Editor: Stanley Omaye

Received: 14 October 2022
Accepted: 29 November 2022
Published: 9 December 2022

Publisher's Note: MDPI stays neutral with regard to jurisdictional claims in published maps and institutional affiliations.

1. Introduction

Amaranth is a promising millennium vegetable with vast diversity [1–7]. It is an alternate source of nutrients because of its richness in vitamin C, minerals [8–15], vitamins [16–20], protein [21,22], dietary fiber [23–25], leaf pigments [26–42], phenolic compounds [43–58], and flavonoids [59–73] with strong antioxidants [74–86]. Amaranth has a noteworthy contribution as an antioxidant in food manufacturing owing to quenching reactive oxygen species (ROS) [87,88]. Wahid and Ghazanfar [89] reported that extreme salt enhanced the secondary plant metabolites, eventually accelerating plant protection apparatuses against ROS. Salinity enhances ROS production, which causes the oxidation of cellular components. ROS [90]. In plants, antioxidants (non-enzymatic), such as proteins, flavonoids, carbohydrates, carotenoids, and phenolic compounds, and enzymatic antioxidants are capable of ROS detoxification [90,91]. Hence, in human life, salt-tolerant plants could be considered a source of potent antioxidants. These compounds have extraordinary benefits to our food owing to quenching ROS and protecting against numerous diseases, such as cancer, cardiovascular diseases, atherosclerosis, cataracts, retinopathy, emphysema, arthritis, and neuron-damaging diseases [88].

Taste, flavor, and color determine the suitability of foods. Recently, consumers are very much interested in coloring food products. These products have much interest in the nutritional, safety, and beautification aspects of customers as foods. The utilization of natural pigments is considerably increasing day by day. The selected *A. gangeticus* genotype had sufficient betalains with bright red-violet color. *Amaranthus* leafy vegetable is an exclusive origin of betalains with significant quenching capacity of free radicals [92]. In low-acid foods, betalains are preferable to be utilized as a food colorant. These have greater stability than anthocyanins for pH [93], have preferential utility in the promotion of health, act as anti-inflammatory compounds, and diminish the risk of cancers of the skin and lungs and cardiovascular diseases.

Amaranth is an extensively acclimated leafy vegetable due to diverse stresses, such as salinity [94–96] and drought [97], as well as having multiple uses. Salinity stress is a pioneer for the rapid augmentation of the quantity and quality of natural antioxidants through diverse factors, such as physiological, environmental, ecological, biological, biochemical, and evolutionary processes [98]. Very limited reports on the effect of salinity stress are available in terms of minerals, proximate, and bioactive compounds in different crops including leafy vegetables. Petropoulos et al. [99] reported the salinity-induced reduction of chlorophylls, fat, sugar, and carbohydrate and the augmentation of flavonoids, ascorbic acid (AsA), phenolics, proteins, and ARP in *Cichorium spinosum*. Different concentrations of sodium chloride enhanced the carotenoid content in buckwheat sprouts in comparison to the control [100]. Alam et al. [101] reported salt-induced amelioration of phenolics, ARP, and flavonoids in purslane. Ahmed et al. [102] recorded a salinity-induced increase in ARP and phenolics in barley. The influence of sodium chloride stress on the phytochemicals, nutrients, ARP, and PA profiles in *A. gangeticus* was studied for the first time. Based on our previous studies, the ARP genotype (accession LS1) along with high yield potential were selected. Therefore, the response of sodium chloride stress was assessed in *A. gangeticus* in terms of phytochemicals, nutrients, ARP, and PA profiles.

2. Materials and Methods

2.1. Experimental Site, Conditions, and Plant Materials

A high-yielding ARP genotype (accession LS1) was selected from among 120 genotypes from the Department of Genetics and Plant Breeding's collection. The seeds were sown in four replicates following a block design with complete randomization (RCBD) in plastic pots at the Bangabandhu Sheikh Mujibur Rahman Agricultural University (24°23′ N, 90°08′ E, 8.4 m.s.l., AEZ-28 [103,104]. Pots were filled with sandy loam soil. P_2O_5:K_2O was applied @ 48:60 kg ha^{-1} during the final land preparation. However, N was applied @ 46 kg ha^{-1} in two equally split doses during the final land preparation and 10 days after the sowing of the seeds. Four salt treatments, 100 (severe sodium chloride stress, SSCS), 50 (moderate sodium chloride stress, MSCS), and 25 (low sodium chloride stress, LSCS) mM NaCl, and a control (normal water) were used in the study. Pots were regularly irrigated with normal water for 10 days after sowing (DAS). At 11 DAS, salt treatments were imposed and sustained until the edible stage (30 DAS). Pots were irrigated once a day using salt water (100, 50, and 25 mM NaCl) and normal water. *A. gangeticus* leaves were harvested at 30 DAS.

2.2. Chemicals

Acetone, $HClO_4$, HNO_3, Sb, dithiothreitol (DTT), CsCl, AsA, 2, 2-dipyridyl, Trolox, PAs, HPLC grade acetonitrile, acetic acid, gallic acid (GAA), NaOH, rutin, DPPH, H_2SO_4, Folin-Ciocalteu reagent, MeOH, ABTS$^+$, $AlCl_3$.6H2O, Na_2CO_3, CH_3CO_2K, and $K_2S_2O_8$. All chemicals were bought from Kanto Chemical Co. Inc. (Tokyo, Japan) and Merck (Germany).

2.3. Ash, Fiber, Moisture, Fat, Gross Energy, Carbohydrate, and Protein Estimation

The ash, fiber, moisture, fat, gross energy, and protein were estimated by the AOAC method [105–107]. The mini-Kjeldahl method was followed to measure nitrogen (N). Protein was calculated by multiplying N with 6.25. Protein, ash, fat, and moisture (%) were deducted from 100 to estimate carbohydrates.

2.4. Elements Estimation

The leaves were dried in an oven at 70 °C temperature for 24 h. Mineral elements were determined from the ground leaf by digesting with HNO_3 and perchloric acid [105,108]. Exactly 0.5 g of the leaf samples were digested with 400 mL HNO_3 (65%), 40 mL $HClO_4$ (70%), and 10 mL H_2SO_4 (96%). The absorbance was read at 213.9 (Zn), 285.2 (Mg), 766.5 (K), 279.5 (Mn), 248.3 (Fe), 258.056 (S), 422.7 (Ca), 880 (P), 589 (Na), 430 (B), 313.3 (Mo), and 324.8 (Cu) nm wavelengths using an atomic absorption spectrophotometer (AAS with flame) (Hitachi, Japan). Macro- and micro-elements were expressed in mg g^{-1} and μg g^{-1} FW.

2.5. Beta-Carotene

In a mortar and pestle, 500 mg leaves (fresh) were thoroughly mixed with 10 mL acetone (80%). The mixture was centrifuged at $10,000\times g$ for 3–4 min for β-carotene determination [109]. After the separation of the filtrate in a flask, the final volume of 20 mL was maintained. The absorbance was taken at 510 and 480 nm by spectrophotometer (Tokyo, Japan). β-Carotene was expressed in fresh weight as mg 100 g^{-1}.

2.6. Ascorbic Acid (AsA) Estimation

Fresh leaves were used to determine AsA and DHA. The sample was pre-incubated by dithiothreitol (DTT), which reduced DHA to AsA. With the reduction of AsA, Fe^{3+} converted to Fe^{2+}. Fe^{2+} complexes were formed by reacting Fe^{2+} and 2, 2-dipyridyl [109]. The absorbance of the complexes was taken at 525 nm by a spectrophotometer (Hitachi, Japan) to measure AsA in mg 100 g^{-1}.

2.7. Samples Extraction and Determination of Total Polyphenols (TP), Total Flavonoids (TF), and Antiradical Potential (ARP)

Leaves were dried in a shady place to avoid direct sunshine. The extraction was performed from both the ground dried and fresh leaves (30 d) separately with a mortar and pestle. Total polyphenols (TP) were measured from fresh leaves, while total flavonoids (TF) content and ARP were determined from dried leaves. A 90% MeOH solution 10 mL was added with 0.25 g samples in a capped bottle tightly. The mixture was placed for 1 h in a shaker (Tokyo, Japan) at 60 °C. The final filtrate was stored for TP, TF, and ARP estimation. TF and TP were estimated by the $AlCl_3$ colorimetric method and the Folin-Ciocalteu reagent, respectively [105,110]. The absorbance at 760 and 415 nm with a spectrophotometer (Hitachi, Japan). TP and TF were measured as GAA and rutin equivalent μg GAE g^{-1} of FW and μg RE g^{-1} DW using standard GAA and rutin curves. The Trolox equivalent antioxidant activity (TEAC) of ARP was estimated by the DPPH reduction and the $ABTS^+$ assay [105,111]. $ABTS^+$ and DPPH reduction percentage equivalent to the control was measured for estimating the ARP using the equation:

$$\text{ARP (\%)} = (Ac - As/Ac) \times 100$$

where Ac denotes the control absorbance (150 μL MeOH for ARP (ABTS) and 10 μL MeOH for ARP (DPPH) instead of leaf extract) and As is the absorbance of the samples. The results were calculated as μg Trolox equivalent g^{-1} DW.

2.8. Samples Extraction and Determination of Phenolic Acids (PAs) by HPLC

Fresh leaves (1 g) were extracted in MeOH (10 mL, 80%) containing CH_3COOH (1%). The thoroughly homogenized mixture was kept in a 50 mL tightly capped test tube and placed in a shaker (Scientific Industries Inc., New York, NY, USA) for 15 h at 400 rpm. It was filtered in a 0.45 µm filter (MA, New York, USA) and centrifuged for 15 min at $10,000 \times g$. The filtrate was used to estimate PAs. All extractions were repeated 3 times. The method of Sarker and Oba [112] was followed to determine PAs using HPLC. Shimadzu HPLC (Kyoto, Japan) was furnished with a binary pump, degasser, and detector. A column (150×4.6 mm, 5 µm; Shinwa Chemical Industries, Ltd., Kyoto, Japan) was used for the separation of PAs. Solvent B and solvent A (acetonitrile and 6% (v/v) acetic acid in water, respectively) were pumped for 70 min at 1 mL min^{-1}. HPLC system was run using a gradient program with 0–15% acetonitrile for 45 min, 15–30% for 15 min, 30–50% for 5 min, and 50–100% for 5 min; 35 °C temperature in the column was maintained with a 10 µL volume of injection. For monitoring Pas continuously, the detector was set at 254 and 280 nm. The retention time and UV–vis spectra with their respective standards were compared for the identification of the compound. Pas was estimated as µg g^{-1} FW.

Each PA was quantified using the corresponding standards of calibration curves. A total of 16 PAs were dissolved in MeOH (80%) 100 mg mL^{-1} as stock solutions. Individual PAs were quantified using corresponding standard curves (10, 20, 40, 60, 80, and 100 µg mL^{-1}) with external standards. Retention times, co-chromatography of samples spiked with commercially available standards, and UV spectral characteristics were utilized for identification and matching the PA.

2.9. Statistical Analysis

All the sample data of a trait were averaged for each treatment to obtain a replication mean [113–115]. The mean data of various traits were statistically and biometrically analyzed [116–118]. Data analysis and ANOVA were performed using Statistix software version 8.0, Tallahassee FL 32312, USA [119–121]. The means were compared at a 1% level of probability using Duncan's Multiple Range Test (DMRT). The results were reported as the mean $\pm$ SD of four separate replicates [122–124].

3. Results and Discussion

3.1. The Response of Proximate Compositions to Sodium Chloride Stress

Figure 1 represents the nutritional compositions of *A. gangeticus* under different salinity stresses. *A. gangeticus* leaves had a high moisture content like most leafy vegetables. Nevertheless, our study revealed that *A. gangeticus* leaves have copious ash, carbohydrates, dietary fiber, moisture, and protein. The constituents of these components were several times greater than *C. spinosum* [99]. The maximum moisture and fat were exhibited under the control treatment, whereas the minimum moisture and fat were observed under SSCS. Petropoulos et al. [99] reported a similar reduction in fat with the increase in salinity stress in *C. spinosum*. Moisture and fat were significantly reduced in the order: (control > LSCS > MSCS > SSCS) and (control > LSCS > MSCS = SSCS), respectively. Higher leaf dry matter obtained from leaves ensure lower moisture content. Hence, salt-stressed *A. gangeticus* leaves confirmed greater dry matter in comparison to the control. The maximum dietary fiber, ash, carbohydrates, energy, and protein were recorded at SSCS, while the minimum dietary fiber, ash, carbohydrates, energy, and protein were noticed under the control. Similarly, Petropoulos et al. [99] reported higher ash and protein at the maximum and 8.0 and 6.0 dS m^{-1}, than the control and minimum salinity in *C. spinosum*. Energy, protein, and dietary fiber contents were sharply augmented in the following order: control < LSCS < MSCS < SSCS, whereas ash and carbohydrates contents were statistically similar in the control and LSS levels and progressively augmented from MSCS to SSCS levels.

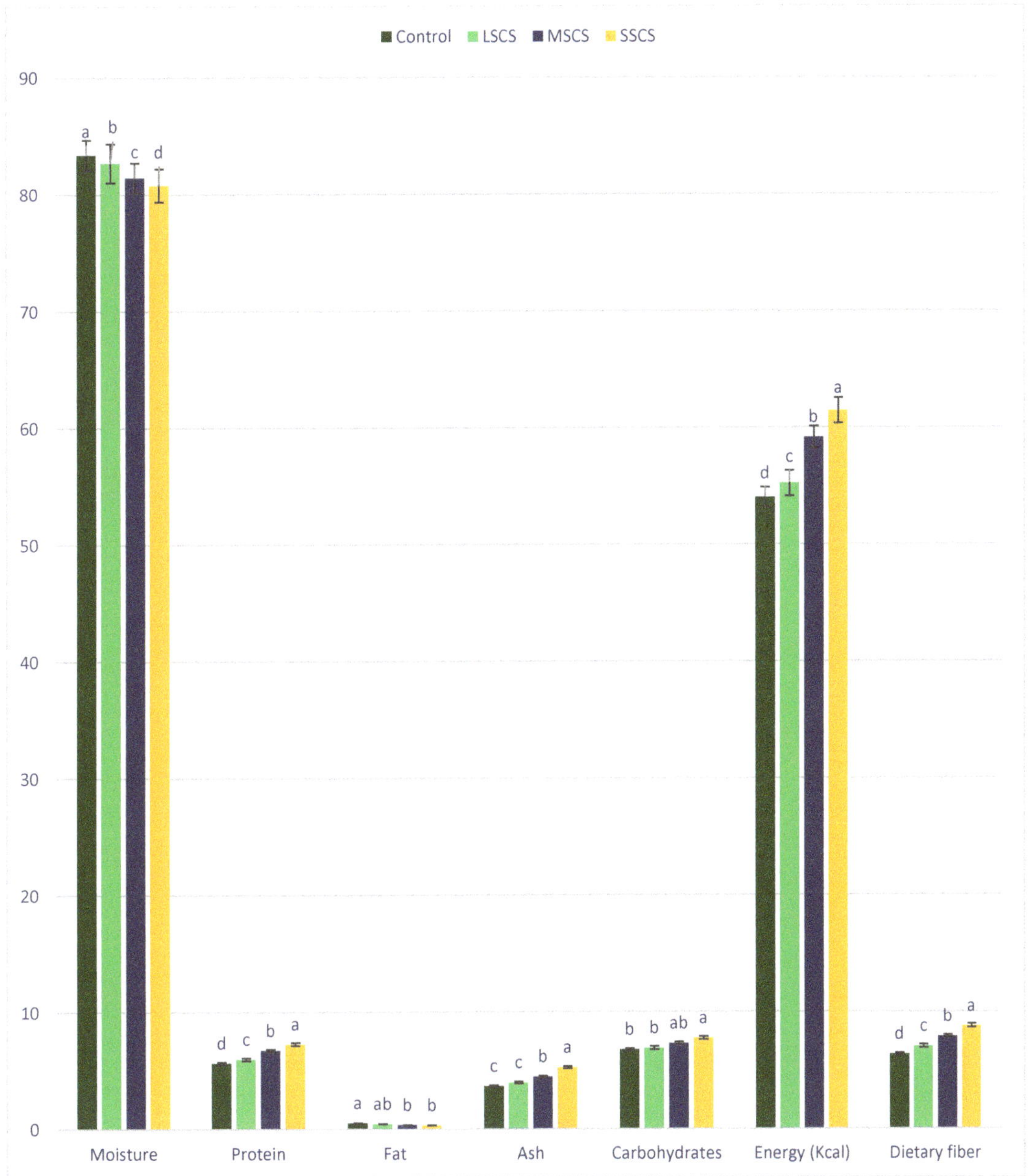

Figure 1. The response of ash, fiber, moisture, fat, gross energy, carbohydrate, and protein (g 100 g^{-1}) to control, LSCS, MSCS, and SSCS in *A. gangeticus* accession; (n = 6), different letters in columns are varied significantly by Duncan Multiple Range Test (DMRT) ($p < 0.01$).

In LSCS, MSCS, and SSCS, dietary fiber, energy, carbohydrates, ash, and protein were increased by 17%, 2%, 2%, 9%, and 4%; 6%, 10%, 8%, 14%, and 19%; and 23%, 14%, 9%, 16%, and 29%, respectively, in comparison with the control condition (Figure 2).

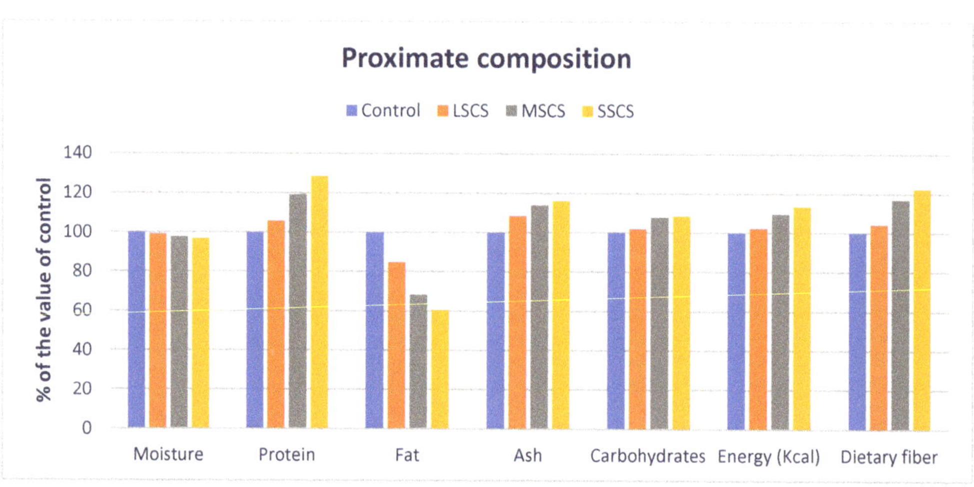

Figure 2. Changes of ash, fiber, moisture, fat, gross energy, carbohydrate, and protein over control in *A. gangeticus* accession.

Dietary fiber has significantly acted in the remedy of constipation, increased digestibility, and palatability. Vegetarians and deprived rural communities in underdeveloped countries mostly trust *A. gangeticus* for protein. Since the low amounts consumed in a daily diet, the increments of energy content in the order of control < LSCS < MSCS < SSCS had no substantial influence on the energy balance in humans. The findings of *A. gangeticus* conformed with the outcomes of AT [91] and leaves of *Ipomoea batata* [125], respectively. They specified that it influences cell function, the fat covering the body's organs, and continues the temperature of the body. The fats of vegetables are prime sources of crucial fatty acids, such as Ω-6 and Ω-3. Fats perform a noteworthy contribution to the absorption, digestion, and transportation of vitamins A, E, K, and D.

3.2. Sodium Chloride Impact on Minerals (Macroelements and Microelements) Composition

A. gangeticus has abundant minerals (macroelements and microelements) (Figure 3). High levels of minerals were observed and corroborated with *A. tricolor* under normal cultivation practice in an open field [126]. *A. gangeticus* had higher Fe and Zn than *Manihot esculenta* leaves [127] and *Lathyrus japonicus* [128]. Jimenez-Aguiar and Grusak [129] also found abundant Zn, Cu, Mn, and Fe in different *A. spp.* They also found higher iron and copper compared with kale and higher Zn compared with leaf cabbage, *Spinacia oleracea*, and *Solanum nigrum*. The maximum Zn, Ca, Mo, Mg, Na, S, Cu, B, Mn, and Fe was noticed under the SSCS level, while the minimum levels Zn, Ca, Mo, Mg, Na, Cu, B, Mn, and Fe were reported under control conditions, and the lowest sulfur content was observed under the LSCS level. Zn, Ca, Mo, Mg, Na, Cu, B, and Mn were progressively augmented in the order control < LSCS < MSCS < SSCS. In contrast, potassium and phosphorus contents were drastically reduced in the order control > LSCS > MSCS > SSCS.

In LSCS, MSCS, and SSCS, Zn, Ca, Cu, Mo, Mg, Mn, B, and Na were augmented by −1%, 0.8%, 13%, −1%, 10%, 4%, 1%, and 6%; 21%, 16%, 29%, 24%, 46%, 67%, 24%, and 12%; and 30%, 34%, 67%, 52%, 72%, 100%, 81%, and 36%, respectively, in comparison with the control condition (Figure 4). In LSCS, MSCS, and SSCS, potassium and phosphorus content declined to 5%, 14%, 25%, and 3%, 36%, 42%, respectively, in comparison with the control condition (Figure 4).

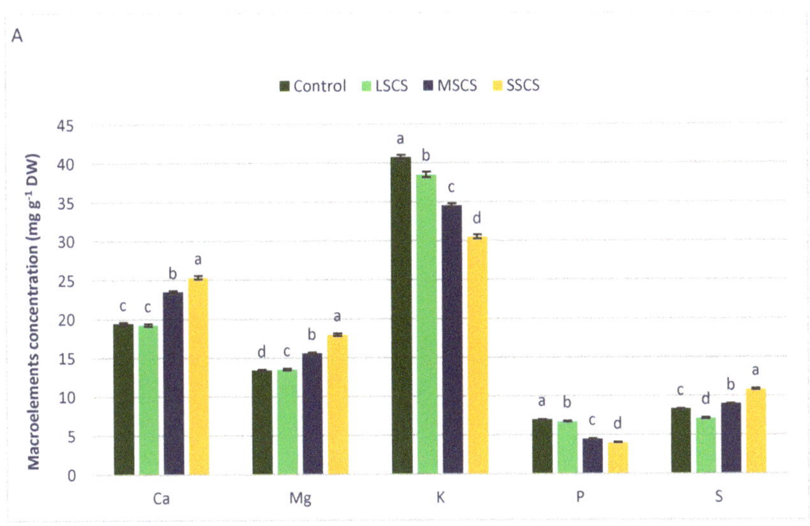

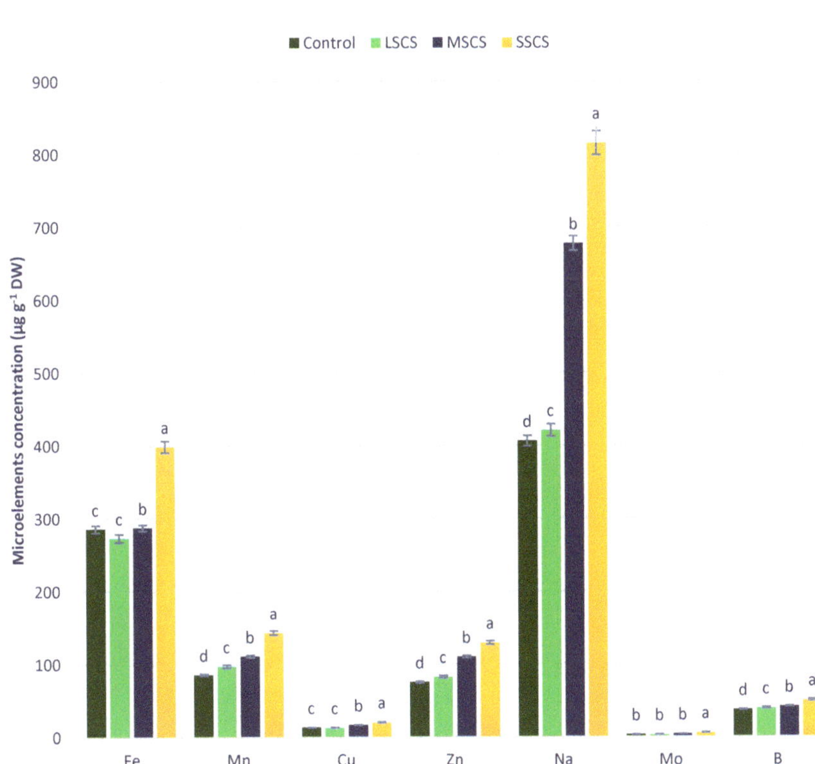

Figure 3. Response of minerals concentration (**A**) macroelements and (**B**) microelements under control, LSCS, MSCS, and SSCS in *A. gangeticus* accession; (n = 6), different letters in columns are varied significantly by DMRT (*p* < 0.01).

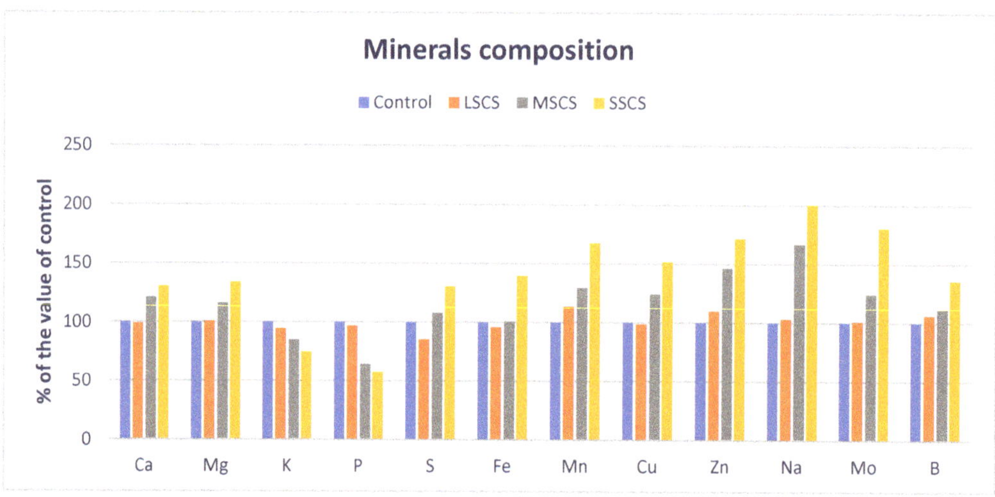

Figure 4. Response of minerals (macroelements and microelements) over control in *A. gangeticus* accession.

Most of the minerals increased under different salt levels compared with control conditions, which were corroborated with minerals of *C. spinosum* under salinity stress [99]. Petropoulos et al. [99] reported sharp augmentation in calcium, magnesium, iron, manganese, zinc, and sodium content and a reduction in potassium content in *C. spinosum*. They stated that the application of fertilizer and treatments of salinity could be the reason for the amelioration of sodium content and suggested that the species utilized accumulated sodium to cope with the adverse effects of salinity. Iron content was statistically similar to the value of the control and LSCS levels, while iron content was progressively augmented under MSCS and SSCS levels by 12% and 62%, respectively. The lowest sulfur content was obtained from the LSCS levels, which differed significantly from the control condition. The sulfur content was gradually augmented under MSCS and SSCS levels by 20% and 51%, respectively (Figure 4).

3.3. Impact of Salinity on Phytochemicals and ARP

Polyphenols, beta-carotene, AsA, flavonoids, and ARP varied significantly under different sodium chloride stresses (Figure 5). Sodium chloride stress progressively augmented polyphenols, beta-carotene, AsA, flavonoids, and ARP in the following order: control < LSCS < MSCS < SSCS.

Beta-carotene, AsA, polyphenols, flavonoids, and ARP (DPPH and ABTS$^+$) under LSCS, MSCS, and SSCS were predominately augmented by 12%, 4%, 5%, 7%, 6%, and 3%; 28%, 18%, 22%, 22%, 20%, and 19%; and 47%, 52%, 54%, 45%, 38%, and 41% than control, respectively (Figure 6).

The maximum polyphenols, beta-carotene, flavonoids, AsA, and ARP (DPPH and ABTS$^+$) were recorded under SSCS. Conversely, the lowest polyphenols, beta-carotene, flavonoids, AsA, and ARP (DPPH and ABTS$^+$) were confirmed under the control. Petropoulos et al. [99] reported the salinity-induced augmentation of flavonoids, ARP, AsA, and phenolics in *C spinosum*. Different concentrations of sodium chloride enhanced the carotenoid content in buckwheat sprouts in comparison with the control (Lim et al. [100]. Alam et al. [101] reported salt-induced amelioration of phenolics, ARP, and flavonoids in purslane. In barley, a similar salinity-induced increase of ARP and phenolics were stated.

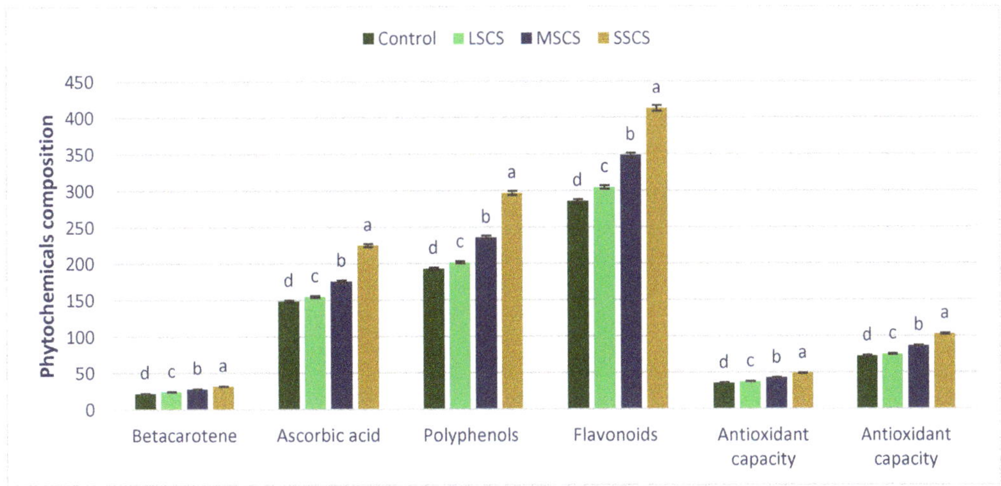

Figure 5. Effect of salinity treatments (control, LSCS, MSCS, and SSCS) on phytochemicals composition in *A. gangeticus* accession. Flavonoids (µg RE g^{-1} DW), AsA and beta-carotene (mg 100 g^{-1} FW), ARP (DPPH and ABTS$^+$) (µg TEAC g^{-1} DW), and polyphenols (µg GAE g^{-1} FW), (n = 6); different letters in columns are varied significantly by DMRT (*p* < 0.01).

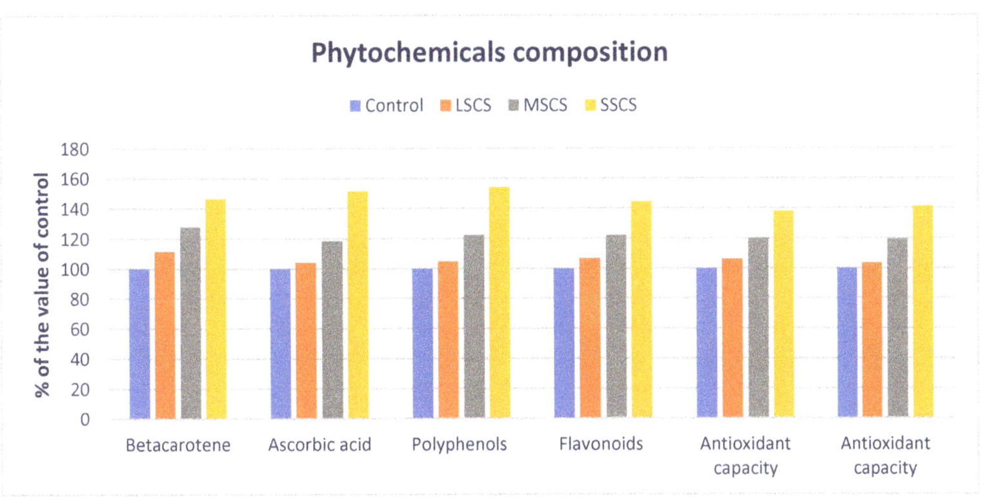

Figure 6. Comparison of phytochemicals over control in *A. gangeticus* accession.

3.4. Response of Salinity on PA Profiles

The HPLC-identified PA values of *A. gangeticus* (accession LS11) under four sodium chloride stress were compared with PAs using the respective peaks of the compounds (Table 1). Sixteen PAs, including seven CAs and nine Bas, were confirmed in *A. gangeticus*. Three BAs [protocatechuic acid (PCA), β-resorcylic acid (β-RA), and gentisic acid (GA)] were identified as new compounds for the first time in *Amaranthus* leaves.

Table 1. Wavelengths of maximum absorption in the visible region (λ_{max}), mass spectral data, retention time (Rt), and tentative identification of PAs in *A. gangeticus*.

Peak No	Rt (min)	λ_{max} (nm)	Molecular Ion [M-H]$^-$ (*m/z*)	MS^2 (*m/z*)	Identity of Tentative Phenolic Acids
1	9.1	254	169.1142	169.1563	3,4,5 Trihydroxybenzoic acid
2	30.6	254	167.1214	167.1564	4-Hydroxy-3-methoxybenzoic acid
3	34.8	254	197.1132	197.1104	3,5-Dimethoxy-4-hydroxybenzoic acid
4	31.5	254	137.0213	137.1574	4-Hydroxybenzoic acid
5	48.2	254	137.2113	137.1582	2-Hydroxybenzoic acid
6	52.5	254	301.0423	301.0643	2,3,7,8-Tetrahydroxy-chromeno [5,4,3-cde] chromene-5,10-dione
7	2.2	280	154.1212	154.1157	3,4-Dihydroxybenzoic acid
8	4.0	280	154.1212	154.0156	2,4-Dihydroxybenzoic acid
9	3.7	280	154.1212	154.1157	2,5- Dihydroxybenzoic acid
10	32.0	280	179.0821	179.0687	3,4-Dihydroxy-trans-cinnamate
11	31.1	280	353.1253	353.1542	3-(3,4-Dihydroxy cinnamoyl) quinic acid
12	42.0	280	163.0658	163.1241	4-Hydroxy cinnamic acid
13	47.9	280	193.1726	193.1649	3-Methoxy-4-hydroxy cinnamic acid
14	49.6	280	163.2547	163.2872	3-Hydroxy cinnamic acid
15	49.0	280	223.1568	223.1748	4-Hydroxy-3,5-dimethoxy cinnamic acid
16	67.3	280	147.1142	147.1103	3-Phenyl acrylic acid

BAs were the amplest among the two categories of acids, thereafter CAs in *A. gangeticus* (Figures 7 and 9). Salicylic acid (SA) was the most copious PAs across BAs thereafter GAA, GA, PCA, vanillic acid (VA), *p*-hydroxybenzoic acid (*p*-HBA), β-RA, and syringic acid (SYA) (Figure 7). BA contents in the *A. gangeticus* genotype under control conditions were superior to the BA content of *A. tricolor* [130]. Chlorogenic acid (CHA) was the most noticeable compound across CAs thereafter ferulic acid (FA), sinapic acid (SIA), *m*-coumaric acid (*m*-COA), *trans*-cinnamic acid (*Trans*-CA), and caffeic acid (CFA) (Figure 7). *A. gangeticus* had abundant CAs under control conditions. Seven CAs obtained were confirmed superior to CAs of *A. tricolor* [130]. Phenylalanine is the most extensively distributed PA in plant tissues, which are finally synthesized into CAs [131]. Identified Benzoic acids (BAs) have important biological activities. For instance, gallic acid and its ester derivatives ARE flavoring agents and preservatives in the food industry. There are diverse scientific reports on the biological and pharmacological activities of these phytochemicals, with emphasis on antioxidant, antimicrobial, anti-inflammatory, anticancer, cardioprotective, gastroprotective, and neuroprotective effects [132]. Vanillic acid exerts diverse bioactivity against cancer, diabetes, obesity, neurodegenerative, cardiovascular, and hepatic diseases by inhibiting the associated molecular pathways. Its derivatives also possess the therapeutic potential to treat autoimmune diseases, as well as fungal and bacterial infections [133]. Syringic acid shows a wide range of therapeutic applications in the prevention of diabetes, CVDs, cancer, and cerebral ischemia, as well as antioxidant, antimicrobial, anti-inflammatory, antiendotoxic, neurologic, and hepatoprotective activities [134]. High salicylate in diets has proven health benefits, such as lower risks of cancer, heart disease, and diabetes. Ellagic acid has been reported to have antimutagenic on bacteria and in mammalian systems as well. It has also shown strong antioxidant, anti-inflammatory, and anticarcinogenic activities, as well as a better preservative effect against oxidative stress when compared with vitamin E [135]. PCA is a major metabolite of anthocyanin. The pharmacological actions of PCA have been shown to include strong in vitro and in vivo antioxidant activity. In in vivo experiments using rats and mice, PCA has been shown to exert anti-inflammatory as well as antihyperglycemic and antiapoptotic activities [136]. β-resorcylic acid has antimicrobial activity [137]. Finally, gentisic acid possesses fibro growth factor inhibition, antimicrobial, antioxidant, anti-inflammatory, hepatoprotective, and neuroprotective activities [138].

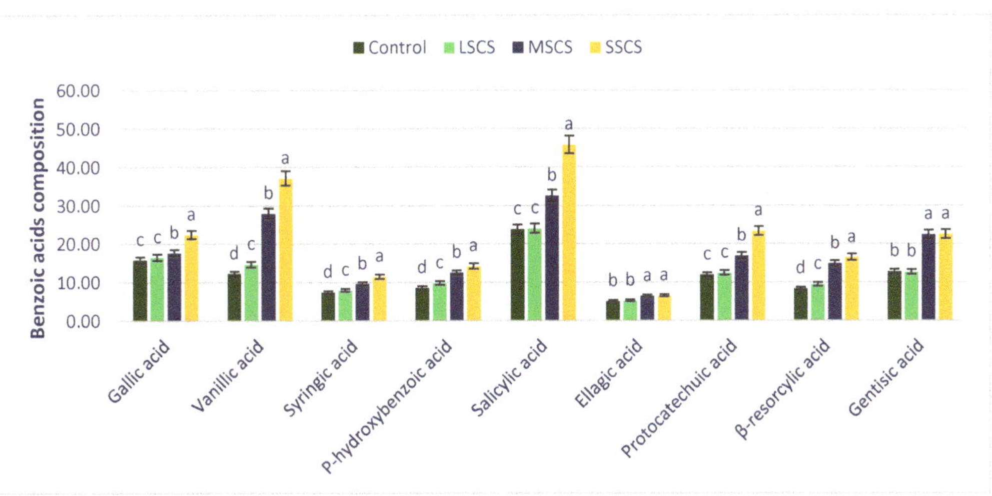

Figure 7. Impact of BAs concentrations (μg g^{-1} FW) under control, LSCS, MSCS, and SSCS in *A. gangeticus* accession; (n = 6), different letters in columns are varied significantly by DMRT ($p < 0.01$).

Sodium chloride stress predominately augmented all the BA compositions. At SSCS, all the BAs displayed the maximum contents, while the minimum BA contents were obtained from the control treatment. From control to SSCS, VA, β-RA, *p*-HBA, and SYA ranged from 12.24 to 37.15, 8.26 to 16.48, 8.55 to 14.23, and 7.36 to 11.52 μg g^{-1} FW, respectively (Figure 7). VA, β-RA, *p*-HBA, and SYA progressively augmented in the order: Control < LSCS < MSCS < SSCS (Figure 7). VA, β-RA, *p*-HBA, and SYA under LSCS, MSCS, and SSCS were predominately augmented by 20%, 28%, 204%; 14%, 79%, 100%; 15%, 46%, and 66%; and 8%, 30%, and 57% than control, respectively (Figure 8).

SA, GAA, and PCA contents had no statistical variations at the control and LSCS level; however, three acids were augmented remarkably from LSCS to SSCS with a range from 23.83 to 45.82, 15.76 to 22.46, and 11.95 to 23.42 μg g^{-1} FW, respectively (Figure 7). In MSCS and SSCS, SA, GAA, and PCA contents were augmented by (37% and 92%), (12% and 43%), and (41% and 96%), respectively (Figure 8). GA and ellagic acid (EA) ranged from 12.68 to 22.58 and 5.08 to 6.55 μg g^{-1} FW. GA and EA had no statistical variations between control and LSCS levels and between MSCS and SSCS levels; however, the contents of these acids were augmented remarkably from control condition or LSCS to MSCS or SSCS level (26% and 77%) (Figures 7 and 8).

All the CA contents were sharply augmented under sodium chloride levels. All the CAs showed the highest contents under the SSCS level, whereas the control treatment exhibited the lowest CA contents. From control to SSCS, CHA, *m*-COA, and *p*-coumaric acid (*p*-COA) ranged from 14.38 to 27.35, 7.87 to 21.36, and 4.16 to 8.75 μg g^{-1} FW, respectively, (Figure 9). Identified Cinnamic acids (CAs) have important biological activities. For instance, Caffeic acid (CA) and its derivatives have antioxidant, anti-inflammatory, and anticarcinogenic activity [139]. Chlorogenic acid was effective in preventing weight gain, inhibiting the development of liver steatosis, and blocking insulin resistance induced by a high-fat diet [140]. *p*-coumaric acid decreases low-density lipoprotein (LDL) peroxidation, shows antioxidant and antimicrobial activities, and plays an important role in human health [141]. Ferulic acid has low toxicity and possesses many physiological functions (anti-inflammatory, antioxidant, antimicrobial activity, anticancer, and antidiabetic effects). It has been widely used in the pharmaceutical, food, and cosmetics industries [142]. Sinapic acid shows antioxidant, antimicrobial, anti-inflammatory, anticancer, and anti-anxiety activity [143]. Cinnamic acids have been identified as interesting compounds with antioxidant, anti-inflammatory, and cytotoxic properties [144].

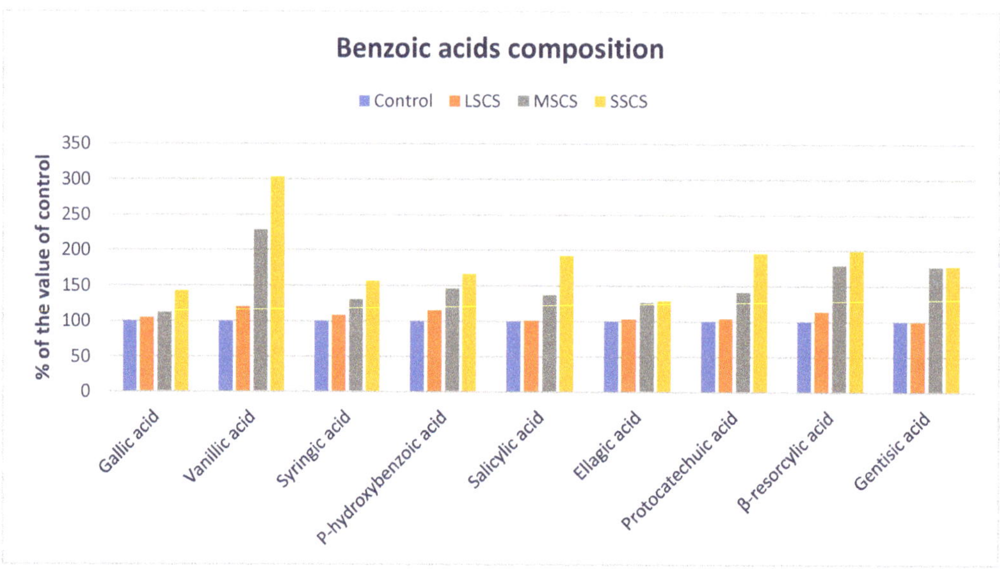

Figure 8. Comparison of BAs composition over control in *A. gangeticus* accession.

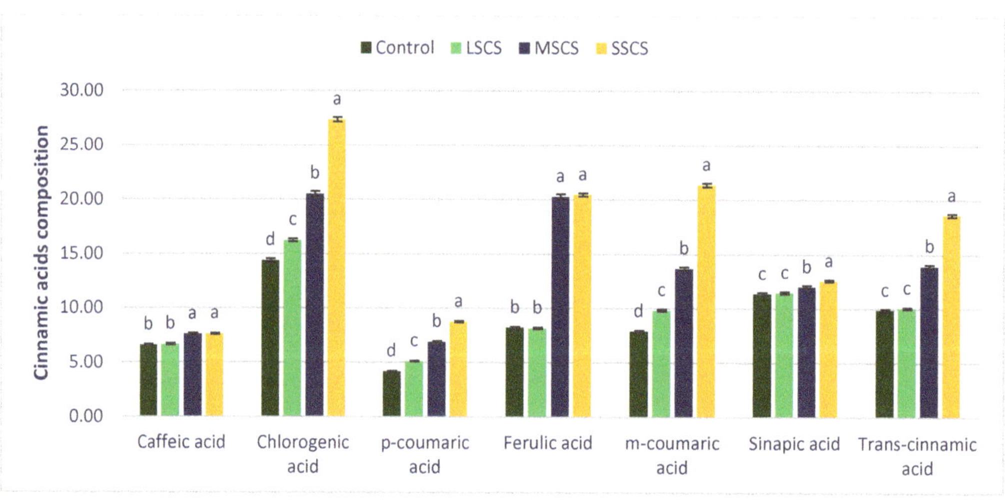

Figure 9. Response of CAs composition (µg g^{-1} FW) under control, LSCS, MSCS, and SSCS in *A. gangeticus* accession; (n = 6), different letters in columns are varied significantly by DMRT ($p < 0.01$).

CHA, *m*-COA acid, and *p*-COA were progressively augmented in the order of control < LSCS < MSCS < SSCS (Figure 9). In LSCS, MSCS, and SSCS, CHA, *m*-COA acid, and *p*-COA were predominately augmented by 13%, 42%, 90%; 25%, 74%, 171%; and 23%, 65%, 110% compared with the control condition, respectively (Figure 10). *Trans*-CA and SIA contents at control condition were statistically similar to the LSCS level; however, these two acids' contents were remarkably augmented from LSCS to SSCS with a range from 9.85 to 18.62 and 11.35 to 12.56 µg g^{-1} FW, respectively (Figure 9). In MSCS and SSCS, *Trans*-CA and SIA contents were augmented by 41% and 89%; and 6% and 11%, respectively (Figure 10). FA and CFA ranged from 8.20 to 20.45 and 6.56 to 7.62 µg g^{-1} FW, respectively. FA and CFA contents at the control condition were statistically similar to the LSCS level, and at the MSCS level were statistically similar to the SSCS level. However, the contents of these acids were remarkably augmented from the control condition or LSCS to MSCS or SSCS level (47% and 16%) (Figures 9 and 10).

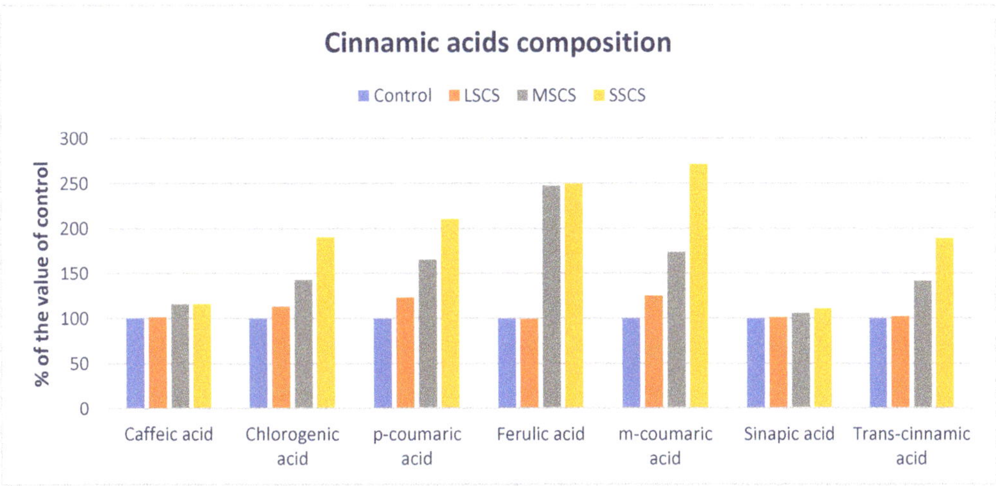

Figure 10. Comparison of CAs over control in *A. gangeticus* accession.

All the PA fractions were sharply and remarkably augmented under sodium chloride stress. All the PA fractions exhibited the highest contents under SSCS level, whereas the control treatment had the lowest PA fractions. From control to SSCS, total BAs, total CAs, and total PAs ranged from 105.71 to 200.21, 62.37 to 116.71, and 168.08 to 316.92, $\mu g\ g^{-1}$ FW, respectively (Figure 11).

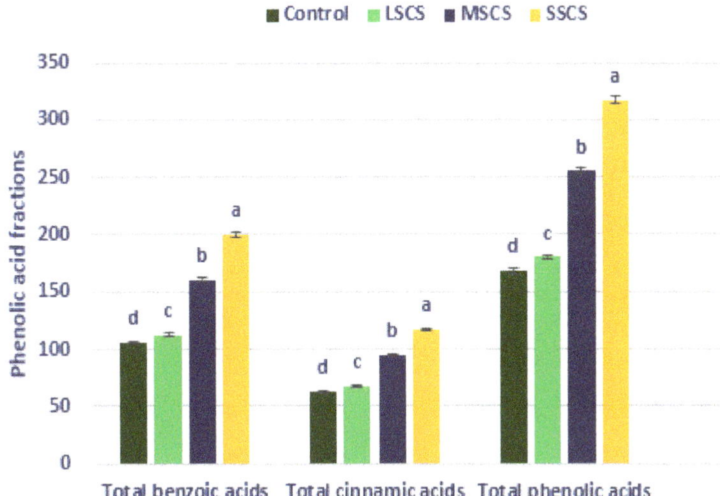

Figure 11. Increase of PA fractions ($\mu g\ g^{-1}$ FW) (total BAs, total CAs, and total PAs) under control, LSCS, MSCS, and SSCS in *A. gangeticus* accession; (n = 6); different letters in columns are varied significantly by DMRT ($p < 0.01$).

Total BAs, total CAs, and total PAs were progressively augmented in the order control < LSCS < MSCS < SSCS (Figure 11). In LSCS, MSCS, and SSCS, total BAs, total CAs, and total PAs were predominately augmented by 7%, 52%, and 89%), (8%, 52%, and 87%), and (7%, 52%, and 89%), compared with control condition, respectively (Figure 12).

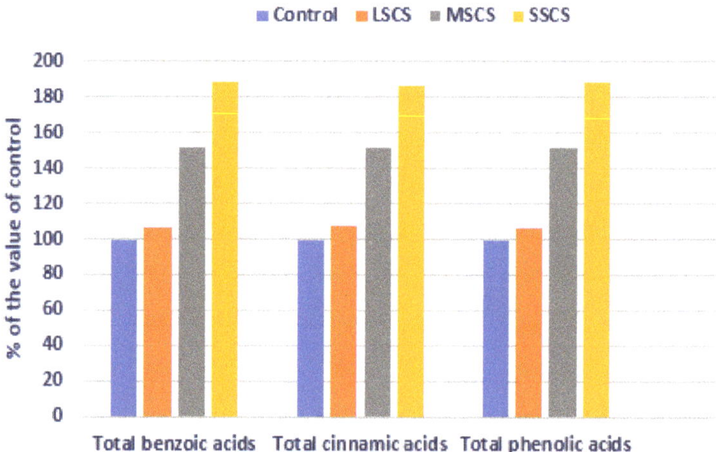

Figure 12. Comparison of phenolics acid fractions (total BA, total CAs, and total PAs) over control in *A. gangeticus* accession.

Petropoulos et al. [99] reported the salinity-induced augmentation of PAs in *C. spinosum*. Klados and Tzortzakis [145] showed a progressive increment of total PAs under increased sodium chloride stress in *C. spinosum*. Alam et al. [101] reported salt-induced amelioration of phenolics in purslane. Ahmed et al. [103] reported a salinity-induced increment of PA profiles in barley. In contrast, Neffati et al. [146] stated the reduction of PA profiles with an increment of sodium chloride concentrations in coriander.

The cost is very low to maintain salt stress by adding sodium chloride to the plants. Furthermore, we suggested cultivating in salt-prone areas where there are no salt susceptible crops grown successfully. So, those areas will be efficiently utilized for amaranth leafy vegetable cultivation to meet the demand for the leafy vegetable of that locality, as leafy vegetables are too susceptible to salinity stress as amaranth is a salinity-tolerant leafy vegetable with up to 200 mM salt concentration. It can produce enough biomass and perform optimal photosynthesis at 100 mM saline stress. Amaranth is highly tolerant to salinity. It can tolerate 200 mM NaCl [147]. As amaranth is salt tolerant, it increases all enzymatic and non-enzymatic antioxidants, and metabolites, to detoxify ROS and cope with salt stress.

4. Conclusions

Sodium chloride stress remarkably augmented the energy, ash, carbohydrates, protein, calcium, dietary fiber, magnesium, S, Fe, Mo, Mn, Na, Cu, B, Zn, and ARP of *A. gangeticus* leaves. All the nutrients, phytochemicals, PA profiles, and ARP of *A. gangeticus* leaves under MSCS and SSCS levels were superior to the control. It can be utilized as a valued product for human consumption and health benefits. Salt-treated *A. gangeticus* leaves had abundant nutrients, phytochemicals, PA profiles, and ARP. Phytochemicals, PA profiles, and ARP scavenge ROS that would be advantageous for human health benefits as these bioactive compounds have potent antioxidants. Furthermore, sodium chloride-stressed *A. gangeticus* contributed with excellent quality in the end users for nutrients, phytochemicals, PA profiles, and ARP. It can be cultivated as a promising substitute crop in sodium chloride-affected areas of the world.

Author Contributions: Conceptualization, U.S.; methodology, U.S.; software, U.S.; validation U.S. and S.E.; formal analysis, U.S.; investigation, U.S.; resources, U.S.; data curation, U.S.; writing—original draft preparation, U.S. writing—review and editing, U.S. and S.E.; visualization, U.S.; supervision, U.S. All authors have read and agreed to the published version of the manuscript.

Funding: This research received no external funding.

Institutional Review Board Statement: Not applicable.

Informed Consent Statement: Not applicable.

Data Availability Statement: The data that are recorded in the current study are available in all of the tables and figures of the manuscript.

Acknowledgments: For its support of the research work, the author acknowledges the Department of Genetics and Plant Breeding, the Bangabandhu Sheikh Mujibur Rahman Agricultural University, Gazipur 1706, Bangladesh.

Conflicts of Interest: The author declares no conflict of interest.

References

1. Rastogi, A.; Shukla, S. Amaranth: A New Millennium Crop of Nutraceutical Values. *Crit. Rev. Food Sci. Nutr.* **2013**, *53*, 109–125. [CrossRef] [PubMed]
2. Promising Das, S. Amaranths: The Crop of Great Prospect. In *Amaranthus: A Promising Crop of Future*; Springer: Singapore, 2016; pp. 13–48.
3. Sreelathakumary, I.; Peter, K.V. Amaranth: *Amaranthus* spp. In *Genetic Improvement of Vegetable Crops*; Elsevier: Amsterdam, The Netherlands, 1993; pp. 315–323.
4. Sauer, J.D. The Grain Amaranths and Their Relatives: A Revised Taxonomic and Geographic Survey. *Ann. Mo. Bot. Gard.* **1967**, *54*, 103. [CrossRef]
5. Anu, R.; Mishra, B.K.; Mrinalini, S.; Ameena, S.; Rawli, P.; Nidhi, V.; Sudhir, S. Identification of Heterotic Crosses Based on Combining Ability in Vegetable Amaranthus (*Amaranthus tricolor* L.). *Asian J. Agric. Res.* **2015**, *9*, 84–94.
6. Nguyen, D.C.; Tran, D.S.; Tran, T.T.H.; Ohsawa, R.; Yoshioka, Y. Genetic diversity of leafy amaranth (*Amaranthus tricolor* L.) resources in Vietnam. *Breed. Sci.* **2019**, *69*, 640–650. [CrossRef]
7. Shukla, S.; Bhargava, A.; Chatterjee, A.; Srivastava, A.; Singh, S.P. Genotypic variability in vegetable amaranth (*Amaranthus tricolor* L for foliage yield and its contributing traits over successive cuttings and years. *Euphytica* **2006**, *151*, 103–110. [CrossRef]
8. Alvarez-Jubete, L.; Arendt, E.K.; Gallagher, E. Nutritive Value of Pseudocereals and Their Increasing Use as Functional Gluten-Free Ingredients. *Trends Food Sci. Technol.* **2010**, *21*, 106–113. [CrossRef]
9. Achigan-Dako, E.G.; Sogbohossou, O.E.D.; Maundu, P. Current knowledge on *Amaranthus* spp.: Research avenues for improved nutritional value and yield in leafy amaranths in sub-Saharan Africa. *Euphytica* **2014**, *197*, 303–317. [CrossRef]
10. Akin-Idowu, P.E.; Odunola, O.A.; Gbadegesin, M.A.; Ademoyegun, O.T.; Aduloju, A.O.; Olagunju, Y.O. Nutritional evaluation of Five Species of Grain Amaranth—An Underutilized Crop. *Int. J. Sci.* **2017**, *3*, 18–27. [CrossRef]
11. Alegbejo, J.O. Nutritional Value and Utilization of Amaranthus (*Amaranthus* spp.)—A Review. *Bayero J. Pure Appl. Sci.* **2014**, *6*, 136–143. [CrossRef]
12. Shukla, S.; Bhargava, A.; Chatterjee, A.; Pandey, A.C.; Mishra, B.K. Diversity in Phenotypic and Nutritional Traits in Vegetable Amaranth (*Amaranthus tricolor*), A Nutritionally Underutilised Crop. *J. Sci. Food Agric.* **2010**, *90*, 139–144. [CrossRef]
13. Soriano-García, M.; Arias-Olguín, I.I.; Montes, J.P.C.; Ramírez, D.G.R.; Figueroa, J.S.M.; Flores-Valverde, E.; Valladares-Rodríguez, M.R. Nutritional functional value and therapeutic utilization of Amaranth. *J. Anal. Pharm. Res.* **2018**, *7*, 596–600. [CrossRef]
14. Sarker, U.; Oba, S. Nutraceuticals, antioxidant pigments, and phytochemicals in the leaves of Amaranthus spinosus and Amaranthus viridis weedy species. *Sci. Rep.* **2019**, *9*, 20413. [CrossRef] [PubMed]
15. Sarker, U.; Oba, S. Protein, dietary fiber, minerals, antioxidant pigments and phytochemicals, and antioxidant activity in selected red morph Amaranthus leafy vegetable. *PLoS ONE* **2019**, *14*, e0222517. [CrossRef]
16. Sarker, U.; Islam, T.; Rabbani, G.; Oba, S. Variability, heritability and genetic association in vegetable amaranth (*Amaranthus tricolor* L.). *Span. J. Agric. Res.* **2015**, *13*, e0702. [CrossRef]
17. Andini, R.; Yoshida, S.; Ohsawa, R. Variation in Protein Content and Amino Acids in the Leaves of Grain, Vegetable and Weedy Types of Amaranths. *Agronomy* **2013**, *3*, 391–403. [CrossRef]
18. Manólio Soares, R.A.; Mendonça, S.; Andrade de Castro, L.I.; Cardoso Corrêa Carlos Menezes, A.C.; Gomes Arêas, J.A. Major Peptides from Amaranth (*Amaranthus cruentus*) Protein Inhibit HMG-CoA Reductase Activity. *Int. J. Mol. Sci.* **2015**, *16*, 4150. [CrossRef]

19. Písaríková, B.; Krácmar, S.; Herzig, I. Amino acid Contents and Biological Value of Protein Amaranth. *Czech J. Anim. Sci.* **2005**, *50*, 169–174. [CrossRef]

20. López, D.N.; Galante, M.; Raimundo, G.; Spelzini, D.; Boeris, V. Functional properties of amaranth, quinoa and chia proteins and the biological activities of their hydrolyzates. *Food Res. Int.* **2019**, *116*, 419–429. [CrossRef]

21. Sarker, U.; Oba, S. Nutraceuticals, phytochemicals, and radical quenching ability of selected drought-tolerant advance lines of vegetable amaranth. *BMC Plant Biol.* **2020**, *20*, 564. [CrossRef]

22. Sarker, U.; Oba, S. Nutritional and bioactive constituents and scavenging capacity of radicals in Amaranthus hypochondriacus. *Sci. Rep.* **2020**, *10*, 19962. [CrossRef]

23. Sarker, U.; Oba, S.; Daramy, M.A. Nutrients, minerals, antioxidant pigments and phytochemicals, and antioxidant capacity of the leaves of stem amaranth. *Sci. Rep.* **2020**, *10*, 3892. [CrossRef] [PubMed]

24. Sarker, U.; Hossain, M.M.; Oba, S. Nutritional and antioxidant components and antioxidant capacity in green morph Amaranthus leafy vegetable. *Sci. Rep.* **2020**, *10*, 1336. [CrossRef] [PubMed]

25. Sarker, U.; Hossain, N.; Iqbal, A.; Oba, S. Bioactive Components and Radical Scavenging Activity in Selected Advance Lines of Salt-Tolerant Vegetable Amaranth. *Front. Nutr.* **2020**, *7*, 587257. [CrossRef]

26. Cai, Y.; Corke, H. Amaranthus Betacyanin Pigments Applied in Model Food Systems. *J. Food Sci.* **1999**, *64*, 869–873. [CrossRef]

27. Cai, Y.; Sun, M.; Corke, H. Identification and Distribution of Simple and Acylated Betacyanins in the Amaranthaceae. *J. Agric. Food Chem.* **2001**, *49*, 1971–1978. [CrossRef]

28. Cai, Y.; Sun, M.; Corke, H. HPLC Characterization of Betalains from Plants in the Amaranthaceae. *J. Chromatogr. Sci.* **2005**, *43*, 454–460. [CrossRef] [PubMed]

29. Cai, Y.; Sun, M.; Wu, H.; Huang, R.; Corke, H. Characterization and Quantification of Betacyanin Pigments from Diverse *Amaranthus* Species. *J. Agric. Food Chem.* **1998**, *46*, 2063–2070. [CrossRef]

30. Cai, Y.; Sun, M.; Corke, H. Characterization and Application of Betalain Pigments from Plants of the Amaranthaceae. *Trends Food Sci. Technol.* **2005**, *16*, 370–376. [CrossRef]

31. Slimen, I.B.; Najar, T.; Abderrabba, M. Chemical and Antioxidant Properties of Betalains. *J. Agric. Food Chem.* **2017**, *65*, 675–689. [CrossRef]

32. Esatbeyoglu, T.; Wagner, A.E.; Motafakkerazad, R.; Nakajima, Y.; Matsugo, S.; Rimbach, G. Free radical scavenging and antioxidant activity of betanin: Electron spin resonance spectroscopy studies and studies in cultured cells. *Food Chem. Toxicol.* **2014**, *73*, 119–126. [CrossRef]

33. Gandía-Herrero, F.; Escribano, J.; García-Carmona, F. The Role of Phenolic Hydroxy Groups in the Free Radical Scavenging Activity of Betalains. *J. Nat. Prod.* **2009**, *72*, 1142–1146. [CrossRef] [PubMed]

34. Gandía-Herrero, F.; Escribano, J.; García-Carmona, F. Structural Implications on Color, Fluorescence, and Antiradical Activity in Betalains. *Planta* **2010**, *232*, 449–460. [CrossRef] [PubMed]

35. Gandía-Herrero, F.; Escribano, J.; García-Carmona, F. Purification and Antiradical Properties of the Structural Unit of Betalains. *J. Nat. Prod.* **2012**, *75*, 1030–1036. [CrossRef] [PubMed]

36. Gandía-Herrero, F.; Escribano, J.; García-Carmona, F. Biological Activities of Plant Pigments Betalains. *Crit. Rev. Food Sci. Nutr.* **2016**, *56*, 937–945. [CrossRef] [PubMed]

37. Khan, M.I. Plant Betalains: Safety, Antioxidant Activity, Clinical Efficacy, and Bioavailability. *Compr. Rev. Food Sci. Food Saf.* **2016**, *15*, 316–330. [CrossRef] [PubMed]

38. Khan, M.I.; Giridhar, P. Plant betalains: Chemistry and biochemistry. *Phytochemistry* **2015**, *117*, 267–295. [CrossRef]

39. Stintzing, F.C.; Carle, R. Functional properties of anthocyanins and betalains in plants, food, and in human nutrition. *Trends Food Sci. Technol.* **2004**, *15*, 19–38. [CrossRef]

40. Taira, J.; Tsuchida, E.; Katoh, M.C.; Uehara, M.; Ogi, T. Antioxidant capacity of betacyanins as radical scavengers for peroxyl radical and nitric oxide. *Food Chem.* **2015**, *166*, 531–536. [CrossRef]

41. Sarker, U.; Lin, Y.-P.; Oba, S.; Yoshioka, Y.; Hoshikawa, K. Prospects and potentials of underutilized leafy Amaranths as vegetable use for health-promotion. *Plant Physiol. Biochem.* **2022**, *182*, 104–123. [CrossRef]

42. Miguel, M.G. Betalains in Some Species of the Amaranthaceae Family: A Review. *Antioxidants* **2018**, *7*, 53. [CrossRef]

43. Sarker, U.; Oba, S. Nutrients, minerals, pigments, phytochemicals, and radical scavenging activity in Amaranthus blitum leafy vegetables. *Sci. Rep.* **2020**, *10*, 3868. [CrossRef] [PubMed]

44. Sarker, U.; Oba, S. Phenolic profiles and antioxidant activities in selected drought-tolerant leafy vegetable amaranth. *Sci. Rep.* **2020**, *10*, 18287. [CrossRef] [PubMed]

45. Sarker, U.; Oba, S. Polyphenol and flavonoid profiles and radical scavenging activity in leafy vegetable Amaranthus gangeticus. *BMC Plant Biol.* **2020**, *20*, 499. [CrossRef] [PubMed]

46. Sarker, U.; Oba, S. The Response of Salinity Stress-Induced A. tricolor to Growth, Anatomy, Physiology, Non-Enzymatic and Enzymatic Antioxidants. *Front. Plant Sci.* **2020**, *11*, 559876. [CrossRef]

47. Sarker, U.; Oba, S. Antioxidant constituents of three selected red and green color Amaranthus leafy vegetable. *Sci. Rep.* **2019**, *9*, 18233. [CrossRef]

48. Sarker, U.; Oba, S. Leaf pigmentation, its profiles and radical scavenging activity in selected *Amaranthus tricolor* leafy vegetables. *Sci. Rep.* **2020**, *10*, 18617. [CrossRef]

49. Sarker, U.; Oba, S. Color attributes, betacyanin, and carotenoid profiles, bioactive components, and radical quenching capacity in selected Amaranthus gangeticus leafy vegetables. *Sci. Rep.* **2021**, *11*, 11559. [CrossRef]
50. Pasko, P.; Bartón, H.; Zagrodzki, P.; Gorinstein, S.; Folta, M.; Zachwieja, Z. Anthocyanins, Total Polyphenols and Antioxidant Activity in Amaranth and Quinoa Seeds and Sprouts During Their Growth. *Food Chem.* **2009**, *115*, 994–998. [CrossRef]
51. Li, H.; Deng, Z.; Liu, R.; Zhu, H.; Draves, J.; Marcone, M.; Sun, Y.; Tsao, R. Characterization of phenolics, betacyanins and antioxidant activities of the seed, leaf, sprout, flower and stalk extracts of three Amaranthus species. *J. Food Compos. Anal.* **2015**, *37*, 75–81. [CrossRef]
52. Barba de la Rosa, A.P.; Fomsgaard, I.S.; Laursen, B.; Mortensen, A.G.; Olvera-Martínez, L.; Silva-Sánchez, C.; Mendoza-Herrera, A.; González-Castañeda, J.; De León-Rodrígueza, A. Amaranth (*Amaranthus hypochondriacus*) as An Alternative Crop for Sustainable Food Production: Phenolic Acids and Flavonoids with Potential Impact on Its Nutraceutical Quality. *J. Cereal Sci.* **2009**, *49*, 117–121. [CrossRef]
53. Peiretti, P.G.; Meineri, G.; Gai, F.; Longato, E.; Amarowicz, R. Antioxidative activities and phenolic compounds of pumpkin (*Cucurbita pepo*) seeds and amaranth (*Amaranthus caudatus*) grain extracts. *Nat. Prod. Res.* **2017**, *31*, 2178–2182. [CrossRef]
54. Stintzing, F.C.; Kammerer, D.; Schieber, A.; Adama, H.; Nacoulma, O.G.; Carle, R. Betacyanins and Phenolic Com-pounds from *Amaranthus spinosus* L. and *Boerhavia erecta* L. Z. *Naturforsch. C* **2004**, *59*, 1–8. [CrossRef] [PubMed]
55. Kalinova, J.; Dadakova, E. Rutin and Total Quercetin Content in Amaranth (*Amaranthus* spp.). *Plant Foods Hum. Nutr.* **2009**, *64*, 68–74. [CrossRef] [PubMed]
56. Asao, M.; Watanabe, K. Functional and Bioactive Properties of Quinoa and Amaranth. *Food Sci. Technol. Res.* **2010**, *16*, 163–168. [CrossRef]
57. Tang, Y.; Xiao, Y.; Tang, Z.; Jin, W.; Wang, Y.; Chen, H.; Yao, H.; Shan, Z.; Bu, T.; Wang, X. Extraction of Polysaccha-rides from *Amaranthus hybridus* L. by Hot Water and Analysis of Their Antioxidant Activity. *Peer J.* **2019**, *7*, e7149. [CrossRef] [PubMed]
58. Ozsoy, N.; Yilmaz, T.; Kurt, O.; Can, A.; Yanardag, R. In vitro antioxidant activity of *Amaranthus lividus* L. *Food Chem.* **2009**, *116*, 867–872. [CrossRef]
59. Sarikurkcu, C.; Sahinler, S.S.; Tepe, B. Astragalus gymnolobus, A. leporinus var. hirsutus, and A. onobrychis: Phyto-chemical Analysis and Biological Activity. *Ind. Crop. Prod.* **2020**, *150*, 112366. [CrossRef]
60. Tatiya, A.U.; Surana, S.J.; Khope, S.D.; Gokhale, S.B.; Sutar, M.P. Phytochemical Investigation and Immunomodulatory Activity of *Amaranthus spinosus* L. *Indian J. Pharm. Educ. Res.* **2007**, *444*, 337–341.
61. Pamela, E.A.I.; Olufemi, T.A.; Yemisi, O.O.; Aduloju, O.A.; Usifo, G.A. Phytochemical Content and Antioxidant Activity of Five Grain Amaranth Species. *Am. J. Food Sci. Technol.* **2017**, *5*, 249–255.
62. Pasko, P.; Barton, H.; Folta, M.; Gwizdz, J. Evaluation of Antioxidant Activity of Amaranth Amaranthus cruentus Grain and by-Products Flour, Popping, Cereal. *Rocz. Pánstwowego Zakładu Hig.* **2007**, *581*, 35–40.
63. Nsimba, R.Y.; Kikuzaki, H.; Konishi, Y. Antioxidant Activity of Various Extracts and Fractions of Chenopodium quinoa and Amaranthus spp. Seeds. *Food Chem.* **2008**, *106*, 760–766. [CrossRef]
64. Tang, Y.; Tsao, R. Phytochemicals in quinoa and amaranth grains and their antioxidant, anti-inflammatory, and potential health beneficial effects: A review. *Mol. Nutr. Food Res.* **2017**, *61*, 1600767. [CrossRef] [PubMed]
65. Barku, V.Y.A.; Opoku-Boahen, Y.; Owusu-Ansah, E.; Mensah, E.F.; Barku, V.Y.A.; Opoku-Boahen, Y.; Owusu-Ansah, E.; Mensah, E.F. Antioxidant Activity and The Estimation of Total Phenolic and Flavonoid Contents of The Root Extract of Amaranthus spinosus. *Asian J. Plant Sci. Res.* **2013**, *3*, 69–74.
66. Bulbul, I.J.; Nahar, L.; Ripa, F.A.; Haque, O. Antibacterial, Cytotoxic and Antioxidant Activity of Chloroform, N-Hexane and Ethyl Acetate Extract of Plant Amaranthus spinosus. *Int. J. PharmTech Res.* **2011**, *33*, 1675–1680.
67. Kumar, B.S.A.; Lakshman, K.; Jayaveera, K.; Shekar, D.S.; Kumar, A.A.; Manoj, B. Antioxidant and antipyretic properties of methanolic extract of Amaranthus spinosus leaves. *Asian Pac. J. Trop. Med.* **2010**, *3*, 702–706. [CrossRef]
68. Ishtiaq, S.; Ahmad, M.; Hanif, U.; Akbar, S.; Kamran, S.H. Phytochemical and in-vitro Antioxidant Evaluation of Dif-ferent Fractions of Amaranthus graecizan subsp. Silvestris Vill. Brenan. *Asian Pac. J. Trop. Biomed.* **2014**, *412*, 965–971.
69. Karamać, M.; Gai, F.; Longato, E.; Meineri, G.; Janiak, M.A.; Amarowicz, R.; Peiretti, P.G. Antioxidant Activity and Phenolic Composition of Amaranth (*Amaranthus caudatus*) during Plant Growth. *Antioxidants* **2019**, *8*, 173. [CrossRef]
70. Kraujalis, P.; Venskutonis, P.R.; Kraujaliene, V.; Pukalskas, A. Antioxidant properties and preliminary evaluation of phytochemical composition of different anatomical parts of amaranth. *Plant Foods Hum. Nutr.* **2013**, *68*, 322–328. [CrossRef]
71. Kumari, S.; Elancheran, R.; Devi, R. Phytochemical Screening, Antioxidant, Antityrosinase, And Antigenotoxic Potential of Amaranthus viridis Extract. *Indian J. Pharmacol.* **2018**, *50*, 130–138.
72. Lopez-Mejía, O.A.; Lopez-Malo, A.; Palou, E. Antioxidant Capacity of Extracts from Amaranth Amaranthus hypochondriacus L. Seeds or Leaves. *Ind. Crop. Prod.* **2014**, *53*, 55–59. [CrossRef]
73. Lucero-Lopez, V.R.; Razzeto, G.S.; Gimenez, M.S.; Escudero, N.L. Antioxidant Properties of Amaranthus hypochondri-acus Seeds and Their Effect on The Liver of Alcohol-Treated Rats. *Plant Foods Hum. Nutr.* **2011**, *66*, 157–162. [CrossRef] [PubMed]
74. Salvamani, S.; Gunasekaran, B.; Shukor, M.Y.; Shaharuddin, N.A.; Sabullah, M.K.; Ahmad, S.A. Anti-HMG-CoA Re-ductase, An-tioxidant, and Anti-Inflammatory Activities of Amaranthus viridis Leaf Extract as A Potential Treatment for Hypercholesterolemia. *Evid. Based Complement. Altern. Med.* **2016**, *2016*, 8090841. [CrossRef]

75. Sandoval-Sicairos, E.S.; Milán-Noris, A.K.; Luna-Vital, D.A.; Milán-Carrillo, J.; Montoya-Rodríguez, A. Anti-Inflammatory and Antioxidant Effects of Peptides Released from Germinated Amaranth During In Vitro Simulated Gastrointestinal Digestion. *Food Chem.* **2021**, *1*, 128394. [CrossRef] [PubMed]

76. Medoua, G.N.; Oldewage-Theron, W.H. Effect of Drying and Cooking on Nutritional Value and Antioxidant Capac-ity of Morogo (*Amaranthus hybridus*) A Traditional Leafy Vegetable Grown in South Africa. *J. Food Sci. Technol.* **2014**, *51*, 736–742. [CrossRef]

77. Tesoriere, L.; Allegra, M.; Gentile, C.; Livrea, M.A. Betacyanins as phenol antioxidants. Chemistry and mechanistic aspects of the lipoperoxyl radical-scavenging activity in solution and liposomes. *Free Radic. Res.* **2009**, *43*, 706–717. [CrossRef] [PubMed]

78. Repo-Carrasco-Valencia, R.; Peña, J.; Kallio, H.; Salminen, S. Dietary fiber and other functional components in two varieties of crude and extruded kiwicha (*Amaranthus caudatus*). *J. Cereal Sci.* **2009**, *49*, 219–224. [CrossRef]

79. Jo, H.-J.; Chung, K.-H.; Yoon, J.A.; Lee, K.-J.; Song, B.C.; An, J.H. Radical Scavenging Activities of Tannin Extracted from Amaranth (*Amaranthus caudatus* L.). *J. Microbiol. Biotechnol.* **2015**, *25*, 795–802. [CrossRef]

80. Subhasree, B.; Baskar, R.; Keerthana, R.L.; Susan, R.L.; Rajasekaran, P. Evaluation of antioxidant potential in selected green leafy vegetables. *Food Chem.* **2009**, *115*, 1213–1220. [CrossRef]

81. Lacatusu, I.; Arsenie, K.L.V.; Badea, G.; Popa, O.; Oprea, O.; Badea, N. New Cosmetic Formulations with Broad Pho-toprotective and Antioxidative Activities Designed by Amaranth and Pumpkin Seed Oils Nanocarriers. *Ind. Crops Prod.* **2018**, *123*, 424–433. [CrossRef]

82. Steffensen, S.K.; Pedersen, H.A.; Labouriau, R.; Mortensen, A.G.; Laursen, B.; de Troiani, R.M.; Noellemeyer, E.J.; Janovska, D.; Stavelikova, H.; Taberner, A.; et al. Variation of Polyphenols and Betaines in Aerial Parts of Young, Field-Grown Amaranthus Genotypes. *J. Agric. Food Chem.* **2011**, *59*, 12073–12082. [CrossRef]

83. Niveyro, S.L.; Mortensen, A.G.; Fomsgaard, I.S.; Salvo, A. Differences among five amaranth varieties (*Amaranthus* spp.) regarding secondary metabolites and foliar herbivory by chewing insects in the field. *Arthropod-Plant Interact.* **2013**, *7*, 235–245. [CrossRef]

84. Bao, X.; Han, X.; Du, G.; Wei, C.; Zhu, X.; Ren, W.; Zeng, L.; Zhang, Y. Antioxidant Activities and Immunomodulatory Effects in Mice of Betalain in vivo. *Food Sci.* **2019**, *40*, 196–201.

85. Venskutonis, P.R.; Kraujalis, P. Nutritional Components of Amaranth Seeds and Vegetables: A Review on Composition, Properties, and Uses. *Comp. Rev. Food Sci. Food Saf.* **2013**, *12*, 381–412. [CrossRef] [PubMed]

86. Repo-Carrasco-Valencia, R.; Hellstrom, J.K.; Philava, J.M.; Mattila, P.H. Flavonoids and Other Phenolic Compounds in Andean Indigenous Grains: Quinoa (*Chenopodium quinoa*), Kaniwa (*Chenopodium pallidicaule*) and Kiwicha (*Amaranthus caudatus*). *Food Chem.* **2010**, *120*, 128–133. [CrossRef]

87. Esatbeyoglu, T.; Wagner, A.E.; Schiniatbeyo, V.B.; Rimbach, G. Betanin-A food colorant with biological activity. *Mol. Nutr. Food Res.* **2015**, *59*, 36–47. [CrossRef]

88. Isabelle, M.; Lee, B.L.; Lim, M.T.; Koh, W.-P.; Huang, D.; Ong, C.N. Antioxidant activity and profiles of common fruits in Singapore. *Food Chem.* **2010**, *123*, 77–84. [CrossRef]

89. Wahid, A.; Ghazanfar, A. Possible involvement of some secondary metabolites in salt tolerance of sugarcane. *J. Plant Physiol.* **2006**, *163*, 723–730. [CrossRef]

90. Gill, S.S.; Tuteja, N. Reactive oxygen species and antioxidant machinery in abiotic stress tolerance in crop plants. *Plant Physiol. Biochem.* **2010**, *48*, 909–930. [CrossRef]

91. Sarker, U.; Oba, S. Drought Stress Effects on Growth, ROS Markers, Compatible Solutes, Phenolics, Flavonoids, and Antioxidant Activity in *Amaranthus tricolor*. *Appl. Biochem. Biotechnol.* **2018**, *186*, 999–1016. [CrossRef]

92. Cai, Y.; Sun, M.; Corke, H. Antioxidant Activity of Betalains from Plants of the Amaranthaceae. *J. Agric. Food Chem.* **2003**, *51*, 2288–2294. [CrossRef]

93. Stintzing, F.C.; Carle, R. Betalains—Emerging prospects for food scientists. *Trends Food Sci. Technol.* **2007**, *18*, 514–525. [CrossRef]

94. Sarker, U.; Islam, T.; Oba, S. Salinity stress accelerates nutrients, dietary fiber, minerals, phytochemicals and antioxidant activity in *Amaranthus tricolor* leaves. *PLoS ONE* **2018**, *13*, e0206388. [CrossRef] [PubMed]

95. Sarker, U.; Oba, S. Salinity stress enhances color parameters, bioactive leaf pigments, vitamins, polyphenols, flavonoids and antioxidant activity in selected Amaranthus leafy vegetables. *J. Sci. Food Agric.* **2019**, *99*, 2275–2284. [CrossRef] [PubMed]

96. Sarker, U.; Oba, S. Augmentation of leaf color parameters, pigments, vitamins, phenolic acids, flavonoids and antioxidant activity in selected *Amaranthus tricolor* under salinity stress. *Sci. Rep.* **2018**, *8*, 12349. [CrossRef]

97. Sarker, U.; Oba, S. Catalase, superoxide dismutase and ascorbate-glutathione cycle enzymes confer drought tolerance of *Amaranthus tricolor*. *Sci. Rep.* **2018**, *8*, 16496. [CrossRef]

98. Selmar, D.; Kleinwächter, M. Influencing the product quality by deliberately applying drought stress during the cultivation of medicinal plants. *Ind. Crop. Prod.* **2013**, *42*, 558–566. [CrossRef]

99. Petropoulos, S.A.; Levizou, E.; Ntatsi, G.; Fernandes, Â.; Petrotos, K.; Akoumianakis, K.; Barros, L.; Ferreira, I.C. Salinity effect on nutritional value, chemical composition and bioactive compounds content of *Cichorium spinosum* L. *Food Chem.* **2017**, *214*, 129–136. [CrossRef] [PubMed]

100. Lim, J.H.; Park, K.-J.; Kim, B.-K.; Jeong, J.-W.; Kim, H.-J. Effect of salinity stress on phenolic compounds and carotenoids in buckwheat (*Fagopyrum esculentum* M.) sprout. *Food Chem.* **2012**, *135*, 1065–1070. [CrossRef]

101. Alam, A.; Juraimi, A.S.; Rafii, M.Y.; Hamid, A.A.; Aslani, F.; Alam, M.Z. Effects of salinity and salinity-induced augmented bioactive compounds in purslane (*Portulaca oleracea* L.) for possible economical use. *Food Chem.* **2015**, *169*, 439–447. [CrossRef]

102. Ahmed, I.M.; Cao, F.; Han, Y.; Nadira, U.A.; Zhang, G.; Wu, F. Differential changes in grain ultrastructure, amylase, protein and amino acid profiles between Tibetan wild and cultivated barleys under drought and salinity alone and combined stress. *Food Chem.* **2013**, *141*, 2743–2750. [CrossRef]

103. Hossain, N.; Sarker, U.; Raihan, S.; Al-Huqail, A.A.; Siddiqui, M.H.; Oba, S. Influence of Salinity Stress on Color Parameters, Leaf Pigmentation, Polyphenol and Flavonoid Contents, and Antioxidant Activity of *Amaranthus lividus* Leafy Vegetables. *Molecules* **2022**, *27*, 1821. [CrossRef] [PubMed]

104. Sarker, U.; Rabbani, G.; Oba, S.; Eldehna, W.M.; Al-Rashood, S.T.; Mostafa, N.M.; Eldahshan, O.A. Phytonutrients, Colorant Pigments, Phytochemicals, and Antioxidant Potential of Orphan Leafy *Amaranthus* Species. *Molecules* **2022**, *27*, 2899. [CrossRef] [PubMed]

105. Sarker, U.; Oba, S. Response of nutrients, minerals, antioxidant leaf pigments, vitamins, polyphenol, flavonoid and antioxidant activity in selected vegetable amaranth under four soil water content. *Food Chem.* **2018**, *252*, 72–83. [CrossRef] [PubMed]

106. Dola, D.B.; Mannan, A.; Sarker, U.; Al Mamun, A.; Islam, T.; Ercisli, S.; Saleem, M.H.; Ali, B.; Pop, O.L.; Marc, R.A. Nano-iron oxide accelerates growth, yield, and quality of Glycine max seed in water deficits. *Front. Plant Sci.* **2022**, *13*, 992535. [CrossRef] [PubMed]

107. Yasmin, A.; Mannan, M.A.; Sarker, U.; Dola, D.B.; Higuchi, H.; Ercisli, S.; Ali, B.; Saleem, M.H.; Babalola, O.O. Foliar application of seaweed extracts enhances yield and drought tolerance of soybean. *Front. Plant Sci.* **2022**, *13*, 992880. [CrossRef]

108. Sarker, U.; Azam, M.; Alam Talukder, Z. Genetic variation in mineral profiles, yield contributing agronomic traits, and foliage yield of stem amaranth. *Genetika* **2022**, *54*, 91–108. [CrossRef]

109. Sarker, U.; Oba, S.; Ercisli, S.; Assouguem, A.; Alotaibi, A.; Ullah, R. Bioactive Phytochemicals and Quenching Activity of Radicals in Selected Drought-Resistant *Amaranthus tricolor* Vegetable Amaranth. *Antioxidants* **2022**, *11*, 578. [CrossRef]

110. Sarker, U.; Oba, S.; Alsanie, W.F.; Gaber, A. Characterization of Phytochemicals, Nutrients, and Antiradical Potential in Slim Amaranth. *Antioxidants* **2022**, *11*, 1089. [CrossRef]

111. Sarker, U.; Iqbal, A.; Hossain, N.; Oba, S.; Ercisli, S.; Muresan, C.C.; Marc, R.A. Colorant Pigments, Nutrients, Bioactive Components, and Antiradical Potential of Danta Leaves (*Amaranthus lividus*). *Antioxidants* **2022**, *11*, 1206. [CrossRef]

112. Sarker, U.; Oba, S. Drought stress enhances nutritional and bioactive compounds, phenolic acids and antioxidant capacity of Amaranthus leafy vegetable. *BMC Plant Biol.* **2018**, *18*, 258. [CrossRef]

113. Faysal, A.S.M.; Ali, L.; Azam, G.; Sarker, U.; Ercisli, S.; Golokhvast, K.S.; Marc, R.A. Genetic Variability, Character Association, and Path Coefficient Analysis in Transplant Aman Rice Genotypes. *Plants* **2022**, *11*, 2952. [CrossRef] [PubMed]

114. Hassan, J.; Jahan, F.; Rajib, M.R.; Sarker, U.; Miyajima, I.; Ozaki, Y.; Ercisli, S.; Golokhvast, K.S.; Marc, R.A. Color and physiochemical attributes of pointed gourd (Trichosanthes dioica Roxb.) influenced by modified atmosphere packaging and postharvest treatment during storage. *Front. Plant Sci.* **2022**, *13*, 1016324. [CrossRef] [PubMed]

115. Kulsum, U.; Sarker, U.; Rasul, G. Genetic variability, heritability and interrelationship in salt-tolerant lines of T. Aman rice. *Genetika* **2022**, *54*, 761–776. [CrossRef]

116. Hasan, M.J.; Kulsum, M.U.; Sarker, U.; Matin, M.Q.I.; Shahin, N.H.; Kabir, M.S.; Ercisli, S.; Marc, R.A. Assessment of GGE, AMMI, Regression, and Its Deviation Model to Identify Stable Rice Hybrids in Bangladesh. *Plants* **2022**, *11*, 2336. [CrossRef] [PubMed]

117. Rahman, M.; Sarker, U.; Swapan, A.H.; Raihan, M.S.; Oba, S.; Alamri, S.; Siddiqui, M.H. Combining Ability Analysis and Marker-Based Prediction of Heterosis in Yield Reveal Prominent Heterotic Combinations from Diallel Population of Rice. *Agronomy* **2022**, *12*, 1797. [CrossRef]

118. Azam, G.; Sarker, U.; Hossain, A.; Iqbal, S.; Islam, R.; Hossain, F.; Ercisli, S.; Kul, R.; Assouguem, A.; Al-Huqail, A.A.; et al. Genetic Analysis in Grain Legumes [*Vigna radiata* (L.) Wilczek] for Yield Improvement and Identifying Heterotic Hybrids. *Plants* **2022**, *11*, 1774. [CrossRef]

119. Hassan, J.; Rajib, M.R.; Sarker, U.; Akter, M.; Khan, N.-E.; Khandaker, S.; Khalid, F.; Rahman, G.K.M.M.; Ercisli, S.; Muresan, C.C.; et al. Optimizing textile dyeing wastewater for tomato irrigation through physiochemical, plant nutrient uses and pollution load index of irrigated soil. *Sci. Rep.* **2022**, *12*, 10088. [CrossRef]

120. Prodhan, M.; Sarker, U.; Hoque, A.; Biswas, S.; Ercisli, S.; Assouguem, A.; Ullah, R.; Almutairi, M.H.; Mohamed, H.R.H.; Najda, A. Foliar Application of GA₃ Stimulates Seed Production in Cauliflower. *Agronomy* **2022**, *12*, 1394. [CrossRef]

121. Azad, A.K.; Sarker, U.; Ercisli, S.; Assouguem, A.; Ullah, R.; Almeer, R.; Sayed, A.A.; Peluso, I. Evaluation of Combining Ability and Heterosis of Popular Restorer and Male Sterile Lines for the Development of Superior Rice Hybrids. *Agronomy* **2022**, *12*, 965. [CrossRef]

122. Hasan, M.J.; Kulsum, M.U.; Majumder, R.R.; Sarker, U. Genotypic Variability for Grain Quality Attributes in Restorer Lines of Hybrid Rice. *Genetika* **2020**, *52*, 973–989 doi 102298/GENSR2003973H. [CrossRef]

123. Hasan-Ud-Daula, M.; Sarker, U. Variability, Heritability, Character Association, and Path Coefficient Analysis in Advanced Breeding Lines of Rice (*Oryza sativa* L.). *Genetika* **2020**, *52*, 711–726. [CrossRef]

124. Rashad, M.M.I.; Sarker, U. Genetic Variations in Yield and Yield Contributing Traits of Green Amaranth. *Genetika* **2020**, *52*, 393–407. [CrossRef]

125. Sun, H.; Mu, T.; Xi, L.; Zhang, M.; Chen, J. Sweet potato (*Ipomoea batatas* L.) leaves as nutritional and functional foods. *Food Chem.* **2014**, *156*, 380–389. [CrossRef]

126. Shukla, S.; Bhargava, A.; Chatterjee, A.; Srivastava, J.; Singh, N.; Singh, S.P. Mineral profile and variability in vegetable amaranth (*Amaranthus tricolor*). *Plant Food. Hum. Nutri.* **2006**, *61*, 23–28. [CrossRef]

127. Madruga, M.S.; Camara, F.S. The chemical composition of "Multimistura" as a food supplement. *Food Chem.* **2000**, *68*, 41–44. [CrossRef]

128. Shahidi, F.; Chavan, U.; Bal, A.; McKenzie, D. Chemical composition of beach pea (*Lathyrus maritimus* L.) plant parts. *Food Chem.* **1999**, *64*, 39–44. [CrossRef]

129. Jiménez-Aguilar, D.M.; Grusak, M.A. Minerals, vitamin C, phenolics, flavonoids and antioxidant activity of Amaranthus leafy vegetables. *J. Food Compos. Anal.* **2017**, *58*, 33–39. [CrossRef]

130. Khanam, U.K.S.; Oba, S.; Yanase, E.; Murakami, Y. Phenolic acids, flavonoids and total antioxidant capacity of selected leafy vegetables. *J. Funct. Foods.* **2012**, *4*, 979–987. [CrossRef]

131. Robbins, R.J. Phenolic Acids in Foods: An Overview of Analytical Methodology. *J. Agric. Food Chem.* **2003**, *51*, 2866–2887. [CrossRef]

132. Kahkeshani, N.; Farzaei, F.; Fotouhi, M.; Alavi, S.S.; Bahramsoltani, R.; Naseri, R.; Momtaz, S.; Abbasabadi, Z.; Rahimi, R.; Farzaei, M.H.; et al. Pharmacological effects of gallic acid in health and disease: A mechanistic review. *Iran J. Basic Med. Sci* **2019**, *22*, 225–237. [CrossRef]

133. Kaur, J.; Gulati, M.; Singh, S.K.; Kuppusamy, G.; Kapoor, B.; Mishra, V.; Gupta, S.; Arshad, M.F.; Porwal, O.; Jha, N.K.; et al. Discovering multifaceted role of vanillic acid beyond flavours: Nutraceutical and therapeutic potential. *Trends Food Sci. Technol.* **2022**, *122*, 187–200. [CrossRef]

134. Cheemanapalli, S.; Mopuri, R.; Golla, R.; Anurudha, C.M.; Chitta, S.K. Syringic acid (SA)—A review of its occurrence, biosynthesis, pharmacological and industrial importance. *Biomed. Pharmacother.* **2018**, *108*, 547–557. [CrossRef]

135. Sharifi-Rad, J.; Quispe, C.; Castillo, C.M.S.; Caroca, R.; Lazo-Vélez, M.A.; Antonyak, H.; Polishchuk, A.; Lysiuk, R.; Oliinyk, P.; De Masi, L.; et al. Ellagic Acid: A Review on Its Natural Sources, Chemical Stability, and Therapeutic Potential. *Oxidative Med. Cell. Longev.* **2022**, *2022*, 3848084. [CrossRef] [PubMed]

136. Semaming, Y.; Pannengpetch, P.; Chattipakorn, S.C.; Chattipakorn, N. Pharmacological Properties of Protocatechuic Acid and Its Potential Roles as Complementary Medicine. *Evid. Based Complement. Altern. Med.* **2015**, *11*, 593902. [CrossRef]

137. Wagle, B.R.; Upadhyay, A.; Arsi, K.; Shrestha, S.; Venkitanarayanan, K.; Donoghue, A.M.; Donoghue, D.J. Application of β-Resorcylic Acid as Potential Antimicrobial Feed Additive to Reduce Campylobacter Colonization in Broiler Chickens. *Front. Microbiol.* **2017**, *8*, 599. [CrossRef]

138. Abedi, F.; Razavi, B.M.; Hosseinzadeh, H. A review on gentisic acid as a plant derived phenolic acid and me-tabolite of aspirin: Comprehensive pharmacology, toxicology, and some pharmaceutical aspects. *Phytother. Res.* **2020**, *34*, 729–734. [CrossRef]

139. Espíndola, K.M.M.; Ferreira, R.G.; Narvaez, L.E.M.; Rosario, A.C.R.S.; Da Silva, A.H.M.; Silva, A.G.B.; Vieira, A.P.O.; Monteiro, M.C. Chemical and Pharmacological Aspects of Caffeic Acid and Its Activity in Hepatocarcinoma. *Front. Oncol.* **2019**, *9*, 541. [CrossRef]

140. Santana-Gálvez, J.; Cisneros-Zevallos, L.; Jacobo-Velázquez, D.A. Chlorogenic Acid: Recent Advances on Its Dual Role as a Food Additive and a Nutraceutical against Metabolic Syndrome. *Molecules* **2017**, *22*, 358. [CrossRef]

141. Boz, H. *p*-Coumaric acid in cereals: Presence, antioxidant and antimicrobial effects. *Int. J. Food Sci. Technol.* **2015**, *50*, 2323–2328. [CrossRef]

142. Zduńska, K.; Dana, A.; Kolodziejczak, A.; Rotsztejn, H. Antioxidant Properties of Ferulic Acid and Its Possible Application. *Ski. Pharmacol. Physiol.* **2018**, *31*, 332–336. [CrossRef]

143. Nićiforović, N.; Abramovič, H. Sinapic Acid and Its Derivatives: Natural Sources and Bioactivity. *Compr. Rev. Food Sci. Food Saf.* **2014**, *13*, 34–51. [CrossRef] [PubMed]

144. Pontiki, E.; Hadjipavlou-Litina, D.; Litinas, K.; Geromichalos, G. Novel Cinnamic Acid Derivatives as Antioxidant and Anticancer Agents: Design, Synthesis and Modeling Studies. *Molecules* **2014**, *19*, 9655–9674. [CrossRef] [PubMed]

145. Klados, E.; Tzortzakis, N. Effects of Substrate and Salinity in Hydroponically Grown *Cichorium spinosum*. *J. Soil Sci. Plant Nutr.* **2014**, *14*, 211–222. [CrossRef]

146. Neffati, M.; Sriti, J.; Hamdaoui, G.; Kchouk, M.E.; Marzouk, B. Salinity Impact on Fruit Yield, Essential Oil Composition and Antioxidant Activities of *Coriandrum sativum* Fruit Extracts. *Food Chem.* **2011**, *124*, 221–225. [CrossRef]

147. Omamt, E.N.; Hammes, P.S.; Robbertse, P.J. Differences in Salinity Tolerance for Growth and Water-Use Efficiency in Some Amaranth (*Amaranthus* spp.) genotypes. *New Zealand J. Crop Hortic. Sci.* **2006**, *34*, 11–22. [CrossRef]

 *antioxidants*

Article

Romanian Wild-Growing *Armoracia rusticana* L.—Untargeted Low-Molecular Metabolomic Approach to a Potential Antitumoral Phyto-Carrier System Based on Kaolinite

Adina-Elena Segneanu [1,*], Gabriela Vlase [1,2], Liviu Chirigiu [3], Daniel Dumitru Herea [4], Maria-Alexandra Pricop [5], Patricia-Aida Saracin [3] and Ștefania Eliza Tanasie [3]

[1] Institute for Advanced Environmental Research, West University of Timisoara (ICAM-WUT), Oituz nr. 4, 300086 Timisoara, Romania; gabriela.vlase@e-uvt.ro

[2] Research Center for Thermal Analysis in in Environmental Problems, West University of Timisoara, Pestalozzi St. 16, 300115 Timisoara, Romania

[3] Faculty of Pharmacy, University of Medicine and Pharmacy Craiova, 2, Petru Rareș, 200349 Craiova, Romania; liviu.chirigiu@umfcv.ro (L.C.); ada_patricia62@yahoo.com (P.-A.S.); eliza_tanasie@yahoo.com (Ș.E.T.)

[4] National Institute of Research and Development for Technical Physics, 47 Mangeron Blvd, 700050 Iasi, Romania; dherea@phys-iasi.ro

[5] OncoGen Centre, Clinical County Hospital "Pius Branzeu", Blvd. Liviu Rebreanu 156, 300723 Timisoara, Romania; alexandra.pricop@oncogen.ro

* Correspondence: adina.segneanu@e-uvt.ro

Citation: Segneanu, A.-E.; Vlase, G.; Chirigiu, L.; Herea, D.D.; Pricop, M.-A.; Saracin, P.-A.; Tanasie, Ș.E. Romanian Wild-Growing *Armoracia rusticana* L.—Untargeted Low-Molecular Metabolomic Approach to a Potential Antitumoral Phyto-Carrier System Based on Kaolinite. *Antioxidants* 2023, 12, 1268. https://doi.org/10.3390/ antiox12061268

Academic Editors: Antonella D'Anneo and Marianna Lauricella

Received: 24 May 2023
Revised: 24 May 2023
Accepted: 6 June 2023
Published: 13 June 2023

Abstract: Horseradish is a globally well-known and appreciated medicinal and aromatic plant. The health benefits of this plant have been appreciated in traditional European medicine since ancient times. Various studies have investigated the remarkable phytotherapeutic properties of horseradish and its aromatic profile. However, relatively few studies have been conducted on Romanian horseradish, and they mainly refer to the ethnomedicinal or dietary uses of the plant. This study reports the first complete low-molecular-weight metabolite profile of Romanian wild-grown horseradish. A total of ninety metabolites were identified in mass spectra (MS)-positive mode from nine secondary metabolite categories (glucosilates, fatty acids, isothiocyanates, amino acids, phenolic acids, flavonoids, terpenoids, coumarins, and miscellaneous). In addition, the biological activity of each class of phytoconstituents was discussed. Furthermore, the development of a simple target phyto-carrier system that collectively exploits the bioactive properties of horseradish and kaolinite is reported. An extensive characterization (FT-IR, XRD, DLS, SEM, EDS, and zeta potential) was performed to investigate the morpho-structural properties of this new phyto-carrier system. The antioxidant activity was evaluated using a combination of three in vitro, non-competitive methods (total phenolic assay, 2,2-Diphenyl-1-picrylhydrazyl (DPPH) radical-scavenging assay, and phosphomolybdate (total antioxidant capacity)). The antioxidant assessment indicated the stronger antioxidant properties of the new phyto-carrier system compared with its components (horseradish and kaolinite). The collective results are relevant to the theoretical development of novel antioxidant agent fields with potential applications on antitumoral therapeutic platforms.

Keywords: secondary metabolites; horseradish; mass spectra; kaolinite; phyto-carrier system; antioxidant activity

1. Introduction

Armoracia rusticana G. Gaertn., B. Mey. & Scherb (*Armoracia rusticana* L.) from the *Brassicaceae* family has been part of traditional European medicine since ancient times. The first mention of the healing effects of this plant (analgesic, diuretic, and antiparasitic) occurs in *De Materia Medica* [1]. Dacian medicine recommends horseradish as an anti-inflammatory cure for colds, coughs, and migraines [1]. Currently, horseradish root is used globally and on a large scale in food, food preservation, and traditional medicine [1].

It is known that there is an interdependence between the content of phytoconstituents in horseradish and different abiotic factors (pH, humidity, temperature, light, etc.) [2,3]. Furthermore, various studies reported that the profiles of metabolites considered responsible for the aroma of horseradish differ, depending on the genotype and plant maturity [3,4].

Recent research has shown that horseradish has collective therapeutic properties: antimicrobial, antifungal, anti-inflammatory, antiviral, and antitumor activity [5–8]. This herb's notable pharmacological activity is due to the combined and synergistic action of its numerous secondary metabolites: glucosinolates, isothiocyanates, organo-sulfur compounds, flavonoids, terpenoids, phenolic acids, coumarins, amino acids, and fatty acids [6–10].

Recently, particular consideration has been given to advanced materials based on natural compounds, which feature extended-release, site-specific delivery and outperform the alternatives in terms of therapeutic activity (anti-tumor, antioxidant, antiviral, antimicrobial, neuroprotective, and anti-inflammatory) [11]. Various studies have investigated the isolation of glucosinolates, isothiocyanates, and organo-sulfur compounds, the main bioactive compounds of horseradish. It has been reported that their chemical stability and implicit bioavailability are influenced by time and temperature [12].

Among the foremost challenges related to new drug discovery from natural products are the composition and proportion differences of secondary metabolites resulting from the influence of biotic and abiotic factors [11,13]. Furthermore, the total synthesis of some phytoconstituents with sophisticated chemical structures and numerous chiral centers is demanding [14].

The use of natural products, especially those based on medicinal plants, has seen an upward trend across the whole world in recent years [15–17]. Although, for the majority of the population, herbal medicine is the primary strategy used in various ailments, recently, particularly in developed countries, natural products have begun to take an increasingly important place in modern civilization due to their high level of biocompatibility and weak side effects [15–17].

Several studies have reported the possible toxic effects of the different herbal products available on the market, which are mainly due to self-administration and exceeding the dosage [15–17]. The pandemic also contributed to this situation, and numerous deaths and severe complications were registered due to the effective lack of medicines [15–17]. Therefore, the most recent studies address the development of new plant-based materials with high performance, binding-site specificity, and controlled release [11,13]. Particular consideration is given to secondary metabolites with high levels of antioxidant, antimicrobial, antiviral, anti-inflammatory, neuroprotective, and antitumor activity [11].

On the other hand, clay therapeutic, food, and protective applications are an integral parts of human culture [18]. In Mesopotamia, Ancient Egypt, and Ancient Greece, clay was used for its anti-inflammatory, antiseptic, and wound-healing properties [18]. The great scholars of the ancient world, Hippocrates and Aristotle, were the first to create a classification of therapeutic clays according to their origin, chemical composition, and biological activity [18].

Recent studies have demonstrated that due to its outstanding physico-chemical properties, including small grain size (in the micrometric order), and large specific surface area (of approximately 100 m^2/g), ensuring high adsorption, swelling, intercalation, and cation-exchange capacity, mineral clays can be used as carrier materials or drug-delivery-system substrates or supports [19–21].

In addition, various studies have confirmed the biological properties of clay minerals and reported high chemical stability and the absence of toxicity in vivo. Currently, clay mineral applications are used as active agents or excipients in numerous pharmaceutical and dermato-cosmetic preparations [19–23].

Kaolinite, $Al_2Si_2O_5(OH)_4$, with a ratio of SiO_2 to Al_2O_3 of approximately 1.18:1, consists of a two-dimensional layer of silica groups linked to a layer of aluminum groups. The distance between the two layers is about 7.2 Å, and it has minor cation-exchange capacity. Furthermore, hydrogen bonds restrict the possibility of expansion or swelling between layers. The surface area is 10–30 m^2/g. Due to its high chemical stability and inertness in vivo, kaolinite has numerous pharmaceutical applications (anti-inflammatory, antiviral, detoxification, hemostatic, antitumoral, protection against gastro-intestinal problems and skin damage, pelotherapy, detoxification, and others) [23–26].

Kaolinite increases the bioavailability of the drug through a controlled release and an oral administration route [23,25]. Many studies reported different drug-delivery systems based on clay minerals for use as in antioxidant, anti-inflammatory, antibiotic, antitumor, antimycotic, anticoagulant, antidiabetic, osteoporosis, and cardioprotective, applications, among others. The main benefits are the prolonged release, increased bioavailability, and minimized toxicity [24,25,27].

The most recent studies addressed the development of antitumoral and immunomodulation drug-delivery systems [24,25,27,28].

It is well-known that the excessive generation of reactive oxygen species (ROS) causes the onset of serious pathologies, including cancer [29–32]. Numerous studies investigated the use of antioxidants as a novel and potent approach to cancer prevention and treatment [29–32]. It is acknowledged that the excessive generation of reactive oxygen species (ROS) causes the onset of serious pathologies, including cancer [29–32]. Consequently, many studies have investigated the antioxidant function as a novel and robust approach to cancer prevention and treatment [29–32]. However, there are still many controversies regarding the effectiveness of antioxidants in cancer therapy [29–32]. Nevertheless, the most recent studies reported some possible factors that can significantly reduce their beneficial effects, such as low bioavailability and low transmembrane permeability, the absence of an adequate dosage, uneven distribution, and others [32].

Furthermore, the biological activity of a plant is the result of the synergistic action of the mixture or complex of secondary metabolites in different proportions [9,33,34].

The antioxidant activities of phytoconstituents are determined by several factors: diversity, climatic factors (temperature, humidity, pH, and soil chemical composition), and harvest maturity stage [35]. Additionally, antioxidant agents are grouped into several categories depending on their mechanism of action (direct or indirect), their source, and the physical-chemical properties of the biomolecule (size, solubility, and others) [36–38]. The efficiency of an antioxidant agent is influenced by several criteria: metabolism pathway, bioavailability, rate constant, concentration, the chemical structure of the biomolecule, and others [36,39].

The development of a successful phyto-carrier assembly relies upon the complementary and synergistic action of the carrier and the secondary metabolites. Furthermore the morpho-structural characteristics, chemical and thermal stability, and biological properties of the carrier have an essential role [13].

Consequently, the high-performance carrier system based on kaolinite development represents a novel multifunctional strategy that will overcome the limitations of the current therapeutic approach related to the drug resistance of cancer cells and ensure site-specific targeting and controlled release.

This study investigates, for the first time, the development and characterization of a phyto-engineered carrier system that accumulates the biological properties of horseradish and kaolinite. Furthermore, to the best of our knowledge, another novelty of this study is the identification of a complete low-molecular-weight metabolite profile of *Armoracia rusticana*, grown in the wild in Romania.

2. Materials and Methods

All used reagents were analytical grade. Methanol, chloroform, dichloromethane, and ethanol were acquired from Sigma-Aldrich (München, Germany) and used without further purification. The DPPH (2,2-diphenyl-1-picrylhydrazyl), β-carotene Type II, synthetic (≥95%), ascorbic acid, AgNO$_3$, sodium citrate, sodium carbonate, Folin–Ciocalteu phenol reagent (2 N), potassium persulfate, sodium phosphate, ammonium molybdate, and potassium chloride of 99% purity or higher were purchased from Sigma-Aldrich (München, Germany). Propyl gallate (purum) was purchased from Fluka (Buchs, Switzerland). The horseradish sample (leaves (28 cm in height) and roots (lengths of about 35 cm) were collected in November 2022 from the area of Timis County, Romania (geographic coordinates 45°45′59.99″ N 21°17′60.00″ E) and taxonomically authenticated at the University of Medicine and Pharmacy Craiova, Romania. Kaolinite was purchased from local market in Timisoara, Romania. The double distilled water (DDW) was used throughout the experiments.

2.1. Phyto-Carrier-System Components' Preparation

2.1.1. Plant-Sample Preparation for Chemical Screening

The plant samples (roots and leaves) were cut and then quickly frozen in liquid nitrogen (180 °C). Subsequently, they were ground and sieved to obtain a particle size lower than 0.45 mm and then stored at -38 °C to prevent enzyme-mediated degradation of phytoconstituents, in a 100 mL conical flask containing 1.5 g dried plant sample and 15 mL of solvent (methanol/chloroform = 1:1). Subsequently, the mixture was subjected to sonication extraction for 30 min at 35 °C with a frequency of 60 kHz. The resulting solution was concentrated using a rotary evaporator, and the obtained residue was dissolved in 10 mL MeOH. The obtained extract was centrifuged (10,000 rot/min, 10 min), and the supernatant was filtered through a 0.2 μm syringe filter and stored at -25 °C until further analysis. All samples were prepared in triplicate.

2.1.2. GC-MS Analysis

Gas chromatography was carried out on a GCMS-QP2020NX Shimadzu apparatus with a ZB-5MS capillary column (30 m × 0.25 mm id × 0.25 μm) (Agilent Technologies, Santa Clara, CA, USA), helium, flow of 1 mL/min.

2.1.3. GC–MS Separation Conditions

The oven-temperature program started from 50 °C to 300 °C with a rate of 5 °C/minute, and it was finally kept at this temperature for 3 min. The temperature of the injector was 280 °C and the temperature at the interface was 230 °C. The compounds' mass was registered at 70 eV ionization energy starting after 3 min of solvent delay. The source of the mass spectrometer was heated at 235 °C and the MS quad was heated at 165 °C. The mass values of identified compounds were scanned from 50 amu to 570 amu. Compounds were identified based on their mass spectra, which were compared to the NIST0.2 mass-spectra-library database (USA National Institute of Science and Technology Software, (NIST, Gaithersburg, MD, USA). Furthermore, the calculated retention indices (RIs) for each compound were compared with the Adams indices in the literature (Table 1) [40].

2.1.4. Mass Spectrometry

The MS experiments were performed using EIS-QTOF-MS (Bruker Daltonics, Bremen, Germany). The mass spectra were acquired in the positive ion mode in a mass range of 100–3000 m/z, scan speed was 2.0 scans/s, collision energy was 25–85 eV, and the temperature of source block was 80 °C. The identification of phytoconstituents was based on standard library NIST/NBS-3 (National Institute of Standards and Technology/National Bureau of Standards) (NIST, Gaithersburg, MD, USA). The obtained mass-spectra values and the identified secondary metabolites are presented in Table 2.

2.1.5. Phyto- Carrier System Preparation

For each analysis, 2.5 g of sample was prepared from dried horseradish, and kaolinite powder was added (horseradish/kaolinite nanoparticles = 1:3) at room temperature (22 °C), ground, and homogenized for 10 min using a pestle and mortar.

2.2. Characterization of the Phyto- Carrier System

2.2.1. Fourier-Transform Infrared (FTIR) Spectroscopy

Data collection was conducted after 20 recordings at a resolution of 4 cm^{-1}, in the range of 4000–400 cm^{-1}, on Shimadzu AIM-9000 with ATR devices (Shimadzu, Kyoto, Japan).

2.2.2. XDR Spectroscopy

The X-ray powder diffraction (XRD) was performed using a Bruker AXS D8-Advance X-ray diffractometer (Bruker AXS GmbH, Karlsruhe, Germany) equipped with a rotating sample stage, Anton Paar TTK low-temperature cell (-180 °C $\div$ 450 °C), high vacuum, inert atmosphere, relative humidity control, and Anton Paar TTK high-temperature cell (up to 1600 °C). The XRD patterns were compared with those from the ICDD Powder Diffraction Database (ICDD file 04-015-9120). The average crystallite size and the phase content were determined using the whole-pattern profile-fitting method (WPPF).

2.2.3. Scanning-Electron Microscopy (SEM)

The SEM micrographs were obtained with a SEM–EDS system (QUANTA INSPECT F50) equipped with a field-emission gun (FEG), 1.2 nm resolution, and energy-dispersive X-ray spectrometer (EDS) with a MnK resolution of 133 eV.

2.2.4. Dynamic Light Scattering (DLS) Particle-Size-Distribution Analysis

The DLS analysis was carried on a Microtrac/Nanotrac 252 (Montgomeryville, PA, USA). Each sample was analyzed in triplicate at room temperature (22 °C) at a scattering angle of 172°.

2.2.5. Zeta-Potential Analysis

The zeta-potential analysis was conducted using an AMERIGO particle-size and zeta-potential analyzer (Pessac, France), with six measurements/s. The main experimental conditions were as follows. Electrode distance: 5 mm; temperature: 25 °C; conductivity: 5.10 V; carrier frequency: 8210 Hz; reference intensity: 2660 kcps; applied field: 20.27 V/cm; and scattering intensity: 2850 kcps.

2.2.6. Antioxidant Activity

The antioxidant activity of the newly phyto-carrier system was estimated using three different assays: a 2,2-diphenyl-1-picrylhydrazyl (DPPH) radical-scavenging assay, a Folin–Ciocalteu assay, and phosphomolybdate assay (total antioxidant capacity).

The phyto-carrier system (0.25 g) and horseradish (0.3 g) samples were dissolved in methanol (10 mL and 12 mL, respectively). The mixtures were stirred at room temperature (22 °C) for 8 h, and then centrifuged at 10,000 rpm for 10 min. The supernatant was then collected for use in the antioxidant assays (2,2-diphenyl-1-picrylhydrazyl (DPPH) radical-scavenging assay, Folin–Ciocalteu assay, and phosphomolybdate assay (total antioxidant capacity).

2.2.7. Determination of Total Phenolic Content

The total phenolic contents in the newly phyto-carrier system and horseradish samples were determined spectrophotometrically according to the Folin–Ciocalteu procedure adapted from the literature [41].

A volume of 2 mL of Folin–Ciocalteu reagent (0.2 N) and 0.2 mL of each sample were vortexed and stored at room temperature (22 °C) for 8 min, in the dark. Sequentially, 2 mL sodium carbonate (7.5%) was added. Next, after two h of incubation at room temperature (vortexed in the dark) the absorbance was measured at 725 nm using a Tecan i-control, 1.10.4.0 infinite 200 Pro spectrophotometer with Corning 96 flat-bottomed clear polystyrol plates (Tecan, Männedorf, Switzerland). The phenol content was expressed in gallic acid equivalents (mg GAE/g sample) using a propyl gallate standard calibration curve between 1 mg/mL and 12.5 µg/mL in methanol [42].

Sample extract concentrations were calculated based on the linear equation obtained from the standard curve (y = 0.9873x − 0.0989).

2.2.8. DPPH Radical-Scavenging Assay

The stock solution was prepared by dissolving 2 mg DPPH in 20 mL MeOH followed by dilutions for a calibration curve with a range of concentrations between 3.12 µg/mL and 0.1 mg/mL. Serial dilutions of ascorbic acid and β-carotene were used as positive standards and MeOH as a vehicle control sample. The ratio (v/v) of DPPH to samples was of 1:1. All samples were placed, in triplicate, in a 96-well plate and stored at 22 °C for 30 min in the dark. At 515 nm, the absorbance was determined on a Tecan i-control, 1.10.4.0 infinite 200 Pro spectrophotometer (Tecan Group Ltd., Männedorf, Switzerland).

The obtained results were used to calculate the average and the inhibition percentage (Inh%) (Equation (1)).

$$Inh\% = (A0 − As)/A0 \times 100 \tag{1}$$

where:

A0 = vehicle control absorbance;

As—sample absorbance.

Further, the IC_{50} value was obtained from the inhibition percentage using the equation of a calibration curve generated for each sample and standard. The results were presented as Inh% versus concentration (µg/mL) [43].

2.2.9. Phosphomolybdate Assay (Total Antioxidant Capacity)

The total-antioxidant-capacity assay of the new phyto-carrier system and horseradish samples was carried out by the phosphomolybdenum procedure using ascorbic acid as standard [44].

A volume of 5 mL reagent solution (0.6 M sulfuric acid, 28 mM sodium phosphate and 4 mM ammonium molybdate) and 0.5 mL of each sample were placed into a water bath at 95 °C for 120 min. Next, the mixed solutions were cooled at room temperature (22 °C). The absorbance was measured at 765 nm using a UV-VIS Perkin-Elmer Lambda 35 (Perkin Elmer, Waltham, MA, USA).

A blank solution was used (5 mL reagent was added in 0.5 mL methanol, and then the mixture was incubated in the same experimental conditions (at 95 °C for 120 min, and then cooled at room temperature (22 °C)). Total antioxidant capacity was determined according to the following equation (Equation (2))

$$\text{Total antioxidant capacity (\%)} = [(\text{Abs. of control} − \text{Abs. of sample})/(\text{Abs. of control}] \times 100 \tag{2}$$

The results are presented as µg/mL of ascorbic acid equivalents (AAE).

2.2.10. Statistical Analysis

All results were obtained with Microsoft Office Excel 2019. Data were used to calculate the average of three replicates for all samples, and all calibration curves and concentrations.

3. Results and Discussion

Plants contain an extensive range of categories of secondary metabolites, with complex chemical compositions [45,46].

In recent years, numerous studies addressed the phytochemical composition and pharmacological activities of metabolites from horseradish roots [4–8,47–54]. There are relatively few studies related to the phytoconstituents from horseradish leaves [1,7,8,55].

Nevertheless, a specific plant's biological activity is the synergistic action of whole phytoconstituent result. Furthermore, researchers have reported that various biotic or abiotic factors (stress, pathogens, and others) altered the metabolite balance and, implicitly, their variability and interrelation [56–58]. In addition, several other elements (the part of the plant used, the extraction process, and the solvent used) influence the type and proportion of bioactive compounds collected from plants [58–61]. Therefore, a plant extract's pharmacological activity differs from the experimental conditions, making it difficult to evaluate the relationship between chemical composition and therapeutic effect [58].

The chemical screening of the phytoconstituents from the horseradish sample was carried out via gas chromatography coupled with mass spectroscopy (GC-MS) and electrospray ionization–quadrupole time-of-flight mass spectrometry (ESI–QTOF–MS) analysis.

The gas-chromatography method coupled with mass spectroscopy (GC–MS) is the most convenient technique for secondary metabolites with relatively low molecular mass (volatile compounds, fatty acids, etc.), providing efficient separation and identification [62].

The GC–MS analysis (Figure 1) revealed the separation of several low-molecular-weight metabolites from the horseradish sample.

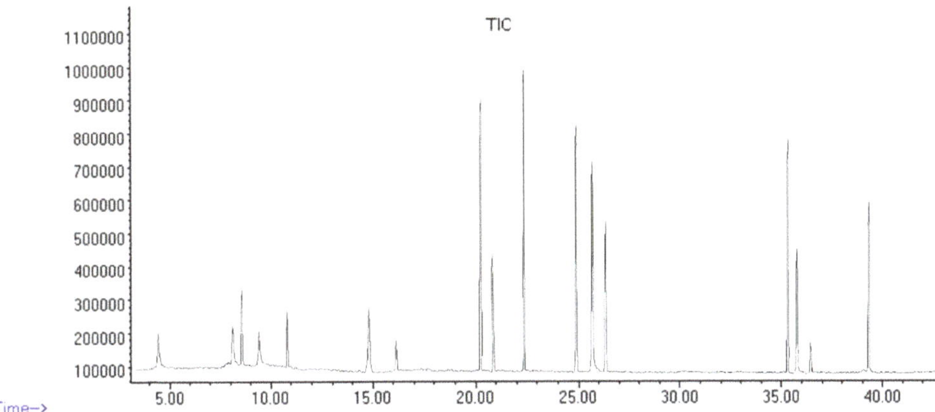

Figure 1. TIC chromatogram of horseradish extract.

The results are summarized in Table 1, which presents the tentative compound identification from the horseradish sample using GC–MS.

Table 1. Main compounds identified by GC–MS analysis of horseradish sample.

No	Retention Time (RT)	Retention Index (RI) Determined	Adams Indices (AI)	Area%	Compound Name	Ref
1	4.23	503	517	0.67	dimethyl sulfide	[47]
2	8.36	1005	1175	0.84	α-phellandrene	[7,48]
3	8.79	1387	1397	1.13	junipene	[7,48]
4	9.44	275	279	0.69	carbonyl sulphide	[7,48]
5	10.76	1169	1173	1.13	menthol	[7,48]
6	14.88	557	574	2.24	carbon disulfide	[7,48]
7	16.02	1198	1202	0.56	3-phenylpropionitrile	[47]
8	20.09	1713	1699	10.47	isobutyl isothiocyanate	[48]
9	20.97	652	706	4.76	2-ethylfuran	[63]
10	22.38	889	1349	13.82	allyl isothiocyanate	[48]
11	24.97	949	963	12.59	3-butenyl isothiocyanate	[48]
12	25.67	1075	1113	11.88	2-pentyl isothiocyanate	[48]

Table 1. *Cont.*

No	Retention Time (RT)	Retention Index (RI) Determined	Adams Indices (AI)	Area%	Compound Name	Ref
13	26.47	1287	1303	10.77	cyclopentyl isothiocyanate	[48]
14	35.08	1363	1317	9.74	benzylisothiocyanate	[7,48]
15	35.97	1317	1435	5.41	erucin	[64]
16	36.49	1215	1231	0.55	2-pentylfuran	[63]
17	39.37	1165	1267	7.88	phenylisothiocyanate	[48]

RI—retention indices calculated based upon a calibration curve of a C8–C20 alkane standard mixture.

The GC–MS analysis showed the presence of seventeen major components, accounting for 95.13% of the total peak area in the horseradish samples (Figure 1).

However, thermally unstable biomolecules require additional procedures (for instance, derivatization). Therefore, the mass-spectrometry method was selected for the metabolite-profile screening [65].

3.1. Mass-Spectrometric Analysis of Horseradish Sample

The spectra revealed a complex combination of low-molecular-weight components, of which some were detected. The mass spectra of the identified metabolites were compared with those of the NIST/EPA/NIH Mass Spectral Library 3.0 database, in addition to a literature review [7,48,55,66]. The mass spectrum and the phytoconstituents identified by the ESI–QTOF–MS analysis are presented in Figure 2 and Table 2, respectively.

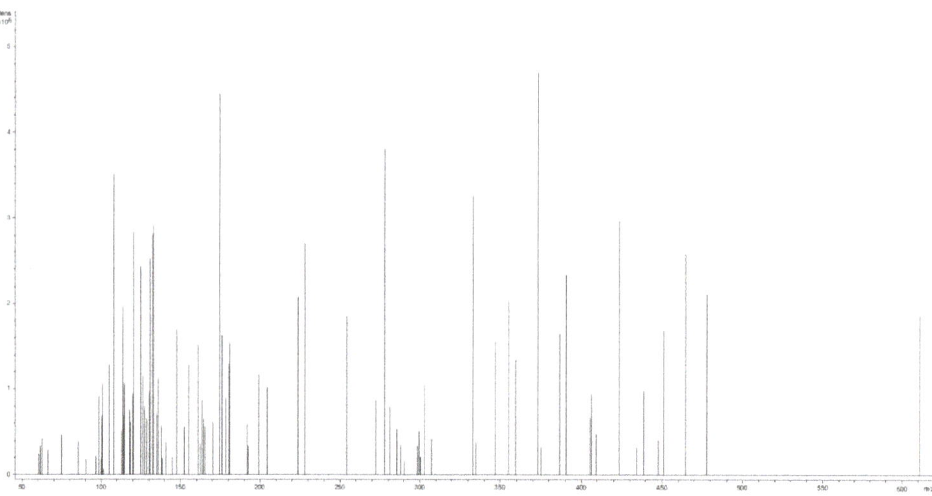

Figure 2. The mass spectrum of *Armoracia rusticana* L.

Table 2. The molecules identified through electrospray-ionization–quadrupole time-of-flight mass spectrometry (ESI–QTOF–MS) analysis.

No	*m/z* Detected	Theoretic *m/z*	Formula	Tentative of Identification	Category	Ref
1	61.05	60.05	$C_2H_4O_2$	acetic acid	organic acid	[48]
2	61.08	60.08	COS	carbonyl sulphide	sulfur compound	[48]
3	63.14	62.14	C_2H_6S	dimethyl sulfide	sulfur compound	[7]

Table 2. *Cont.*

No	*m/z* Detected	Theoretic *m/z*	Formula	Tentative of Identification	Category	Ref
4	68.09	67.09	C_4H_5N	allyl cyanide	miscellaneous	[48]
5	77.15	76.15	CS_2	carbon disulfide	sulfur compound	[7]
6	87.11	86.13	$C_5H_{10}O$	pentanal	aldehyde	[7]
7	91.01	90.03	$C_2H_2O_4$	oxalic acid	organic acid	[48]
8	97.11	96.13	C_6H_8O	2-ethylfuran	furans	[7]
9	99.13	98.14	$C_6H_{10}O$	3-hexenal	aldehyde	[49]
10	100.15	99.16	C_4H_5NS	allyl isothiocyanate	isothiocyanates	[7,48]
11	101.14	100.16	$C_6H_{12}O$	hexanal	aldehyde	[48]
12	103.13	102.13	$C_5H_{10}O_2$	isovaleric acid	organic acid	[49]
13	104.11	103.12	$C_4H_9NO_2$	γ-aminobutyric acid	organic acid	[50]
14	107.11	106.12	C_7H_6O	benzaldehyde	aldehyde	[48]
15	109.13	108.14	C_7H_8O	benzyl alcohol	organic acid	[48]
16	114.17	113.18	C_5H_7NS	3-butenyl isothiocyanate	isothiocyanates	[48]
17	116.11	115.13	$C_5H_9NO_2$	proline	amino acid	[50]
18	116.18	115.20	C_5H_9NS	isobutyl isothiocyanate	isothiocyanates	[48]
19	117.06	116.07	$C_4H_4O_4$	fumaric acid	organic acid	[48]
20	119.07	118.09	$C_4H_6O_4$	succinic acid	organic acid	[48]
21	120.11	119.12	$C_4H_9NO_3$	threonine	amino acid	[51]
22	121.14	120.15	C_8H_8O	phenylacetaldehyde	aldehyde	[49]
23	122.17	121.16	$C_3H_7NO_2S$	cysteine	amino acid	[52]
24	127.19	126.20	$C_8H_{14}O$	vinyl amyl ketone	ketone	[66]
25	128.19	127.21	C_6H_9NS	cyclopentyl isothiocyanate	isothiocyanate	[48]
26	129.15	128.17	$C_{10}H_8$	naphthalene	miscellaneous	[7]
27	130.21	129.23	$C_6H_{11}NS$	2-pentyl isothiocyanate	isothiocyanate	[7,48]
28	132.19	131.17	C_9H_9N	3-phenylpropionitrile	miscellaneous	[7]
29	133.11	132.12	$C_4H_8N_2O_3$	asparagine	amino acid	[8]
30	135.07	134.09	$C_4H_6O_5$	malic acid	organic acid	[56]
31	135.15	134.17	$C_9H_{10}O$	4-ethylbenzaldehyde	aldehyde	[7]
32	137.13	136.15	$C_8H_8O_2$	anisaldehyde	aldehyde	[49]
33	137.25	136.23	$C_{10}H_{16}$	α-phellandrene	terpenoid	[56]
34	139.11	138.12	$C_7H_6O_3$	p-salicylic acid	organic acid	[56]
35	139.19	138.21	$C_9H_{14}O$	2-pentylfuran	furans	[7]
36	143.21	142.24	$C_9H_{18}O$	nonanal	aldehyde	[7]
37	147.17	146.19	$C_6H_{14}N_2O_2$	lysine	amino acid	[50]
38	150.19	149.21	C_8H_7NS	benzyl isothiocyanate	isothiocyanate	[6,48,53–55]
39	153.13	152.15	$C_8H_8O_3$	vanillin	aldehyde	[50]
40	157.25	156.26	$C_{10}H_{20}O$	menthol	terpenoid	[7]
41	162.23	161.3	$C_6H_{11}NS_2$	erucin	isothiocyanate	[49]
42	164.21	163.24	C_9H_9NS	phenethyl isothiocyanate	isothiocyanate	[48]
43	165.15	164.16	$C_9H_8O_3$	coumarinic acid	phenolic acid	[55]
44	166.21	165.19	$C_9H_{11}NO_2$	phenylalanine	amino acid	[51]

Table 2. *Cont.*

No	*m/z* Detected	Theoretic *m/z*	Formula	Tentative of Identification	Category	Ref
45	167.23	166.22	$C_9H_{14}N_2O$	2-sec-butyl-methoxy-pyrazine	miscellaneous	[49]
46	171.13	170.12	$C_7H_6O_5$	gallic acid	phenolic acid	[50]
47	175.19	174.20	$C_6H_{14}N_4O_2$	arginine	amino acid	[50]
48	177.13	176.12	$C_6H_8O_6$	ascorbic acid	organic acid	[55]
49	179.15	178.14	$C_9H_6O_4$	esculetin	coumarin	[8]
50	181.15	180.16	$C_9H_8O_4$	caffeic acid	phenolic acid	[50]
51	182.17	181.19	$C_9H_{11}NO_3$	tyrosine	amino acid	[52]
52	193.11	192.12	$C_6H_8O_7$	citric acid	organic acid	[55]
53	193.17	192.17	$C_{10}H_8O_4$	scopoletin	coumarin	[8]
54	199.19	198.17	$C_9H_{10}O_5$	syringic acid	phenolic acid	[55]
55	205.33	204.35	$C_{15}H_{24}$	junipene	terpenoid	[7]
56	225.19	224.21	$C_{11}H_{12}O_5$	sinapinic acid	phenolic acid	[55]
57	229.36	228.37	$C_{14}H_{28}O_2$	myristic acid	fatty acid	[50]
58	255.40	254.41	$C_{16}H_{30}O_2$	13-hexadecenoic acid	fatty acid	[50]
59	273.43	272.42	$C_{16}H_{32}O_3$	beta-hydroxypalmitic acid	fatty acid	[50]
60	279.39	278.4	$C_{18}H_{30}O_2$	pinolenic acid	fatty acid	[50]
61	281.38	280.4	$C_{18}H_{32}O_2$	linoleic acid	fatty acid	[6]
62	287.25	286.24	$C_{15}H_{10}O_6$	kaempferol	flavonoid	[55]
63	289.41	288.42	$C_{16}H_{32}O_4$	9,10-dihydroxypalmitic acid	fatty acid	[50]
64	291.25	290.27	$C_{15}H_{14}O_6$	catechin	flavonoid	[55]
65	299.49	298.5	$C_{18}H_{34}O_3$	17-hydroxyoleic acid	fatty acid	[50]
66	300.51	299.5	$C_{18}H_{37}NO_2$	sphingosine	miscellaneous	[50]
67	301.49	300.5	$C_{18}H_{36}O_3$	14-hydroxystearic acid	fatty acid	[50]
68	303.25	302.23	$C_{15}H_{10}O_7$	quercetin	flavonoid	[66]
69	309.49	308.5	$C_{20}H_{36}O_2$	eicosadienoic acid	fatty acid	[50]
70	317.51	316.5	$C_{18}H_{36}O_4$	10,11-dihydroxy stearic acid	fatty acid	[50]
71	333.25	332.26	$C_{13}H_{16}O_{10}$	glucogallin	tannin	[50]
72	334.31	333.3	$C_8H_{15}NO_9S_2$	glucocapparin	glucosinolates	[6]
73	348.39	347.4	$C_9H_{17}NO_9S_2$	glucolepidiin	glucosinolates	[6]
74	355.29	354.31	$C_{16}H_{18}O_9$	chlorogenic acid	phenolic acid	[55]
75	360.39	359.4	$C_{10}H_{17}NO_9S_2$	sinigrin	glucosinolates	[8]
76	374.39	373.4	$C_{11}H_{19}NO_9S_2$	gluconapin	glucosinolates	[6,8]
77	376.41	375.4	$C_{11}H_{21}NO_9S_2$	glucocochlearin	glucosinolates	[67]
78	388.39	387.4	$C_{12}H_{21}NO_9S_2$	glucobrassicanapin	glucosinolates	[6,8]
79	392.41	391.4	$C_{11}H_{21}NO_{10}S_2$	glucoconringiin	glucosinolates	[7]
80	407.51	406.5	$C_{11}H_{20}NO_9S_3$	glucoiberverin	glucosinolates	[6,8]
81	408.49	407.5	$C_{11}H_{21}NO_9S_3$	glucosativin	glucosinolates	[7]
82	410.39	409.4	$C_{14}H_{19}NO_9S_2$	glucotropaeolin	glucosinolates	[6,8]

Table 2. *Cont.*

No	*m/z* Detected	Theoretic *m/z*	Formula	Tentative of Identification	Category	Ref
83	424.52	423.5	$C_{11}H_{21}NO_{10}S_3$	glucoiberin	glucosinolates	[6,8]
84	436.49	435.5	$C_{13}H_{25}NO_9S_3$	glucoberteroin	glucosinolates	[6,8]
85	440.51	439.5	$C_{11}H_{21}NO_{11}S_3$	glucocheirolin	glucosinolates	[6,8]
86	449.52	448.5	$C_{16}H_{20}N_2O_9S_2$	glucobrassicin	glucosinolates	[7]
87	452.49	451.5	$C_{13}H_{25}NO_{10}S_3$	glucoalyssin	glucosinolates	[7]
88	465.52	464.5	$C_{16}H_{20}N_2O_{10}S_2$	5-hydroxyglucobrassicin	glucosinolates	[7]
89	479.51	478.5	$C_{17}H_{22}N_2O_{10}S_2$	4-methoxyglucobrassicin	glucosinolates	[7]
90	611.49	610.5	$C_{27}H_{30}O_{16}$	rutin	flavonoid	[55]

The metabolite profile from the horseradish sample conducted through the GC–MS and mass spectroscopy corroborated the data reported in the literature [6–8,48–55,63–65].

3.2. Screening and Classification of the Differential Metabolites

The 90 secondary metabolites identified through mass spectroscopy were assigned to different chemical classes: glucosilates (18.9%), fatty acids (11.12%), isothiocyanates (8.9%), amino acids (8.9%), phenolic acids (6.67%), flavonoids (4.45%), terpenoids (3.34%), coumarins (2.23%), and miscellaneous. The assignment of the identified secondary metabolites into different chemical categories is presented in Table 3.

Table 3. Classification of bioactive secondary metabolites from the *Armoracia rusticana* L. sample in different chemical categories.

Chemical Class	Metabolite Name
glucosinolates	glucocapparin
	glucolepidiin
	sinigrin
	gluconapin
	glucocochlearin
	glucobrassicanapin
	glucoconringiin
	glucoiberverin
	glucosativin
	glucotropaeolin
	glucoiberin
	glucoberteroin
	glucocheirolin
	glucobrassicin
	glucoalyssin
	5-hydroxyglucobrassicin
	4-methoxyglucobrassicin

Table 3. *Cont.*

Chemical Class	Metabolite Name
isothiocyanates	allyl isothiocyanate
	3-butenyl isothiocyanate
	isobutyl isothiocyanate
	cyclopentyl isothiocyanate
	2-pentyl isothiocyanate
	benzyl isothiocyanate
	erucin
	phenethyl isothiocyanate
fatty acids	myristic acid
	13-hexadecenoic acid
	beta-hydroxypalmitic acid
	pinolenic acid
	9,10-dihydroxypalmitic acid
	17-hydroxyoleic acid
	14-hydroxystearic acid
	eicosadienoic acid
	10,11-dihydroxy stearic acid
	linoleic acid
amino acids	proline
	threonine
	cysteine
	asparagine
	lysine
	phenylalanine
	arginine
	tyrosine
	proline
phenolic acids	coumarinic acid
	gallic acid
	caffeic acid
	syringic acid
	sinapinic acid
	chlorogenic acid
flavonoids	kaempferol
	catechin
	quercetin
	rutin

Table 3. *Cont.*

Chemical Class	Metabolite Name
terpenoids	α-phellandrene
	menthol
	junipene
coumarins	esculetin
	scopoletin
aldehyde & ketone	pentanal
	3-hexenal
	hexanal
	benzaldehyde
	phenylacetaldehyde
	vinyl amyl ketone
	4-ethylbenzaldehyde
	anisaldehyde
	nonanal
	vanillin
organic acids	acetic acid
	oxalic acid
	isovaleric acid
	γ-aminobutyric acid
	benzyl alcohol
	fumaric acid
	succinic acid
	malic acid
	p-salicylic acid
	ascorbic acid
	citric acid
furans	2-ethylfuran
	2-pentylfuran
miscellaneous	sphingosine
	glucogallin
	allyl cyanide
	naphthalene
	3-phenylpropionitrile
	2-sec-butyl-3 methoxypyrazine

Figure 3 presents the classification chart of the phytoconstituents from the horseradish sample based on the data analysis reported in Table 3.

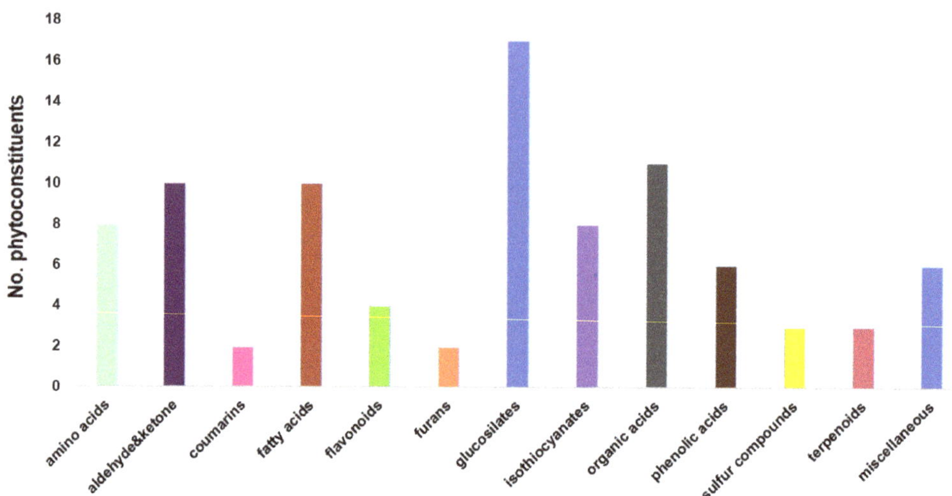

Figure 3. Phytoconstituent-classification bar chart for *Armoracia rusticana*.

According to Figure 3, *glucosinolates* are the largest category of phytochemicals, comprising about 19% of the total found in the horseradish sample. Recent studies demonstrated their antioxidant, anti-inflammatory, and antitumoral properties [7,66,67].

Isothiocyanates are a category of metabolites characteristic of cruciferous plants, with remarkable anti-cancer, anti-inflammatory, and neuroprotective effects [67,68].

Organo-sulfur phytoconstituents represented over 30% of all the metabolites identified in the horseradish sample. Various studies have shown that sulfur phytochemicals possess antioxidant, antiviral, antifungal, antibacterial, and antitumor properties [7,54,69,70].

Fatty acids represented more than 11% of the phytoconstituents identified in the horseradish sample. These secondary metabolites exhibit antioxidant, anti-inflammatory, cardio, and neuroprotective activities [71,72].

Amino acids: eight compounds were identified in the sample extract; the proportions of non-essential amino acids (proline, cysteine, asparagine, and tyrosine) and essential amino acids (threonine, lysine, phenylalanine, and arginine) were equal [73–75].

About one-third of the amino acids identified in the horseradish sample (arginine, phenylalanine, and proline) act as antitumor, neuroprotective, antiproliferative, and immunomodulating agents [71,74–77].

Phenolic acids are another class of phytochemicals with outstanding therapeutic properties (antioxidants, anti-inflammatory, antimicrobial, antidiabetic, antitumor, neuroprotective) [78,79].

The *Terpenoids* found in the horseradish samples were α-phellandrene, junipene, and menthol. Studies have reported that these have antitumor properties. Furthermore, menthol also acts as an antibacterial, antifungal, antipruritic, and analgesic agent [80–82].

Flavonoids are other category of secondary metabolites identified in the horseradish sample with notable pharmacological proprieties, including antioxidant, anti-inflammatory, antitumoral, and antimicrobial properties, as well as activities against neurodegenerative diseases (Alzheimer's) [73,79,83].

The two *coumarins* identified in the horseradish sample, scopoletin and esculetin, show exceptional therapeutic activity, with antioxidant, anti-inflammatory, antitumor, hepatoprotective, and antidiabetic properties, as well as activities against neurodegenerative diseases (Alzheimer's) [84,85].

Among the *miscellaneous* compounds identified in the horseradish sample, sphingosine exerts antitumoral, immunomodulatory, and neuroprotective activities [86–88]. Furthermore, glucogallin possesses antioxidant, anti-inflammatory, and antidiabetic properties [87].

The aromatic compounds of volatile metabolites (VOCs) identified in the horseradish sample are shown in Table 4 and Figure 4.

Table 4. Aromatic compounds identified in *Armoracia rusticana* using ATOF-MS.

No	Name	Odor
1	acetic acid	vinegar
2	carbonyl sulphide	sulphuric
3	dimethylsulfide	cabbage, sulphurous onion
4	allyl cyanide	onion
5	carbon disulphide	sulphuric
6	pentanal	acrid
7	2-ethylfuran	ethereal rum, cocoa
8	3-hexenal	fruity, green, vegetable
9	allyl isothiocyanate	purgent, sulfuric, mustard, garlic
10	hexanal	green, woody, grassy
11	isovaleric acid	cheesy
12	benzaldehyde	almond
13	benzyl alcohol	floral, berry
14	3-butenyl isothiocyanate	purgent
15	isobutyl isothiocyanate	purgent
16	phenylacetaldehyde	green, floral, honey
17	cysteine	sulphur
18	vinyl amyl ketone	earthy, mushroom
19	naphthalene	mothballs
20	2-pentyl isothiocyanate	purgent
21	malic acid	apple, cherry
22	4-ethylbenzaldehyde	sweet, almond, cherry
23	anisaldehyde	sweet, floral, aniseed
24	α-phellandrene	peppery, woody, grassy
25	p-salicylic acid	phenolic
26	2-pentylfuran	green
27	nonanal	citrus, rose
28	benzyl isothiocyanate	pungent green
29	vanillin	vanilla, sweet
30	menthol	minthy
31	erucin	purgent raddish, cabbage
32	phenethyl isothiocyanate	sulfurous
33	2-sec-butyl-3-methoxypyrazine	bell pepper, galbanum
34	junipene	pine, woody
35	kaempferol	bitter
36	quercetin	bitter
37	glucocapparin	purgent, horseradish-

Table 4. *Cont.*

No	Name	Odor
38	gluconapin	pungent, green, cabagge
39	glucobrassicanapin	acrid, purgent, mustard, horseradish
40	glucobrassicin	purgent
41	glucosativin	rocket
42	glucoiberverin	purgent, radish
43	sinigrin	pungent, sulfurous, mustard
44	glucotropaeolin	purgent

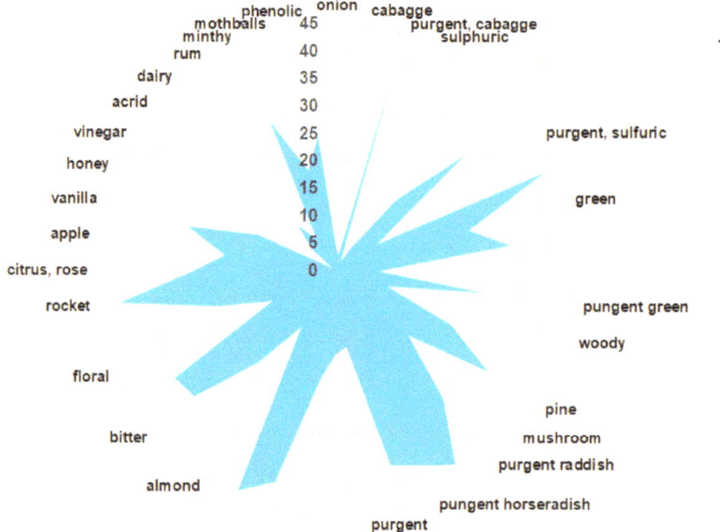

Figure 4. VOC-aroma profile of phytoconstituients identified in horseradish sample.

The predominant aromatic components of the investigated Romanian horseradish depend on different conditions (climatic conditions, maturity soil parameters, varieties, harvest time, and others) [4,7,47–49].

Their fragrances are unique, encompassing a purgent aroma with rocket and sulfuric, green, sweet-vanilla, and floral notes [4,7,47,48].

3.3. Phyto-Carrier System

The main challenges in the novel therapeutic approaches to cancer are the drug resistance of cancer cells, determined by the reduced retention interval, low permeability, the triggering of inactivation by the immune system, and the lack of specificity [89,90].

Hence, the development of an innovative phyto-carrier target system with cumulative and synergistic kaolinite and horseradish biological activity could make it possible to overcome the limitations related to vectorization, site-specific distribution, prolonged release, and membrane permeability.

3.4. FT-IR Spectroscopy

The use of FTIR is one of the most common analytical techniques, and it is considered fundamental in the analysis of complex carrier systems due to its features (sensitivity, flexibility, robustness, and specificity), allowing the investigation of interactions between biomolecules and mineral components [90].

The incorporation of the horseradish phytoconstituents into the pores of the kaolinite particles was successfully achieved and confirmed through FT-IR spectroscopy. Figure 5A presents the spectra of the horseradish, the kaolinite particles, and the new phyto-carrier system.

Table 5. The characteristic absorption bands attributed to secondary metabolites identified in *Armoracia rusticana*.

Secondary Metabolite	Wavenumber (cm^{-1})	Ref
glucosinolates	990–1090, 1433–1470, 1695, 1730, 1920, 1990–2150, 2270	[91,92]
isothiocyanates	2060–2190. 2269–2275, 1990 − 2150, 2034, 925–1250, 680, 520–570, 425–440, 464	[91,92]
flavonoids	4000–3125, 3140–3000, 1670–1620, 1650–1600, 1600–1500, 1450–1490	[93,94]
amino acids	3400; 3330–3130; 2530–2760; 2130; 1724–1754 1687, 1675, 1663, 1652, 1644, 1632, 1621, 1611, 1500–1600	[95]
terpenoids	2939, 1740, 1651, 810	[96]
phenolic acids	1800–1650, 1734, 1720, 1627, 1522, 1440, 1410, 1420–1300, 1367, 1315, 1255, 1170–1100	[97]
fatty acids	3020–3010, 2924–2915, 2855–2847, 2800–2900, 1746, 1710, 1250, 720	[97,98]
coumarins	2963, 3061, 3381, 1608, 1715, 1489, 1450, 1254, 1028, 600–900	[99]

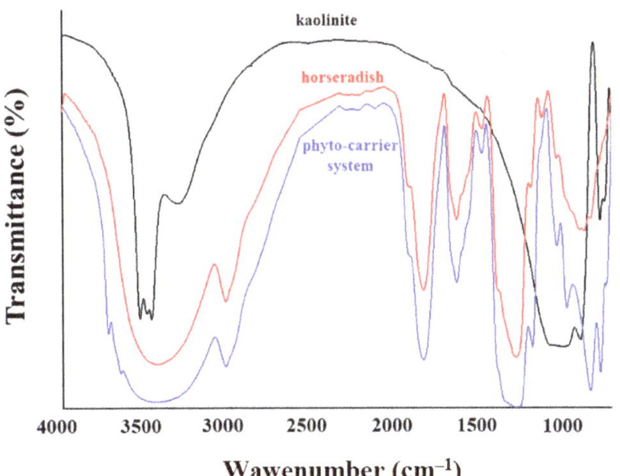

(**A**)

Figure 5. *Cont.*

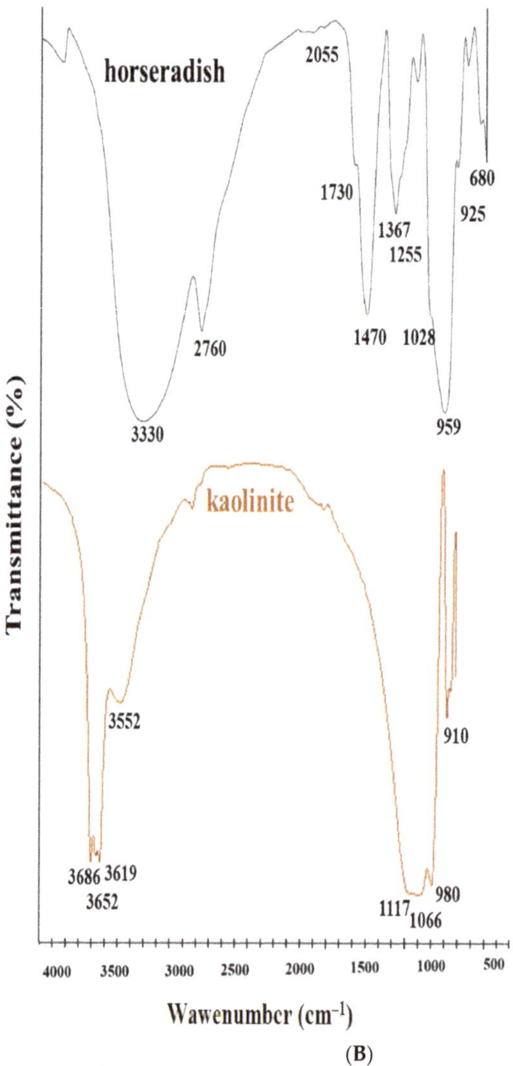

Figure 5. (**A**) FTIR spectra of kaolinite, horseradish, and phyto-carrier system. (**B**) FTIR spectra of FT-IR absorption bands identified in the horseradish sample are presented in the following table (Table 5).

The FTIR peak of the kaolinite (Figure 5B) presented vibrational bands characteristic at 3686, 3652, 3619, and 3552 cm^{-1} (attributed to the OH stretching vibrations), and 1117, 1066, 980, and 912 cm^{-1} (associated with the Si-O stretching vibration) [100–103].

The data obtained and presented in Figure 5 confirm the successful development of the new phyto-carrier system.

The obtained IR spectra of the new phyto-carrier system incorporated peaks specific to the secondary metabolites from the horseradish at the following: 3330 cm^{-1}, assigned to the OH group; 2760 cm^{-1}, attributed to the O-H stretching in the amino acids; 2055 cm^{-1} (N=C=S stretching of isothiocyanate); 1730 cm^{-1} (C-H stretching by methylene groups); 1470 cm^{-1} (C-H bending); 1367 cm^{-1} (O-H bending); 1255 cm^{-1} (C-O stretching); 1028 cm^{-1} (C-N stretching); 959 and 925 cm^{-1} (symmetric N-C-S stretch); and 680 cm^{-1} (aromatic ring); and the characteristic absorption bands of the kaolinite [90–103].

In addition, the kaolinite absorption bands at 3686, 3652, 3619, and 3552 cm^{-1} (attributed to O-H stretching vibrations) and the vibrational bands at 1117, 1066, 980, and 912 cm^{-1} (associated with Si-O stretching vibration) were shifted to lower wavenumbers, indicating that this functional group was involved in the binding of the O-H, C-N, N-H, and C-O functional groups from the horseradish (Figure 5, Table 5) [90–103].

Moreover, several detectable changes occurred in the horseradish spectra, particularly the hydroxyl vibrations (O-H stretching and O-H bending), indicating that this functional group is involved in the binding of kaolinite [23–27,104].

3.5. X-ray-Diffraction Spectroscopy

The XRD technique was used to obtain information about the atomic structure of the phyto-carrier system and the raw materials.

Figure 6 displays the XRD patterns of the horseradish sample and the new phytocarrier system.

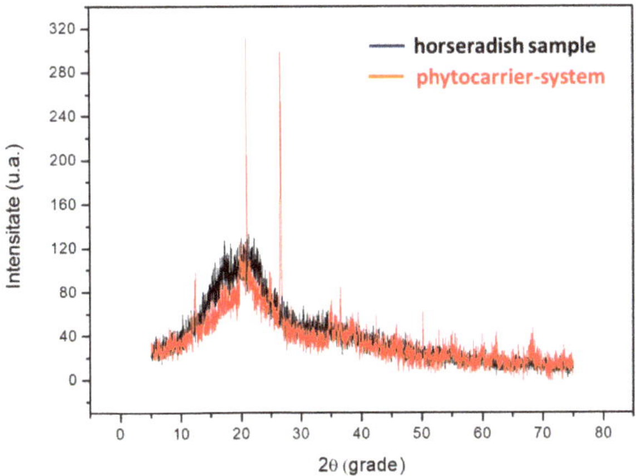

Figure 6. The overlapping XRD spectra of horseradish sample and new phyto-carrier system.

In the XRD spectrum of the new phyto-carrier system, the characteristic XRD peaks of the kaolinite and horseradish samples are easily observable. Hence, the absorption peaks at 2θ (degrees) values of 12°, 25°, 34°, 36°, and 51° can be assigned to a triclinic structure [105].

The XRD pattern of the horseradish sample (Figure 6) was in the range of 11.8–34.6°, with large bands and weak peaks characteristic of amorphous phases, which can be attributed to the phytoconstituents from the horseradish (minerals, hydroxides, and fibers).

3.6. Scanning-Electron Microscopy–Energy-Dispersive X-ray (SEM–EDX)

Scanning-electron microscopy–energy-dispersive X-ray (SEM–EDX) is a versatile technique to investigate the morphologies, compositions, and microstructures of materials. In some complex materials, it allows the identification of the component phases through qualitative chemical analysis [106].

The morphological changes (the size, shape, and distribution of the particles) in the horseradish and kaolinite samples before and after the preparation of the new phyto-carrier system were investigated by using the SEM–EDX technique.

To acquire insights, the SEM micrographs were recorded at different magnifications. The obtained two-dimensional images are shown in Figure 7.

Figure 7. SEM images of kaolinite (**A**,**B**), horseradish (**C**,**D**), and phyto-carrier system (**E**,**F**).

The SEM micrograph of the kaolinite sample (Figure 7A,B) exhibited a heterogeneous size distribution of small anhedral and pseudo-hexagonal particles up to 5 μm in size [25].

It appears that the horseradish micrographs (Figure 7C,D) indicated the presence of a heterogeneous fibrous structure, with a thickness of about a few μm, with porous regions with irregular shapes. These porous regions allowed the arrest of the kaolinite particles.

The morphology of the phyto-carrier system (Figure 7E,F) indicated the presence of kaolinite particles both on the surface and in the porous areas of the horseradish sample. Changes in the sizes of the horseradish and kaolinite particles (reduction) were observed, which can be explained by the experimental conditions of the new phyto-carrier system preparation.

Accompanying the SEM spectra are EDX analyses on the elemental composition of the kaolinite and phyto-carrier investigated (Figure 8A,B).

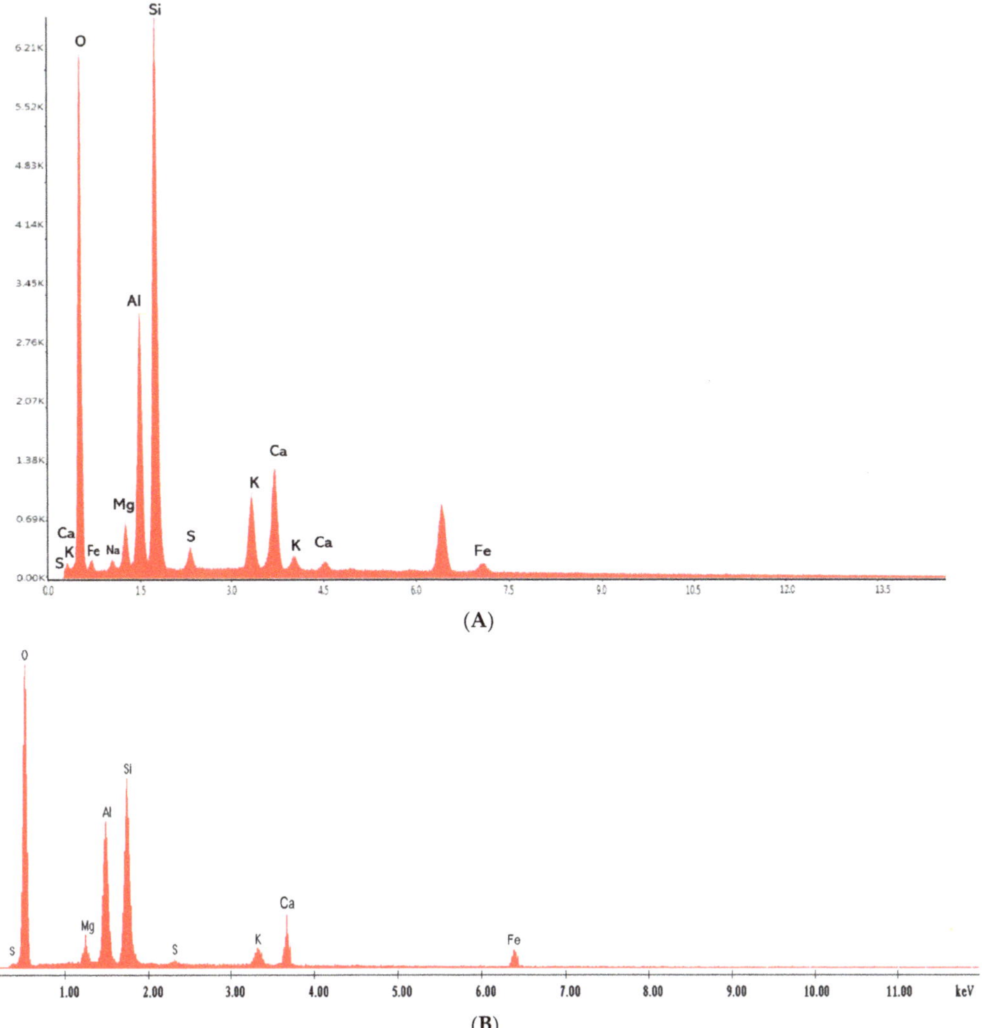

Figure 8. (**A**) EDX composition of kaolinite sample; (**B**) EDX composition of the new phyto-carrier system.

According to the data from the EDX (Figure 7A), the predominant element contents in the kaolinite sample were silica, aluminum, magnesium, calcium, potassium, iron, sodium, oxygen, and sulphur. Overall, the chemical analysis revealed the significant oxides SiO_2, Al_2O_3, Fe_2O_3, MgO, CaO, Na_2O, K_2O_5, and SO^{-3}, which was in good agreement with the data reported in the literature [25].

The comparative analysis in Figure 8B highlights the presence of peaks corresponding to the kaolinite (Figure 8A) in the new phyto-carrier system. The EDX results confirmed the preparation of the new phyto-carrier system.

3.7. Dynamic Light Scattering (DLS)

Dynamic light scattering (DLS) is a fast and very efficient method for determining the sizes of particles and the particle-size distribution (PSD) in suspensions [107]. Particle-size measurement is established indirectly by using the intensity of the light-scattered fluctuations, yielding the rate of the Brownian motion [107].

The DLS method was used to obtain information about the average mean particle size of the phyto-carrier system and its raw components. The DLS results are displayed in Figure 9.

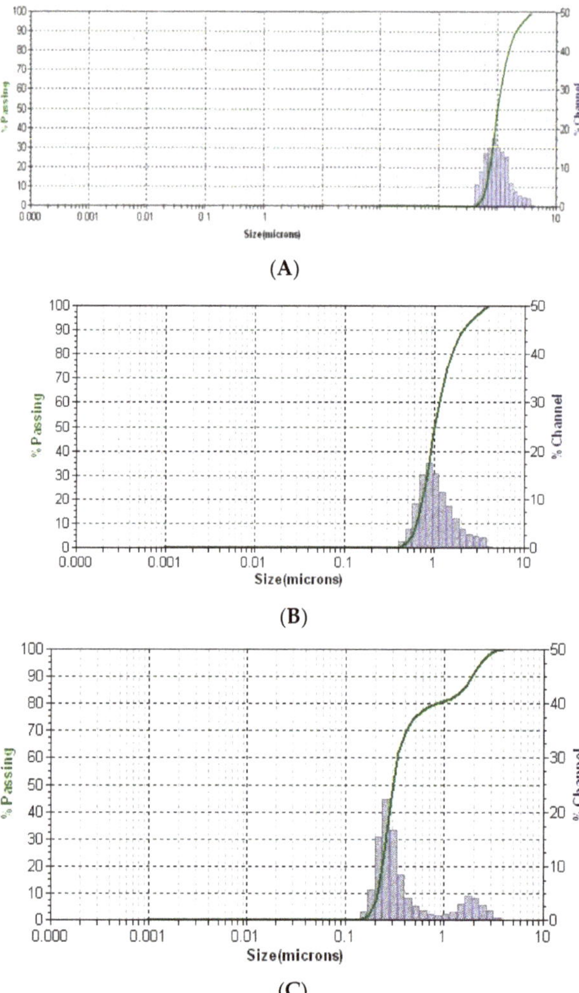

Figure 9. DLS patterns of kaolinite (**A**), horseradish sample (**B**), and phyto-carrier system (**C**).

The average diameter size of the kaolinite particles was 500.03 nm (Figure 9A), corroborating the SEM results. In the horseradish sample (Figure 9B), the average diameter of the particles was 100.2 nm.

The DLS pattern of the new phyto-carrier system (Figure 9C) exhibited two peaks that can be attributed to the kaolinite and horseradish particles, distributed in a narrow range. The mean diameter of the kaolinite in the phyto-carrier system was 277.5 nm. The second mean of the hydrodynamic diameter, associated with the horseradish particles, was about 186.4 nm. The fact that the average diameter of the horseradish particles in the phyto-carrier increased compared to that determined in the horseradish sample can be attributed to the loading of the pores on the plant surfaces with the kaolinite particles, which was confirmed by the results of the SEM analysis.

Furthermore, the reduction in the mean size of the kaolinite particles from 500.03 nm (Figure 9A) to 277.5 nm (Figure 9C) was attributed to the experimental conditions for the preparation of the new phyto-carrier. In addition, Figure 9C shows well-dispersed particles of horseradish and kaolinite, which indicates the high stability of the new phyto-carrier system.

3.8. Zeta Potential

The zeta-potential method determines the charge of a particle in a suspension, providing an estimation of interactions between particles and the suspension stability.

The zeta-potential value of the kaolinite particles was -35.09 mV, indicating the high stability of the suspension, in good agreement with the data reported in the literature [108].

The zeta potential changed to -23.12 mV for the phyto-carrier system, indicating high biocompatibility.

3.9. Screening of Antioxidant Activity

For a specific herb, the total antioxidant capacity (TAC) is the outcome of the cumulative action of entire antioxidant classes from its composition [37]. The adequate investigation of the antioxidant activity of a plant requires an appropriate variety of tests to address the mechanism of action characteristic of each category of phytochemicals [37,38,109,110].

Various chemical (spectrometric, chromatographic, and electrochemical) and biochemical methods have been developed for the assessment of the antioxidant capacities of different biomolecules [37,38,109,110]. The most common are the in vitro tests, divided based on the reaction-mechanism type into hydrogen-atom transfer (HAT) and electron transfer (ET) methods [37,38,109,110].

The first category, HAT methods, includes the oxygen-radical-absorbance capacity (ORAC), the total radical-trapping-antioxidant parameter (TRAP), the total radical-scavenging-capacity assay (TOSCA), the chemiluminescent assay, β-carotene bleaching assays, and the inhibition of induced LDL oxidation [37,38,111–113].

The main ET methods (based on electron transfer) are the total phenolics assay (Folin–Ciocalteu reagent assay), the 2,2-Diphenyl-1-picrylhydrazyl radical-scavenging assay (DPPH•), the Trolox equivalence antioxidant-capacity assay (TEAC), the ferric-ion-reducing antioxidant-power assay (FRAP), the cupric reducing antioxidant capacity (CUPRAC) assay, the N,N-Dimethyl-p-phenylenediamine radical-scavenging assay (DMPD•+), and the 2,2-Azinobis 3-ethylbenzthiazoline-6-sulfonic acid radical-scavenging assay (ABTS•+) [37,38,109,110].

The choice of a particular method depends on criteria related to simplicity, sensitivity, associated costs, and reproducibility [37–39,111,112].

The biological activity of a plant varies depending on the complexity of the chemical composition and, implicitly, on the collective, complementary, and the synergistic actions of a variety of secondary metabolites. Moreover, the antioxidant activities of plants differ, depending on morphological parts, degree of maturity, and exogenous parameters (temperature, pH, humidity, and others) [37].

Hence, the antioxidant activity of the phyto-carrier system is a combined result of the complementary and synergistic actions of its components (horseradish and kaolinite). A total amount of ninety secondary metabolites from nine different chemical classes were identified in the horseradish sample. Consequently, to consider the antioxidant properties of the new phyto-carrier system more precisely, three different in vitro, non-competitive methods were used (DPPH, Folin–Ciocalteu, and phosphomolybdate (total antioxidant capacity).

3.9.1. DPPH (1,1-diphenyl-2-picrylhydrazyl) Free-Radical-Scavenging Assay

The DPPH (2,2-diphenyl-1-picrylhydrazyl) is a fast, simple, low-cost, and accurate method based on a single electron transfer (ET)-type mechanism for the antioxidant assessment of plant extracts or other complex matrices. Furthermore, it is a highly frequently used assay to determine the free scavenging capacity of antioxidants based on the ability of compounds to act as free-radical scavengers or hydrogen donors [37–39,110–113].

Hence, the antioxidant activity of the new phyto-carrier system and its components were evaluated in relation to the antioxidant standards of β-carotene and ascorbic acid. It is noteworthy that different studies reported the presence of β-carotene and ascorbic acid in the chemical composition of horseradish [114,115]. The data obtained are presented in Table 6 and Figure 10.

Table 6. IC50 values for horseradish, the new phyto-carrier system, ascorbic acid, and beta-carotene.

Sample Name	Horseradish	Phyto-Carrier System	Ascorbic Acid	Beta-Carotene
IC50 (_g/mL)	8.21.00 ± 0.06	4.68 ± 0.11	28.17 ± 0.02	2.11 ± 0.017

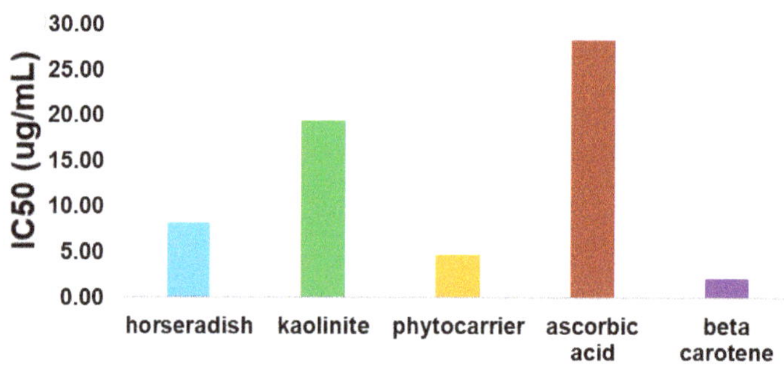

Figure 10. Graphic representation of DPPH results expressed as IC50 (μg/mL).

The obtained IC50 values indicated that the antioxidant activity of the new phyto-carrier system was higher than that of the horseradish sample, the kaolinite, and the ascorbic acid. For the new phyto-carrier system, the IC50 value was about half that of the horseradish sample. The increase in the antioxidant activity of the phyto-carrier system compared to the horseradish and kaolinite was in good agreement with the literature data [116,117]. The IC50 value for the beta-carotene standard can be explained by the experimental conditions (the low solubility of beta carotene in methanol) [118].

3.9.2. Folin–Ciocalteu Assay

This assay is widely used as a fast, simple, precise, and inexpensive measure of total phenolics from natural products based on an oxidation/reduction-reaction mechanism (electron transfer) [38,39,41,119,120].

The total polyphenolic contents (TPCs) of the horseradish sample and phyto-carrier system were determined and the obtained results are presented in Table 7.

Table 7. Total polyphenolic contents in horseradish and the phytocarrier system.

Sample Name	Total Phenolic Content (μg/mL)
horseradish	13.79667
phyto-carrier system	35.18658

According to the results, the total polyphenolic content identified in the new phyto-carrier system was more than 39% higher than that of the horseradish sample. The higher antioxidant capacity of the phyto-carrier system compared to the horseradish sample can be attributed to the synergistic action of the kaolinite and corresponds to the data reported in the literature [121].

3.9.3. Phosphomolybdate Assay (Total Antioxidant Capacity)

Phosphomolybdate (total antioxidant capacity) is a frequently used and precise assay used to evaluate the total antioxidant potentials of plant extracts or other complex mixtures of biomolecules. It is based on the Mo(VI)-to-Mo(V) reduction of the presence of antioxidants [44].

The phosphomolybdate assay (total antioxidant capacity) was used to determine the total antioxidant potential of the prepared phyto-carrier system compared to those of the horseradish and ascorbic acid. The obtained experimental results are displayed in Table 8 and Figure 11.

Table 8. Total antioxidant potentials of phyto-carrier system and horseradish sample.

Sample Name	
horseradish	171.82 ± 0.00343
phyto-carrier system	248.96 ± 0.014

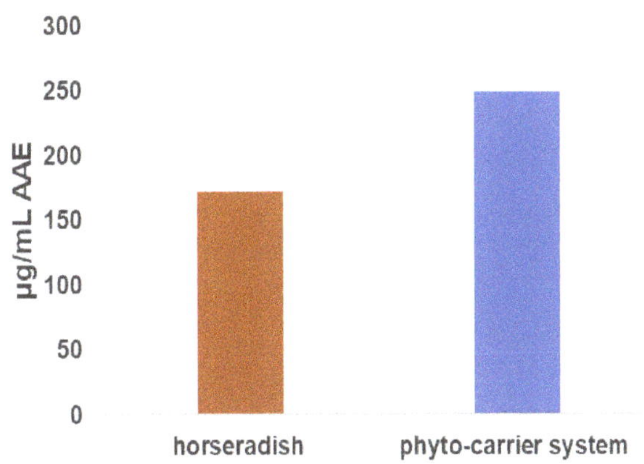

Figure 11. Graphic representation of phosphomolybdate (total antioxidant capacity) results expressed as µg/mL AAE.

The phyto-carrier system displayed a higher antioxidant activity than the horseradish sample. This result can be attributed to the synergistic and complementary action of the phytoconstituents in the horseradish and the antioxidant mechanism of the kaolinite [28]. In addition, the kaolinite potentiated the antioxidant activities of secondary metabolites in the horseradish sample [121].

4. Conclusions

In this study, a new phyto-carrier system with particular morpho-structural properties and high antioxidant activity was prepared. The low-molecular-mass-metabolite profiling and the VOC-aroma profile of the *Armoracia rusticana* grown in the wild in Romania were determined. The biological activities of each identified phytoconstituent category in the horseradish were discussed. The development of the horseradish–kaolinite carrier system was confirmed through FTIR, EDX, XRD, DLS, zeta-potential, and SEM studies. The size distributions of the kaolinite and horseradish particles were investigated through a DSL analysis. The kaolinite and the phyto-carrier system's stability levels in aqueous suspensions were determined using a zeta-potential analysis. A combination of assays

(DPPH, Folin–Ciocalteu, and phosphomolybdate (total antioxidant capacity)) was used to evaluate the antioxidant properties of the proposed phyto-carrier system. The results demonstrated the significantly higher antioxidant activity of the phyto-carrier compared with its components (horseradish and kaolinite). However, further studies are required to investigate the biological activity, bioavailability, and biocompatibility of the new phyto-carrier system. This study may motivate future research on therapies in the area of advanced antitumoral agents.

Author Contributions: Conception and study design: A.-E.S.; methodology: A.-E.S.; data acquisition: G.V., M.-A.P., D.D.H., P.-A.S. and Ş.E.T.; analysis and data interpretation: L.C., G.V., M.-A.P. and D.D.H.; writing—original draft preparation: A.-E.S.; writing—review and editing: A.-E.S. and D.D.H.; investigation: G.V., M.-A.P., Ş.E.T. and D.D.H. All authors have read and agreed to the published version of the manuscript.

Funding: This research received no external funding.

Institutional Review Board Statement: Not applicable.

Informed Consent Statement: Not applicable.

Data Availability Statement: All data are contained within the article.

Acknowledgments: National Center for Micro and Nanomaterials (the Center is part of the Department of Science and Engineering of Oxide and Nanomaterials Materials of the Faculty of Applied Chemistry and Materials Science of the Politehnica University of Bucharest). This work was supported by a grant of the Ministry of Research, Innovation and Digitization CNCS–UEFISCDI, project number PN-III-P4-PCE-2021-1081 within PNCDI III (Contract no. 75/2022).

Conflicts of Interest: The authors declare no conflict of interest.

References

1. Courter, J.W.; Rhodes, A.M. Historical notes on horseradish. *Econ. Bot.* **1969**, *23*, 156–164. [CrossRef]
2. Segneanu, A.-E.; Cepan, M.; Bobica, A.; Stanusoiu, I.; Dragomir, I.C.; Parau, A.; Grozescu, I. Chemical Screening of Metabolites Profile from Romanian *Tuber* spp. *Plants* **2021**, *10*, 540. [CrossRef]
3. Rivelli, A.R.; De Maria, S. Exploring the physiological and agronomic response of *Armoracia rusticana* grown in rainfed Mediterranean conditions. *Ital. J. Agron.* **2019**, *14*, 133–141. [CrossRef]
4. Kroener, E.-M.; Buettner, A. Sensory-Analytical Comparison of the Aroma of Different Horseradish Varieties (*Armoracia rusticana*). *Front. Chem.* **2018**, *6*, 149. [CrossRef] [PubMed]
5. Papp, N.; Gonda, S.; Kiss-Szikszai, A.; Plaszkó, T.; Lőrincz, P.; Vasas, G. Ethnobotanical and ethnopharmacological data of *Armoracia rusticana* P. Gaertner, B. Meyer et Scherb. in Hungary and Romania: A case study. *Genet. Resour. Crop Evol.* **2018**, *65*, 1893–1905. [CrossRef]
6. Anam, Y.; Qayyum, F.; Ahmad Mugha, A.; Amjad, O. Phytochemical and biological investigation of *Armoracia rusticana*. *Int. J. Pharm. Integr. Health Sci.* **2020**, *1*, 58.
7. Agneta, R.; Möllers, C.; Rivelli, A.R. Horseradish (*Armoracia rusticana*), a neglected medical and condiment species with a relevant glucosinolate profile: A review. *Genet. Resour. Crop. Evol.* **2013**, *60*, 1923–1943. [CrossRef]
8. Nguyen, N.M.; Gonda, S.; Vasas, G. A Review on the Phytochemical Composition and Potential Medicinal Uses of Horseradish (*Armoracia rusticana*) Root. *Food Rev. Int.* **2013**, *29*, 261–275. [CrossRef]
9. Zhao, Q.; Luan, X.; Zheng, M.; Tian, X.-H.; Zhao, J.; Zhang, W.-D.; Ma, B.-L. Synergistic mechanisms of constituents in herbal extracts during intestinal absorption: Focus on natural occurring nanoparticles. *Pharmaceutics* **2020**, *12*, 128. [CrossRef] [PubMed]
10. Hussein, R.A.; El-Anssary, A.A. Plants Secondary Metabolites: The Key Drivers of the Pharmacological Actions of Medicinal Plants. In *Herbal Medicine*; Builders, P.F., Ed.; IntechOpen: London, UK, 2019.
11. Nath, R.; Roy, R.; Barai, G.; Bairagi, S.; Manna, S.; Chakraborty, R. Modern developments of nano based grug delivery system by combined with phytochemicals-presenting new aspects. *Int. J. Sci. Res. Sci. Technol.* **2021**, *8*, 107–129.
12. Barba, F.J.; Nikmaram, N.; Roohinejad, S.; Khelfa, A.; Zhu, Z.; Koubaa, M. Bioavailability of Glucosinolates and Their Breakdown Products: Impact of Processing. *Front. Nutr.* **2016**, *3*, 24. [CrossRef] [PubMed]
13. Patra, J.K.; Das, G.; Fraceto, L.F.; Campos, E.V.R.; del Pilar Rodriguez-Torres, M.; Acosta-Torres, L.S.; Diaz-Torres, L.A.; Grillo, R.; Swamy, M.K.; Sharma, S.; et al. Nano based drug delivery systems: Recent developments and future prospects. *J. Nanobiotechnol.* **2018**, *16*, 71. [CrossRef]
14. Singh, I.P.; Ahmad, F.; Chatterjee, D.; Bajpai, R.; Sengar, N. *Natural Products: Drug Discovery and Development in Drug Discovery and Development from Targets and Molecules to Medicines*; Poduri, R., Ed.; Springer Nature: Singapore, 2021; ISBN 978-981-15-5533-6.

15. Pal, S.K.; Shukla, Y. Herbal medicine: Current status and the future. *Asian Pac. J. Cancer Prev.* **2003**, *4*, 281–288. [PubMed]

16. Sánchez, M.; González-Burgos, E.; Iglesias, I.; Lozano, R.; Gómez-Serranillos, M.P. Current uses and knowledge of medicinal plants in the Autonomous Community of Madrid (Spain): A descriptive cross-sectional study. *BMC Complement. Med. Ther.* **2020**, *20*, 306. [CrossRef]

17. Alam, S.; Sarker, M.R.; Afrin, S.; Richi, F.T.; Zhao, C.; Zhou, J.-R.; Mohamed, I.N. Traditional Herbal Medicines, Bioactive Metabolites, and Plant Products Against COVID-19: Update on Clinical Trials and Mechanism of Actions. *Front. Pharmacol.* **2021**, *12*, 671498. [CrossRef] [PubMed]

18. Stojiljković, S.T.; Stojiljković, M.S. Application of bentonite clay for human use. In *Proceedings of the IV Advanced Ceramics and Applications Conference*; Atlantis Press: Paris, France, 2017; pp. 349–356.

19. Rebitski, E.P.; Darder, M.; Sainz-Diaz, C.I.; Carraro, R.; Aranda, P.; Ruiz-Hitzky, E. Theoretical and experimental investigation on the intercalation of metformin into layered clay minerals. *Appl. Clay Sci.* **2020**, *186*, 105418. [CrossRef]

20. Behroozian, S.; Zlosnik, J.E.A.; Xu, W.; Li, L.Y.; Davies, J.E. Antibacterial Activity of a Natural Clay Mineral against *Burkholderia cepacia* Complex and Other Bacterial Pathogens Isolated from People with Cystic Fibrosis. *Microorganisms* **2023**, *11*, 150. [CrossRef]

21. Martsouka, F.; Papagiannopoulos, K.; Hatziantoniou, S.; Barlog, M.; Lagiopoulos, G.; Tatoulis, T.; Tekerlekopoulou, A.G.; Lampropoulou, P.; Papoulis, D. The Antimicrobial Properties of Modified Pharmaceutical Bentonite with Zinc and Copper. *Pharmaceutics* **2021**, *13*, 1190. [CrossRef] [PubMed]

22. Williams, L.B. Natural Antibacterial Clays: Historical Uses and Modern Advances. *Clays Clay Miner.* **2019**, *67*, 7–24. [CrossRef]

23. Massaro, M.; Colletti, C.G.; Lazzara, G.; Riela, S. The Use of Some Clay Minerals as Natural Resources for Drug Carrier Applications. *J. Funct. Biomater.* **2018**, *9*, 58. [CrossRef] [PubMed]

24. Awad, M.E.; López-Galindo, A.; Setti, M.; El-Rahmany, M.M.; Iborra, C.V. Kaolinite in pharmaceutics and biomedicine. *Int. J. Pharm.* **2017**, *533*, 34–48. [CrossRef] [PubMed]

25. Awad, M.; López-Galindo, A.; El-Rahmany, M.; El-Desoky, H.; Viseras, C. Characterization of Egyptian kaolins for health-care uses. *Appl. Clay Sci.* **2017**, *135*, 176–189. [CrossRef]

26. Saad, H.; Ayed, A.; Srasra, M.; Attia, S.; Srasra, E.; Charrier-El Bouhtoury, F.; Tabbene, O. New trends in clay-based nanohybrid applications: Essential oil encapsulation strategies to improve their biological activity. In *Nanoclay—Recent Advances, New Perspectives and Applications*; Oueslati, W., Ed.; IntechOpen: London, UK, 2022; ISBN 978-1-80356-558-3.

27. Gianni, E.; Avgoustakis, K.; Papoulis, D. Kaolinite group minerals: Applications in cancer diagnosis and treatment. *Eur. J. Pharm. Biopharm.* **2020**, *154*, 359–376. [CrossRef]

28. García-Tojal, J.; Iriarte, E.; Palmero, S.; Pedrosa, M.R.; Rad, C.; Sanllorente, S.; Zuluaga, M.C.; Cavia-Saiz, M.; Rivero-Perez, D.; Muñiz, P. Phyllosilicate-content influence on the spectroscopic properties and antioxidant capacity of Iberian Cretaceous clays. *Spectrochim. Acta Part A Mol. Biomol. Spectrosc.* **2021**, *251*, 119472. [CrossRef] [PubMed]

29. Sardas, S. The Role of Antioxidants in Cancer Prevention and Treatment. *Indoor Built Environ.* **2003**, *12*, 401–404. [CrossRef]

30. Fuchs-Tarlovsky, V. Role of antioxidants in cancer therapy. *Nutrition* **2013**, *29*, 15–21. [CrossRef]

31. Akanji, M.A.; Fatinukun, H.D.; Rotimi, E.D.; Afolabi, B.L.; Adeyemi, O.S. The two sides of dietary antioxidants in cancer therapy. In *Antioxidants-Benefits, Sources, Mechanisms of Action*; Waisundara, V., Ed.; IntechOpen: London, UK, 2021; ISBN 978-1-83968-865-2.

32. Luo, M.; Zhou, L.; Huang, Z.; Li, B.; Nice, E.C.; Xu, J.; Huang, C. Antioxidant Therapy in Cancer: Rationale and Progress. *Antioxidants* **2022**, *11*, 1128. [CrossRef]

33. Yuan, H.; Ma, Q.; Ye, L.; Piao, G. The Traditional Medicine and Modern Medicine from Natural Products. *Molecules* **2016**, *21*, 559. [CrossRef]

34. Yarnell, E. Synergy in Herbal Medicines: Part 1. *J. Restor. Med.* **2015**, *4*, 60–73. [CrossRef]

35. Flieger, J.; Flieger, W.; Baj, J.; Maciejewski, R. Antioxidants: Classification, natural sources, activity/capacity measurements, and usefulness for the synthesis of nanoparticles. *Materials* **2021**, *14*, 4135. [CrossRef]

36. Amorati, R.; Valgimigli, L. Methods To Measure the Antioxidant Activity of Phytochemicals and Plant Extracts. *J. Agric. Food Chem.* **2018**, *66*, 3324–3329. [CrossRef]

37. Dinkova-Kostova, A.T.; Talalay, P. Direct and indirect antioxidant properties of inducers of cytoprotective proteins. *Mol. Nutr. Food Res.* **2008**, *52*, S128–S138. [CrossRef]

38. Losada-Barreiro, S.; Sezgin-Bayindir, Z.; Paiva-Martins, F.; Bravo-Díaz, C. Biochemistry of antioxidants: Mechanisms and pharmaceutical applications. *Biomedicines* **2022**, *10*, 3051. [CrossRef]

39. Shahidi, F.; Zhong, Y. Measurement of antioxidant activity. *J. Funct. Foods* **2015**, *18*, 757–781. [CrossRef]

40. Adams, R.P. *Identification of Essential Oil Components by Gas Chromatography/Mass Spectrometry*; Allured Publishing Corporation: Carol Stream, IL, USA, 2007; p. 456.

41. Swain, T.; Hillis, W.E. The phenolic constituents of Prunus domestica. I.-The quantitative analysis of phenolic constituents. *J. Sci. Food Agric.* **1959**, *10*, 63–68.

42. Jianu, C.; Goleţ, I.; Stoin, D.; Cocan, I.; Lukinich-Gruia, A.T. Antioxidant activity of *Pastinaca sativa* L. ssp. sylvestris [Mill.] Rouy and Camus essential oil. *Molecules* **2020**, *25*, 869. [CrossRef]

43. Rădulescu, M.; Jianu, C.; Lukinich-Gruia, A.T.; Mioc, M.; Mioc, A.; Șoica, C.; Stanca, L.G. Chemical composition, in vitro and in silico antioxidant potential of Melissa officinalis subsp. officinalis essential oil. *Antioxidants* **2021**, *10*, 1081. [CrossRef] [PubMed]
44. Khatoon, M.; Islam, E.; Islam, R.; Rahman, A.A.; Alam, A.K.; Khondkar, P.; Rashid, M.; Parvin, S. Estimation of total phenol and in vitro antioxidant activity of Albizia procera leaves. *BMC Res. Notes* **2013**, *6*, 121. [CrossRef]
45. Teoh, E.S. Secondary metabolites of plants. *Med. Orchid. Asia* **2015**, *5*, 59–73.
46. Twaij, B.M.; Hasan, M.N. Bioactive secondary metabolites from plant sources: Types, synthesis, and their therapeutic uses. *Int. J. Plant Biol.* **2022**, *13*, 4–14. [CrossRef]
47. Mazza, G. Volatiles in distillates of fresh, dehydrated and freez dried horseradish. *Can. Inst. Food Sci. Technol. J.* **1984**, *17*, 18–23. [CrossRef]
48. Tomsone, L.; Kruma, Z.; Galoburda, R.; Talou, T. Composition of volatile compounds of horseradish roots (*Armoracia rusticana* L.) depending on the genotype. *Rural. Sustain. Res.* **2013**, *29*, 1–10. [CrossRef]
49. Kroener, E.-M.; Buettner, A. Unravelling important odorants in horseradish (*Armoracia rusticana*). *Food Chem.* **2017**, *232*, 455–465. [CrossRef]
50. Negro, E.J.; Sendker, J.; Stark, T.; Lipowicz, B.; Hensel, A. Phytochemical and functional analysis of horseradish (*Armoracia rusticana*) fermented and non-fermented root extracts. *Fitoterapia* **2022**, *162*, 105282. [CrossRef] [PubMed]
51. Welinder, K.G. Amino Acid Sequence Studies of Horseradish Peroxidase. Amino and Carboxyl Termini, Cyanogen Bromide and Tryptic Fragments, the Complete Sequence, and Some Structural Characteristics of Horseradish Peroxidase C. *Eur. J. Biochem.* **1979**, *96*, 483–502. [CrossRef]
52. Plaszkó, T.; Szűcs, Z.; Cziáky, Z.; Ács-Szabó, L.; Csoma, H.; Géczi, L.; Vasas, G.; Gonda, S. Correlations Between the Metabolome and the Endophytic Fungal Metagenome Suggests Importance of Various Metabolite Classes in Community Assembly in Horseradish (*Armoracia rusticana*, Brassicaceae) Roots. *Front. Plant Sci.* **2022**, *13*, 921008. [CrossRef]
53. Petrović, S.; Drobac, M.; Ušjak, L.; Filipović, V.; Milenković, M.; Niketić, M. Volatiles of roots of wild-growing and cultivated Armoracia macrocarpa and their antimicrobial activity, in comparison to horseradish, *A. rusticana*. *Ind. Crops Prod.* **2017**, *109*, 398–403. [CrossRef]
54. Herz, C.; Tran, H.T.T.; Márton, M.-R.; Maul, R.; Baldermann, S.; Schreiner, M.; Lamy, E. Evaluation of an Aqueous Extract from Horseradish Root (*Armoracia rusticana* Radix) against Lipopolysaccharide-Induced Cellular Inflammation Reaction. *Evidence-Based Complement. Altern. Med.* **2017**, *2017*, 1950692. [CrossRef] [PubMed]
55. Tomsone, L.; Galoburda, R.; Kruma, Z.; Cinkmanis, I. Characterization of dried horseradish leaves pomace: Phenolic compounds profile and antioxidant capacity, content of organic acids, pigments and volatile compounds. *Eur. Food Res. Technol.* **2020**, *246*, 1647–1660. [CrossRef]
56. Wawrosch, C.; Zotchev, S.B. Production of bioactive plant secondary metabolites through in vitro technologies—Status and outlook. *Appl. Microbiol. Biotechnol.* **2021**, *105*, 6649–6668. [CrossRef]
57. Ashraf, M.A.; Iqbal, M.; Rasheed, R.; Hussain, I.; Riaz, M.; Arif, M.S. Environmental stress and secondary metabolites in plants. In *Plant Metabolites and Regulation under Environmental Stress*; Academic Press: Cambridge, MA, USA, 2018; pp. 153–167.
58. Pirintsos, S.; Panagiotopoulos, A.; Bariotakis, M.; Daskalakis, V.; Lionis, C.; Sourvinos, G.; Karakasiliotis, I.; Kampa, M.; Castanas, E. From Traditional Ethnopharmacology to Modern Natural Drug Discovery: A Methodology Discussion and Specific Examples. *Molecules* **2022**, *27*, 4060. [CrossRef] [PubMed]
59. Segneanu, A.E.; Grozescu, I.; Sfirloaga, P. The influence of extraction process parameters of some biomaterials precursors from Helianthus annuus. *Dig. J. Nanomater. Biostruct.* **2013**, *8*, 1423–1433.
60. Segneanu, A.-E.; Grozescu, I.; Cziple, F.; Berki, D.; Damian, D.; Niculite, C.M.; Florea, A.; Leabu, M. *Helleborus purpurascens*—Amino Acid and Peptide Analysis Linked to the Chemical and Antiproliferative Properties of the Extracted Compounds. *Molecules* **2015**, *20*, 22170–22187. [CrossRef]
61. Segneanu, A.-E.; Damian, D.; Hulka, I.; Grozescu, I.; Salifoglou, A. A simple and rapid method for calixarene-based selective extraction of bioactive molecules from natural products. *Amino Acids* **2016**, *48*, 849–858. [CrossRef]
62. Hill, C.B.; Roessner, U. Metabolic profiling of plants by GC–MS. In *The Handbook of Plant Metabolomics*, 1st ed.; Weckwerth, W., Kahl, G., Eds.; Wiley-VCH Verlag GmbH: Weinheim, Germany, 2013.
63. Shen, M.; Liu, Q.; Jia, H.; Jiang, Y.; Nie, S.; Xie, J.; Xie, M. Simultaneous determination of furan and 2-alkylfurans in heat-processed foods by automated static headspace gas chromatography-mass spectrometry. *LWT—Food Sci. Technol.* **2016**, *72*, 44–54. [CrossRef]
64. Al-Gendy, A.A.; Lockwood, G.B. GC-MS analysis of volatile hydrolysis products from glucosinolates inFarsetia aegyptia var.ovalis. *Flavour Fragr. J.* **2003**, *18*, 148–152. [CrossRef]
65. Andini, S.; Araya-Cloutier, C.; Sanders, M.; Vincken, J.-P. Simultaneous Analysis of Glucosinolates and Isothiocyanates by Reversed-Phase Ultra-High-Performance Liquid Chromatography–Electron Spray Ionization–Tandem Mass Spectrometry. *J. Agric. Food Chem.* **2020**, *68*, 3121–3131. [CrossRef]
66. Maina, S.; Misinzo, G.; Bakari, G.; Kim, H.-Y. Human, Animal and Plant Health Benefits of Glucosinolates and Strategies for Enhanced Bioactivity: A Systematic Review. *Molecules* **2020**, *25*, 3682. [CrossRef]

67. Connolly, E.L.; Sim, M.; Travica, N.; Marx, W.; Beasy, G.; Lynch, G.S.; Bondonno, C.P.; Lewis, J.R.; Hodgson, J.M.; Blekkenhorst, L.C. Glucosinolates From Cruciferous Vegetables and Their Potential Role in Chronic Disease: Investigating the Preclinical and Clinical Evidence. *Front. Pharmacol.* **2021**, *12*, 767975. [CrossRef]

68. Yadav, K.; Dhankhar, J.; Kundu, P. Isothiocyanates—A Review of their Health Benefits and Potential Food Applications. *Curr. Res. Nutr. Food Sci.* **2022**, *10*, 476–502. [CrossRef]

69. Miękus, N.; Marszałek, K.; Podlacha, M.; Iqbal, A.; Puchalski, C.; Świergiel, A.H. Health Benefits of Plant-Derived Sulfur Compounds, Glucosinolates, and Organosulfur Compounds. *Molecules* **2020**, *25*, 3804. [CrossRef]

70. Hill, C.R.; Shafaei, A.; Balmer, L.; Lewis, J.R.; Hodgson, J.M.; Millar, A.H.; Blekkenhorst, L.C. Sulfur compounds: From plants to humans and their role in chronic disease prevention. *Crit. Rev. Food Sci. Nutr.* **2022**, *04*, 1–23. [CrossRef]

71. Dhama, K.; Karthik, K.; Khandia, R.; Munjal, A.; Tiwari, R.; Rana, R.; Khurana, S.K.; Ullah, S.; Khan, R.U.; Alagawany, M.; et al. Medicinal and therapeutic potential of herbs and plant metabolites/extracts countering viral pathogens-current knowledge and future prospects. *Curr. Drug Metab.* **2018**, *19*, 236–263. [CrossRef] [PubMed]

72. Nagy, K.; Tiuca, I.D. Importance of fatty acids. In *Physiopathology of Human Body in Fatty Acids*; Catala, A., Ed.; IntechOpen: Rijeka, Croatia, 2017.

73. Das, K.; Gezici, S. Review article Plant secondary metabolites, their separation, identification and role in human disease prevention. *Ann. Phytomedicine Int. J.* **2018**, *7*, 13–24. [CrossRef]

74. Kim, S.-H.; Roszik, J.; Grimm, E.A.; Ekmekcioglu, S. Impact of l-Arginine Metabolism on Immune Response and Anticancer Immunotherapy. *Front. Oncol.* **2018**, *8*, 67. [CrossRef] [PubMed]

75. Chiangjong, W.; Chutipongtanate, S.; Hongeng, S. Anticancer peptide: Physicochemical property, functional aspect and trend in clinical application (Review). *Int. J. Oncol.* **2020**, *57*, 678–696. [CrossRef] [PubMed]

76. Lieu, E.L.; Nguyen, T.; Rhyne, S.; Kim, J. Amino acids in cancer. *Exp. Mol. Med.* **2020**, *52*, 15–30. [CrossRef]

77. Albaugh, V.L.; Pinzon-Guzman, C.; Barbul, A. Arginine metabolism and cancer. *Surg. Oncol. March* **2017**, *115*, 273–280. [CrossRef] [PubMed]

78. Kumar, N.; Goel, N. Phenolic acids: Natural versatile molecules with promising therapeutic applications. *Biotechnol. Rep.* **2019**, *24*, e00370. [CrossRef]

79. Rasouli, H.; Farzaei, M.H.; Khodarahmi, R. Polyphenols and their benefits: A review. *Int. J. Food Prop.* **2017**, *20*, 1700–1741. [CrossRef]

80. Radice, M.; Durofil, A.; Buzzi, R.; Baldini, E.; Martínez, A.P.; Scalvenzi, L.; Manfredini, S. Alpha-phellandrene and alpha-phellandrene-rich essential oils: A systematic review of biological activities, Pharmaceutical and Food Applications. *Life* **2022**, *12*, 1602. [CrossRef]

81. Kamatou, G.P.; Vermaak, I.; Viljoen, A.M.; Lawrence, B.M. Menthol: A simple monoterpene with remarkable biological properties. *Phytochemistry* **2013**, *96*, 15–25. [CrossRef] [PubMed]

82. Grover, M.; Behl, T.; Virmani, T.; Sanduja, M.; Makeen, H.A.; Albratty, M.; Alhazmi, H.A.; Meraya, A.M.; Bungau, S.G. Exploration of Cytotoxic Potential of Longifolene/Junipene Isolated from *Chrysopogon zizanioides*. *Molecules* **2022**, *27*, 5764. [CrossRef]

83. Kozłowska, A.; Szostak-Wegierek, D. Flavonoids–food sources, health benefits, and mechanisms involved. In *Bioactive Molecules in Food, Reference Series in Phytochemistry*; Mérillon, J.M., Ramawat, K., Eds.; Springer: Cham, Switzerland, 2018.

84. Gnonlonfin, G.J.B.; Sanni, A.; Brimer, L. Review Scopoletin—A Coumarin Phytoalexin with Medicinal Properties. *Crit. Rev. Plant Sci.* **2012**, *31*, 47–56. [CrossRef]

85. Garg, S.S.; Gupta, J.; Sahu, D.; Liu, C.-J. Pharmacological and Therapeutic Applications of Esculetin. *Int. J. Mol. Sci.* **2022**, *23*, 12643. [CrossRef]

86. Wang, X.; Wang, Y.; Xu, J.; Xue, C. Sphingolipids in food and their critical roles in human health. *Crit. Rev. Food Sci. Nutr.* **2020**, *61*, 462–491. [CrossRef]

87. Ghosh, D.; Khan, A.N.; Singh, R.; Bhattacharya, A.; Chakravarti, R.; Roy, S.; Ravichandiran, V. A Short Review on Glucogallin and its Pharmacological Activities. *Mini-Rev. Med. Chem.* **2022**, *22*, 2820–2830. [CrossRef]

88. Emran, T.B.; Shahriar, A.; Mahmud, A.R.; Rahman, T.; Abir, M.H.; Siddiquee, M.F.; Ahmed, H.; Rahman, N.; Nainu, F.; Wahyudin, E.; et al. Multidrug resistance in cancer: Understanding molecular mechanisms, immunoprevention and therapeutic approaches. *Front. Oncol.* **2022**, *12*, 2581. [CrossRef]

89. Xie, P.; Wang, Y.; Wei, D.; Zhang, L.; Zhang, B.; Xiao, H.; Song, H.; Mao, X. Nanoparticle-based drug delivery systems with platinum drugs for overcoming cancer drug resistance. *J. Mater. Chem. B* **2021**, *9*, 5173–5194. [CrossRef]

90. Eid, M.M. Characterization of nanoparticles by FTIR and FTIR-microscopy. In *Handbook of Consumer Nanoproducts*; Springer: Singapore, 2022; ISBN 978-981-16-8697-9.

91. Redha, A.A.; Torquati, L.; Langston, F.; Nash, G.R.; Gidley, M.J.; Cozzolino, D. Determination of glucosinolates and isothiocyanates in glucosinolate-rich vegetables and oilseeds using infrared spectroscopy: A systematic review. *Crit. Rev. Food Sci. Nutr.* **2023**, 1–17. [CrossRef] [PubMed]

92. Lieber, E.; Rao, C.; Ramachandran, J. The infrared spectra of organic thiocyanates and isothiocyanates. *Spectrochim. Acta* **1959**, *13*, 296–299. [CrossRef]

93. Noh, C.H.C.; Azmin, N.F.M.; Amid, A. Principal Component Analysis Application on Flavonoids Characterization. *Adv. Sci. Technol. Eng. Syst. J.* **2017**, *2*, 435–440. [CrossRef]

94. Heneczkowski, M.; Kopacz, M.; Nowak, D.; Kuźniar, A. Infrared spectrum analysis of some flavonoids. *Acta Pol. Pharm. Drug Res.* **2001**, *58*, 415–420.

95. Segneanu, A.; Velciov, S.M.; Olariu, S.; Cziple, F.; Damian, D.; Grozescu, I. Bioactive molecules profile from natural compounds. In *Amino Acid—New Insights and Roles in Plant and Animal*; Asao, T., Asaduzzaman, M., Eds.; IntechOpen: Rijeka, Croatia, 2017.

96. Segneanu, A.-E.; Marin, C.N.; Herea, D.D.; Stanusoiu, I.; Muntean, C.; Grozescu, I. Romanian *Viscum album* L.—Untargeted low-molecular metabolomic approach to engineered Viscum–AuNPs carrier assembly. *Plants* **2022**, *11*, 1820. [CrossRef] [PubMed]

97. Scarsini, M.; Thurotte, A.; Veidl, B.; Amiard, F.; Niepceron, F.; Badawi, M.; Lagarde, F.; Schoefs, B.; Marchand, J. Metabolite Quantification by Fourier Transform Infrared Spectroscopy in Diatoms: Proof of Concept on Phaeodactylum tricornutum. *Front. Plant Sci.* **2021**, *12*, 756421. [CrossRef] [PubMed]

98. Topala, C.M.; Tatarua, L.D.; Ducu, C. ATR-FTIR spectra fingerprinting of medicinal herbs extracts prepared using microwave extraction. *Arab. J. Med. Aromat. Plants* **2017**, *3*, 1–9.

99. Umashankar, T.; Govindappa, M.; Ramachandra, Y.L.; Padmalatha, R.; Channabasava, S. Isolation and characterization of coumarin isolated from endophyte, Alternaria species-1 of Crotalaria pallida and its apoptotic action on HeLa cancer cell Line. *Metabolomics* **2015**, *5*, 158.

100. Frost, R.L.; Kristof, J. Raman and Infrared Spectroscopic Studies of Kaolinite Surfaces Modified by Intercalation. In *Clay Surfaces—Fundamentals and Applications*; Wypych, F., Satyanarayana, K.G., Eds.; Academic Press: Cambridge, MA, USA, 2004; pp. 184–215. [CrossRef]

101. Balan, E.; Saitta, A.M.; Mauri, F.; Calas, G. First-principles modeling of the infrared spectrum of kaolinite. *Am. Miner.* **2001**, *86*, 1321–1330. [CrossRef]

102. Jozanikohan, G.; Abarghooei, M.N. The Fourier transform infrared spectroscopy (FTIR) analysis for the clay mineralogy studies in a clastic reservoir. *J. Pet. Explor. Prod. Technol.* **2022**, *12*, 2093–2106. [CrossRef]

103. Panda, A.K.; Mishra, B.G.; Mishra, D.K.; Singh, R.K. Effect of sulphuric acid treatment on the physico-chemical characteristics of kaolin clay. *Colloids Surf. A Physicochem. Eng. Asp.* **2010**, *363*, 98–104. [CrossRef]

104. Yu, W.H.; Li, N.; Tong, D.S.; Zhou, C.H.; Lin, C.X.; Xu, C.Y. Adsorption of proteins and nucleic acids on clay minerals and their interactions: A review. *Appl. Clay Sci.* **2013**, *80–81*, 443–452. [CrossRef]

105. Sachan, A.; Penumadu, D. Identification of Microfabric of Kaolinite Clay Mineral Using X-ray Diffraction Technique. *Geotech. Geol. Eng.* **2007**, *25*, 603–616. [CrossRef]

106. Zhou, W.; Apkarian, R.; Wang, Z.L.; Joy, D. Fundamentals of Scanning Electron Microscopy (SEM). In *Scanning Microscopy for Nanotechnology*; Springer: Berlin/Heidelberg, Germany, 2006; pp. 1–40. ISBN 978-1-4419-2209-0. [CrossRef]

107. Wang, M.; Shen, J.; Thomas, J.C.; Mu, T.; Liu, W.; Wang, Y.; Pan, J.; Wang, Q.; Liu, K. Particle Size Measurement Using Dynamic Light Scattering at Ultra-Low Concentration Accounting for Particle Number Fluctuations. *Materials* **2021**, *14*, 5683. [CrossRef] [PubMed]

108. Yukselen, Y.; Abidin, K. Zeta potential of kaolinite in the presence of alkali, alkaline earth and hydrolyzable metal ions. *Water Air Soil Pollut.* **2003**, *145*, 155–168. [CrossRef]

109. Gulcin, İ. Antioxidants and antioxidant methods: An updated overview. *Arch. Toxicol.* **2020**, *94*, 651–715. [CrossRef]

110. Dontha, S.A. Review on antioxidant methods. *Asian J. Pharm. Clin. Res.* **2016**, *9*, 14–32.

111. Haida, Z.; Hakiman, M. A comprehensive review on the determination of enzymatic assay and nonenzymatic antioxidant activities. *Food Sci. Nutr.* **2019**, *7*, 1555–1563. [CrossRef]

112. Kedare, S.B.; Singh, R.P. Genesis and development of DPPH method of antioxidant assay. *J. Food Sci. Technol.* **2011**, *48*, 412–422. [CrossRef]

113. Rahman, M.M.; Islam, M.B.; Biswas, M.; Alam, A.H.M.K. In vitro antioxidant and free radical scavenging activity of different parts of Tabebuia pallida growing in Bangladesh. *BMC Res. Notes* **2015**, *8*, 621. [CrossRef]

114. Rivelli, A.R.; Caruso, M.C.; De Maria, S.; Galgano, F. Vitamin C content in leaves and roots of horseradish (*Armoracia rusticana*): Seasonal variation in fresh tissues and retention as affected by storage conditions. *Emir. J. Food Agric.* **2017**, *29*, 799. [CrossRef]

115. Tomsone, L.; Kruma, Z. Spectrophotometric analysis of pigments in horseradish by using various extraction solvents. In Proceedings of the 13th Baltic Conference on Food Science and Technology "Food. Nutrition. Well-Being.", Jelgava, Latvia, 2–3 May 2019. [CrossRef]

116. Dinis, L.-T.; Bernardo, S.; Conde, A.; Pimentel, D.; Ferreira, H.; Félix, L.; Gerós, H.; Correia, C.; Moutinho-Pereira, J. Kaolin exogenous application boosts antioxidant capacity and phenolic content in berries and leaves of grapevine under summer stress. *J. Plant Physiol.* **2016**, *191*, 45–53. [CrossRef] [PubMed]

117. Rudayni, H.A.; Aladwani, M.; Alneghery, L.M.; Allam, A.A.; Abukhadra, M.R.; Bellucci, S. Insight into the Potential Antioxidant and Antidiabetic Activities of Scrolled Kaolinite Single Sheet (KNs) and Its Composite with ZnO Nanoparticles: Synergetic Studies. *Minerals* **2023**, *13*, 567. [CrossRef]

118. Nisar, T.; Iqbal, M.; Raza, A.; Safdar, M.; Iftikhar, F.; Waheed, M. Estimation of total phenolics and free radical scavenging of turmeric (*Curcuma longa*). *Environ. Sci.* **2015**, *15*, 1272–1277.

119. Lamuela-Raventós, R.M. Folin-Ciocalteu method for the measurement of total phenolic content and antioxidant capacity. In *Measurement of Antioxidant Activity & Capacity: Recent Trends and Applications*; Wiley: New York, NY, USA, 2017; pp. 107–115. [CrossRef]
120. Carmona-Hernandez, J.C.; Taborda-Ocampo, G.; González-Correa, C.H. Folin-Ciocalteu Reaction Alternatives for Higher Polyphenol Quantitation in Colombian Passion Fruits. *Int. J. Food Sci.* **2021**, *2021*, 8871301. [CrossRef] [PubMed]
121. Hamdy, A.E.; Abdel-Aziz, H.F.; El-Khamissi, H.; AlJwaizea, N.I.; El-Yazied, A.A.; Selim, S.; Tawfik, M.M.; AlHarbi, K.; Ali, M.S.M.; Elkelish, A. Kaolin Improves Photosynthetic Pigments, and Antioxidant Content, and Decreases Sunburn of Mangoes: Field Study. *Agronomy* **2022**, *12*, 1535. [CrossRef]

Article

Dryopteris juxtapostia Root and Shoot: Determination of Phytochemicals; Antioxidant, Anti-Inflammatory, and Hepatoprotective Effects; and Toxicity Assessment

Abida Rani [1], Muhammad Uzair [1,*], Shehbaz Ali [2,3], Muhammad Qamar [4,*], Naveed Ahmad [5], Malik Waseem Abbas [6] and Tuba Esatbeyoglu [7,*]

[1] Department of Pharmaceutical Chemistry, Faculty of Pharmacy, Bahauddin Zakariya University, Multan 60800, Pakistan
[2] Department of Bioscience and Technology, Khwaja Fareed University of Engineering and Information Technology, Rahim Yar Khan 64200, Pakistan
[3] School of Environment and Safety Engineering, Jiangsu University, Zhenjiang 212013, China
[4] Institute of Food Science and Nutrition, Bahauddin Zakariya University, Multan 60800, Pakistan
[5] Multan College of Food & Nutrition Sciences, Multan Medical and Dental College, Multan 60000, Pakistan
[6] Institute of Chemical Sciences, Bahauddin Zakariya University, Multan 60800, Pakistan
[7] Institute of Food Science and Human Nutrition, Gottfried Wilhelm Leibniz University Hannover, Am Kleinen Felde 30, 30167 Hannover, Germany
*   Correspondence: muhammaduzair@bzu.edu.pk (M.U.); muhammad.qamar44@gmail.com (M.Q.); esatbeyoglu@lw.uni-hannover.de (T.E.)

Citation: Rani, A.; Uzair, M.; Ali, S.; Qamar, M.; Ahmad, N.; Abbas, M.W.; Esatbeyoglu, T. *Dryopteris juxtapostia* Root and Shoot: Determination of Phytochemicals; Antioxidant, Anti-Inflammatory, and Hepatoprotective Effects; and Toxicity Assessment. *Antioxidants* **2022**, *11*, 1670. https://doi.org/10.3390/antiox11091670

Academic Editors: Marianna Lauricella and Antonella D'Anneo

Received: 20 July 2022
Accepted: 24 August 2022
Published: 27 August 2022

Publisher's Note: MDPI stays neutral with regard to jurisdictional claims in published maps and institutional affiliations.

Abstract: An estimated 450 species of *Dryopteris* in the Dryoperidaceae family grow in Japan, North and South Korea, China, Pakistan, and Kashmir. This genus has been reported to have biological capabilities; however, research has been conducted on *Dryopteris juxtapostia*. Therefore, with the present study, we aimed to exploring the biological potential of *D. juxtapostia* root and shoot extracts. We extracted dichloromethane and methanol separately from the roots and shoots of *D. juxtapostia*. Antioxidant activity was determined using DPPH, FRAP, and H_2O_2 assays, and anti-inflammatory activities were evaluated using both in vitro (antiurease activity) and in vivo (carrageenan- and formaldehyde-induced paw edema) studies. Toxicity was evaluated by adopting a brine shrimp lethality assay followed by determination of cytotoxic activity using an MTT assay. Hepatoprotective effects of active crude extracts were examined in rats. Activity-bearing compounds were tentatively identified using LC-ESI-MS/MS analysis. Results suggested that *D. juxtapostia* root dichloromethane extract exhibited better antioxidant (DPPH, IC_{50} of 42.0 μg/mL; FRAP, 46.2 mmol/g; H_2O_2, 71% inhibition), anti-inflammatory (urease inhibition, 56.7% at 50 μg/mL; carrageenan-induced edema inhibition, 61.7% at 200 μg/mL; formaldehyde-induced edema inhibition, 67.3% at 200 μg/mL), brine shrimp % mortality (100% at 1000 μg/mL), and cytotoxic (HeLa cancer, IC_{50} of 17.1 μg/mL; prostate cancer (PC3), IC_{50} of 45.2 μg/mL) effects than *D. juxtapostia* root methanol extract. *D. juxtapostia* shoot dichloromethane and methanol extracts exhibited non-influential activity in all biological assays and were not selected for hepatoprotective study. *D. juxtapostia* root methanol extract showed improvement in hepatic cell structure and low cellular infiltration but, in contrast the dichloromethane extract, did not show any significant improvement in hepatocyte morphology, cellular infiltration, or necrosis of hepatocytes in comparison to the positive control, i.e., paracetamol. LC-ESI-MS/MS analysis showed the presence of albaspidin PP, 3-methylbutyryl-phloroglucinol, flavaspidic acid AB and BB, filixic acid ABA and ABB, tris-desaspidin BBB, tris-paraaspidin BBB, tetra-flavaspidic BBBB, tetra-albaspidin BBBB, and kaempferol-3-*O*-glucoside in the dichloromethane extract, whereas kaempferol, catechin, epicatechin, quinic acid, liquitrigenin, and quercetin 7-*O*-galactoside in were detected in the methanol extract, along with all the compounds detected in the dichloromethane extract. Hence, *D. juxtapostia* is safe, alongside other species of this genus, although detailed safety assessment of each isolated compound is obligatory during drug discovery.

Keywords: cytotoxicity; hepatoprotective effects; HeLa cancer; inflammation; mass spectrometry; oxidation; prostate cancer; phytochemical

1. Introduction

Generation of free radicals in living systems is associated with intrinsic (stress) and extrinsic (alcohol, smoking, and radiation) factors, whereas antioxidant mechanisms help to neutralizing the negative impacts induced by oxidative stress [1]. Oxidative stress is a condition whereby imbalance occurs between reactive oxygen species (ROS) and the body's antioxidant system. Undue generation of ROS disrupts the normal functioning of important organs and can lead to the onset of many ailments, including inflammation, cancer, polygenic disorders, diabetes, and aging [2–4].

Pakistan, regardless of being an agricultural territory, is home to various ecological zones with many indigenous medicinal plant species [5,6]. However, limited research has been conducted to evaluate their pharmaceutical prominence due to the phytochemical potential of secondary metabolites [7]. Of 5700 reported medicinal plant species, Pakistan possesses 500-600 species, with only a few having been probed for biochemical assessment [8]. The *Dryopteris* genus of the Dryoperidaceae family of the North Temperate Zone comprises more than 450 species grown in Japan, North and South Korea, China, Pakistan, and Kashmir [9]. *Dryopteris ramose* and *Dryopteris cochleata* extracts, i.e., species from the aforementioned genus, have been reported to exhibit antioxidant activity [10,11]. *Dryopteris chrysocoma*, *Dryopteris blanfordii*, and *Dryopteris crassirhizoma* were reported to exhibit anti-inflammatory activities [12–14]. In other studies, *Dryopteris fragrans* and *Dryopteris crassirhizoma* extracts were found to possess anticancer potential [15,16]. However, *Dryopteris juxtapostia* extracts have not been explored for biological potential to date. Therefore, the aim of the present study is to explore phytochemical, radical-scavenging, anti-inflammatory, cytotoxic, and hepatoprotective potential of *Dryopteris juxtapostia* root and shoot extracts.

2. Materials and Methods

2.1. Plant Material and Its Preparation

Dryopteris juxtapostia (DJ) roots and shoot were collected from the Sawat area (Tehsil Matta upper swat. KPK., Village, Shukhdara, Biha, Charrma, Fazal Banda), Pakistan, and identified and authenticated at the Institute of Pure and Applied Biology, Bahauddin Zakariya University, Multan, Pakistan. The plant was assigned voucher no. tro-26609785. For the purpose of effective extraction, whole DJ root and shoot material was shade-dried for 15 days. Then, dried plant material was ground in a blender and weighed. Extraction was performed with dichloromethane and methanol in an orbital shaker in the dark for 48 h. The process was repeated three times, and the solvent was evaporated using a rotary evaporator (Heidolph, Schwabach, Germany) in to obtain semisolid plant material.

2.2. Quantification of Total Phenolic and Flavonoid Contents

Total phenolic contents were determined by Folin–Ciocalteu (FC) colorimetric assay using gallic acid as standard [17]. Absorbance was recorded at 765 nm using a spectrophotometer (UV-Vis 3000 ORI, Reinbeker, Germany), and values were recorded in triplicate using ethanol as a blank. Total flavonoid contents were determined using an $AlCl_3$ assay [18]. Sample absorbance was read at 510 nm using a spectrophotometer (UV-Vis 3000 ORI). A quercetin standard curve was plotted; sample results are expressed as mg quercetin equivalents per gram (mg QE/g) of the dried weight.

2.3. Antioxidant Activity

DJ dichloromethane and methanol crude extracts of roots and shoots were evaluated for antioxidant activity using three assays i.e., DPPH, H_2O_2, and FRAP assays, as adopted by Qamar et al. (2021) [19]. Samples were prepared as 1 g/20 mL for all three assays. In the DPPH and H_2O_2 assays, distilled water was used as a blank, with quercetin (125 µg/mL) as standard. The findings are reported as percent inhibition according to the following equation:

$$\%\text{Inhibition} = [(\text{absorbance of control} - \text{absorbance of sample/standard}) \div \text{absorbance of control}] \times 100$$

For the FRAP assay, ferrous sulphate was used for calibration. Results are expressed as Fe mmol/g.

2.4. Anti-Inflammatory Activity

2.4.1. Urease Inhibition Assay (In Vitro)

Weatherburn's indophenol method was used to evaluate the urease activity by determining the ammonia production in the reaction mixture [20]. The reaction mixture comprising 25 µL jack bean urease enzyme, 55 µL buffer (100 mM urea), and 5 µL test compounds (0.5 mM) was incubated for 15 min in a 96-well plate at 30 °C. Urease activity was assessed by Weatherburn's method by measuring ammonia production using indophenol. In brief, 70 µL alkali (0.1% active chloride (NaOCl) and 0.5% NaOH *w/v*) and 45 µL phenol reagents (0.005% *w/v* sodium nitroprusside and 1% *w/v* phenol) were added to each well. After 50 min, the increase in absorbance was measured at 630 nm with a microplate reader (Molecular Device, Ramsey, NJ, USA). The reaction was performed in a triplicate run, with pH 6.8 and a final volume of 200 µL. Absorbance readings were processed with Max Pro software (Molecular Device, USA), and % inhibition was calculated using following equation:

$$\text{Percentage inhibitions } (\%) = 100 - \left(\frac{\text{OD t}}{\text{OD c}} \right) \times 100$$

where OD t is optical density of the test well, and OD c is the optical density of the control. Thiourea was used as the standard urease inhibitor in this study.

2.4.2. Carrageenan- and Formaldehyde-Induced Paw Oedema (In Vivo)

A carrageenan-induced paw inflammation assay was employed to assess the pain-relieving capabilities of DJ dichloromethane and methanol (root and shoot) extracts in rats according to Morris (2003) [21], with some modifications. The study was performed by adopting the parameters mentioned in the guidelines of the National Research Council [22] (NRC, 1996, Washington, DC, USA). The study was also approved by the departmental Committee pf Animal Care at BZU, Pakistan (approval number ACC-10-2019). Rats were divided into six groups (n = 5); animals in group 1 were provided with normal saline and designated the control group. Animals in group 2 were given standard indomethacin at a dose of 100 mg/kg body weight (b.w.) and designated the positive control. Rats in groups 3 and 4 were fed with the dichloromethane extract (200 mg/kg) of *D. juxtapostia* roots and shoots, respectively. Rats in groups 5 and 6 were fed with the methanol extract (200 mg/kg) of *D. juxtapostia* roots and shoots, respectively. One half hour after extract administration, the animals were injected with carrageenan into the plantar aponeurosis surface of the right hind paw. Any change in paw linear circumference was noted after 0, 1, 2, and 3 h using a plethysmometer (UGO-BASILE 7140, Comerio, Italy). An increase in paw circumference was taken as indicator of inflammation.

Likewise, a formaldehyde-induced hind-paw edema assay was used to examine the anti-inflammatory potential of DJ dichloromethane and methanol (root and shoot) extracts in mice, adopting the method of Brownlee with minor changes [23]. We divided animal into a total of six study groups; details are the same as those mentioned above for the carrageenan-induced paw inflammation assay. One half hour after extract administration, formaldehyde (100 µL, 4%) was injected into the plantar aponeurosis of each mouse's right paw, and changes in paw circumference were recorded after 0, 3, 6, 12, and 24 h.

2.5. Brine Shrimp Lethality Assay

The method described by Meyer et al. (1982) [24] was used to perform a brine shrimp lethality assay. Commercial salt was dissolved in distilled water to prepare artificial seawater in a rectangular plastic tray (22 × 32 cm) in the dark. Fifty milligrams of shrimp eggs (*Artemia salina*) obtained from Husein Ebrahim Jamal Research Institute of Chemistry (HEJ, Karachi, Pakistan) was scattered into the artificial seawater. Incubation lasted 48 h at 37 °C. Pasteur pipettes were used to collect hatched larvae. Dichloromethane and methanol extracts from roots and shoots of DJ were prepared at concentrations of 10, 100, and 1000 µg/mL. Samples with varying strengths were separately transferred to clean vials. Each incubation vial contained 1 mL artificial seawater (to a final volume of 5 mL) and 30 shrimp with pH 7.4 adjusted using 1N NaOH and incubated for 24 h at 26 °C. The shrimp survival rate was quantified in each vial, including the positive control (i.e., etoposide).

2.6. Cytotoxic Activity

To assess the cytotoxic potential of DJ dichloromethane and methanol extracts (root and shoot), we adopted the method described by Mosmann et al. (1983) [25]. Experimental samples of varying strengths (0.5–200 µg/mL) were prepared in 100 µL dimethylsulphoxide (1% *v/v*) in 96-well microtiter plates. After incubating the microtiter plates (37 °C, 48 h), 50 µL of the MTT solution (5 mg/mL) was added to each well. A microplate reader was used to check the reduction in MTT after a second incubation (37 °C for 4 h) by recording the absorbance at 570 nm. The untreated cells were used as a control against which to measure the effect of experimental extracts on the cell viability. The percent inhibition exhibited on the cell cultures by the test samples was computed using the following equation:

$$\text{Survival (\%)} = (\text{At} - \text{Ab})/(\text{Ac} - \text{Ab}) \times 100$$

where At, Ab, and Ac indicate the sample, blank (complete media without cells), and control absorbance, respectively.

$$\text{Cell inhibition (\%)} = 100 - \text{cell survival (\%)}$$

2.7. Hepatoprotective Studies

The active crude extracts of DJ roots and parts were subjected to hepatoprotective analysis following the OECD 423 (Organization for Economic Co-operation and Development) guidelines [26]. Preset parameters and guidelines of the National Research Council (1996, Washington, USA) were also considered. The institutional ethical committee of Bahauddin Zakariya University (BZU) Multan Pakistan approved the animal study under the title "Study of hepatoprotective potential of *Dryopteris juxtapostia*". The regimen presented in Table 1 was used to orally administer DJ plant extracts to the groups for ten days, but only the most biologically active extracts were considered for this analysis. All animals were treated according to the regimen presented in Table 2. After last dose, retro-orbital plexus blood was collected. Blood of animals in each treatment group was saved for lipid and protein analysis. Furthermore, serum samples were allowed to clot for 60–70 min at ambient room temperature to test for biochemical liver function markers, followed by centrifugation (2500 rpm at 30 °C) for 15–20 min.

2.7.1. Assessment of Liver Functions

Serum glutamic-pyruvic transaminase (SGPT)/alanine aminotransferase (ALT), serum glutamic-oxaloacetic transaminase (SGOT)/aspartate aminotransferase (AST), alkaline phosphate, total bilirubin, total protein, and lipid profile were analyzed and quantified in each the serum from each group. The activity of serum transaminases (SGPT and SGOT) and blood lipid profile were examined using the Rietman and Frankel method [27]. Total protein was evaluated according to the Lowry procedure [13]. Total bilirubin (TB) and alkaline phosphate (ALP) were estimated using the methodologies described by Keiding et al. (1974) [28] and Tietz et al. (1983) [29], respectively.

Table 1. Hepatoprotective activity of DJ crude extracts.

Group No.	Treatment n = 4	Ten-Day Regimen	
		Days 1–7	Days 8–10
1	Normal saline (negative control)	Normal saline	Normal saline
2	Paracetamol (2 g/kg) (positive control)	Normal saline	Paracetamol (2 g/kg)
3	Silymarin (standard) (10 mg/kg)	10 mg/kg	Silymarin + Paracetamol
4	DJ DCM root (300 mg/kg)	300 mg/kg	DJ DCM root (300 mg/kg) + paracetamol
5	DJ DCM root (500 mg/kg)	500 mg/kg	DJ DCM root (500 mg/kg) + paracetamol
6	DJ MeOH root (300 mg/kg)	300 mg/kg	DJ MeOH root (300 mg/kg) + paracetamol
7	DJ MeOH root (500 mg/kg)	500 mg/kg	DJ MeOH root (500 mg/kg) + paracetamol
8	DJ DCM shoot (300 mg/kg)	300 mg/kg	DJ DCM shoot (300 mg/kg) + paracetamol
9	DJ DCM shoot (500 mg/kg)	500 mg/kg	DJ DCM shoot (500 mg/kg) + paracetamol
10	DJ MeOH shoot (300 mg/kg)	300 mg/kg	DJ MeOH shoot (300 mg/kg) + paracetamol
11	DJ MeOH shoot (500 mg/kg)	500 mg/kg	DJ MeOH shoot (500 mg/kg) + paracetamol

Table 2. Carrageenan-induced edema in rat hind paw.

Type of Extract	Dose mg/kg	0 h % Inhibition	1 h % Inhibition	2 h % Inhibition	3 h % Inhibition
Control	-	-	-	-	-
Indomethacin	100	23.3 *	38.9 **	40.6 **	77.6 ****
DJ root DCM extract	200	16.2 [ns]	18.5 [ns]	39.5 **	61.7 ***
DJ root MeOH extract	200	17.0 [ns]	22.8 *	28.8 *	43.9 **
DJ shoot DCM extract	200	9.30 [ns]	14.2 [ns]	20.4 *	24.4 *
DJ shoot MeOH extract	200	3.10 [ns]	7.4 [ns]	10.4 [ns]	16.3 [ns]

DJ, *Dryopteris juxtapostia*; DCM, dichloromethane; h, hours; MeOH, methanol; ns, non-significant. Values are presented as means $\pm$ S.D. of three measurements. * $p < 0.05$, ** $p < 0.01$, *** $p < 0.001$, **** $p < 0.0001$.

2.7.2. Histopathology of Liver

The liver of all experimental animals (rats) were excised and washed with normal saline. The cleansed liver tissues were separately preserved in 10% formalin solution (neutral) in air-tight, labelled jars. After eight days, the tissues were dehydrated using ethanol solution. The tissues were dried, embedded in paraffin, and sliced into 5 μm length segments. The liver sections were placed on a marked slide and dyed (hematoxylin–eosin (H & E) 400X). The prepared labelled slides were then observed under a photomicro-scope (Olympus-CX23 Upright, Japan) for vacuolar degeneration, cellular infiltration, and necrosis of hepatocytes.

2.8. LC-ESI-MS/MS Analysis of Active Crude Extracts

Crude extracts exhibiting biological potential and outlined hepatoprotective effects were further subjected to mass spectrometry analysis using LC-ESI-MS/MS (Thermo Electron Corporation, Waltham, MA, USA) with the aim of tentative identification of activity-bearing compounds. Detection was carried out by adopting direct-injection-mode ESI (electron spray ionization) in both negative and positive modes. Range of mass, temperature of capillaries, and sample flow rate were maintained at *m/z* 50 to 1000, 280 °C, and 8 μL/min, respectively. Collision-induced energy generated during MS/MS analysis depended upon the nature/type of the parent molecular ion subjected to 10 to 45 eV. Furthermore, in order to ensure sufficient ionization and ion transfer, every compound was optimized for MS parameters. Similarly, for every analyte, the source parameters were unchanged but parent, whereas daughter signals were optimized either by analyte infusion or manually. Moreover, online mass data banks, software, and previously published literature were used for compound identification (www.chemspider.com, accessed on 12 December 2021).

2.9. Statistical Analysis

Study data are expressed as the mean (SEM) of three measurements. ANOVA was used to compare the differences between the control and treatment groups, and Dunnett's test was run using GraphPad Prism (Graph Pad Software V8, San Diego, CA, USA).

3. Results

3.1. Phytochemical Constituents and Antioxidant Activity of Dryopteris juxtapostia (DJ) Crude Extracts

The quantitative investigation recorded the maximum total phenolic contents in 100% DCM extract of *D. juxtapostia* roots, root methanol extract, shoot DCM extract, and shoot methanol extract as 222 ± 0.41 mg GAE/g, 163 ± 0.2 mg GAE/g, 109 ± 0.41 mg GAE/g, and 91.4 ± 0.2 mg GAE/g, respectively (Figure 1). In contrast, total flavonoid contents recorded in methanol extracts of *D. juxtapostia* roots and shoots, i.e., 83.7 ± 0.1 mg QE/g and 43.8 ± 0.3 mg QE/g, respectively, were higher compared to those of DCM root and shoot extracts, i.e., 51 ± 0.2 mg QE/g and 13.2 ± 0.5 mg QE/g, respectively.

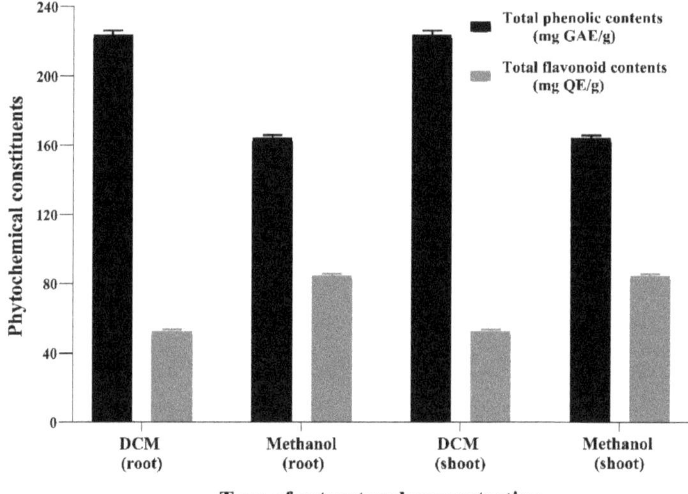

Figure 1. Total phenolic and flavonoid contents of *Dryopteris juxtapostia* root and shoot crude extracts. Values are presented as means $\pm$ S.D. of three measurements.

The radical-scavenging ability of various *D. juxtapostia* crude extracts (shoot DCM extract, shoot methanol extract, root dichloromethane extract, and root methanol extract) was evaluated using various antioxidant assays, such as stable radical assay (DPPH), reducing assay (FRAP), and hydrogen peroxide (H_2O_2) inhibition assay. As shown in Figure 2, among all extracts, *D. juxtapostia* root DCM extract exhibited the lowest IC_{50} of 42.0 µg/mL against stable free radicals (DPPH), followed by root methanol extract, with an IC_{50} of 54.0 µg/mL. In contrast, in the present study, moderate activity was shown by *D. juxtapostia* shoot DCM and methanol extracts, with IC_{50} values of 59.0 µg/mL and 61.4 µg/mL, respectively. Quercetin was used as a standard antioxidant compound and exhibited remarkable activity, with an IC_{50} of 22.3 µg/mL.

D. juxtapostia root DCM extract demonstrated a higher reducing potential of 46.2 mmol/g, followed by shoot dichloromethane extract, root methanol extract, and shoot methanol extract, with a reducing potential of 31.1 mmol/g, 34.6 mmol/g, and 29.4 mmol/g, respectively. Quercetin was observed to have the highest reducing potential of 66.0 mmol/g.

In the hydrogen peroxide inhibition assay, *D. juxtapostia* root dichloromethane extract demonstrated 71.0% inhibition against delineated prominent activity relative to the standard quercetin (87.0% inhibition) and in contrast to root methanol extract (51.0% inhibition), shoot dichloromethane extract (32.1% inhibition), and methanol extract (34.2% inhibition).

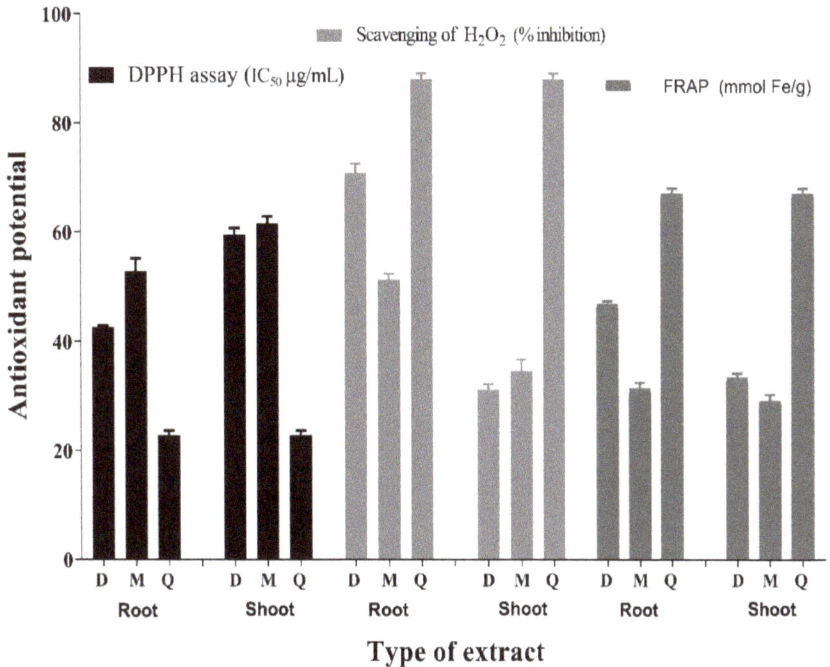

Figure 2. Antioxidant activity of *Dryopteris juxtapostia* root and shoot crude extracts. D, 100% dichloromethane extract; M, 100% methanol extract; Q, quercetin (standard). Values are presented as means ± S.D. of three measurements.

3.2. In Vitro Anti-Inflammatory Activity

D. juxtapostia crude extracts were evaluated for possible antiurease activity at varying concentrations, i.e., 12.5, 25, and 50 µg/mL, using thiourea as a standard anti-inflammatory drug. The results presented in Figure 3 illustrate that *D. juxtapostia* root dichloromethane extract exhibited urease inhibition activity of 56.7% at 50 µg/mL, followed by root methanol extract, with moderate urease inhibition activity of 32.9% at 50 µg/mL. Similarly, the standard drug, i.e., thiourea, exhibited potent inhibition of 88.9% at 50 µg/mL. In contrast, *D. juxtapostia* shoot dichloromethane and methanol extracts evinced non-influential antiurease activity at all concentrations. The anti-inflammatory activity of *D. juxtapostia* root dichloromethane and methanol extracts was found to be consistent with total phenolic contents and antioxidant activity. Statistically, activity outlined by DJ root DCM extract was comparable to that of standard thiourea, with a non-significant difference (ns) observed between their activities, whereas the activity of DJ root MeOH ($p < 0.01$), DJ shoot DCM ($p < 0.001$), and DJ shoot MeOH $p < 0.001$) extracts was significantly lower when compared to standard thiourea (Figure 3).

3.3. In Vivo Anti-Inflammatory Activity

In the present study, the experimental crude extracts, including *D. juxtapostia* root dichloromethane and methanol extracts and *D. juxtapostia* shoot dichloromethane and methanol extracts were evaluated for possible in vivo pain-alleviating properties induced by carrageenan and formaldehyde at various concentrations, i.e., 50, 100, and 200 mg/kg (Table 2).

DJ root dichloromethane extract showed an antiedematous effect in a dose-dependent manner, with a maximum inhibition of 61.7% ($p < 0.001$) at 200 mg/kg after 3 h, in contrast to the control (normal saline). This is comparable to the anti-inflammatory effects of the standard anti-inflammatory drug indomethacin at a dose of 100 mg/kg, which altered inflammation by as much as 77.6% ($p < 0.0001$). In contrast, *D. juxtapostia* root methanol extract evinced moderate inhibition (43.9% at 200 mg/kg), whereas *D. juxtapostia* shoot extracts exhibited non-substantial activity when compared to the control. The in vivo anti-inflammatory properties of *D. juxtapostia* are in line with its in vitro anti-inflammatory, antioxidant, and phytochemical potential.

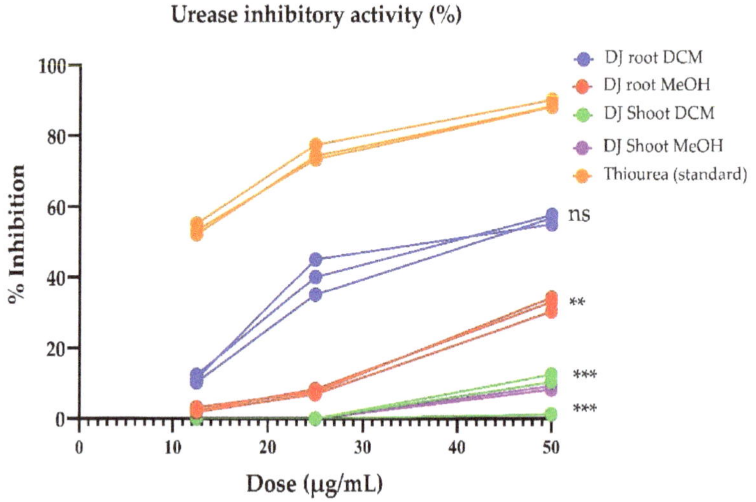

Figure 3. Urease inhibition activity (%) of various *D. juxtapostia* crude extracts. DCM, dichloromethane; DJ, *Dryopteris juxtapostia*; MeOH, methanol; ns, non-significant. Values are presented as means $\pm$ S.D. of three readings. ** $p < 0.01$. *** $p < 0.001$.

Akin to the previous model of inflammation, *D. juxtapostia* root dichloromethane extract, when administrated at a rate of 200 mg/kg, evinced inhibition of 67.3% ($p < 0.001$) after 24 h against formaldehyde-induced pain behavior in contrast to the control, i.e., normal saline. The activity was parallel to the anti-inflammatory effects of the standard anti-inflammatory drug indomethacin at a dose of 100 mg/kg, which altered inflammation by as much as 86.3% ($p < 0.0001$). Moreover, *D. juxtapostia* root methanol extract showed moderate inhibition of 45.1% at 200 mg/kg, whereas both extracts of the shoot portion, i.e., dichloromethane and methanol, exhibited non-substantial activity against formaldehyde-intoxicated pain behavior (Table 3).

Table 3. Formaldehyde-induced edema in mouse hind paw.

Type of Extract	Dose mg/kg	1 h % Inhibition	3 h % Inhibition	6 h % Inhibition	12 h % Inhibition	24 h % Inhibition
Control	-	-	-	-	-	-
Indomethacin	100	68.7 ***	71.2 ***	76.6 ***	80.2 ***	86.3 ****
DJ root DCM extract	200	53.2 **	57.6 **	61.5 **	65.2 ***	67.3 ***
DJ root MeOH extract	200	29.9 *	28.7 *	34.5 *	44.9 **	45.1 **
DJ shoot DCM extract	200	15.5 ns	19.2 ns	23.5 ns	25.5 *	31.9 *
DJ shoot MeOH extract	200	3.08 ns	0.96 ns	14.4 ns	10.8 ns	22.11 ns

DJ, *Dryopteris juxtapostia*; DCM, dichloromethane; MeOH, methanol; ns. non-significant. Values are presented as means $\pm$ S.D. of three measurements. * $p < 0.05$, ** $p < 0.01$, *** $p < 0.001$, **** $p < 0.0001$.

3.4. Brine Shrimp Lethality Assay

Brine shrimp lethality assay is an imperative method for determining the preliminary cytotoxicity of experimental plant extracts and other substances based on their ability to kill laboratory-cultured larvae (nauplii). Such an assay is easy to use, inexpensive, and requires only a small amount of test material. In the present study, the cytotoxicity of various crude extracts of *D. juxtapostia* were evaluated at varying concentrations, i.e., 10, 100, and 1000 µg/mL, to compute % mortality. Dichloromethane extracts were observed to be more cytotoxic in comparison to methanol extracts in a dose-dependent manner (Table 4). In brief, *D. juxtapostia* root dichloromethane extract was found to be the most lethal of all investigated extracts, with 100%, 76%, and 10% mortality at 1000 µg/mL, 100 µg/mL, and 10 µg/mL, respectively, followed by standard etoposide (70% mortality at 10 µg/mL), *D. juxtapostia* shoot dichloromethane extract, *D. juxtapostia* root methanol extract, and *D. juxtapostia* shoot methanol extract. These findings revealed that *D. juxtapostia* root dichloromethane extract may contain some compounds that exert cytotoxic effects on certain cancer cells.

Table 4. Toxicity assessment of various *D. juxtapostia* crude extracts using a brine shrimp lethality assay.

Extract	Dose (µg/mL)	% Mortality	EC_{50}
DJ root DCM	1000	100 ± 0.00	
	100	73.1 ± 2.45	26.74
	10	10 ± 0.00	
DJ root MeOH	1000	20 ± 0.00	
	100	3.3 ± 1.55	391.9
	10	00 ± 0.00	
DJ shoot DCM	1000	49.9 ± 2.73	
	100	20 ± 0.00	69.17
	10	1.1 ± 1.55	
DJ shoot MeOH	1000	12.2 ± 1.55	
	100	5.5 ± 1.55	208.8
	10	00 ± 0.00	
Etoposide (standard drug)	10	71 ± 1.41	19.29

DJ, *Dryopteris juxtapostia*; DCM, dichloromethane; MeOH, methanol. Values are presented as means $\pm$ S.D. of three measurements.

3.5. Cytotoxic Activity of Various D. juxtapostia Crude Extracts Using MTT Assay

D. juxtapostia root dichloromethane extract, root methanol extract, shoot dichloromethane extract, and shoot methanol extract were evaluated for possible anticancer potential using doxorobicin as a standard anticancer drug by MTT assay (Table 5). The MTT (3-[4,5-dimethylthiazol-2-yl]-2,5 diphenyl tetrazolium bromide) assay is based on the conversion of MTT into formazan crystals by living cells, which determines mitochondrial activity. Because for most cell populations, the total mitochondrial activity is related to the number of viable cells, this assay is broadly used to measure the in vitro cytotoxic effects of drugs on cell lines or primary patient cells.

D. juxtapostia root dichloromethane extract was not only found to anticipate a reduction in oxidative stress induced by DPPH, FRAP, and H_2O_2 but also yielded significant inhibition in cancer progression among both investigated cancer cell lines, i.e., HeLa human cervical and prostate cancer cell lines (PC3), with an IC_{50} of 17.1 µg/mL and 45.2 µg/mL, respectively. Moreover, *D. juxtapostia* root methanol extract demonstrated prominent inhibitory activity against the Hela cancer cell line, with an IC_{50} of 36.9 µg/mL, and moderate activity against human prostate cancer cell lines, with an IC_{50} of 98.3 µg/mL. The standard anticancer drug doxorubicin exhibited potent inhibition against both cancer cell lines, with IC_{50} values of 0.90 µg/mL (HeLa human cervical cancer cell line) and 1.90 µg/mL (PC3).

Table 5. Cytotoxic activity of various *D. juxtapostia* crude extracts at 30 µg/mL determined by MTT assay.

Sample	Cell Line	% Inhibition	IC$_{50}$ (µg/mL)
DJ root DCM		76.7 ± 0.5	17.1 ± 1.3
DJ root MeOH		62.9 ± 1.1	36.9 ± 0.9
DJ shoot DCM	HeLa cervical cancer cell line	34.4 ± 1.3	87.2 ± 1.1
DJ shoot MeOH		20.2 ± 0.9	143.6 ± 0.7
Doxorobicin (standard)		98.0 ± 1.1	0.90 ± 0.14
DJ root DCM		56.5 ± 0.1	45.2 ± 0.1
DJ root MeOH		30.6 ± 0.2	98.3 ± 1.1
DJ shoot DCM	Human prostate cancer cell line	28.5 ± 1.4	101.2 ± 2.1
DJ shoot MeOH		18.3 ± 0.6	187.4 ± 0.1
Doxorobicin (standard)		89.9 ± 0.12	1.90 ± 0.3

DJ, *Dryopteris juxtapostia*; DCM, dichloromethane; MeOH, Methanol. Values are presented as means ± S.D. of three measurements.

3.6. Hepatoprotective Activity

Histopathology results revealed normal hepatocytes with no inflammatory changes in the group given silymarin (group 3). However, the group treated with paracetamol (group 2) showed fatty changes, vacuolar degeneration, cellular infiltration, and necrosis of hepatocytes. Groups given *D. juxtapostia* root methanol extract at 300 mg/kg (group 6) and 500 mg/kg (group 7) showed improvement in cell structure, and low cellular infiltration was observed at a higher dose as compared to a lower dose. Groups 4 and 5 given *D. juxtapostia* root dichloromethane extract at doses of 300 mg/kg and 500 mg/kg, respectively, did not show any notable improvement in hepatocyte morphology, cellular infiltration, or necrosis of hepatocytes, as shown in Figure 4. As shown in Table 6, *D. juxtapostia* root methanol extract was found to be more effective in a liver function test, as well as total protein, and total lipid profile tests, in a dose-dependent manner as compared to *D. juxtapostia* root dichloromethane extract.

Table 6. Hepatoprotective analysis of *D. juxtapostia* root dichloromethane and methanol extracts.

Sample	Liver Function Test				Total Protein				Lipid Profile			
	Tb (mg/dL)	SGPT/ALT (IU/L)	SGOT/AST (IU/L)	ALP (IU/L)	SM	SA	Gb	A/G Ratio	Cholesterol (mg/dL)	Triglycerides (mg/dL)	HDL (mg/dL)	LDL (mg/dL)
Control	0.34	44	73.60	164	6.30	3.79	2.99	1.26	169	97	64	76
Paracetamol	0.8	103	151	497	7.96	3.22	4.03	0.97	243	152	53	95
Silymarin	0.49	66	104	241	7.16	3.62	3.01	1. 2	185	71	69	77
DJMR 300 mg	0.69	83	119	301	7.13	3.45	3.8	0.9	183	148	61	83
DJMR 500 mg	0.57	77	113	258	6.8	3.68	3.6	1.02	179	123	64	74
DJDR 300 mg	0.71	96	134	339	7.12	3.10	3.64	0.85	193	146	59	79
DJDR 500 mg	0.67	94	141	321	7.01	3.56	3.28	1.08	186	129	62	71

ALP, alkaline phosphate; SM, serum protein; SA, serum albumin; Gb, globulin; Tb, total bilirubin; SGPT/ALT, serum glutamic-pyruvic transaminase/alanine aminotransferase; SGOT/AST, serum glutamic-oxaloacetic transaminase/aspartate aminotransferase.

3.7. Mass Spectrometry Analysis of Various Extracts

D. juxtapostia root dichloromethane and methanol extracts showing notable biological activities were subjected to mass spectrometry analysis (ESI-MS/MS) to identify (tentative) compounds by comparing the mass spectra and their fragments with mass banks and previously published literature. In detail, albaspidin PP, 3-methylbutyryl-phloroglucinol, flavaspidic acid AB, flavaspidic acid BB, filixic acid ABA, filixic acid ABB, tris-desaspidin BBB, tris-paraaspidin BBB, tetra-flavaspidic BBBB, tetra-albaspidin BBBB, and kaempferol-3-*O*-glucoside were detected in DCM extract. All the aforementioned compounds were also detected in methanol extract, along with kaempferol, catechin, epicatechin, quinic acid, liquitrigenin, and quercetin 7-*O*-galactoside (Table 7). Compound (**A**) was previously identified in another species of the same genus called *Dryopteris crassirhizoma* [30]. Compounds (**B–I**) were also identified in the same genus in a species called *Dryopteris Adanson* [31]. Compounds (**J–M**) were identified according to recent literature reports [19,32,33].

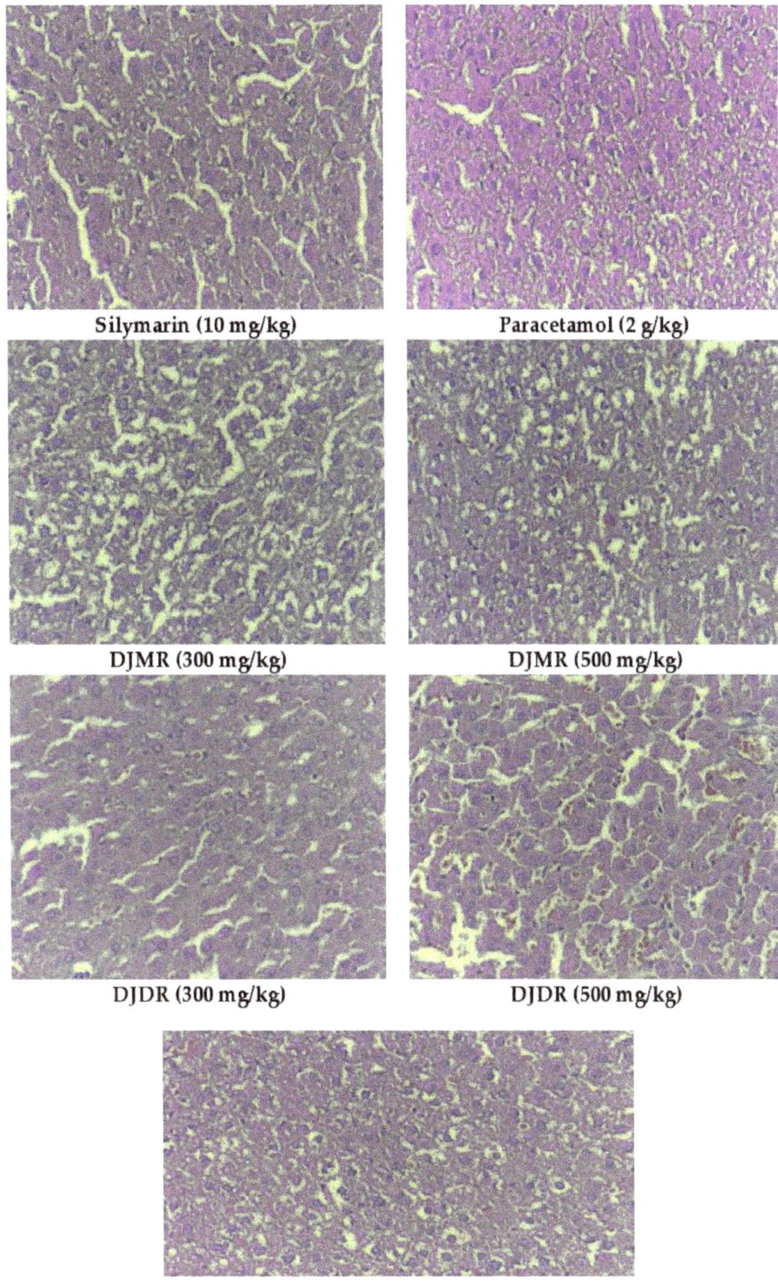

Figure 4. Histopathological analysis of liver. DJDR, *Dryopteris juxtapostia* methanol root; DJMR, *Dryopteris juxtapostia* dichloromethane root.

Table 7. ESI-MS/MS analysis of various crude extracts.

Sample Name	Compound	Average Mass (*m/z*)	ESI-MS/MS^n (*m/z*)	Mode	Identification	References
	A	433	432.3, 236.1, 196	Positive	Albaspidin PP	[30]
	B	445	445.2, 235.08, 223.08, 209.17	Negative	Flavaspidic acid BB	[31]
	C	627	625, 417.2, 403.1, 221,	Negative	Filixic acid ABP	[31]
	D	641	429, 417.2, 403.1, 237	Positive	Filixic acid ABB	[31]
DJDR	E	653	429, 417.2, 221.08	Positive	Tris-desaspidin BBB	[31]
	F	667	666.4, 442.3, 431, 235, 223	Positive	Tris-paraaspidin BBB	[31]
	G	863	866.4, 653.3, 625.3	Positive	Tetra-flavaspidic BBBB	[31]
	G	419	419, 223, 209, 196.9	Positive	Flavaspidic acid AB	[31]
	I	667	667.3, 653.4, 639.3, 431.1	Positive	Tetra-albaspidin BBBB	[31]
	J	290	289.1, 271.08, 247.08	Positive	Catechin	[19]
DJMR	K	291	273.1, 163.3, 139.08	Positive	Epi-catechin	[32]
	L	191	191, 173, 127	Positive	Quinic acid	[32]
	M	267	257, 237.1, 211.1,	Positive	Liquitrigenin	[33]

4. Discussion

D. juxtapostia root dichloromethane and methanol extracts were found to have higher phenolic and flavonoid contents, respectively, as compared to shoot extracts (Figure 1). The identical potential of bioactive metabolites was reported in an previous study by Baloch et al. (2019) [34], wherein dichloromethane and methanol extracts of *Dryopteris ramose* belonging to the same genus as our experimental plant exhibited notable phenolic (184.2–199.2 mg GAE/g) and flavonoid (50.13–73.02 mg rutin equivalent (RE)/g) contents, supporting the findings of the present investigation. Another study revealed that *Dryopteris ramose* was high in total flavonoid contents using various solvents, including ethyl acetate extract (45.28 μg QE/mg), methanol extract (36.94 μg QE/mg), and water extract (25.69 μg QE/mg) [10]. Similarly, successive extracts of another species of the same genus, i.e., *Dryopteris cochleata* leaves, were reported to contain considerable amounts of total phenolic contents, with 17.7 μg GAE/g petroleum ether, 32.9 μg GAE/g chloroform, 43.4 μg GAE/g ethyl acetate, 90.4 μg GAE/g acetone, 30.86 μg GAE/g methanol, and 28.4 μg GAE/g water. Furthermore, successive extracts of *Dryopteris cochleata* leaves were also found to contain a considerable amount of total flavonoid contents, with 9.16 μg catechin-equivalent (CE)/g petroleum ether, 122.5 μg CE/g chloroform, 145.78 μg CE/g ethyl acetate, 146.9 μg CE/g acetone, 77.71 μg CE/g methanol, and 25.74 μg CE/g water [11].

In the present study, radical scavenging potential was found to align with total phenolic and flavonoid contents. *D. juxtapostia* root dichloromethane extract exhibited considerable antioxidant potential in all three assays as compared to other extracts. These findings are supported by the fact that experimental variables, such as the type of solvent, are important with respect to estimation of antioxidant activity [35–37], as in the present study, dichloromethane extract was found to contain a considerable amount of phytochemicals, in addition to considerable radical-scavenging activity. These findings are in line with those reported by Kathirvel and Sujhata (2016) [11], i.e., that acetone extract of *Dryopteris cochleata* leaves exhibited notable radical-scavenging potential as compared to other tested extracts, owing to its total phenolic and flavonoid contents. Numerous reports have highlighted the antioxidant potential of plants due to the presence of phenolic compounds [38,39]. Additionally, it has been determined that the antioxidant activity of phenolics and flavonoids is mainly a result of their redox properties, which can play an important role in absorbing and neutralizing free radicals, quenching singlet and triplet oxygen, and decomposing peroxides Osawa et al. [40]. The radical-scavenging potential of *Dryopteris ramose* crude extract (91.95%), methanol fraction (88.25%), water fraction (87.28%), and ethyl acetate fraction (69.97%) was recently reported by Alam et al. (2021) [10] using a DPPH assay. *Dryopteris affinis*, another species from the same *genus* as DJ was reported to comprise a reasonable amount of total phenolic contents of 112.5 mg GAE/g. *Dryopteris affinis* rhizome extract showed antioxidant potential in DPPH (IC$_{50}$ of 4.60 μg/mL) and ABTS (22.35 μmol Trolox/g) assays, which is even superior to that of standard butylated hydroxytoluene (BHT), with an IC$_{50}$ of 9.96 μg/mL [41]. Another study reported the remarkable antioxidant potential of *Dryopteris ramose* dichloromethane (55.7% inhibition) and methanol extract (72.7% inhibition) in a DPPH assay parallel to standard quercetin (74.54% inhibition), further supporting the antioxidant potential of species belonging to this *genus*. A decade earlier, Kathirvel and Sujhata (2012) [42] reported that several extracts of *Dryopteris cochleata* leaves, i.e., acetone, ethyl acetate, methanol, chloroform, and water, exhibited reducing activities, with EC_{50} values of 243 µg, 327 µg, 378 µg, and 494 µg, respectively, in a dose-dependent manner. The current findings corroborate prior findings of antioxidant activity in components of the genus *Dryopteris*, although varying potency levels have been reported in the literature. Climate, geography, soil conditions, irrigation methods, harvesting timing, storage, transit facilities, drying procedures (shade drying, sun drying, oven drying, or freeze drying), the polarity of solvents, extraction methods, and extraction time could all play a significant role [19].

Urease is an enzyme that mediates the hydrolysis of urea, resulting in the production of ammonia and carbon dioxide, with its primary function being to protect bacteria in the acidic environment of the stomach [43]. Urease inhibitors have the potential to counteract urease's detrimental effects on living organisms. Urease inhibitors are effective against a variety of infections caused by urease secretion by *Helicobacter pylori*, including gastrointestinal disorders, such as gastritis, duodenal ulcers, peptic ulcers, and stomach cancer [44]. Antibiotic treatment can heal ulcers, prevent recurrence of peptic ulcers, and reduce the risk of stomach cancer in high-risk groups. However, resistance to one or more antibiotics, as well as other considerations, such as poor patient compliance, drug side effects, and the considerable expense of combination therapy, have resulted in concerns among consumers with respect to safety, cost-effectiveness, and availability [45]. Figure shows 3 that *D. juxtapostia* root dichloromethane extract exhibited notable urease inhibition activity, followed by root methanol extract, exhibiting moderate urease inhibition activity. The results of the present investigation cannot be compared with previous data, as no species from this genus has previously be explored for antiurease activity, although various studies have reported the anti-inflammatory potential of this genus in other assays. For example, *Dryopteris crassirhizoma* ethanol extract was reported to diminish the mediation of nitric oxide and prostaglandin production in lipopolysaccharide-stimulated RAW264.7 cells. It also downregulated the levels of mRNA expression of pro-inflammatory genes, such as inducible nitric oxide synthase, cyclooxygenase, and TNF-α [46]. Some compounds isolated from water extract of *Dryopteris fragrans* were reported to have nitric oxide production inhibition potential in lipopolysaccharide-induced RAW 264.7 macrophages, with IC_{50} values of 45.8, 65.8, and 49.8 µM, respectively [47]. Another study explored *Dryopteris filixmas* leaves and reported that aqueous extract inhibited the hemolysis of red blood cell membranes (56.45% inhibition at 6 mg/mL) parallel to the inhibition induced by standard drug acetylsalicylic acid (70% inhibition at 6 mg/mL) [48]. These reports support the anti-inflammatory results of the present investigation, indicating that species from the genus *Dryopteris* possess health-promoting potential.

Carrageenan-induced paw edema is considered a credible anti-inflammatory compound screening test. The development of carrageenan-induced paw edema is a biphasic reaction, with the first phase including the release of kinins, histamine, and 5-HT and the second phase involving the release of prostaglandins [49]. Similarly, formaldehyde-induced pain and edema are mediated by bradykinins and substance P during the early phase of formaldehyde injection, whereas a tissue-mediated response in the later phase is associated with the release of histamine, 5-hydroxytryptamine (5-HT), prostaglandins, and bradykinins [50]. In our study, *D. juxtapostia* root dichloromethane extract was found to inhibit inflammation in carrageenan-induced ($p < 0.001$) and formaldehyde-induced ($p < 0.001$) paw edema models in a significant manner, suggesting that anti-inflammatory activity may be accredited to the inhibition of inflammatory mediators during both phases of edema formation. Recently, *S. cumini* extract was recorded to contain a considerable amount of total phenolic, flavonoid, and antioxidant potential, in addition to showing remarkable anti-inflammatory activities against carrageenan-, formalin-, and PGE2-induced intoxication [19]. Another group of researchers reported a direct relationship between phytochemical contents, antioxidant activity, and inhibition of inflammation [32], supporting the findings of current investigation. As previously stated, this is first report on the biological potential of *D. juxtapostia*, although many species of this genus have been found to offer inflammation aversion potential in the past. Ahmad et al. (2011) [12] reported that *Dryopteris chrysocoma* root extract exhibited significant ($p < 0.001$) inhibition of 51.1% when dispensed orally at a rate of 500 mg/kg after administration of formalin comparable to the standard aspirin. In the same study, significant inhibition ($p < 0.001$) of 57% was induced by *Dryopteris chrysocoma* root extract when administered at a rate of 500 mg/kg against carrageenan-induced paw edema, demonstrating the inflammation-averting potential of this genus. The inhibition was dose-dependent, and root extract was found to be more effective than leaf and stem extract. The findings are also in agreement with those reported by Khan et al. (2018) [13], i.e., that *Dryopteris blanfordii* extract showed significant analgesic activity (45% reduction in writhing, $p < 0.01$) when administered at a dose of 300 mg/kg, which is more than standard drug aspirin, exhibiting 40% inhibition at 150 mg/kg. In the same study, ethanolic extract of *Dryopteris blanfordii* exhibited significant ($p < 0.01$) inhibition against carrageenan-induced paw edema, with activity was parallel to the standard drug diclofenac sodium. Another species from the same genus, i.e., *Dryopteris crassirhizoma*, was reported to demonstrate significant antiallergic and anti-inflammatory activities by modulating the T-helper type 1 (Th1) and T-helper type 2 (Th2) response and reducing the allergic inflammatory reaction in phorbol myristate acetate (PMA)- and A23187-stimulated HMC-1 cells via NF-κB signaling in an ovalbumin (OVA)-induced allergic asthma model [51]. A decade earlier, radix methanolic extract of *Dryopteris crassirrhizoma* showed significant ($p < 0.01$) anti-inflammatory effects during the screening of almost 150 medicinal plants [52].

The findings of the brine shrimp assay revealed that *D. juxtapostia* root dichloromethane extract may contain some compounds that may have cytotoxic effects on certain cancer cells. As previously suggested by Mayer et al. (1982) [24], the brine shrimp lethality test can be used to predict compounds or extracts that may exhibit anticancer activity. These findings are in accordance with those reported by Baloch et al. (2019) [34], i.e., that dichloromethane extract of *Dryopteris ramosa* whole plant exhibited a very high potential for cytotoxicity, with an LD_{50} of 0.6903 μl/mL, which is 10 times more potent than etoposide, with an LD_{50} of 7.46 μl/mL. Another study reported *Dryopteris ramosa* to exhibit brine shrimp lethality potential, with an LD_{50} value of 47.64 μg/mL Alam et al. [10]. *Dryopteris affinis* rhizome and leaf methanol extracts were reported to exhibit moderate cytotoxicity in a brine shrimp assay, with LC_{50} values of 323.9 μg/mL and 85.5 μg/mL, respectively [41]. The methanol extract of *Dryopteris filixmas* leaves was also reported to exhibit potent cytotoxic activity, with an LC_{50} of 25.9 μg/mL, whereas the LC_{50} value of standard vincristine sulfate was 10.0 μg/mL [53].

D. juxtapostia root dichloromethane extract not only exhibited antioxidant, anti-inflammatory, and cytotoxic properties but also inhibited the proliferation HeLa human cervical and prostate cancer (PC3) cells (Table 6). Several species of the same genus have been reported to exhibit notable anticancer effects against different cancer cell lines [15,54,55], supporting the findings of the current investigation. The anticancer activity reported in the present study is consistent with total phenolics and flavonoids, anti-inflammatory effects, and cytotoxic potential. Anticancer activity of *Dryopteris cochleat* was previously believed to be associated with its phenolic and flavonoid contents [11,56]. *Dryopteris crassirhizoma* (50 and 100 g/mL) markedly inhibited the proliferation of PC-3 and PC3-MM2 cells without disturbing or inducing cytotoxicity toward normal spleen cells from BALB/C mice through the activation of caspase-3, -8, -9, bid, and PARP in PC3-MM2 cells [53].

5. Conclusions

The findings of the present investigation demonstrate the phytochemical, antioxidant, anti-inflammatory (in vitro and in vivo), and cytotoxic potential of *Dryopteris juxtapostia* (DJ) crude extracts, supporting the use of species belonging to this genus in traditional medicinal systems. DJ root dichloromethane extract exhibited the highest biological potential in all aforementioned assays, followed by DJ root methanol extract. Both extracts were found to exert no toxicity in the livers of the tested animals when administered at dosis 300 and 500 mg/kg according to liver function test, total protein, and lipid profile. Mass spectrometry analysis showed that some phenolic compounds are responsible for the antioxidant, anti-inflammatory, and anticancer potential of DJ. Overall, the current study confirms the potential of DJ with respect to bioactivity and as a possible alternative therapeutic vector. This research also provides a database for future research to optimize extraction methods and solvents for the maximum extraction of polyphenols and/or flavonoids. Furthermore, preserving the bioactivity of polyphenols and optimizing their delivery also represents a future challenge. In conclusion, plant polyphenols and flavonoids may represent a significant complementary medicine for treatment of oxidation- and inflammation-induced physiological dysfunction. However, further clinical trials are required to establish the safety and efficacy of bioactive compounds.

Author Contributions: Conceptualization, M.U., A.R. and M.Q.; methodology, A.R.; software, M.Q.; validation, A.R., S.A. and M.Q.; investigation, A.R.; resources, M.U.; writing—original draft preparation, A.R., T.E., M.Q., N.A., M.W.A. and S.A.; writing—review and editing, T.E., N.A. and S.A.; visualization, T.E.; supervision, M.U.; project administration, A.R. and M.Q.; funding acquisition, T.E. All authors have read and agreed to the published version of the manuscript.

Funding: The publication of this article was funded by the Open Access Fund of the Leibniz Universität Hannover, Germany.

Institutional Review Board Statement: All trials were conducted following preset guidelines of National Research Council (NRC, 1996, Washington, DC, USA). The study was approved by the departmental Animal Care Committee at BZU, Pakistan (approval number ACC-10-2019).

Informed Consent Statement: Not applicable.

Data Availability Statement: The data is contained within the manuscript.

Conflicts of Interest: The authors declare no conflict of interest.

References

1. Jayachitra, A.; Krithiga, N. Study on antioxidant property in selected medicinal plant extract. *Int. J. Med. Aromat. Plants* **2010**, *2*, 495–500.
2. Hiransai, P.; Tangpong, J.; Kumbuar, C.; Hoonheang, N.; Rodpech, O.; Sangsuk, P.; Kajklangdon, U.; Inkaow, W. Anti-nitric oxide production, antiproliferation and antioxidant effects of the aqueous extract from *Tithonia diversifolia*. *Asian Pac. J. Trop. Biomed.* **2016**, *6*, 950–956. [CrossRef]
3. Qamar, M.; Akhtar, S.; Ismail, T.; Wahid, M.; Barnard, R.T.; Esatbeyoglu, T.; Ziora, Z.M. The chemical composition and health-promoting effects of the *Grewia* species—A systematic review and meta-analysis. *Nutrients* **2021**, *13*, 4565. [CrossRef] [PubMed]
4. Qamar, M.; Akhtar, S.; Ismail, T.; Wahid, M.; Abbas, M.W.; Mubarak, M.S.; Esatbeyoglu, T. Phytochemical Profile, Biological Properties, and Food Applications of the Medicinal Plant *Syzygium cumini*. *Foods* **2022**, *11*, 378. [CrossRef]
5. Maehre, H.K.; Jensen, I.J.; Elvevoll, E.O.; Eilertsen, K.E. ω-3 Fatty acids and cardiovascular diseases: Effects, mechanisms and dietary relevance. *Int. J. Mol. Sci.* **2015**, *16*, 22636–22661. [CrossRef] [PubMed]
6. Patel, V.R.; Patel, P.R.; Kajal, S.S. Antioxidant activity of some selected medicinal plants in western region of India. *Adv. Biol. Res.* **2010**, *4*, 23–26.
7. Mustafa, G.; Arif, R.; Atta, A.; Sharif, S.; Jamil, A. Bioactive compounds from medicinal plants and their importance in drug discovery in Pakistan. *Matrix Sci. Pharma* **2017**, *1*, 17–26. [CrossRef]
8. Ahmad, S.S. Medicinal wild plants from Lahore-Islamabad motorway (M-2). *Pak. J. Bot.* **2007**, *39*, 355–375.
9. Sher, H.; Ali, A.; Ullah, Z.; Sher, H. Alleviation of Poverty through Sustainable Management and Market Promotion of Medicinal and Aromatic Plants in Swat, Pakistan: Alleviation of Poverty through Sustainable Management. *Ethnobot. Res. Appl.* **2022**, *23*, 1–19.
10. Alam, F.; Khan, S.H.A.; Asad, M.H.H.B. Phytochemical, antimicrobial, antioxidant and enzyme inhibitory potential of medicinal plant Dryopteris ramosa (Hope) C. Chr. *BMC Complement. Med. Ther.* **2021**, *21*, 197. [CrossRef]
11. Kathirvel, A.; Rai, A.K.; Maurya, G.S.; Sujatha, V. *Dryopteris cochleata* rhizome: A nutritional source of essential elements, phytochemicals, antioxidants and antimicrobials. *Int. J. Pharm. Pharm. Sci.* **2014**, *6*, 179–188.
12. Ahmad, M.; Jahan, N.; Mehjabeen, A.B.R.; Ahmad, M.; Ullah, O.; Mohammad, N. Differential inhibitory potencies of alcoholic extract of different parts of *Dryopteris chrysocoma* on inflammation in mice and rats. *Pak. J. Pharm. Sci.* **2011**, *24*, 559–563.

13. Khan, M.S.; Ullah, S. Analgesic, anti-inflammatory, antioxidant activity and phytochemical screening of *Dryopteris blanfordii* plant. *J. Pharmacogn. Phytochem.* **2018**, *7*, 536–541.

14. Erhirhie, E.O.; Emeghebo, C.N.; Ilodigwe, E.E.; Ajaghaku, D.L.; Umeokoli, B.O.; Eze, P.M.; Okoye, F.B.G.C. *Dryopteris filix-mas* (L.) *Schott* ethanolic leaf extract and fractions exhibited profound anti-inflammatory activity. *Avicenna J. Phytomed.* **2019**, *9*, 396–409. [PubMed]

15. Liu, Z.D.; Zhao, D.D.; Jiang, S.; Xue, B.; Zhang, Y.L.; Yan, X.F. Anticancer phenolics from *Dryopteris fragrans* (L.) *Schott. Molecules* **2018**, *23*, 680. [CrossRef] [PubMed]

16. Lee, J.; Nho, Y.H.; Yun, S.K.; Hwang, Y.S. Anti-invasive and anti-tumor effects of *Dryopteris crassirhizoma* extract by disturbing actin polymerization. *Integr. Cancer Ther.* **2019**, *18*, 1534735419851197. [CrossRef]

17. Singleton, V.L.; Rossi, J.A. Colorimetry of total phenolics with phosphomolybdic phosphotungstic acid reagents. *Am. J. Enol. Vitic.* **1965**, *16*, 144–158.

18. Pekal, A.; Pyrzynska, K. Evaluation of aluminium complexation reaction for flavonoid content assay. *Food Anal. Methods* **2014**, *7*, 1776–1782. [CrossRef]

19. Qamar, M.; Akhtar, S.; Ismail, T.; Yuan, Y.; Ahmad, N.; Tawab, A.; Ziora, Z.M. *Syzygium cumini* (L.), Skeels fruit extracts: In vitro and in vivo anti-inflammatory properties. *J. Ethnopharmacol.* **2021**, *271*, 113805. [CrossRef]

20. Weatherburn, M. Enzymic method for urea in urine. *Anal. Chem.* **1967**, *39*, 971–974. [CrossRef]

21. Morris, C.J. Carrageenan-induced paw edema in the rat and mouse. *Methods Mol. Biol.* **2003**, *225*, 115–121. [PubMed]

22. NRC National Research Council. *Guide for the Care and Use of Laboratory Animals*; National Academy Press: Washington, DC, USA, 1996.

23. Brownlee, G. Effect of deoxycortone and ascorbic acid on formaldehyde-induced arthritis in normal and adrenalectomised rats. *Lancet* **1950**, *255*, 157–159. [CrossRef]

24. Meyer, B.; Ferrigni, N.; Putnam, J.; Jacobsen, L.; Nichols, D.; McLaughlin, J.L. Brine shrimp: A convenient general bioassay for active plant constituents. *Planta Med.* **1982**, *45*, 31–34. [CrossRef] [PubMed]

25. Mosmann, T. Rapid colorimetric assay for cellular growth and survival: Application to proliferation and cytotoxicity assays. *J. Immunol. Methods* **1983**, *65*, 55–63. [CrossRef]

26. OECD. *OECD Guidelines for Testing of Chemicals, Section 4 Test No. 423: Acute Oral Toxicity—Acute Toxic Class Method*; OECD: Paris, France, 2001.

27. Reitman, S.; Frankel, S. A colorimetric method for the determination of serum glutamic oxalacetic and glutamic pyruvic transaminases. *Am. J. Clin. Pathol.* **1957**, *28*, 56–63. [CrossRef]

28. Keiding, R.; Hörder, M.; Denmark, W.G.; Pitkänen, E.; Tenhunen, R.; Strömme, J.H.; Westlund, L. Recommended methods for the determination of four enzymes in blood. *Scand. J. Clin. Lab. Investig.* **1974**, *33*, 291–306. [CrossRef]

29. Tietz, N.W.; Burtis, C.A.; Duncan, P.; Ervin, K.; Petitclerc, C.J.; Rinker, A.D.; Zygowicz, E.R. A reference method for measurement of alkaline phosphatase activity in human serum. *Clin. Chem.* **1983**, *29*, 751–761. [CrossRef]

30. Ren, Q.; Quan, X.G.; Wang, Y.L.; Wang, H.Y. Isolation and identification of phloroglucinol derivatives from *Dryopteris crassirhizoma* by HPLC-LTQ-Orbitrap Mass Spectrometry. *Chem. Nat. Compd.* **2016**, *52*, 1137–1140. [CrossRef]

31. Ren, Q. Mass spectral fragmentation pattern and spectroscopic rules of phloroglucinol constituents in plants of *Dryopteris Adanson*. *Chin. Tradit. Herb. Drugs* **2015**, *24*, 932–937.

32. Qamar, M.; Akhtar, S.; Ismail, T.; Sestili, P.; Tawab, A.; Ahmed, N. Anticancer and anti-inflammatory perspectives of Pakistan's indigenous berry *Grewia asiatica* Linn (Phalsa). *J. Berry Res.* **2020**, *10*, 115–131. [CrossRef]

33. Qamar, M.; Akhtar, S.; Ismail, T.; Wahid, M.; Ali, S.; Nazir, Y.; Murtaza, S.; Abbas, M.W.; Ziora, Z.M. *Syzygium cumini* (L.) Skeels extracts; in vivo anti-nociceptive, anti-inflammatory, acute and subacute toxicity assessment. *J. Ethnopharmacol.* **2022**, *287*, 114919. [CrossRef] [PubMed]

34. Baloch, R.; Uzair, M.; Chauhdary, B. Phytochemical analysis, antioxidant and cytotoxic activities of *Dryopteris ramosa. Biomed. Res. J.* **2019**, *30*, 764–769.

35. Dawidowicz, A.L.; Olszowy, M. The importance of solvent type in estimating antioxidant properties of phenolic compounds by ABTS assay. *Eur. Food Res. Technol.* **2013**, *236*, 1099–1105. [CrossRef]

36. Dawidowicz, A.L.; Olszowy, M. Antioxidant properties of BHT estimated by ABTS assay in systems differing in pH or metal ion or water concentration. *Eur. Food Res. Technol.* **2011**, *232*, 837–842. [CrossRef]

37. Dawidowicz, A.L.; Olszowy, M. Influence of some experimental variables and matrix components in the determination of antioxidant properties by β-carotene bleaching assay: Experiments with BHT used as standard antioxidant. *Eur. Food Res. Technol.* **2010**, *231*, 835–840. [CrossRef]

38. Ahmad, N.; Qamar, M.; Yuan, Y.; Nazir, Y.; Wilairatana, P.; Mubarak, M.S. Dietary Polyphenols: Extraction, Identification, Bioavailability, and Role for Prevention and Treatment of Colorectal and Prostate Cancers. *Molecules* **2022**, *27*, 2831. [CrossRef]

39. Abbas, M.W.; Hussain, M.; Qamar, M.; Ali, S.; Shafiq, Z.; Wilairatana, P.; Mubarak, M.S. Antioxidant and Anti-Inflammatory Effects of Peganum harmala Extracts: An In Vitro and In Vivo Study. *Molecules* **2021**, *26*, 6084. [CrossRef]

40. Osawa, T.; Uritani, I.; Garcia, V.V.; Mendoza, E.M. *Postharvest Biochemistry of Plant Food-Materials in the Tropics*; Japan Scientific Societies Press: Tokyo, Japan, 1994; p. 241.

41. Valizadeh, H.; Sonboli, A.; Kordi, F.M.; Dehghan, H.; Bahadori, M.B. Cytotoxicity, antioxidant activity and phenolic content of eight fern species, from north of Iran. *Pharm. Sci.* **2016**, *21*, 18–24. [CrossRef]

42. Kathirvel, A.; Sujatha, V. Phytochemical studies, antioxidant activities and identification of active compounds using GC-MS of *Dryopteris cochleata* leaves. *Arab. J. Chem.* **2016**, *9*, S1435–S1442. [CrossRef]

43. Stingl, K.; Altendorf, K.; Bakker, E.P. Acid survival of Helicobacter pylori: How does urease activity trigger cytoplasmic pH homeostasis? *Trends Microbiol.* **2002**, *10*, 70–74. [CrossRef]

44. Amtul, Z.; Siddiqui, R.; Choudhary, M. Chemistry and mechanism of urease inhibition. *Curr. Med. Chem.* **2002**, *9*, 1323–1348. [CrossRef] [PubMed]

45. O'Connor, C.J.; Laraia, L.; Spring, D.R. Chemical genetics. *Chem. Soc. Rev.* **2011**, *40*, 4332–4345. [CrossRef] [PubMed]

46. Yang, Y.; Lee, G.J.; Yoon, D.H.; Yu, T.; Oh, J.; Jeong, D.; Lee, J.; Kim, S.H.; Kim, T.W.; Cho, J.Y. ERK1-and TBK1-targeted anti-inflammatory activity of an ethanol extract of *Dryopteris crassirhizoma*. *J. Ethnopharmacol.* **2013**, *145*, 499–508. [CrossRef] [PubMed]

47. Peng, B.; Bai, R.-F.; Li, P.; Han, X.-Y.; Wang, H.; Zhu, C.-C.; Zeng, Z.-P.; Chai, X.-Y. Two new glycosides from *Dryopteris fragrans* with anti-inflammatory activities. *J. Asian Nat. Prod. Res.* **2016**, *18*, 59–64. [CrossRef]

48. Salemcity, A.; Attah, A.; Oladimeji, O.; Olajuyin, A.; Usifo, G.; Audu, T. Comparative study of membrane-stabilizing activities of kolaviron *Dryopteris filix-mas* and *Ocimum gratissimum* extracts. *Egypt. Pharm. J.* **2016**, *15*, 6–9. [CrossRef]

49. Wills, A.L. Release of histamin, kinin and prostaglandins during carrageenin induced inflammation of the rats. *Prostaglandins Pept. Amines* **1969**, 31–48.

50. Wheeler-Aceto, H.; Cowan, A. Neurogenic and tissue-mediated components of formalin-induced edema: Evidence for supraspinal regulation. *Agents Actions* **1991**, *34*, 264–269. [CrossRef]

51. Piao, C.H.; Bui, T.T.; Fan, Y.J.; Nguyen, T.V.; Shin, D.U.; Song, C.H.; Chai, O.H. In vivo and in vitro anti allergic and anti inflammatory effects of *Dryopteris crassirhizoma* through the modulation of the NF kB signaling pathway in an ovalbumin induced allergic asthma mouse model. *Mol. Med. Rep.* **2020**, *22*, 3597–3606. [CrossRef]

52. Lee, K.K.; Kim, J.H.; Cho, J.J.; Choi, J.D. Inhibitory effects of 150 plant extracts on elastase activity, and their anti-inflammatory effects. *Int. J. Cosmet. Sci.* **1999**, *21*, 71–82. [CrossRef]

53. Ali, M.S.; Mostafa, K.; Raihan, M.O.; Rahman, M.K.; Aslam, M. Antioxidant and Cytotoxic activities of Methanolic extract of *Dryopteris filix-mas* (L.) Schott Leaves. *Int. J. Drug Dev. Res.* **2012**, *4*, 223–229.

54. Chang, S.-H.; Bae, J.-H.; Hong, D.-P.; Choi, K.-D.; Kim, S.-C.; Her, E.; Kim, S.-H.; Kang, C.-D. *Dryopteris crassirhizoma* has anti-cancer effects through both extrinsic and intrinsic apoptotic pathways and G0/G1 phase arrest in human prostate cancer cells. *J. Ethnopharmacol.* **2010**, *130*, 248–254. [CrossRef] [PubMed]

55. Zhang, T.; Wang, L.; Duan, D.-H.; Zhang, Y.-H.; Huang, S.-X.; Chang, Y. Cytotoxicity-guided isolation of two new phenolic derivatives from *Dryopteris fragrans* (L.) Schott. *Molecules* **2018**, *23*, 1652. [CrossRef] [PubMed]

56. Thakur, R.S.; Ahirwar, B. Ethno pharmacological evaluation of medicinal plants for cytotoxicity against various cancer cell lines. *Int. J. Pharm. Pharm. Sci.* **2017**, *9*, 198–202. [CrossRef]

Article

Echinacea purpurea Fractions Represent Promising Plant-Based Anti-Inflammatory Formulations

Sara F. Vieira [1,2], Samuel M. Gonçalves [2,3], Virgínia M. F. Gonçalves [4,5], Carmen P. Llaguno [6], Felipe Macías [6], Maria Elizabeth Tiritan [4,7,8], Cristina Cunha [2,3], Agostinho Carvalho [2,3], Rui L. Reis [1,2], Helena Ferreira [1,2,*] and Nuno M. Neves [1,2,*]

[1] 3B's Research Group, I3BS—Research Institute on Biomaterials, Biodegradables and Biomimetics, University of Minho, Headquarters of the European Institute of Excellence on Tissue Engineering and Regenerative Medicine, AvePark, Parque de Ciência e Tecnologia, Zona Industrial da Gandra, Barco, 4805-017 Guimarães, Portugal
[2] ICVS/3B's–PT Government Associate Laboratory, Braga, Guimarães, Portugal
[3] Life and Health Sciences Research Institute (ICVS), School of Medicine, University of Minho, Campus de Gualtar, 4710-057 Braga, Portugal
[4] TOXRUN—Toxicology Research Unit, University Institute of Health Sciences, CESPU, CRL, 4585-116 Gandra, Portugal
[5] UNIPRO—Oral Pathology and Rehabilitation Research Unit, University Institute of Health Sciences (IUCS), CESPU, CRL, 4585-116 Gandra, Portugal
[6] Departamento de Edafoloxía e Química Agrícola, Facultade de Bioloxía, Universidade de Santiago de Compostela, 15782 Santiago de Compostela, Spain
[7] Interdisciplinary Centre of Marine and Environmental Research (CIIMAR), University of Porto, Terminal de Cruzeiros do Porto de Leixões, Avenida General Norton de Matos, S/N, 4450-208 Matosinhos, Portugal
[8] Laboratório de Química Orgânica e Farmacêutica, Departamento de Ciências Químicas, Faculdade de Farmácia da Universidade do Porto, Rua Jorge de Viterbo Ferreira 228, 4050-313 Porto, Portugal
[*] Correspondence: helenaferreira@i3bs.uminho.pt (H.F.); nuno@i3bs.uminho.pt (N.M.N.); Tel.: +351-253-510-913 (H.F.); +351-253-510-905 (N.M.N.); Fax: +351-253-510-909 (H.F. & N.M.N.)

Citation: Vieira, S.F.; Gonçalves, S.M.; Gonçalves, V.M.F.; Llaguno, C.P.; Macías, F.; Tiritan, M.E.; Cunha, C.; Carvalho, A.; Reis, R.L.; Ferreira, H.; et al. *Echinacea purpurea* Fractions Represent Promising Plant-Based Anti-Inflammatory Formulations. *Antioxidants* **2023**, *12*, 425. https://doi.org/10.3390/antiox12020425

Academic Editors: Antonella D'Anneo and Marianna Lauricella

Received: 16 January 2023
Revised: 6 February 2023
Accepted: 7 February 2023
Published: 9 February 2023

Abstract: *Echinacea purpurea* is traditionally used in the treatment of inflammatory diseases. Therefore, we investigated the anti-inflammatory capacity of *E. purpurea* dichloromethanolic (DE) and ethanolic extracts obtained from flowers and roots (R). To identify the class of compounds responsible for the strongest bioactivity, the extracts were fractionated into phenol/carboxylic acid (F1) and alkylamide fraction (F2). The chemical fingerprint of bioactive compounds in the fractions was evaluated by LC-HRMS. *E. purpurea* extracts and fractions significantly reduced pro-inflammatory cytokines (interleukin 6 and/or tumor necrosis factor) and reactive oxygen and nitrogen species (ROS/RNS) production by lipopolysaccharide-stimulated primary human monocyte-derived macrophages. Dichloromethanolic extract obtained from roots (DE-R) demonstrated the strongest anti-inflammatory activity. Moreover, fractions exhibited greater anti-inflammatory activity than whole extract. Indeed, alkylamides must be the main compounds responsible for the anti-inflammatory activity of extracts; thus, the fractions presenting high content of these compounds presented greater bioactivity. It was demonstrated that alkylamides exert their anti-inflammatory activity through the downregulation of the phosphorylation of p38, ERK 1/2, STAT 3, and/or NF-κB signaling pathways, and/or downregulation of cyclooxygenase 2 expression. *E. purpurea* extracts and fractions, mainly DE-R-F2, are promising and powerful plant-based anti-inflammatory formulations that can be further used as a basis for the treatment of inflammatory diseases.

Keywords: *Echinacea purpurea* extracts; fractions; phenols/carboxylic acids; alkylamides; inflammation; human primary macrophages

1. Introduction

Inflammation is crucial for the survival and maintenance of human health [1]. The inflammatory response is coordinated by the activation of several inflammatory signaling pathways in tissue-resident and recruited immune cells [2]. The main inflammatory signaling pathways associated with the initiation and progression of inflammation are nuclear factor-kappa B (NF-κB) [3], mitogen-activated protein kinase (MAPK) family (extracellular signal-regulated kinase (ERK), C-Jun N-terminal kinase/stress-activated protein kinase (JNK/SAPK), and p38 kinase) [4], cyclooxygenase (COX)-2 expression [5], and Janus kinase/signal transducers and activators of transcription (JAK/STAT) [6].

The dysregulation of the magnitude or duration of the inflammatory response can lead to chronic inflammation, which is characterized by the continuous infiltration of immune cells into the injured tissue [7]. Particularly, macrophages are key mediators of inflammation, orchestrating the immune response. Those cells are responsible for engulfing damaged cells and invading pathogens and present antigens to the adaptive immune system [8]. Once activated, macrophages release high levels of pro-inflammatory mediators, including reactive oxygen and nitrogen species (ROS/RNS) and cytokines (e.g., interleukin (IL)-6 and tumor necrosis factor (TNF)-α) [9–11]. These molecules allow for the communication between immune cells, regulating the intensity and duration of the inflammatory response. Hence, their suppression can be a valuable hallmark in the therapy of chronic inflammation where the immune system is overactivated.

The most severe and deleterious outcome of chronic inflammation is the continuous damage and destruction of tissues and organs, which leads to an increased risk of several pathologies (e.g., autoimmune disorders) [12,13]. The current treatment for chronic inflammation-associated diseases varies with their severity, but often it focuses on reducing the overactivity of the immune system. Available anti-inflammatory drugs include nonsteroidal anti-inflammatory drugs (NSAIDs, e.g., celecoxib), corticosteroids (e.g., dexamethasone), conventional disease-modifying anti-rheumatic drugs (cDMARDs, e.g., methotrexate), and biological (b) DMARDs (e.g., anti-IL-6 and anti-TNF-α) [14–16]. However, the prolonged administration of these drugs is frequently associated with several serious side effects. Those include disturbances in the gastrointestinal tract and an increased incidence of opportunistic infections and cancer [17,18]. Therefore, there is an urgent need to discover effective and safe anti-inflammatory drugs.

Plants have been an excellent resource of unique compounds with an important role in the development of many therapeutics [19]. Particularly, *Echinacea purpurea* formulations, recognized as safe by the World Health Organization, have been traditionally used as a potent immunomodulatory medicines [20]. Indeed, *E. purpura* extracts are employed to reduce oxidative stress and inflammation, as well as to prevent cold and flu. Moreover, the ability of *E. purpurea* to interact with immune cells is leading to new insights about its anticancer properties [21]. Other biological properties, such as antifungal, antiviral, and antibacterial activities, have also been reported [22]. Particularly, the antioxidant and anti-inflammatory activities of *E. purpurea* have been associated with its ability to reduce the production of ROS/RNS and pro-inflammatory mediators [23], decrease the infiltration of inflammatory cells [24], and block the receptors of the immune cell [25]. The anti-inflammatory properties have been attributed to alkylamides [26–31], polysaccharides [32–36], and caffeic acid derivatives [37,38]. More recently, sesquiterpenes have also been proposed as bioactive principles of *E. purpurea* [39]. However, other studies suggested that the anti-inflammatory activity arises from the synergy between the different bioactive classes of compounds present in the *E. purpurea* extracts [40]. Additionally, the mechanism through which *E. purpurea* extracts exert anti-inflammatory activity is still unclear. Although few studies report the cellular mechanism of *E. purpurea* extracts, they are mainly developed in mouse-derived immune cells [32,39,40]. To the best of our knowledge, only two studies evaluated the mechanism of action of *E. purpurea* extracts using human-derived immune cells. Fast et al. prepared an aqueous extract that reduced the TNF-α production via the inhibition of Toll-like receptor (TLR) 1/2 in Pam3Csk4-stimulated human macrophages [33].

Chicca et al. reported that the standardized commercial tincture Echinaforce decreased the TNF-α production in part via cannabinoid type 2 (CB2) receptor signaling in lipopolysaccharide (LPS)-stimulated human peripheral blood mononuclear cells (PBMCs) [41]. Thus, the understanding of how a particular bioactive class of compounds present in *E. purpurea* extracts produces its effects in human-derived primary cells, which mimic the human cell environment, is urgently needed. Bioactivity-guided fractionation assays will help in the identification of substances responsible for the biological activity.

In a previous study, we demonstrated the potential of several *E. purpurea* extracts to reduce cytokine production and ROS/RNS levels in an LPS-stimulated macrophage cell line [42]. In this work, we aim to corroborate their anti-inflammatory effects with human primary monocyte-derived macrophages (hMDMs), investigate the bioactive principles, and explore the therapeutic targets. The three most promising extracts in the previous study—dichloromethanolic extracts obtained from roots (DE-R), dichloromethanolic extracts obtained from flowers (DE-F), and ethanolic extracts obtained from flowers (EE-F)—were selected for this work [42]. The *E. purpurea* extracts prepared using a green and innovative extraction technique, the Accelerated Solvent Extractor (ASE), were fractionated by semi-preparative high-performance liquid chromatography (HPLC) into a phenol/carboxylic acid rich fraction (F1) and an alkylamide rich fraction (F2) to identify the class of compounds responsible for the strongest bioactivity. Moreover, the chemical fingerprint of the bioactive compounds in the fractions was also evaluated by liquid chromatography–high-resolution mass spectrometry (LC-HRMS). The reduction in the production of pro-inflammatory cytokines (IL-6 and TNF-α), the decrease in intracellular ROS/RNS generation, and the downregulation of inflammatory signaling pathways (NF-κB, ERK1/2, p38, JNK/SAPK, STAT3, COX-2) were investigated in LPS-stimulated primary human monocyte-derived macrophages (hMDMs). LPS is an exogenous stimulus derived from the cell wall of Gram-negative bacteria that promotes the release of pro-inflammatory mediators (e.g., cytokines and ROS/RNS) [43]. After fractionation of *E. purpurea* extracts, it was observed that F2 enhanced the anti-inflammatory activity, suggesting that alkylamides are the bioactive compounds mainly responsible for this bioactivity. Interestingly, the further fractionation of alkylamides fraction demonstrated the existence of a possible synergistic effect between them. To the best of our knowledge, this is the first study demonstrating the anti-inflammatory effects of *E. purpurea*, mainly of dichloromethanolic extracts and their fractions, in LPS-stimulated hMDM, through the suppression of ERK1/2, p38, STAT3, and COX-2 inflammatory signaling pathways.

2. Materials and Methods

A scheme detailing the sequence of the methodology used in this work is illustrated in Figure 1.

2.1. Reagents and Chemicals

E. purpurea was purchased from Cantinho das Aromáticas (Vila Nova de Gaia, Portugal) in May 2017. The plants were transferred to soil and grown following a sustainable agriculture procedure (41°37′04.5″ N, 7°16′14.4″ W). After two years of cultivation, the flowers were collected in a full bloom phase (June and July 2019), and the roots, including rhizomes, were harvested in the autumn (October 2019). Flowers and roots were dried in the dark and stored at room temperature (RT) and protected from light and humidity until further use. HPLC-grade dichloromethane, acetonitrile (ACN) and HPLC-grade methanol were obtained from Fisher Scientific, Portugal. Dimethyl sulfoxide (DMSO) was purchased from VWR, Portugal. Roswell Park Memorial Institute (RPMI)-1640 media, 4-(2-hydroxyethyl)-1-piperazineethanesulfonic acid (HEPES) buffer solution 1 M, penicillin–streptomycin (10,000 U/mL), Dulbecco's phosphate-buffered saline (DPBS), formalin 10% (*v/v*), Quant-iT PicoGreen dsDNA Kit, Pierce Phosphatase Inhibitor Mini Tablets, PageRuler Plus Prestained Protein Ladder (10 to 250 kDa), Bolt Sample Reducing Agent, Bolt LDS Sample Buffer, Bis-Tris Bolt 8%, Bolt MES SDS Running Buffer, and iBlot 2 Transfer Stacks (polyvinylidene fluoride, PVDF) were purchased from Thermo Fisher Scientific, Lisbon, Portugal. OctoMACS separator, human CD14 microbeads, MS columns, and human recombinant granulocyte–macrophage colony-stimulating factor (GM-CSF) were obtained from Miltenyi Biotec, Bergisch Gladbach, Germany. AlamarBlue, Bio-Rad Protein Assay Dye Reagent Concentrate, and Tween 20 were purchased from Bio-Rad, Lisbon, Portugal. Human IL-6 and TNF-α DuoSet Enzyme-linked immunosorbent assay (ELISA) and DuoSet ELISA Ancillary Reagent Kit 2 were purchased from R&D Systems, Minneapolis, MN, USA. Ethanol, formic acid analytical grade, dexamethasone, Histopaque-1077, human serum, lipopolysaccharide (LPS, *Escherichia coli* O26:B6), radioimmunoprecipitation assay (RIPA) buffer, complete mini protease inhibitor cocktail tablets, bovine serum albumin (BSA), Tris-base, and high-purity standards of echinacoside, chicoric acid, caftaric acid, caffeic acid, chlorogenic acid, and cynarin were obtained from Sigma-Aldrich, Lisbon, Portugal. Echinacea isobutylamide standards kit, composed of undeca-2E/Z-ene-8,10-diynoic acid isobutylamide, dodeca-2E-ene-8,10-diynoic acid isobutylamide, and dodeca-2E,4E-dienoic acid isobutylamide, was acquired from ChromaDex, Los Angeles, CA, USA. High-purity standard dodeca-2E,4E,8Z,10E/Z-tetraenoic acid isobutylamide was obtained from Biosynth Carbosynth, Spain. Cellular ROS/Superoxide ($O_2^{\bullet-}$) detection assay kit and rabbit glyceraldehyde-3-phosphate dehydrogenase (GAPDH) were acquired from Abcam, Boston, MA, USA. IRDye 800CW Goat anti-Rabbit IgG and IRDye 680RD Goat anti-Rabbit IgG secondary antibodies were obtained from LI-COR Biosciences GmbH, Bad Homburg, Germany. Rabbit NF-κB p65, rabbit p44/42 MAPK (ERK 1/2), rabbit p38 MAPK, rabbit SAPK/JNK, rabbit STAT3, rabbit COX-2, rabbit inducible nitric oxide synthase (iNOS), rabbit phospho-NF-κB p65, rabbit phospho-p38 MAPK, rabbit phospho-STAT3, rabbit phospho-SAPK/JNK, and rabbit phospho-p44/42 MAPK (ERK 1/2) were purchased from Cell Signaling Technology, Lisbon, Portugal. Sodium chloride was acquired from PanReac AppliChem, Lisbon, Portugal. Celecoxib was obtained from abcr GmbH, Karlsruhe, Germany. DAPI (4',6-diamidino-2-phenylindole) was purchased from Biotium, Fremont, CA, USA. Ultra-pure water was obtained from a Milli-Q® Direct Water Purification System (Milli-Q Direct 16, Millipore, Molsheim, France).

2.2. Bioactive Compounds Extraction

Dried flowers (F) or roots (R) were ground using an Analytical Sieve Shaker (AS200 Digit, Retsch, Haan, Germany) before extraction. Dichloromethanolic extracts (DE) and ethanolic extracts (EE) were prepared using an Accelerated Solvent Extractor 200 (ASE, Dionex Corp. Vigo, Spain), as previously described by Vieira et al. [42]. Briefly, the mixture of the plant material (2–5 g) with diatomaceous earth was placed and pressed into stainless-steel extraction cells, presenting cellulose filters in the bottom. Two extraction cycles were carried out at constant pressure (1500 psi) for 30 min at the minimal operation temperature of the equipment (40 °C). The extract solutions were collected in vials, and then the organic solvent was evaporated using nitrogen. Once dried, all the extracts were stored at −80 °C until further use.

2.2.1. Fractionation of Extracts

The chromatographic separation of the phenols/carboxylic acids and alkylamides was first optimized with standards by analytical HPLC. A stock solution of 1 mg/mL of all standards was prepared and stored in amber bottles at −80 °C. All standards were prepared in methanol, except the caffeic acid solution, which was prepared in ethanol. A standard mixture was prepared at a final concentration of 100 μg/mL for each. A LaChrom Merck Hitachi system equipped with a D-7000 Interface, an L-7100 Pump, an L-7200 autosampler, an L-7455 diode array detector (DAD), and an HPLC System Manager HSMD-7000 (Merck Hitachi, Tokyo, Japan), version 3.0, was used in the chromatographic analysis. The chromatographic separation was performed on a LiChrocart LiChrosphere 100 RP-18 (250 mm × 4 mm, 5 μm, Merck, Darmstadt, Germany). The gradient elution was optimized following the previous method reported by Pellati et al. [44], the mobile phase being composed of water containing 0.1% formic acid and ACN (Supplementary Table S1). The flow rate was 1 mL/min, and the column was set at RT. The injection volume was 20 μL. The UV spectra were acquired in the range of 190 to 450 nm, and the peak integration was performed at 254 nm for alkylamides and 330 nm for caffeic acid and its derivatives.

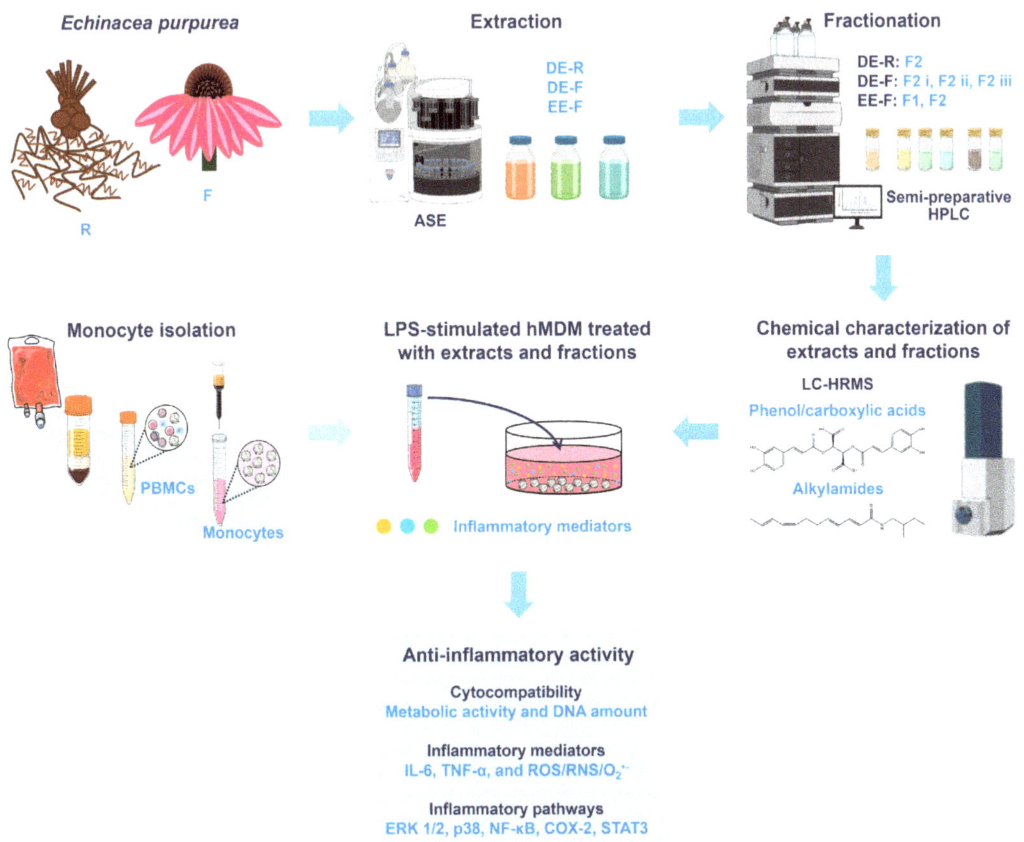

Figure 1. Scheme of the experimental procedure used in this work. *Echinacea purpurea* root (R) and flower (F) extracts were obtained using an Accelerated Solvent Extractor (ASE). Three extracts were prepared: dichloromethanolic extracts obtained from roots (DE-R), dichloromethanolic extracts obtained from flowers (DE-F), and ethanolic extracts obtained from flowers (EE-F). Then, the extracts were fractionated into phenol/carboxylic acid fractions (F1) and alkylamide fractions (F2) by semi-preparative high-performance liquid chromatography (HPLC). Both extracts and fractions were chemically characterized by liquid chromatography–high-resolution mass spectrometry (LC–HRMS). After, the whole extracts and fractions, at different concentrations, were added to lipopolysaccharide (LPS)-stimulated human monocyte-derived macrophages (hMDMs). These cells were isolated from human blood. Their cytocompatibility was evaluated through the metabolic activity and DNA amount determination. Their anti-inflammatory activity was validated by the decrease in interleukin (IL)-6 and tumor necrosis factor (TNF)-α levels in the cell culture medium, as well as by the reduction in the intracellular generation of ROS/RNS/O$_2^{\bullet-}$. Moreover, inflammatory pathways, including ERK1/2, p38, NF-κB, COX-2, and STAT3, were analyzed to determine their mechanisms of action.

The optimized separation method was adapted for the fractionation of *E. purpurea* extracts by semi-preparative HPLC. In order to reduce the time consumed, the previous gradient method was optimized (Table 1). The flow rate was set at 2 mL/min. The injection volume of *E. purpurea* extracts varied between 200 and 400 μL. A Uptisphere WOD homemade semi-preparative column (250 mm × 10 mm, 5 μm, interchrom, Interchim, Montluçon, France) was used.

Table 1. Parameters of the optimized gradient method for semi-preparative HPLC.

Time (Min)	Water with 0.1% Formic Acid (%)	ACN (%)
0	50	50
7	5	95
20	5	95
21	50	50
25	50	50

The dry residues of DE and EE were dissolved in methanol (5 to 25 mg/mL) and centrifuged (10,000× g, 5 min; ScanSpeed Mini, Labogene, Lillerød, Denmark) to collect the supernatant. The DE and EE were fractionated into two main fractions: Fraction 1 (F1, 2–11 min) and Fraction 2 (F2, 11–20 min), defined as phenol and carboxylic acid fraction and alkylamide fractions, respectively. Fractions were obtained through the eluent collection. Briefly, the supernatants were injected in the LaChrom Merck Hitachi system equipped with a D-7000 Interface, an L-7100 Pump, an L-7200 autosampler, an L-7455 diode array detector (DAD) and an HPLC System Manager HSMD-7000, version 3.0. The chromatographic separation was performed on an Uptisphere WOD homemade semi-preparative column (250 mm × 10 mm, 5 μm, interchrom, Interchim, France). The mobile phase was composed of (A) water containing 0.1% formic acid and (B) ACN (Table 1). Only F2 (12.2–21.5 min) was recovered from DE-R. DE-F was fractionated into F2-i (12–14.6 min), F2-ii (14.6–16 min), and F2-iii (16–20 min). EE-F was fractionated into F1 (2–11 min) and F2 (11–20 min). The organic solvent was evaporated in a rotavapor (R210 Buchi, Switzerland), and then the fractions were freeze-dried (LyoQuest Plus Eco, Telstar, Terrassa, Spain) to remove the water. The crude fractions were stored at −80 °C until further use.

2.2.2. Characterization of Fractions Composition by LC-HRMS Analysis

The LC-HRMS analysis of the fractions was performed according to the method described by Vieira et al. [42]. Briefly, the LC-HRMS analysis was performed on UltiMate 3000 Dionex ultra-high-performance liquid chromatography (UHPLC, Thermo Scientific, Lisbon, Portugal), coupled to an ultra-high-resolution quadrupole–quadrupole time-of-flight (UHR-QqTOF) mass spectrometer (Impact II, Bruker, Lisbon, Portugal). The chromatographic separation was performed on an Acclaim RSLC 120 C18 analytical column (100 mm × 2.1 mm i.d.; 2.2 μm, Dionex, Lisbon, Portugal). The mobile phase was composed of (A) water containing 0.1% formic acid and (B) ACN containing 0.1% formic acid. The gradient program was as follows: 0 min, 95% A; 10 min, 79% A; 14 min, 73% A; 18.3 min, 42% A; 20 min, 0% A; 24 min, 0% A; 26 min, 96% A. The LC-HRMS acquired data were processed using Bruker Compass DataAnalysis 5.1 software (Bruker, Lisbon, Portugal) to extract the mass spectral features from the sample raw data. Echinacoside, chicoric acid, caftaric acid, caffeic acid, chlorogenic acid, cynarin, undeca-2E/Z-ene-8,10-diynoic acid isobutylamide, dodeca-2E-ene-8,10-diynoic acid isobutylamide, dodeca-2E,4E-dienoic acid isobutylamide, and dodeca-2E,4E,8Z,10E/Z-tetraenoic acid isobutylamide were the standards used to confirm the identity of the compounds present in the fractions. The identification of these compounds in the *E. purpurea* fractions was confirmed by their retention time (t_R, min), the mass-to-charge ratio (m/z) of the molecular ion, and MS/MS fragmentation patterns. Supplementary Table S2 presents the characteristics of standards obtained by LC-HRMS. The potential identity of the compound in which t_R and MS data did not match with the available standards were assigned by comparing the MS/MS spectra with the theoretical data MS/MS fragments and data in the literature [44–50].

2.3. Preparation of E. purpurea Extracts and Fractions Solutions

Stock solutions of DE-R, DE-F, and EE-F (30.0 mg/mL) and of F1 and F2 (60.0 mg/mL) were prepared in DMSO. Then, serial dilutions were made with complete RPMI (cRPMI, RPMI-1640 culture medium with 2 mM glutamine supplemented with 10% human serum, 1% penicillin/streptomycin, and 1% HEPES). The final concentrations of the samples in the well were 10, 50, and 100 µg/mL. The fractions were only tested in the highest concentration. The maximum concentration of DMSO in the well (0.33%) did not affect the cell viability.

2.4. Human Monocytes

2.4.1. Ethics Statement

The in vitro experiments involved cells isolated from the peripheral blood of healthy volunteers at the Hospital of Braga, Portugal, approved by the Ethics Subcommittee for Life and Health Sciences (SECVS) of the University of Minho, Braga, Portugal (no. 014/015). Experiments were conducted according to the principles expressed in the Declaration of Helsinki, and participants provided written informed consent.

2.4.2. Monocyte Isolation and Differentiation

Monocytes were isolated from the PBMCs of three different donors, as previously described by Gonçalves et al. [51]. Briefly, PMBCs were first subjected to a density gradient centrifugation using a Histopaque-1077 solution. The PBMC ring was carefully collected and washed twice with PBS. Then, the monocytes were isolated from PBMCs using positive magnetic bead separation with CD14 microbeads, according to the manufacturer's instructions. Isolated monocytes were resuspended in cRPMI. After, monocytes were seeded at a density of 1×10^6 cells/mL in adherent 24-well culture plates for 7 days in the presence of 20 ng/mL of recombinant human GM-CSF, at 37 °C, in a humidified atmosphere of 5% CO_2. The culture medium was replaced every 3 days, and the acquisition of macrophage morphology was confirmed by visualization under an inverted microscope (Axiovert 40, Zeiss, Göttingen, Germany).

2.4.3. Evaluation of Anti-Inflammatory Activity

The hMDMs were stimulated with 100 ng/mL of LPS in a fresh cRPMI medium. After 2 h, all *E. purpurea* extracts and fractions, at different concentrations (see Section 2.3), were added to the LPS-stimulated hMDMs and incubated for 22 h at 37 °C, in a humidified atmosphere of 5% CO_2. Afterward, the culture medium was harvested (the triplicates were mixed and homogenized) and stored, aliquoted at −80 °C until cytokine quantification. The cells were washed with warm sterile DPBS, and the metabolic activity and DNA content were determined (see Section 2.4.4). Controls containing the same percentage of DMSO (see Section 2.3) in the maximal concentration of extracts/fractions were also tested. hMDM cultures stimulated or not with LPS were used as negative and positive controls for the production of pro-inflammatory mediators, respectively. Dexamethasone and celecoxib, prepared in ethanol (20 mM) and diluted with cRPMI (10 µM in the well), were used as positive controls of inhibition of the production of the pro-inflammatory mediators.

2.4.4. Metabolic Activity and DNA Quantification

The metabolic activity of hMDM incubated with *E. purpurea* extracts and fractions was determined by the reduction of resazurin (blue) to resorufin (pink) by living macrophages using the alamarBlue assay [43]. These results are expressed in percentages related to the positive control.

The DNA concentration of macrophages was quantified using a fluorometric dsDNA quantification kit, according to the instructions of the manufacturer, as previously described by Vieira et al. [43]. DNA contents are expressed in relative concentrations of the positive control.

2.4.5. Cytokine Measurement

The amounts of IL-6 and TNF-α were assayed using ELISA kits, according to the instructions of the manufacturer. The obtained values were normalized by the respective DNA concentration. The results are expressed in percentage relative to the positive control.

2.4.6. Cellular ROS/RNS/$O_2^{\bullet-}$ Detection Assay

Oxidative stress in the presence or absence of *E. purpurea* extracts and fractions was investigated using a cellular ROS/$O_2^{\bullet-}$ detection assay kit, as previously described by Vieira et al. [42]. Briefly, LPS-stimulated hMDMs were treated with *E. purpurea* extracts and fractions, at 100 µg/mL, as mentioned before (see Section 2.4.3). After, the hMDMs were washed and labeled with the oxidative stress detection reagent (green, Ex/Em 490/525 nm) for the determination of total ROS/RNS, and $O_2^{\bullet-}$ detection reagent (orange, Ex/Em 550/620 nm) for 1 h at 37 °C in the dark. These nonfluorescent detection reagents diffuse into cells, where they can be oxidized by ROS/RNS and $O_2^{\bullet-}$, converting to fluorescent probes. Then, the cells were fixed with 10% of formalin and the nucleus was labeled with DAPI in a ratio of 1:1000 in DPBS, for 10 min. The fluorescent samples were analyzed using a Fluorescence Inverted Microscope with Incubation (Axio Observer, Zeiss, Germany). The fluorescence intensity values, analyzed using ImageJ software (version 1.52a, Wayne Rasband, National Institutes of Health, Bethesda, MD, USA), were normalized against the number of nuclei. Changes in the fluorescence intensity relative to the positive control were related to an increase or decrease in the generation of intracellular ROS/RNS and/or $O_2^{\bullet-}$.

2.4.7. Western Blot Analysis

LPS-stimulated hMDMs (5×10^5/well in 24-well plates) were treated with *E. purpurea* extracts and fractions, at 100 µg/mL, as previously described (see Section 2.4.3). After 24 h, the medium was removed, and the cells were washed with ice DPBS. Then, the cells were lysed in RIPA buffer containing a mixture of protease and phosphatase inhibitors at 4 °C for 30 min under shaking. Samples were collected and centrifuged (2000 rpm, 20 min). The supernatant was transferred to a new Eppendorf flask and the protein content was determined using the Bio-Rad Protein Assay, based on the method of Bradford. Bolt sample reducing agent and bolt LDS sample buffer were added to 30 µg of protein. Then, the samples were heated and denatured at 70 °C (20 min) and 95 °C (5 min). The centrifuged samples were loaded and separated on 8% precast polyacrylamide gels set on a Mini Gel Tank (Invitrogen, Thermo Fisher Scientific, Lisbon, Portugal). The proteins were transferred from the gel to a PVDF membrane using the iBlot 2 Gel Transfer Device (Invitrogen, Thermo Fisher Scientific, Lisbon, Portugal).

After blocking for 30 min at RT with 5% BSA in Tris-buffered saline Tween 20 (TBST), the membranes were incubated overnight at 4 °C with the following primary antibodies diluted in blocking solution: rabbit NF-κB p65 (1:1000), rabbit p44/42 MAPK (ERK1/2; 1:1000), rabbit p38 MAPK (1:1000), rabbit SAPK/JNK (1:1000), rabbit STAT3 (1:1000), rabbit COX-2 (1:500), rabbit iNOS (1:500), rabbit phospho-NF-κB p65 (1:1000), rabbit phospho-p38 MAPK (1:1000), rabbit phospho-STAT3 (1:2000), rabbit phospho-SAPK/JNK (1:1000), rabbit phospho-p44/42 MAPK (p-ERK1/2; 1:1000), and rabbit GAPDH (1:10,000). Afterwards, the membranes were washed three times for 5 min with TBST, and then IRDye 800CW Goat anti-Rabbit IgG or IRDye 680RD Goat anti-Rabbit IgG secondary antibodies, both diluted in TBST (1:15,000), were added and the samples were incubated for 1 h at RT in the dark. The Odyssey Fc Imaging System (LI-COR Inc., 2800, Lincoln, NE, USA) was used for image acquisition of the Western blots using near-infrared wavelengths of 700 or 800 nm. The intensity of the bands was quantified with Image Studio software (LI-COR, Inc. software version, Lincoln, NE, USA). The data were normalized to the housekeeping GAPDH.

2.5. Statistical Analysis

Results are expressed as mean ± standard deviation (SD) of 3 independent experiments with a minimum of 3 replicates for each condition. Statistical analyses were performed using GraphPad Prism 8.0.1 software (Boston, MA, USA). Two-way analysis of variance (ANOVA) and Dunnett's multiple comparisons or Sidak's multiple comparisons test was used for cell assays. Differences between experimental groups were considered significant with a confidence interval of 99% when $p < 0.01$.

3. Results

3.1. Fractionation of the E. purpurea Extracts

The optimized analytical method led to the successful separation of the ten studied standards (Figure 2A-i,ii). It was possible to clearly distinguish between phenols and alkylamides. The phenols, due to their high polarity, eluted first under reversed-phase conditions (from ≈ 5 to 28 min), while alkylamides, which are less polar, eluted later (from ≈ 30 to 37 min). To fractionate the extracts, a semi-preparative HPLC method was employed. The analytical method conditions were optimized to reduce the run time and increase the injection volume while maintaining the baseline separation of the two different fractions of interest. In the chromatogram obtained from the standard mixture (Figure 2B-i,ii), it is possible to observe a robust gap between phenol/acids and alkylamide fractions (from 10.6 min to 12.2 min), ensuring the successful separation between these two types of compounds.

The whole *E. purpurea* extracts were fractioned into F1 (phenol and carboxylic acid fraction) and F2 (alkylamide fraction). Different chromatogram profiles were observed for DE-R, DE-F, and EE-F (Figure 3). Accordingly, DE-R and DE-F showed higher absorbance values for alkylamides than phenols/carboxylic acids (Figure 3A-i,ii,B-i,ii). Moreover, the first extract seems to be more enriched with alkylamides. EE-F also presented phenols/carboxylic acids and alkylamides in their composition (Figure 3C-i,ii). Based on their chromatographic profiles, the *E. purpurea* extracts were fractionated. As phenols/carboxylic acids were not detected in DE-R, only the F2 fraction was harvested. DE-F presented defined alkylamide peaks, being possible their fractionation into three sections: F2 i showing three clear peaks, F2 ii presenting one perfect peak, and F2 iii displaying three main peaks. Finally, EE-F was fractionated into F1 and F2.

3.2. Chemical Composition of the E. purpurea Fractions

The identification of the bioactive compounds present in the *E. purpurea* fractions was performed by LC-HRMS (Table 2). Both product ions and relative intensities for standard fragments perfectly matched those obtained for the compounds in *E. purpurea* fractions. Supplementary Tables S3 and S4 include the retention times (t_R), the precursor ions, and the product ions for phenols/carboxylic acids and alkylamides, respectively. Each extract and fraction exhibited different patterns of phenols/carboxylic acids and alkylamides. The identification of phenols/carboxylic acid compounds and alkylamides in whole DE-R, DE-F, and EE-F was comparable to our previous study [42]. Five phenols/carboxylic acids and twenty-three alkylamides were identified in all *E. purpurea* fractions. As expected, F1 only presented phenols/carboxylic acids, while alkylamides are just observed in F2. EE-F-F1 exhibited five phenols/carboxylic acids in its composition. DE-R-F2 presented the highest number of identified alkylamides (19 compounds), followed by EE-F-F2 (18 compounds), DE-F-F2 i (10 compounds), and DE-F-F2 ii and DE-F-F2 iii (4 compounds).

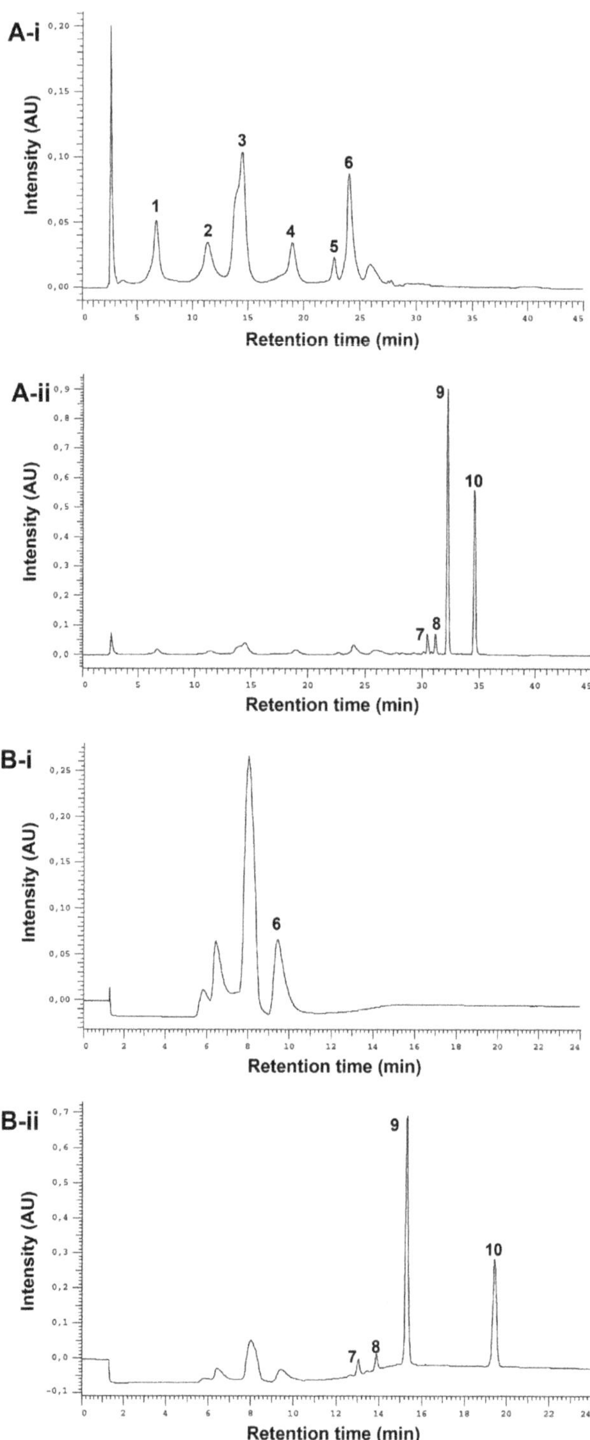

Figure 2. Analytical (**A**) and semi-preparative HPLC chromatograms (**B**) of standard mixture detected at 330 nm (phenols, **i**) and 254 nm (alkylamides, **ii**). Peak numbering on the chromatograms refers

to the following compounds: 1—caftaric acid (t_R-A = 6.77 min); 2—chlorogenic acid (t_R-A = 11.36 min); 3—caffeic acid (t_R-A = 14.51 min); 4—cynarin (t_R-A = 18.99 min); 5—echinacoside (t_R-A = 22.72 min); 6—chicoric acid (t_R-A = 24.03 min, t_R-B = 9.47 min); 7—undeca-2E/Z-ene-8,10-diynoic acid isobutylamide (t_R-A = 30.48 min, t_R-B = 13.07 min); 8—dodeca-2E-ene-8,10-diynoic acid isobutylamide (t_R-A = 31.17 min, t_R-B = 13.89 min); 9—dodeca-2E,4E,8Z,10E/Z-tetraenoic acid isobutylamide (t_R-A = 32.24 min, t_R-B = 15.33 min); 10—dodeca-2E,4E-dienoic acid isobutylamide (t_R-A = 34.59 min, t_R-B = 19.44 min).

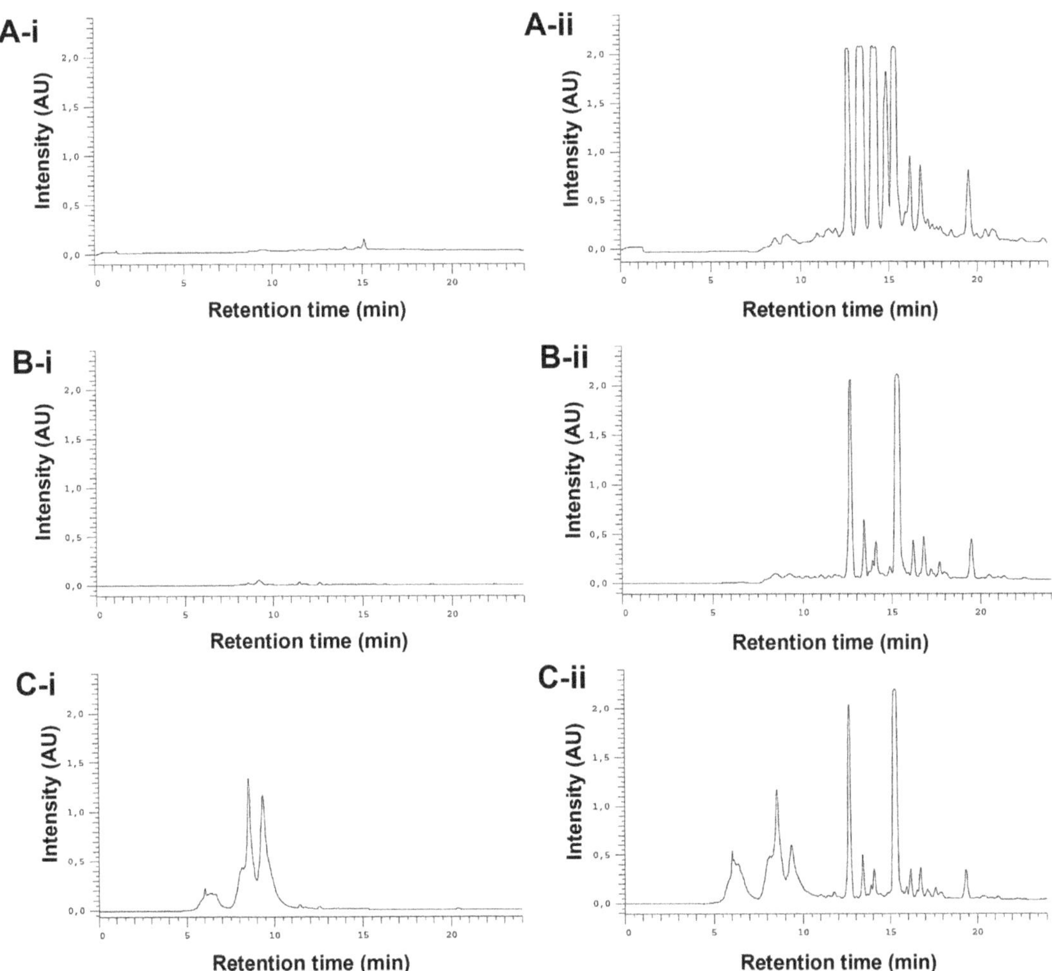

Figure 3. Semi-preparative HPLC chromatograms of DE-R (**A**), DE-F (**B**), and EE-F (**C**) at 19–24.6 µg/mL (100 µL), detected at 330 nm (phenols/carboxylic acids; **A-i,B-i,** and **C-i**) and 254 nm (alkylamides; **A-ii, B-ii,** and **C-ii**).

Table 2. Overview of the identified compounds (phenols/carboxylic acids and alkylamides) in *E. purpurea* extracts and fractions by LC-HRMS.

Compounds	DE		EE	DE-R	DE-F			EE-F	
	R	F	F	F2	F2 i	F2 ii	F2 iii	F1	F2
Malic Acid	+	-	+	-	-	-	-	-	-
Vanillic acid	-	-	+	-	-	-	-	-	-
Protocatechuic acid	-	-	+	-	-	-	-	+	-
Caftaric acid [a]	-	-	+	-	-	-	-	-	-
Chlorogenic acid [a]	-	-	+	-	-	-	-	+	-
Quinic acid	-	-	-	-	-	-	-	-	-
Vanillin	-	-	-	-	-	-	-	-	-
Caffeic acid [a]	+	+	+	-	-	-	-	+	-
Benzoic acid	+	+	+	-	-	-	-	-	-
Cynarin [a]	-	-	-	-	-	-	-	-	-
Echinacoside [a]	-	-	-	-	-	-	-	-	-
p-coumaric acid	+	-	+	-	-	-	-	-	-
Chicoric acid [a]	-	+	+	-	-	-	-	+	-
Rutin	-	-	+	-	-	-	-	+	-
Quercetin	-	-	+	-	-	-	-	-	-
Dodeca-2E,4Z,10E-triene-8-ynoic acid isobutylamide	+	+	+	+	+	-	-	-	+
Dodeca-2E,4Z,10Z-triene-8-ynoic acid isobutylamide	+	+	+	+	+	-	-	-	+
Dodeca-2,4,10-triene-8-ynoic acid isobutylamide (isomer 1)	+	+	-	-	-	+	-	-	-
Dodeca-2E,4E,10Z-triene-8-ynoic acid isobutylamide	+	+	+	+	-	-	-	-	+
Dodeca-2Z,4E,10Z-triene-8-ynoic acid isobutylamide	+	-	-	+	-	-	-	-	-
Dodeca-2E,4E,10E-triene-8-ynoic acid isobutylamide	+	+	+	+	+	-	-	-	+
Undeca-2E,4Z-diene-8,10-diynoic acid isobutylamide	+	+	+	+	+	-	-	-	+
Undeca-2E/Z-ene-8,10-diynoic acid isobutylamide[a]	-	+	+	-	+	-	-	-	+
Undeca-2Z,4E-diene-8,10-diynoic acid isobutylamide	+	-	-	+	-	-	-	-	-
Undeca-2E/Z,4Z/E-diene-8,10-diynoic acid 2-methylbutylamide	-	-	-	-	-	-	-	-	-
Pentadeca-2E,9Z-diene-12,14-diynoic acid 2-hydroxyisobutylamide	-	+	+	-	-	-	-	-	+
Dodeca-2E,4Z-diene-8,10-diynoic acid isobutylamide	+	+	+	+	+	-	-	-	+
Undeca-2E,4E-diene-8,10-diynoic acid isobutylamide	+	-	-	-	-	-	-	-	-

Table 2. *Cont.*

Compounds	DE		EE	DE-R	DE-F			EE-F	
	R	F	F	F2	F2 i	F2 ii	F2 iii	F1	F2
Dodeca-2Z,4E-diene-8,10-diynoic acid isobutylamide	-	-	-	-	-	-	-	-	-
Dodeca-2E-ene-8,10-diynoic acid isobutylamide [a]	+	+	+	+	+	-	-	-	+
Trideca-2E,7Z-diene-10,12-diynoic acid isobutylamide	+	+	+	+	+	-	-	-	+
Dodeca-2,4-diene-8,10-diynoic acid 2-methylbutylamide	+	+	+	+	+	-	-	-	+
Dodeca-2Z,4Z,10Z-triene-8-ynoic acid isobutylamide	+	-	-	+	-	-	-	-	-
Trideca-2E,7Z-diene-10,12-diynoic acid 2-methylbutylamide	+	+	+	+	-	+	-	-	+
Dodeca-2E,4E,8Z,10E/Z-tetraenoic acid isobutylamide [a]	+	+	+	+	-	+	-	-	+
Dodeca-2E,4Z,10E-triene-8-ynoic acid 2-methylbutylamide OR Dodeca-2E-ene-8,10-diynoic acid 2-methylbutylamide	+	+	+	+	+	-	-	-	+
Dodeca-2E,4E,8Z-trienoic acid isobutylamide (isomer 1)	-	+	+	-	-	-	+	-	-
Dodeca-2E,4E-dienoic acid isobutylamide (isomer 1)	-	-	-	-	-	-	-	-	-
Pentadeca-2E,9Z-diene-12,14-diynoic acid isobutylamide	+	+	+	+	-	+	-	-	+
Dodeca-2E,4E,8Z-trienoic acid isobutylamide	+	+	+	+	-	-	+	-	+
Trideca-2Z,7Z-diene-10,12-diynoic acid 2-methylbutylamide	+	-	-	-	-	-	-	-	-
Dodeca-2E,4E,8Z,10E/Z-tetraenoic acid 2-methylbutylamide	+	+	+	+	-	-	+	-	+
Hexadeca-2E,9Z-diene-12,14-diynoic acid isobutylamide	+	-	-	-	-	-	-	-	-
Dodeca-2E,4E,8Z-trienoic acid isobutylamide (isomer 2)	+	-	-	-	-	-	-	-	-
Dodeca-2E,4E-dienoic acid isobutylamide [a]	+	+	+	+	-	-	+	-	+

DE: dichloromethanolic extracts; EE: ethanolic extracts; R: roots; F: flowers; F1: phenol/carboxylic acid fraction; F2: alkylamide fraction; symbol "+" represents the presence of compound; symbol "-" represents the absence of compound. [a] Injected standards. E/Z stereochemistry is indicated here in accordance to existing literature [44–50], but it should be highlighted that without NMR spectra, it is not possible to conclusively distinguish between E and Z isomers.

DE-R-F2 presented the following alkylamides: dodeca-2E,4Z,10E-triene-8-ynoic acid isobutylamide, dodeca-2E,4Z,10Z-triene-8-ynoic acid isobutylamide, dodeca-2E,4E,10Z-triene-8-ynoic acid isobutylamide, dodeca-2Z,4E,10Z-triene-8-ynoic acid isobutylamide, dodeca-2E,4E,10E-triene-8-ynoic acid isobutylamide, undeca-2E,4Z-diene-8,10-diynoic acid isobutylamide, undeca-2Z,4E-diene-8,10-diynoic acid isobutylamide, dodeca-2E,4Z-diene-8,10-diynoic acid isobutylamide, dodeca-2E-ene-8,10-diynoic acid isobutylamide, trideca-2E,7Z-diene-10,12-diynoic acid isobutylamide, dodeca-2,4-diene-8,10-diynoic acid

2-methylbutylamide, dodeca-2Z,4Z,10Z-triene-8-ynoic acid isobutylamide, trideca-2E,7Z-diene-10,12-diynoic acid 2-methylbutylamide, dodeca-2E,4E,8Z,10E/Z-tetraenoic acid isobutylamide, dodeca-2E,4Z,10E-triene-8-ynoic acid 2-methylbutylamide or dodeca-2E-ene-8,10-diynoic acid 2-methylbutylamide, pentadeca-2E,9Z-diene-12,14-diynoic acid isobutylamide, dodeca-2E,4E,8Z-trienoic acid isobutylamide, dodeca-2E,4E,8Z,10E/Z-tetraenoic acid 2-methylbutylamide, and dodeca-2E,4E-dienoic acid isobutylamide.

The fractionation of DE-F originated three different alkylamide fractions, namely (i) DE-F-F2 i composed of dodeca-2E,4Z,10E-triene-8-ynoic acid isobutylamide, dodeca-2E,4Z,10Z-triene-8-ynoic acid isobutylamide, dodeca-2E,4E,10E-triene-8-ynoic acid isobutylamide, undeca-2E,4Z-diene-8,10-diynoic acid isobutylamide, undeca-2E/Z-ene-8,10-diynoic acid isobutylamide, dodeca-2E,4Z-diene-8,10-diynoic acid isobutylamide, dodeca-2E-ene-8,10-diynoic acid isobutylamide, trideca-2E,7Z-diene-10,12-diynoic acid isobutylamide, dodeca-2,4-diene-8,10-diynoic acid 2-methylbutylamide, and dodeca-2E,4Z,10E-triene-8-ynoic acid 2-methylbutylamide or dodeca-2E-ene-8,10-diynoic acid 2-methylbutylamide; (ii) DE-F-F2 ii constituted by dodeca-2,4,10-triene-8-ynoic acid isobutylamide isomer 1, trideca-2E,7Z-diene-10,12-diynoic acid 2-methylbutylamide, dodeca-2E,4E,8Z,10E/Z-tetraenoic acid isobutylamide, and pentadeca-2E,9Z-diene-12,14-diynoic acid isobutylamide; and (iii) DE-F-F2 iii that presented dodeca-2E,4E,8Z-trienoic acid isobutylamide isomer 1, dodeca-2E,4E,8Z-trienoic acid isobutylamide, dodeca-2E,4E,8Z,10E/Z-tetraenoic acid 2-methylbutylamide, and dodeca-2E,4E-dienoic acid isobutylamide. It is important to highlight that none of the alkylamides was repeated in each sub-fraction of DE-F-F2, empathizing the efficiency of the separation method.

EF-F1 was composed of protocatechuic acid, chlorogenic acid, caffeic acid, chicoric acid, rutin, and rutin derivative. EF-F2 was constituted by dodeca-2E,4Z,10E-triene-8-ynoic acid isobutylamide, dodeca-2E,4Z,10Z-triene-8-ynoic acid isobutylamide, dodeca-2E,4E,10Z-triene-8-ynoic acid isobutylamide, dodeca-2E,4E,10E-triene-8-ynoic acid isobutylamide, undeca-2E,4Z-diene-8,10-diynoic acid isobutylamide, undeca-2E/Z-ene-8,10-diynoic acid isobutylamide, pentadeca-2E,9Z-diene-12,14-diynoic acid 2-hydroxyisobutylamide, dodeca-2E,4Z-diene-8,10-diynoic acid isobutylamide, dodeca-2E-ene-8,10-diynoic acid isobutylamide, trideca-2E,7Z-diene-10,12-diynoic acid isobutylamide, dodeca-2,4-diene-8,10-diynoic acid 2-methylbutylamide, trideca-2E,7Z-diene-10,12-diynoic acid 2-methylbutylamide, dodeca-2E,4E,8Z,10E/Z-tetraenoic acid isobutylamide, dodeca-2E,4Z,10E-triene-8-ynoic acid 2-methylbutylamide or dodeca-2E-ene-8,10-diynoic acid 2-methylbutylamide, pentadeca-2E,9Z-diene-12,14-diynoic acid isobutylamide, dodeca-2E,4E,8Z-trienoic acid isobutylamide, dodeca-2E,4E,8Z,10E/Z-tetraenoic acid 2-methylbutylamide, and dodeca-2E,4E-dienoic acid isobutylamide.

3.3. Cytotoxicity of E. purpurea Extracts and Fractions

The metabolic activity and the relative DNA concentration of LPS-stimulated hMDM in the absence or presence of *E. purpurea* extracts and fractions at different concentrations are presented in Figure 4. The cell metabolic activity and the DNA concentration were not affected by the presence of the DE, EE, and fractions at any tested concentration (Figure 4).

3.4. Anti-Inflammatory Activity of E. purpurea Extracts and Fractions
3.4.1. Cytokine Production

The anti-inflammatory activity of *E. purpurea* extracts and fractions was evaluated by the decreased amounts of pro-inflammatory cytokines, namely IL-6 and TNF-α, in the cell culture supernatant of LPS-stimulated hMDM (Figure 5). Non-stimulated hMDM produced basal amounts of IL-6 (8.0 ± 10.5 pg/mL) and TNF-α (60.0 ± 36.0 pg/mL). As expected, LPS stimulation of hMDM led to a significant increase in the levels of these pro-inflammatory cytokines (IL-6: $19{,}139.0 \pm 7850.8$ pg/mL, TNF-α: $21{,}773.4 \pm 9425.9$ pg/mL). Dexamethasone (10 μM) effectively reduced the IL-6 and TNF-α production by $51.2 \pm 6.5\%$ and $38 \pm 5.7\%$, respectively (Figure 5). As expected, celecoxib (10 μM) did not considerably decrease the IL-6 and TNF-α production ($21.2 \pm 17.2\%$ and $4.6 \pm 1.8\%$, respectively).

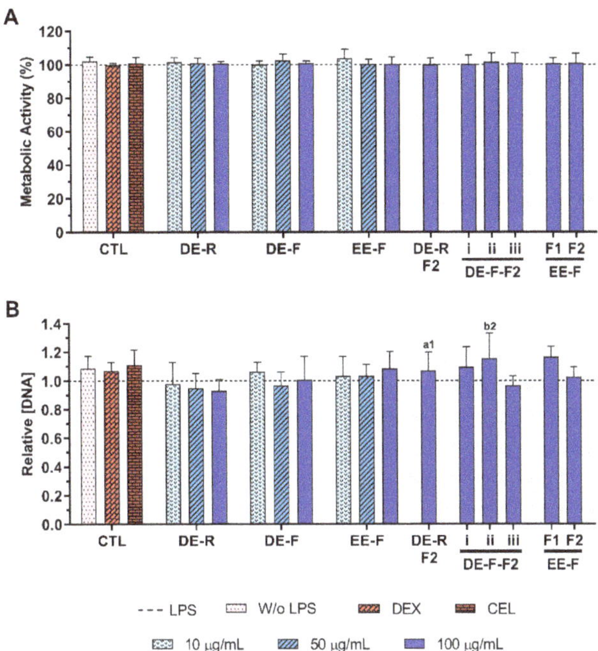

Figure 4. Metabolic activity (**A**) and relative DNA concentration (**B**) of LPS-stimulated human monocyte-derived macrophages (hMDMs) cultured in the presence of different concentrations of the *E. purpurea* extracts, fractions, and clinically used anti-inflammatory drugs (dexamethasone (DEX) and celecoxib (CEL), 10 µM) for 24 h. The dotted line represents the metabolic activity and DNA concentration of positive control (LPS-stimulated hMDM without treatment). There are no statistically significant differences in comparison to the positive control for each tested extract, fraction, DEX, and CEL. Statistically significant differences are 1 ($p < 0.0133$) and 2 ($p < 0.0035$) in comparison with a (DE-R vs. DE-R-F2) and b (DE-F vs. DE-F-F2) at the same concentration. CTL: control; DE: dichloromethanolic extracts; EE: ethanolic extracts; R: roots; F: flowers; F1: phenol/-carboxylic acid fraction; F2: alkylamide fraction.

When LPS-stimulated hMDMs were treated with the whole *E. purpurea* extracts, a significant decrease in the IL-6 amount in the culture supernatant was observed in a concentration-dependent manner (Figure 5A). Particularly, DE showed a higher ability to decrease IL-6 and TNF-α levels than EE. Moreover, the extracts obtained from roots more significantly reduced these two pro-inflammatory cytokines in comparison with the ones obtained from flowers. Indeed, 50 µg/mL of DE-R efficiently decreased the IL-6 production, being even more effective at 100 µg/mL (69.5 ± 10.0%). DE-F was only able to significantly decrease the IL-6 levels by 47.3 ± 6.2% at 100 µg/mL. A comparable significant IL-6 reduction was observed for EE-F over all tested concentrations, with the highest tested concentration displaying greater activity (35.2 ± 12.1%). DE-R was ≈1.5 and 2 times stronger than DE-F and EE-F, respectively. As extracts, all the fractions strongly decreased the IL-6 production, except the EE-F-F1 (25.1 ± 15.8%). Moreover, the bioactivity of DE-F and EE-F was significantly improved with their fractionation into F2. DE-R-F2 led to a more marked IL-6 reduction (84.3 ± 9.1%). DE-F-F2 i, DE-F-F2 ii, DE-F-F2 iii, and EE-F-F2 demonstrated similar bioactivity (62.7 ± 11.5%, 71.2 ± 12.3%, 68.5 ± 14.3%, and 71.6 ± 6.1%, respectively). Besides DE-R-F2 exhibiting a higher efficacy in IL-6 reduction, no significant differences were observed. Analyzing all the *E. purpurea* extracts and fractions, DE-R-F2 led to the strongest reduction in IL-6 production, followed by EE-F-F2 ≈ DE-F-F2 ii, DE-R ≈ DE-F-F2 iii, DE-F-F2 i, DE-F, EE-F, and EE-F-F1. Moreover, DE-R, DE-F, and all the F2 fractions demonstrated similar or higher bioactivity than dexamethasone.

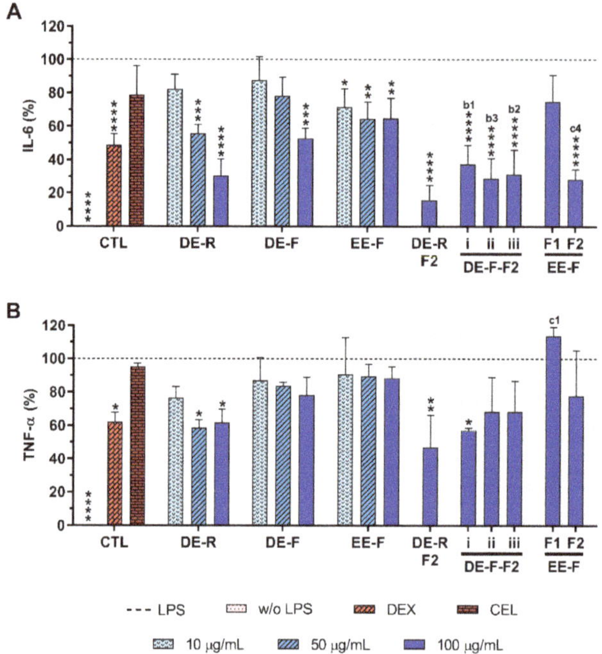

Figure 5. IL-6 (**A**) and TNF-α (**B**) production by LPS-stimulated human monocyte-derived macrophages (hMDMs) cultured in the presence of different concentrations of the *E. purpurea* extracts, fractions, and clinically used anti-inflammatory drugs (dexamethasone (DEX) and celecoxib (CEL), 10 µM) for 24 h. Statistically significant differences are * ($p < 0.0363$), ** ($p < 0.0047$), *** ($p < 0.0003$), and **** ($p < 0.0001$) in comparison to the positive control (LPS-stimulated hMDM without treatment) for all tested *E. purpurea* extracts, fractions, DEX, and CEL and 1 ($p < 0.0428$), 2 ($p < 0.0018$), 3 ($p < 0.0003$), and 4 ($p < 0.0001$) in comparison with b (DE-F vs. DE-R-F2) and c (EE-F vs. EE-F-F) at the same concentration. CTL: control; DE: dichloromethanolic extracts; EE: ethanolic extracts; R: roots; F: flowers; F1: phenol/carboxylic acid fraction; F2: alkylamide fraction.

Some extracts and fractions were also able to significantly decrease the TNF-α levels in LPS-stimulated hMDM cultures (Figure 5B). DE-R, at 50 µg/mL, significantly decreased the TNF-α production by 41.4 ± 4.9%. Conversely, DE-F and EE-F did not demonstrate the capacity to markedly decrease the TNF-α production (21.7 ± 10.5% and 11.6 ± 6.9%, respectively). DE-R was ≈1.7 and 3.3 times stronger than DE-F and EE-F, respectively. Regarding the fractions, only DE-R-F2 and DE-F-F2 i significantly reduced the TNF-α production by 53.1 ± 19.5% and 42.7 ± 1.4%, respectively. Indeed, the bioactivity of the DE-R and DE-F extracts was not significantly improved with their fractionation. DE-F-F2 ii, DE-F-F2 iii, and EE-F-F2 did not show an ability to significantly reduce the TNF-α production (31.6 ± 20.6%, 31.6 ± 18.2%, and 22.1 ± 27.0%, respectively). EE-F-F1 increased the TNF-α amount by 13.9 ± 5.7% in comparison to LPS-stimulated hMDM. Comparing the data obtained for all *E. purpurea* extracts and fractions, it is possible to conclude that DE-R-F2 exhibited the strongest reduction in TNF-α production, followed by DE-F-F2 i, DE-R, DE-F-F2 ii ≈ DE-F-F2 iii, EE-F-F2, DE-F, EE-F, and EE-F1. Moreover, DE-R, DE-R-F2, and DE-F-F2 demonstrated similar or higher bioactivity than dexamethasone.

3.4.2. ROS/RNS/$O_2^{\bullet-}$ Generation

The reduction in the intracellular levels of ROS/RNS and $O_2^{\bullet-}$ in LPS-stimulated hMDM incubated with *E. purpurea* extracts and fractions at the maximal tested concentration are present in Figure 6 and Supplementary Figures S1 and S2. Non-stimulated hMDM produced basal levels of ROS and $O_2^{\bullet-}$, which were significantly increased by the stimulation with LPS (Figure 6 and Supplementary Figure S1). Dexamethasone (10 μM) effectively reduced the ROS/RNS generation, but no differences were observed with the positive control in the reduction in $O_2^{\bullet-}$ (Figure 6). Conversely, celecoxib (10 μM) considerably decreased both intracellular ROS/RNS and $O_2^{\bullet-}$ generation.

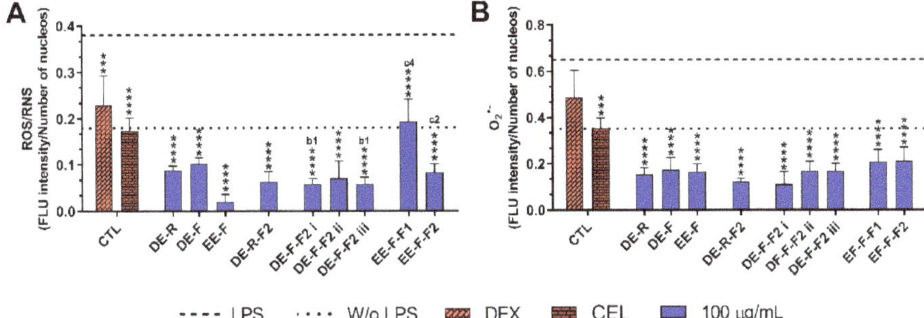

Figure 6. Fluorescence intensity of intracellular ROS/RNS (**A**) and $O_2^{\bullet-}$ (**B**) production in LPS-stimulated human monocyte-derived macrophages (hMDMs) in the absence or presence of *E. purpurea* extracts, fractions, and clinically used anti-inflammatory drugs (dexamethasone (DEX) and celecoxib (CEL), 10 μM) cultured for 24 h. Fluorescence intensity was measured using ImageJ software. The dotted line represents the basal levels of ROS/RNS and $O_2^{\bullet-}$ in non-stimulated hMDM (negative control) and the dashed line corresponds to the amounts of ROS/RNS and $O_2^{\bullet-}$ produced by LPS-stimulated hMDM (positive control). Statistically significant differences are *** ($p < 0.0002$) and **** ($p < 0.0001$) in comparison to the positive control (LPS-stimulated hMDM without treatment) for each tested *E. purpurea* extract, fraction, DEX, and CEL, as well as 1 ($p < 0.0453$), 2 ($p < 0.0023$), and 4 ($p < 0.0001$) in comparison with b (DE-F vs. DE-R-F2) and c (EE-F vs. EE-F-F) at the same concentration. CTL: control; DE: dichloromethanolic extracts; EE: ethanolic extracts; R: roots; F: flowers; F1: phenol/carboxylic acid fraction; F2: alkylamide fraction.

The treatment of LPS-stimulated hMDM with the *E. purpurea* extracts drastically decreased the intracellular levels of ROS/RNS (Figure 6A and Supplementary Figure S2). EE-F demonstrated a strong capacity to decrease intracellular ROS/RNS production. EE-F was ≈4 and 5 times stronger than DE-R and DE-F, respectively. Regarding the fractions, all of them strongly decreased the intracellular ROS/RNS production. The fractionation of DE-F into DE-F-F2 i and DE-F-F2 iii significantly enhanced the reduction in intracellular ROS/RNS generation. However, the same trend was not observed for EE-F and its fractions. Indeed, the antioxidant activity was markedly decreased with the fractionation of EE-F into EE-F-F1 and EE-F-F2. DE-R-F2 and DE-F-F2 ii seem to present greater bioactivity than the whole extract, but no significant differences were observed. Analyzing all the *E. purpurea* extracts and fractions, EE-F demonstrated the most potent bioactivity, followed by DE-F-F2 i ≈ DE-F-F2 iii, DE-R-F2, DE-F-F2 ii, DE-R ≈ EE-F-F2, DE-F, and EE-F-F1. Moreover, all extracts and fractions were able to reestablish or decrease the levels to those observed in the non-stimulated hMDM.

As observed for ROS/RNS, the treatment of LPS-stimulated hMDM with the *E. purpurea* extracts significantly decreased the intracellular levels of $O_2^{\bullet-}$ (Figure 6B and Supplementary Figure S2). The three extracts showed similar bioactivity in the reduction in $O_2^{\bullet-}$ amounts. Additionally, all the fractions were also able to reduce the intracellular $O_2^{\bullet-}$ a generation with comparable efficacy. Consequently, the fractionation of the extracts did not significantly improve their ability to reduce the intracellular $O_2^{\bullet-}$ generation.

Comparing all *E. purpurea* extracts and fractions, DE-F-F2 i exhibited the most powerful activity, followed by DE-R-F2, DE-R, EE-F, DE-F-F2 ii $\approx$ DE-F-F2 iii, DE-F, EE-F-F1, and EE-F-F2. Moreover, LPS-stimulated hMDM in the presence of all the extracts and fractions were able to reach similar or inferior levels of intracellular $O_2^{\bullet-}$ to those observed in the non-stimulated macrophages.

3.4.3. Therapeutic Targets

To understand the therapeutic targets responsible for the anti-inflammatory activity of the *E. purpurea* extracts and fractions, several pro-inflammatory signaling pathways were investigated by Western blot (Figure 7). Non-stimulated hMDM showed basal levels of ERK 1/2 (Figure 7A), p38 (Figure 7B), and NF-κB p65 phosphorylation (Figure 7C), but COX-2 (Figure 7D) and STAT3 (Figure 7E) expressions were not observed. The phosphorylation of all the studied inflammatory proteins was significantly enhanced in LPS-stimulated hMDM. Dexamethasone (10 μM) was able to significantly decrease the phosphorylation of all studied inflammatory proteins. Celecoxib (10 μM) also significantly reduced the phosphorylation of p38, STAT3, and the expression of COX-2, but no significant activity was observed for ERK 1/2 and NF-κB p65.

When LPS-stimulated hMDMs were treated with the whole *E. purpurea* extracts, a marked decrease in the activation of the ERK 1/2 signaling pathway was observed (Figure 7A). DE-R also efficiently decreased the phosphorylation of ERK 1/2, being its activity $\approx$8 and 9.5 times higher than DE-F and EE-F, respectively. As observed in extracts, the fractions reduced the phosphorylation of ERK 1/2, but DE-R-F2 and EE-F-F2 strongly suppressed the phosphorylation of this inflammatory protein. DE-F-F2 and EE-F-F1 demonstrated similar activity. Although F2 exhibited stronger bioactivity, no significant differences were observed compared to the whole extracts. Analyzing all the *E. purpurea* extracts and fractions, DE-R-F2 strongly suppressed the ERK 1/2 signaling pathway, followed by DE-R, EE-F-F2, DE-F-F2 iii, DE-F-F2 ii, DE-F-F2 i, DE-F, EE-F, and EE-F-F1. Moreover, LPS-stimulated hMDM in the presence of DE-R, DE-R-F2, DE-F-F2 ii, DE-F-F2 iii, and EE-F-F2 reached similar or lower levels of ERK 1/2 phosphorylation than non-stimulated macrophages.

Only DE-R was able to significantly reduce the activity of the p38 signaling pathway, being its bioactivity $\approx$2 and 2.7 times stronger than DE-F and EE-F, respectively (Figure 7B). DE-F and EE-F also led to decreased p38 phosphorylation, but no significant differences were observed. All the fractions significantly reduced the phosphorylation of p38, being DE-R-F2, DE-F-F2 iii, and EE-F-F1 the most promising. DE-F-F2 i, DE-F-F2 ii, and EE-F-F2 exhibited comparable bioactivity. Only the fractionation of EE-F into EE-F-F1 markedly reduced the p38 signaling pathway. Although other fractions presented increased bioactivity, no significant differences were observed in comparison with the whole extract. The comparison of all the *E. purpurea* extracts and fractions demonstrates a strong potential for DE-R-F2, DE-F-F2 iii, and EE-F-F1 in the reduction in p38 phosphorylation, followed by DE-F-F2 i $\approx$ DE-F-F2 ii $\approx$ EE-F-F2, DE-R, and DE-F $\approx$ EE-F.

Only EE-F was able to significantly reduce the activity of the NF-κB p65 signaling pathway, being its efficacy $\approx$1.1 times higher than DE-R and DE-F (Figure 7C). DE-F and EE-F showed a small ability to decrease the NF-κB p65 phosphorylation, with no significant differences. The fractionation of the whole *E. purpurea* extracts into fractions strongly improved the reduction in the NF-κB p65 signaling pathway. Indeed, all fractions markedly reduced the phosphorylation of NF-κB p65, being DE-R-F2, DE-F-F2 iii, and EE-F-F1 the most promising. DE-F-F2 i, DE-F-F2 ii, and EE-F-F2 demonstrated equivalent bioactivity. Comparing all the *E. purpurea* extracts and fractions, DE-R-F2, DE-F-F2 iii, and EE-F-F1 exhibited the most powerful bioactivity in the reduction in NF-κB p65 phosphorylation, followed by DE-F-F2 ii $\approx$ DE-F-F2 i $\approx$ EE-F-F1, EE-F, and DE-R $\approx$ DE-F. Moreover, all the fractions enabled LPS-stimulated hMDM to reach lower levels of ERK 1/2 phosphorylation than non-stimulated macrophages.

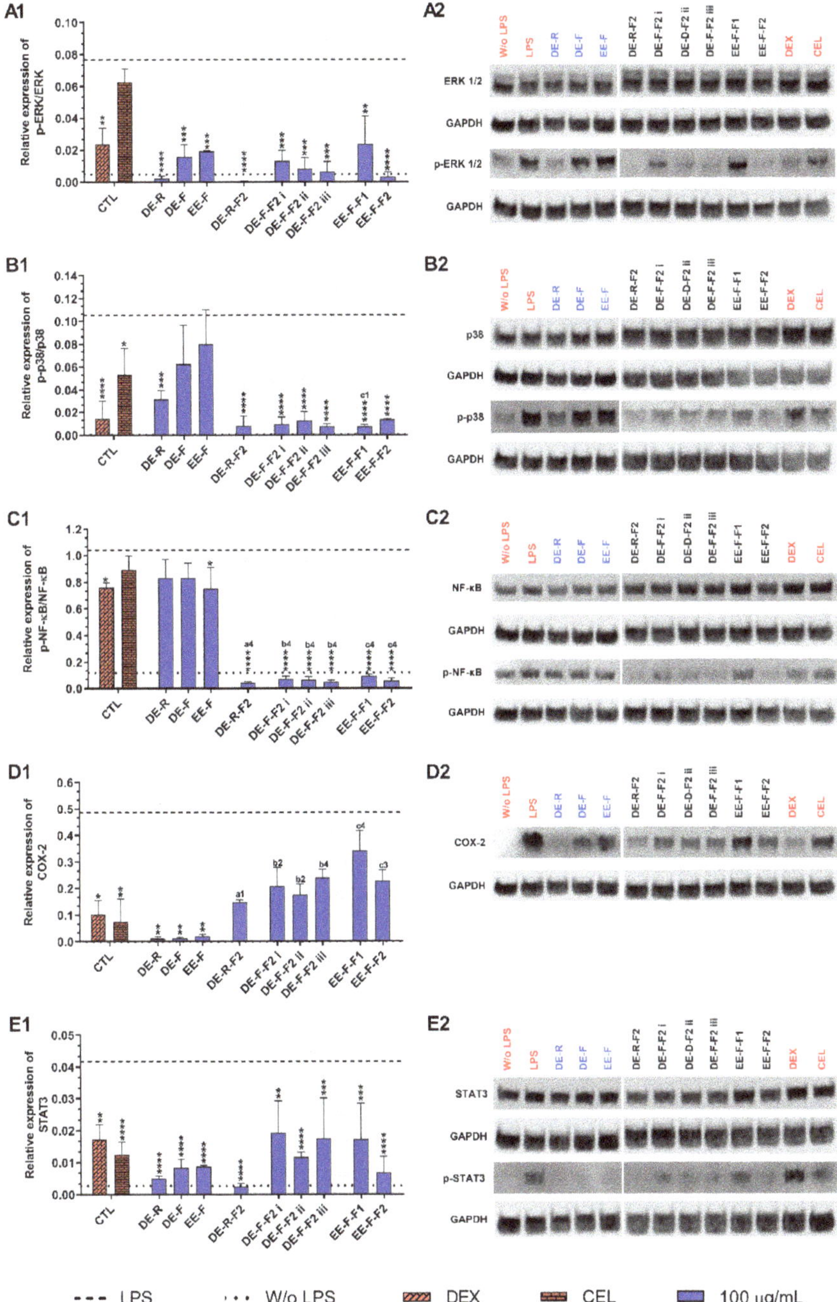

Figure 7. ERK 1/2 (**A1**,**A2**), p38 (**B1**,**B2**), NF-κB p65 (**C1**,**C2**), COX-2 (**D1**,**D2**), and STAT3 (**E1**,**E2**) signaling pathways downregulation of LPS-stimulated human monocyte-derived macrophages (hMDMs) cultured in the presence of *E. purpurea* extracts, factions, and clinically used anti-inflammatory drugs (dexamethasone (DEX) and celecoxib (CEL), 10 µM) for 24 h. Statistically significant differences are

* ($p < 0.0381$), ** ($p < 0.0071$), *** ($p < 0.0009$), and **** ($p < 0.0001$) in comparison to the positive control (LPS-stimulated hMDM without treatment) for each tested *E. purpurea* extract and fraction, as well as DEX and CEL and 1 ($p < 0.0358$), 2 ($p < 0.0023$), 3 ($p < 0.0002$), and 4 ($p < 0.0001$) in comparison with a (DE-R vs. DE-R-F2), b (DE-F vs. DE-R-F2), and c (EE-F vs. EE-F-F) at the same concentration. CTL: control; DE: dichloromethanolic extracts; EE: ethanolic extracts; R: roots; F: flowers; F1: phenol/carboxylic acid fraction; F2: alkylamide fraction.

The treatment of LPS-stimulated hMDM with the whole *E. purpurea* extracts significantly suppressed the COX-2 expression (Figure 7D). Despite the similar bioactivity of the three extracts, DE-R showed to be ≈1.2 and 1.9 times stronger than DE-F and EE-F. The fractions decreased the COX-2 expression, but no significant differences were observed. Nevertheless, DE-R-F2, DE-F-F2, and EE-F-F2 presented comparable bioactivity, while EE-F-F1 showed a lower ability to reduce COX-2 expression. In this case, the fractionation of the whole *E. purpurea* extracts into fractions did not enhance the reduction in COX-2 expression by LPS-stimulated hMDM. Comparing all the *E. purpurea* extracts and fractions, DE-R and DE-F led to the most potent COX-2 suppression, followed by EE-F, DE-R-F2, DE-F-F2 ii, DE-F-F2 i, DE-F-F2 iii, EE-F-F2, DE-F-F2 iii, and EE-F-F1.

E. purpurea extracts also showed a strong capacity to reduce the activation of the STAT3 signaling pathway in LPS-stimulated hMDM (Figure 7E). DE-R efficiently decreased the phosphorylation of STAT3, being its activity ≈1.6 and 1.8 times higher than DE-F and EE-F, respectively. All the fractions also significantly reduced the phosphorylation of STAT3, but DE-R-F2, followed by EE-F-F2 and DE-F-F2 ii, strongly suppressed the phosphorylation of this inflammatory protein. DE-F-F2 i, DE-F-F2 iii, and EE-F-F1 presented less but similar bioactivity. Nevertheless, only DE-R-F2 and EE-F-F2 improved the bioactivity of the extracts, but no significant differences were observed. Analyzing all the *E. purpurea* extracts and fractions, DE-R-F2 was the most promising formulation in the reduction in STAT3 phosphorylation, followed by DE-R, EE-F-F2, DE-F, EE-F, DE-F-F2 ii, and DE-F-F2 iii ≈ EE-F-F1 ≈ DE-F-F2 i. Moreover, LPS-stimulated hMDM in the presence of DE-R and DE-R-F2 achieved similar or lower levels of STAT3 phosphorylation compared with non-stimulated macrophages.

The inflammatory proteins JNK/p-JNK and iNOS were also investigated, but no phosphorylation expression was detected in this study at 24 h of culture.

4. Discussion

The development of chronic pathologies, such as rheumatoid arthritis and osteoarthritis, is strongly correlated with persistent inflammation [12,13]. Additionally, most of the current treatments are associated with significant side effects, and thus, effective and safe therapies are urgently needed. As the initiation and progression of inflammation involve several inflammatory signaling pathways, new and safe entities that effectively modulate different molecular mechanisms are required.

This study used DE-R, DE-F, and EE-F since they exhibited the strongest anti-inflammatory properties in our previous work [42]. These three extracts presented different patterns of phenolic compounds, carboxylic acids, and alkylamides in their composition, comparable to our previous study [42]. Briefly, we identified a higher number of phenolic and carboxylic acids in EE-F (11 compounds) than in DE-R (4 compounds) or DE-F (3 compounds). Relative to the alkylamides, DE-R was the extract where more alkylamides were identified (24 compounds), followed by DE-F (20 compounds) and EE-F (19 compounds). As phenolic compounds, carboxylic acids, and alkylamides have different polarities, the extracts were fractionated into F1 (phenols/carboxylic acids) and F2 (alkylamides) fractions. Afterward, the anti-inflammatory activity of the cytocompatible extracts and fractions was investigated by their ability to decrease IL-6, TNF-α, and ROS/RNS levels in LPS-stimulated hMDM.

The three whole extracts drastically reduced the IL-6 levels in LPS-stimulated hMDM. This ability is important since IL-6 induces hematopoiesis, promotes the expansion and activation of T cells, stimulates B cell differentiation, and regulates neutrophil-activating chemokines, among other processes [6,11]. When the extracts were fractionated into F2, the anti-inflammatory activity was considerably enhanced. On the other hand, the F1 obtained from EE-F did not significantly reduce the IL-6 amount. Therefore, it is possible to conclude that alkylamides are the main class of compounds responsible for the decrease in the IL-6 production by LPS-stimulated hMDM. Moreover, the bioactivity was increased when a high number of alkylamides was present, which involves possible synergistic effects. Specifically, DE-R-F2, composed of 19 alkylamides, demonstrated greater IL-6 reduction ($84.3 \pm 9.1\%$) than DE-F-F2 i (10 alkylamides, $62.7 \pm 11.5\%$), DE-F-F2 ii (4 alkylamides, $71.2 \pm 12.3\%$), or DE-F-F2 iii (4 alkylamides, $68.5 \pm 14.3\%$).

DE-R was the only extract that significantly decreased TNF-α levels ($41.4 \pm 4.9\%$). Although the fractionation of extracts into fractions enhanced the bioactivity of DE-R-F2 ($53.1 \pm 19.5\%$) and DE-F-F2 i ($42.7 \pm 1.4\%$), no significant differences between whole extract and fractions were observed. Conversely, EE-F-F1 failed to decrease TNF-α production, and, consequently, its ability to recruit and enhance the differentiation and proliferation of the immune cells, as well as induce the transcription of several inflammatory genes [52]. These results strengthen the role of alkylamides as promising anti-inflammatory candidates. Furthermore, the bioactive pattern and the decrease in IL-6 and TNF-α levels obtained for LPS-stimulated hMDM are in agreement with our previous study with LPS-stimulated THP-1-derived macrophages [42]. These results support the correlation and similar behavior between the human cell line and primary cells. Nonetheless, it is important to stress that the bioactivity of the extracts was slightly lower in primary macrophages.

All the studied extracts and fractions strongly reduced the intracellular levels of ROS/RNS/$O_2^{\bullet -}$ in LPS-stimulated hMDM, reaching similar or inferior levels to those of the non-stimulated macrophages. Particularly, the fractionated DE-F-F2 i and DE-F-F2 ii significantly reduced the intracellular ROS/RNS generation, suggesting that the alkylamides present in these fractions are directly involved in this bioactivity. Moreover, alkylamide fractions showed, in general, lower intracellular levels of ROS/RNS/$O_2^{\bullet -}$, suggesting that these compounds may be the main compounds responsible for this bioactivity. In this study, the fractionation of EE-F into EE-F-F1 dramatically diminished the capacity of intracellular ROS/RNS reduction, besides phenols/carboxylic acids present in *E. purpurea* extracts are considered strong antioxidants in in vitro assays [53,54]. These results are in agreement with our previous study, where *E. purpurea* extracts enriched in alkylamides presented the strongest intracellular ROS/RNS reduction [42]. It can be hypothesized that alkylamides may inhibit the direct production, mainly in the mitochondria, of these inflammatory mediators [55–57]. Alkylamides may also interfere with the transcription of antioxidant enzymes, such as superoxide dismutase (SOD), catalase (CAT), and glutathione peroxidase (GPx) [58], as well as target the nitric oxide synthases (NOS) [59].

Different patterns were observed for the suppression of the inflammatory signaling pathways, demonstrating that extracts and fractions modulate different inflammatory mechanisms, including ERK 1/2, p38, NF-κB p65, COX-2, and STAT3, to diverse extents. The alkylamide fractions demonstrated stronger potential to drastically inhibit the inflammatory pathways, pointing out the main role of alkylamides in reducing inflammation. The synergistic effect between alkylamides was also observed since DE-R-F2 demonstrated strong bioactivity in general. We also demonstrated the downregulation of COX-2 expression in the presence of extracts, but not of fractions, in LPS-stimulated hMDM. Thus, a synergistic effect between the classes of compounds present in the whole extract should be required for the inhibitory effect of COX-2 expression.

The activation of the STAT3 signaling pathway in LPS-stimulated hMDM was strongly suppressed by extracts. DE-R-F2 was the only fraction that demonstrated higher bioactivity than the whole extract; however, no significant differences were observed. Nevertheless, once again, the alkylamides were the main compounds responsible for the reduction in this inflammatory pathway in LPS-stimulated hMDM. In this study, it was not possible to observe the inflammatory proteins p-JNK and iNOS, perhaps due to the occurrence of their expression at early time points [27,40].

The anti-inflammatory activity of *E. purpurea* preparations has been reported to be due to different alkylamides. Indeed, the major alkylamide found in *E. purpurea*, dodeca-2E,4E,8Z,10Z-tetraenoic acid isobutylamide, demonstrated minor anti-inflammatory effect compared to the alkylamide fraction [60]. Similarly to our results, Hou et al. reported that isolated chicoric acid did not show strong effects in the reduction in TNF-α levels in LPS-stimulated macrophages [27]. On the other hand, the isolated dodeca-2E,4E,8Z,10E/Z-tetraenoic acid isobutylamide potentially decreased the expression of this protein in LPS-stimulated primary human monocyte/macrophage-enriched PBMCs [61]. In fact, as previously reported, alkylamide fractions led to the robust inhibition of $^{\bullet}$NO production in LPS-stimulated RAW 264.7 macrophages [27,40,60]. Indeed, alkylamides, including dodeca-2E,4Z-diene-8,10-diynoic acid isobutylamide (present in DE-R-F2, DE-F-F2 i, and EE-F-F2), dodeca-2E,4E,8Z,10E/Z-tetraenoic acid isobutylamide (present in DE-R-F2, DE-F-F2 ii, and EE-F-F2), dodeca-2E,4E-dienoic acid isobutylamide (present in DE-R-F2, DE-F-F2 iii, and EE-F-F2), dodeca-2E,4Z,10Z-triene-8-ynoic acid isobutylamide (present in DE-R-F2, DE-F-F2 i, and EE-F-F2), dodeca-2E,4E,8Z-trienoic acid isobutylamide (present in DE-R-F2, DE-F-F2 iii, and EE-F-F2), and undeca-2Z,4E-diene-8,10-diynoic acid isobutylamide (present in DE-R-F2), decreased the $^{\bullet}$NO production in RAW 264.7 macrophages [29]. To the best of our knowledge, only isolated phenols obtained from *E. purpurea* have been reported to have an anti-inflammatory effect. Chicoric acid was able to decrease the TNF-α, IL-1β, and IL-6 levels and the infiltration of inflammatory cells in streptozotocin (STZ)-induced diabetic C57BL/6J mice [24,37]. MTX-induced liver injury or chronic kidney disease in male Wistar rats pre-treated with chicoric acid reduced TNF-α, ROS, $^{\bullet}$NO, and malondialdehyde (MDA) levels [62,63].

Our results, together with the currently available evidence, suggest that alkylamides are powerful plant-based drugs, exhibiting strong pharmaceutical advantages to ameliorate the inflammatory process related to chronic diseases.

5. Conclusions

E. purpurea extracts efficiently decreased pro-inflammatory mediators (IL-6, TNF-α, and/or ROS/RNS) in LPS-stimulated hMDM, corroborating their anti-inflammatory effects. The fractionation of the whole extracts into alkylamide fractions drastically enhanced the bioactivity, evidencing these compounds as the main active principles. This study also showed that the combination of different phytochemical compounds exhibited high pharmacological properties. Particularly, an increased number of alkylamides demonstrated greater bioactivity. Moreover, alkylamides exert their anti-inflammatory activity through the reduction in ERK1/2, p38, NF-κB, and STAT3 inflammatory signaling pathways, and the downregulation of COX-2 expression. Therefore, *E. purpurea* extracts and fractions can revert and stop the hyperactivation of macrophages, reaching the desired homeostasis in chronic diseases and preventing damage of the surrounding cells and tissues. Consequently, these results point out the efficiency of *E. purpurea* extracts and fractions, particularly DE-R-F2, an alkylamide extract, as new, innovative, and powerful plant-based anti-inflammatory formulations in the modulation of the fate of macrophages in cases where the immune system is overactive. To the best of our knowledge, the anti-inflammatory activity of DE and DE fractions are studied here for the first time in LPS-stimulated hMDM, in which their therapeutic targets are reported. As the immune response involves both specific and non-specific mechanisms, further studies supporting the role of *E. purpurea* extracts and fractions in complex models of inflammation should be explored.

Supplementary Materials: The following supporting information can be downloaded at: https://www.mdpi.com/article/10.3390/antiox12020425/s1. Figure S1: Intracellular ROS/RNS and $O_2^{\bullet-}$ production in LPS-stimulated human monocyte-derived macrophages in the absence or in the presence of clinically used anti-inflammatory drugs; Figure S2: Intracellular ROS/RNS and $O_2^{\bullet-}$ production in LPS-stimulated human monocyte-derived macrophages in the presence of *E. purpurea* extracts or fractions for 22 h.; Table S1: Parameters of the optimized gradient method for analytical separation; Table S2: Properties of standards determined by LC-HRMS.; Table S3:

Phenolic/carboxylic acidic compounds tentatively identified in *E. purpurea* fractions by LC-HRMS (negative mode).; Table S4: Alkylamides compounds tentatively identified in *E. purpurea* fractions by LC-HRMS (positive mode).

Author Contributions: Conceptualization, S.F.V., H.F. and N.M.N.; methodology, S.F.V., S.M.G., V.M.F.G., C.P.L., F.M., M.E.T., C.C., A.C., H.F. and N.M.N.; software, S.F.V.; validation, S.F.V., H.F. and N.M.N.; formal analysis, S.F.V.; investigation, S.F.V.; resources, S.F.V., H.F., R.L.R. and N.M.N.; data curation, S.F.V.; writing—original draft preparation, S.F.V.; writing—review and editing, S.M.G., V.M.F.G., C.P.L., F.M., M.E.T., C.C., A.C., R.L.R., H.F. and N.M.N.; visualization, S.F.V., H.F. and N.M.N.; supervision, H.F. and N.M.N.; project administration, H.F. and N.M.N.; funding acquisition, H.F., R.L.R. and N.M.N. All authors have read and agreed to the published version of the manuscript.

Funding: This work was financially supported by the FCT to the Ph.D. grants of S.F.V. (PD/BD/135246/2017 and COVID/BD/152012/2021), S.M.G. (SFRH/BD/136814/2018), and C.C. (CEECIND/04058/2018), and the projects PATH (PD/00169/2013), HEALTH-UNORTE (NORTE-01-0145-FEDER-000039) and the NORTE 2020 Structured Project, co-funded by Norte2020 (NORTE-01-0145-FEDER-000021).

Institutional Review Board Statement: The study was conducted in accordance with the Declaration of Helsinki, and approved by the Ethics Subcommittee for Life and Health Sciences (SECVS) of the University of Minho, Braga, Portugal (no. 014/015). Approval on 14 December 2018.

Informed Consent Statement: Informed consent was obtained from all subjects involved in the study.

Data Availability Statement: The data presented in this study are available in the article and supplementary materials.

Acknowledgments: We acknowledge the professionals and donors at the Hospital de Braga (Braga, Portugal) for kindly providing buffy coats.

Conflicts of Interest: The authors declare no conflict of interest.

References

1. Furman, D.; Campisi, J.; Verdin, E.; Carrera-Bastos, P.; Targ, S.; Franceschi, C.; Ferrucci, L.; Gilroy, D.W.; Fasano, A.; Miller, G.W.; et al. Chronic inflammation in the etiology of disease across the life span. *Nat. Med.* **2019**, *25*, 1822–1832. [CrossRef]
2. Medzhitov, R. Inflammation 2010: New adventures of an old flame. *Cell* **2010**, *140*, 771–776. [CrossRef] [PubMed]
3. Li, Q.; Verma, I.M. NF-kappaB regulation in the immune system. *Nat. Rev. Immunol.* **2002**, *2*, 725–734. [CrossRef] [PubMed]
4. Arthur, J.S.C.; Ley, S.C. Mitogen-activated protein kinases in innate immunity. *Nat. Rev. Immunol.* **2013**, *13*, 679–692. [CrossRef]
5. Dennis, E.A.; Norris, P.C. Eicosanoid storm in infection and inflammation. *Nat. Rev. Immunol.* **2015**, *15*, 511–523. [CrossRef]
6. Hunter, C.A.; Jones, S.A. IL-6 as a keystone cytokine in health and disease. *Nat. Immunol.* **2015**, *16*, 448. [CrossRef] [PubMed]
7. Fullerton, J.N.; Gilroy, D.W. Resolution of inflammation: A new therapeutic frontier. *Nat. Rev. Drug Discov.* **2016**, *15*, 551–567. [CrossRef]
8. Hirayama, D.; Iida, T.; Nakase, H. The Phagocytic Function of Macrophage-Enforcing Innate Immunity and Tissue Homeostasis. *Int. J. Mol. Sci.* **2017**, *19*, 92. [CrossRef]
9. Mittal, M.; Siddiqui, M.R.; Tran, K.; Reddy, S.P.; Malik, A.B. Reactive oxygen species in inflammation and tissue injury. *Antioxid. Redox Signal.* **2014**, *20*, 1126–1167. [CrossRef]
10. Conner, E.M.; Grisham, M.B. Inflammation, free radicals, and antioxidants. *Nutrition* **1996**, *12*, 274–277. [CrossRef]
11. Turner, M.D.; Nedjai, B.; Hurst, T.; Pennington, D.J. Cytokines and chemokines: At the crossroads of cell signalling and inflammatory disease. *Biochim. Biophys. Acta-Mol. Cell Res.* **2014**, *1843*, 2563–2582. [CrossRef]
12. Smolen, J.S.; Aletaha, D.; Barton, A.; Burmester, G.R.; Emery, P.; Firestein, G.S.; Kavanaugh, A.; McInnes, I.B.; Solomon, D.H.; Strand, V.; et al. Rheumatoid arthritis. *Nat. Rev. Dis. Prim.* **2018**, *4*, 18001. [CrossRef]
13. Sanchez-Lopez, E.; Coras, R.; Torres, A.; Lane, N.E.; Guma, M. Synovial inflammation in osteoarthritis progression. *Nat. Rev. Rheumatol.* **2022**, *18*, 258–275. [CrossRef] [PubMed]
14. Steinmeyer, J. Pharmacological basis for the therapy of pain and inflammation with nonsteroidal anti-inflammatory drugs. *Arthritis Res.* **2000**, *2*, 379–385. [CrossRef] [PubMed]
15. Barnes, P.J. How corticosteroids control inflammation: Quintiles prize lecture 2005. *Br. J. Pharmacol.* **2006**, *148*, 245–254. [CrossRef] [PubMed]
16. Baumgart, D.C.; Misery, L.; Naeyaert, S.; Taylor, P.C. Biological therapies in immune-mediated inflammatory diseases: Can biosimilars reduce access inequities? *Front. Pharmacol.* **2019**, *10*, 279. [CrossRef] [PubMed]
17. Chamoun-Emanuelli, A.M.; Bryan, L.K.; Cohen, N.D.; Tetrault, T.L.; Szule, J.A.; Barhoumi, R.; Whitfield-Cargile, C.M. NSAIDs disrupt intestinal homeostasis by suppressing macroautophagy in intestinal epithelial cells. *Sci. Rep.* **2019**, *9*, 14534. [CrossRef]

18. Shivaji, U.N.; Sharratt, C.L.; Thomas, T.; Smith, S.C.L.; Iacucci, M.; Moran, G.W.; Ghosh, S.; Bhala, N. Review article: Managing the adverse events caused by anti-TNF therapy in inflammatory bowel disease. *Aliment. Pharmacol. Ther.* **2019**, *49*, 664–680. [CrossRef]

19. Newman, D.J.; Cragg, G.M. Natural products as sources of new drugs over the nearly four decades from 01/1981 to 09/2019. *J. Nat. Prod.* **2020**, *83*, 770–803. [CrossRef]

20. WHO. *WHO Monographs on Selected Medicinal Plants-Volume 1*; World Health Organization: Geneva, Switzerland, 1999; ISBN 9241545178 (v.1).

21. Miller, S.C. Echinacea: A miracle herb against aging and cancer? Evidence in vivo in mice. *Evid. Based. Complement. Alternat. Med.* **2005**, *2*, 309–314. [CrossRef]

22. Burlou-Nagy, C.; Bănică, F.; Jurca, T.; Vicaș, L.G.; Marian, E.; Muresan, M.E.; Bácskay, I.; Kiss, R.; Fehér, P.; Pallag, A. Echinacea purpurea (L.) Moench: Biological and Pharmacological Properties. A Review. *Plants* **2022**, *11*, 1244. [PubMed]

23. Šutovská, M.; Capek, P.; Kazimierová, I.; Pappová, L.; Jošková, M.; Matulová, M.; Fraňová, S.; Pawlaczyk, I.; Gancarz, R. Echinacea complex–chemical view and anti-asthmatic profile. *J. Ethnopharmacol.* **2015**, *175*, 163–171. [CrossRef] [PubMed]

24. Zhu, D.; Zhang, N.; Zhou, X.; Zhang, M.; Liu, Z.; Liu, X. Cichoric acid regulates the hepatic glucose homeostasis via AMPK pathway and activates the antioxidant response in high glucose-induced hepatocyte injury. *RSC Adv.* **2017**, *7*, 1363–1375. [CrossRef]

25. Dong, G.-C.; Chuang, P.-H.; Forrest, M.D.; Lin, Y.-C.; Chen, H.M. Immuno-suppressive effect of blocking the CD28 signaling pathway in T-cells by an active component of Echinacea found by a novel pharmaceutical screening method. *J. Med. Chem.* **2006**, *49*, 1845–1854. [CrossRef]

26. Gulledge, T.V.; Collette, N.M.; Mackey, E.; Johnstone, S.E.; Moazami, Y.; Todd, D.A.; Moeser, A.J.; Pierce, J.G.; Cech, N.B.; Laster, S.M. Mast cell degranulation and calcium influx are inhibited by an Echinacea purpurea extract and the alkylamide dodeca-2E,4E-dienoic acid isobutylamide. *J. Ethnopharmacol.* **2018**, *212*, 166–174.

27. Hou, C.-C.; Chen, C.-H.; Yang, N.-S.; Chen, Y.-P.; Lo, C.-P.; Wang, S.-Y.; Tien, Y.-J.; Tsai, P.-W.; Shyur, L.-F. Comparative metabolomics approach coupled with cell- and gene-based assays for species classification and anti-inflammatory bioactivity validation of Echinacea plants. *J. Nutr. Biochem.* **2010**, *21*, 1045–1059. [CrossRef]

28. Sasagawa, M.; Cech, N.B.; Gray, D.E.; Elmer, G.W.; Wenner, C.A. Echinacea alkylamides inhibit interleukin-2 production by Jurkat T cells. *Int. Immunopharmacol.* **2006**, *6*, 1214–1221. [CrossRef]

29. Chen, Y.; Fu, T.; Tao, T.; Yang, J.; Chang, Y.; Wang, M.; Kim, L.; Qu, L.; Cassady, J.; Scalzo, R.; et al. Macrophage activating effects of new alkamides from the roots of Echinacea species. *J. Nat. Prod.* **2005**, *68*, 773–776. [CrossRef]

30. Raduner, S.; Majewska, A.; Chen, J.-Z.; Xie, X.-Q.; Hamon, J.; Faller, B.; Altmann, K.-H.; Gertsch, J. Alkylamides from Echinacea are a new class of cannabinomimetics. Cannabinoid type 2 receptor-dependent and -independent immunomodulatory effects. *J. Biol. Chem.* **2006**, *281*, 14192–14206. [CrossRef]

31. Hou, C.-C.; Huang, C.-C.; Shyur, L.-F. Echinacea alkamides prevent lipopolysaccharide/D-galactosamine-induced acute hepatic injury through JNK pathway-mediated HO-1 expression. *J. Agric. Food Chem.* **2011**, *59*, 11966–11974. [CrossRef]

32. Zhang, H.; Lang, W.; Wang, S.; Li, B.; Li, G.; Shi, Q. Echinacea polysaccharide alleviates LPS-induced lung injury via inhibiting inflammation, apoptosis and activation of the TLR4/NF-κB signal pathway. *Int. Immunopharmacol.* **2020**, *88*, 106974. [CrossRef] [PubMed]

33. Fast, D.J.; Balles, J.A.; Scholten, J.D.; Mulder, T.; Rana, J. Echinacea purpurea root extract inhibits TNF release in response to Pam3Csk4 in a phosphatidylinositol-3-kinase dependent manner. *Cell. Immunol.* **2015**, *297*, 94–99. [CrossRef] [PubMed]

34. Jiang, W.; Zhu, H.; Xu, W.; Liu, C.; Hu, B.; Guo, Y.; Cheng, Y.; Qian, H. Echinacea purpurea polysaccharide prepared by fractional precipitation prevents alcoholic liver injury in mice by protecting the intestinal barrier and regulating liver-related pathways. *Int. J. Biol. Macromol.* **2021**, *187*, 143–156. [CrossRef]

35. Li, Q.; Yang, F.; Hou, R.; Huang, T.; Hao, Z. Post-screening characterization of an acidic polysaccharide from Echinacea purpurea with potent anti-inflammatory properties in vivo. *Food Funct.* **2020**, *11*, 7576–7583. [CrossRef]

36. Hou, R.; Xu, T.; Li, Q.; Yang, F.; Wang, C.; Huang, T.; Hao, Z. Polysaccharide from Echinacea purpurea reduce the oxidant stress in vitro and in vivo. *Int. J. Biol. Macromol.* **2020**, *149*, 41–50. [CrossRef] [PubMed]

37. Zhu, D.; Zhang, X.; Niu, Y.; Diao, Z.; Ren, B.; Li, X.; Liu, Z.; Liu, X. Cichoric acid improved hyperglycaemia and restored muscle injury via activating antioxidant response in MLD-STZ-induced diabetic mice. *Food Chem. Toxicol.* **2017**, *107*, 138–149. [CrossRef] [PubMed]

38. Dong, G.-C.; Chuang, P.-H.; Chang, K.; Jan, P.; Hwang, P.-I.; Wu, H.-B.; Yi, M.; Zhou, H.-X.; Chen, H.M. Blocking effect of an immuno-suppressive agent, cynarin, on CD28 of T-cell receptor. *Pharm. Res.* **2009**, *26*, 375–381. [CrossRef]

39. Cheng, Z.-Y.; Sun, X.; Liu, P.; Lin, B.; Li, L.-Z.; Yao, G.-D.; Huang, X.-X.; Song, S.-J. Sesquiterpenes from Echinacea purpurea and their anti-inflammatory activities. *Phytochemistry* **2020**, *179*, 112503. [CrossRef]

40. Zhai, Z.; Solco, A.; Wu, L.; Wurtele, E.S.; Kohut, M.L.; Murphy, P.A.; Cunnick, J.E. Echinacea increases arginase activity and has anti-inflammatory properties in RAW 264.7 macrophage cells, indicative of alternative macrophage activation. *J. Ethnopharmacol.* **2009**, *122*, 76–85. [CrossRef]

41. Chicca, A.; Raduner, S.; Pellati, F.; Strompen, T.; Altmann, K.-H.; Schoop, R.; Gertsch, J. Synergistic immunomopharmacological effects of N-alkylamides in Echinacea purpurea herbal extracts. *Int. Immunopharmacol.* **2009**, *9*, 850–858. [CrossRef]

42. Vieira, S.F.; Gonçalves, V.M.F.; Llaguno, C.P.; Macías, F.; Tiritan, M.E.; Reis, R.L.; Ferreira, H.; Neves, N.M. On the Bioactivity of Echinacea purpurea Extracts to Modulate the Production of Inflammatory Mediators. *Int. J. Mol. Sci.* **2022**, *23*, 13616. [CrossRef] [PubMed]

43. Vieira, S.F.; Ferreira, H.; Neves, N.M. Antioxidant and anti-Inflammatory activities of cytocompatible Salvia officinalis extracts: A comparison between traditional and soxhlet extraction. *Antioxidants* **2020**, *9*, 1157. [CrossRef] [PubMed]

44. Pellati, F.; Epifano, F.; Contaldo, N.; Orlandini, G.; Cavicchi, L.; Genovese, S.; Bertelli, D.; Benvenuti, S.; Curini, M.; Bertaccini, A.; et al. Chromatographic methods for metabolite profiling of virus- and phytoplasma-infected plants of Echinacea purpurea. *J. Agric. Food Chem.* **2011**, *59*, 10425–10434. [CrossRef] [PubMed]

45. Cech, N.B.; Eleazer, M.S.; Shoffner, L.T.; Crosswhite, M.R.; Davis, A.C.; Mortenson, A.M. High performance liquid chromatography/electrospray ionization mass spectrometry for simultaneous analysis of alkamides and caffeic acid derivatives from Echinacea purpurea extracts. *J. Chromatogr. A* **2006**, *1103*, 219–228. [CrossRef]

46. Spelman, K.; Wetschler, M.H.; Cech, N.B. Comparison of alkylamide yield in ethanolic extracts prepared from fresh versus dry Echinacea purpurea utilizing HPLC–ESI-MS. *J. Pharm. Biomed. Anal.* **2009**, *49*, 1141–1149. [CrossRef]

47. Thomsen, M.O.; Fretté, X.C.; Christensen, K.B.; Christensen, L.P.; Grevsen, K. Seasonal variations in the concentrations of lipophilic compounds and phenolic acids in the roots of Echinacea purpurea and Echinacea pallida. *J. Agric. Food Chem.* **2012**, *60*, 12131–12141. [CrossRef]

48. He, X.; Lin, L.; Bernart, M.W.; Lian, L. Analysis of alkamides in roots and achenes of Echinacea purpurea by liquid chromatography–electrospray mass spectrometry. *J. Chromatogr. A* **1998**, *815*, 205–211. [CrossRef]

49. Mudge, E.; Lopes-Lutz, D.; Brown, P.; Schieber, A. Analysis of alkylamides in Echinacea plant materials and dietary supplements by ultrafast liquid chromatography with diode array and mass spectrometric detection. *J. Agric. Food Chem.* **2011**, *59*, 8086–8094. [CrossRef]

50. Bauer, R.; Remiger, P. TLC and HPLC analysis of alkamides in Echinacea drugs. *Planta Med.* **1989**, *55*, 367–371. [CrossRef]

51. Gonçalves, S.M.; Duarte-Oliveira, C.; Campos, C.F.; Aimanianda, V.; ter Horst, R.; Leite, L.; Mercier, T.; Pereira, P.; Fernández-García, M.; Antunes, D.; et al. Phagosomal removal of fungal melanin reprograms macrophage metabolism to promote antifungal immunity. *Nat. Commun.* **2020**, *11*, 2282. [CrossRef]

52. Faustman, D.; Davis, M. TNF receptor 2 and disease: Autoimmunity and regenerative medicine. *Front. Immunol.* **2013**, *4*, 478. [CrossRef] [PubMed]

53. Pellati, F.; Benvenuti, S.; Magro, L.; Melegari, M.; Soragni, F. Analysis of phenolic compounds and radical scavenging activity of Echinacea spp. *J. Pharm. Biomed. Anal.* **2004**, *35*, 289–301. [CrossRef] [PubMed]

54. Tsai, Y.-L.; Chiou, S.-Y.; Chan, K.-C.; Sung, J.-M.; Lin, S.-D. Caffeic acid derivatives, total phenols, antioxidant and antimutagenic activities of Echinacea purpurea flower extracts. *LWT-Food Sci. Technol.* **2012**, *46*, 169–176. [CrossRef]

55. Dan Dunn, J.; Alvarez, L.A.J.; Zhang, X.; Soldati, T. Reactive oxygen species and mitochondria: A nexus of cellular homeostasis. *Redox Biol.* **2015**, *6*, 472–485. [CrossRef] [PubMed]

56. Zeeshan, H.M.A.; Lee, G.H.; Kim, H.-R.; Chae, H.-J. Endoplasmic Reticulum Stress and Associated ROS. *Int. J. Mol. Sci.* **2016**, *17*, 327. [CrossRef] [PubMed]

57. Fransen, M.; Nordgren, M.; Wang, B.; Apanasets, O. Role of peroxisomes in ROS/RNS-metabolism: Implications for human disease. *Biochim. Biophys. Acta-Mol. Basis Dis.* **2012**, *1822*, 1363–1373. [CrossRef]

58. Burton, G.W.; Foster, D.O.; Perly, B.; Slater, T.F.; Smith, I.C.P.; Ingold, K.U.; Willson, R.L.; Scott, G.; Norman, R.O.C.; Hill, H.A.O.; et al. Biological antioxidants. *Philos. Trans. R. Soc. London. B Biol. Sci.* **1985**, *311*, 565–578. [CrossRef]

59. Porasuphatana, S.; Tsai, P.; Rosen, G.M. The generation of free radicals by nitric oxide synthase. *Comp. Biochem. Physiol. Part C Toxicol. Pharmacol.* **2003**, *134*, 281–289. [CrossRef]

60. Matthias, A.; Banbury, L.; Stevenson, L.M.; Bone, K.M.; Leach, D.N.; Lehmann, R.P. Alkylamides from echinacea modulate induced immune responses in macrophages. *Immunol. Investig.* **2007**, *36*, 117–130. [CrossRef]

61. Gertsch, J.; Schoop, R.; Kuenzle, U.; Suter, A. Echinacea alkylamides modulate TNF-alpha gene expression via cannabinoid receptor CB2 and multiple signal transduction pathways. *FEBS Lett.* **2004**, *577*, 563–569. [CrossRef]

62. Hussein, O.E.; Hozayen, W.G.; Bin-Jumah, M.N.; Germoush, M.O.; Abd El-Twab, S.M.; Mahmoud, A.M. Chicoric acid prevents methotrexate hepatotoxicity via attenuation of oxidative stress and inflammation and up-regulation of PPARγ and Nrf2/HO-1 signaling. *Environ. Sci. Pollut. Res.* **2020**, *27*, 20725–20735. [CrossRef]

63. Abd El-Twab, S.M.; Hussein, O.E.; Hozayen, W.G.; Bin-Jumah, M.; Mahmoud, A.M. Chicoric acid prevents methotrexate-induced kidney injury by suppressing NF-κB/NLRP3 inflammasome activation and up-regulating Nrf2/ARE/HO-1 signaling. *Inflamm. Res.* **2019**, *68*, 511–523. [CrossRef] [PubMed]

Article

"Golden" Tomato Consumption Ameliorates Metabolic Syndrome: A Focus on the Redox Balance in the High-Fat-Diet-Fed Rat

Giuditta Gambino [1,*,†], Giuseppe Giglia [1,2,†], Mario Allegra [3,4], Valentina Di Liberto [1], Francesco Paolo Zummo [1], Francesca Rappa [1], Ignazio Restivo [4], Filippo Vetrano [5], Filippo Saiano [5], Eristanna Palazzolo [5], Giuseppe Avellone [4,6], Giuseppe Ferraro [1,3], Pierangelo Sardo [1,3,‡] and Danila Di Majo [1,3,‡]

1   Department of Biomedicine, Neuroscience and Advanced Diagnostics (BIND), University of Palermo, 90127 Palermo, Italy; giuseppe.giglia@unipa.it (G.G.); valentina.diliberto@unipa.it (V.D.L.); francescopaolo.zummo@unipa.it (F.P.Z.); francesca.rappa@unipa.it (F.R.); giuseppe.ferraro@unipa.it (G.F.); pierangelo.sardo@unipa.it (P.S.); danila.dimajo@unipa.it (D.D.M.)

2   Euro Mediterranean Institute of Science and Technology (IEMEST), 90139 Palermo, Italy

3   Postgraduate School of Nutrition and Food Science, University of Palermo, 90100 Palermo, Italy; mario.allegra@unipa.it

4   Department of Biological, Chemical and Pharmaceutical Sciences and Technologies (STEBICEF), University of Palermo, Viale delle Scienze, 90128 Palermo, Italy; giuseppe.avellone@unipa.it (G.A.)

5   Dipartimento Scienze Agrarie, Alimentari e Forestali, Università degli Studi di Palermo, Viale delle Scienze Ed.4, 90128 Palermo, Italy; filippo.vetrano@unipa.it (F.V.); filippo.saiano@unipa.it (F.S.); eristanna.palazzolo@unipa.it (E.P.)

6   ATeN (Advanced Technologies Network) Center, Viale delle Scienze, 90128 Palermo, Italy

*   Correspondence: giuditta.gambino@unipa.it

†   Co-first authorship.

‡   Co-last authorship.

Citation: Gambino, G.; Giglia, G.; Allegra, M.; Di Liberto, V.; Zummo, F.P.; Rappa, F.; Restivo, I.; Vetrano, F.; Saiano, F.; Palazzolo, E.; et al. "Golden" Tomato Consumption Ameliorates Metabolic Syndrome: A Focus on the Redox Balance in the High-Fat-Diet-Fed Rat. *Antioxidants* **2023**, *12*, 1121. https://doi.org/10.3390/antiox12051121

Academic Editor: Stanley Omaye

Received: 20 April 2023
Revised: 13 May 2023
Accepted: 16 May 2023
Published: 18 May 2023

Abstract: Tomato fruits defined as "golden" refer to a food product harvested at an incomplete ripening stage with respect to red tomatoes at full maturation. The aim of this study is to explore the putative influence of "golden tomato" (GT) on Metabolic Syndrome (MetS), especially focusing on the effects on redox homeostasis. Firstly, the differential chemical properties of the GT food matrix were characterized in terms of phytonutrient composition and antioxidant capacities with respect to red tomato (RT). Later, we assessed the biochemical, nutraceutical and eventually disease-modifying potential of GT in vivo in the high-fat-diet rat model of MetS. Our data revealed that GT oral supplementation is able to counterbalance MetS-induced biometric and metabolic modifications. Noteworthy is that this nutritional supplementation proved to reduce plasma oxidant status and improve the endogenous antioxidant barriers, assessed by strong systemic biomarkers. Furthermore, consistently with the reduction of hepatic reactive oxygen and nitrogen species (RONS) levels, treatment with GT markedly reduced the HFD-induced increase in hepatic lipid peroxidation and hepatic steatosis. This research elucidates the importance of food supplementation with GT in the prevention and management of MetS.

Keywords: tomato-based products; metabolic syndrome; HFD; antioxidant capacity; phytonutrients

1. Introduction

Tomato fruits (*Lycopersicon esculentum Mill.*) have increasingly grabbed attention as this food product, largely cultivated and consumed throughout the world, represents an invaluable source of bioactive compounds [1]. Noteworthy is that a wide plethora of nutritional substances are encountered in this food, i.e., antioxidants, such as flavonoids and naringenin, but also macronutrients, micronutrients, and organic and phenolic acids. Though the differential composition and quantity of discrete molecules depend on the cultivating conditions and could, hence, influence health-promoting activities, this still needs to be fully unveiled in terms of biochemical and nutritional characterization. In this context, "Golden Tomato" (GT) is a food product harvested at different degrees of ripeness and defined as "golden" due to the degree of coloring it possesses. A complete characterization in phytonutrients of this food product is deserved to fully explore its biological potential. As a matter of fact, previous research suggested a protective role for tomato-based products, modulating lipid profiles and positively influencing the development of cardiovascular diseases [2–5]. In particular, several clinical studies supported a role for different phytoconstituents of red tomato (RT) that can, for instance, improve the levels of antioxidant enzymes reducing lipid peroxidation rate in diabetic syndrome [6–9].

Metabolic syndrome (MetS) is a widely-diffused clustering of risk factors associated with obesity, cardiovascular disease, and alteration of oxidative status, conditions that could severely impact other comorbidities [10–14]. MetS is established if three or more of the following clinical conditions are present: hypertension, atherogenic dyslipidemia, increased visceral obesity, and hyperglycemia/insulin resistance [15,16]. From a mechanistic perspective, MetS development strongly relies on a vicious self-feeding cycle between chronic, low-grade inflammation and oxidative stress that predisposes individuals to cardiovascular diseases and type II diabetes. At the cellular level, oxidative stress consists of an increased production of reactive oxygen and nitrogen species (RONS) through the intervention of enzymatic, non-enzymatic, and/or mitochondrial pathways. In the background of the metabolic syndrome, NADPH oxidase emerges as a key ROS-producing enzyme [17]. Relevantly, RONS can interact with polyunsaturated fatty acids (PUFA), producing reactive lipid by-products (lipid peroxides and byproducts such as aldehydes) prone to interact with macromolecules such as proteins and DNA, leading to cellular dysfunction. At the same time, RONS overproduction modulates specific, intracellular kinase activities such as Jun N-terminal kinase and those related to NF-kB activation. These molecular events result in the increased phosphorylation of Ser/Thr residues of insulin receptor substrate (IRS-1) and lead to the impairment of both insulin signaling and glucose transport, frequently associated to MetS.

In animal models, MetS can be induced by employing a special diet regimen, i.e., a high-fat diet (HFD), that reproduces the complete clinical manifestations in terms of increased body weight, reduced food intake, glucose tolerance, and dyslipidemia [18–20]. Recently, our previous research revealed that specific systemic biomarkers of redox homeostasis are robustly predictive of the development of metabolic dysfunctions, strengthening the impact of oxidative-based alterations in MetS. A growing interest has arisen for the use of nutraceuticals contained in foods that can counteract MetS [21,22]. In this context, tomato stands out as a food of the Mediterranean diet representing an excellent source of nutrients and bioactive compounds, the concentrations of which are related to the prevention of chronic degenerative diseases such as cardiovascular disorders, cancer, and neurodegenerative diseases [11,23]. Numerous studies have already demonstrated the beneficial effect of tomato consumption on the lipid and glycemic profile [23,24]. Along these lines, we here evaluated the impact of the oral supplementation with GT, at an incomplete ripening stage, on the development of a multifactorial metabolic dysfunction, i.e., MetS, focusing on redox homeostasis. Since no studies so far have assessed the specific composition of GT as it is harvested at the ripening stage, we first investigated the different phytonutrients and antioxidant properties with respect to tomatoes harvested at full maturation. In detail, we characterized the food matrices of the tomato samples assessing the amount of phytonutrients present, but also the antioxidant properties via evaluation of radical scavenger activity for the purpose of discriminating between GT and RT. Secondly, we tested the impact of GT extracts in eventually counterbalancing MetS by exploiting a HFD rat model in order to evaluate the impact of this food product on oxidative-based impairments. Thus, we evaluated the influence of GT supplementation on altered body weight gain, glucose tolerance, lipid profile, and, even more importantly, on robust systemic and hepatic biomarkers of oxidative balance in MetS. This research could shed a novel light on the importance of food supplementation with GT in the prevention and management of MetS.

2. Materials and Methods

2.1. Treatment of Tomato Samples

The food matrix named "Golden Tomato" is the sample employed in the present research, which refers to a tomato harvested at the veraison stage and defined as "golden" on the basis of the degree of coloring it possesses. It is a tomato from the cultivar Brigade, grown in a sandy soil type with a northwest exposure and 650 m of altitude. Golden tomatoes were kindly provided by the manufacturer, Mr. Fabrizio Gioia (Company: "Azienda Agricola Fabrizio Gioia", Montemaggiore Belsito, PA, Italy).

The nutritional characterization of the red and golden tomato samples was carried out by determination of the following parameters: macronutrients (proteins, lipids, and carbohydrates) and micronutrients (mineral salts). The degree of ripeness was established by means of the colorimetric analysis of the fruit and based on the following analytical parameters: pH, Brix degree, acidity, and polyphenolic content.

Both tomato samples, after harvesting, were divided into aliquots (~1 kg), freeze-dried and vacuum-preserved at $-20\,^{\circ}$C while waiting to be analyzed and used for animal treatment.

2.1.1. Chemical–Physical Analysis of Tomato Samples

The tomato samples were subdivided according to colorimetric grade and certain chemical–physical parameters. The colorimetric analyses were carried out using a colorimeter (CR-400, Minolta corporation, Ltd., Osaka, Japan) based on the CIELAB color space, also referred to as L*a*b*, is a color space defined by the International Commission on Illumination (abbreviated CIE) in 1976. It expresses color as three values: L* for perceptual lightness and a* and b* for the four unique colors of human vision: magenta, green, blue and yellow. These components were used to calculate hue angle (h°) and chroma (C^*) as

$$h^\circ = 180^\circ + \arctan(b^* | a^*)$$
$$C^* = \sqrt{\left(a^{*2} + b^{*2}\right)}$$

$h^\circ = 180^\circ + \arctan\left(b^*/a^*\right)$ and $C^* = \left(a^{*2} + b^{*2}\right)1/2$

following established procedures [25,26].

The basic analytical parameters characterizing the red and golden tomatoes and their different degree of ripeness were moisture, ash, pH, Brix degree, and acidity [Metodi ufficiali di analisi per le conserve vegetali—Parte generale, Supplemento ordinario alla Gazzetta Ufficiale n. 168 del 20 luglio 1989]. In particular, the pH was determined with the help of a pHmeter instrument, as is stated in the Italian regulations and the Official Methods of Analysis, AOAC.

2.1.2. Determination of Phytonutrients in Golden and Red Tomatoes by HPLC System

The analysis of certain phytonutrients that could differentiate the red from the golden tomato involved a phase of extraction with an organic solvent from the food matrix, a phase of separation, and, successively, identification and quantification in the HPLC system. A total of 200 mg of powdered dried GT and RT samples were weighed and extracted with the addition of 20 mL of Tetrahydrofuran (THF). The samples were then processed with Ultra Turrax for about 30" at 17,500 rpm, filtered with filter paper and, subsequently, with regenerated cellulose syringe filters (0.20 μm), and then loaded into autosampler vials and analyzed in HPLC.

For the HPLC analysis of tomato extracts, we used the UPLC-Q Exactive Orbitrap-HRMS system (Thermo Fisher Scientific™, Bremen, Germany) composed of a Dionex Ultimate 3000 liquid chromatograph coupled to a Q Exactive™ Plus Hybrid Quadrupole-Orbitrap™ Mass Spectrometer equipped with a heated electrospray ionization (HESI) ion source (detailed procedures are reported in Supplementary Materials).

2.1.3. Determination of Total Polyphenolic Content and Antioxidant Properties

In general, the methods used to assess the antioxidant capacity of food can be divided according to their mechanism of action into two categories: methods based on hydrogen atom transfer (HAT) and those based on single-electron transfer (SET).

SET-based methods assess the ability of a compound defined as an antioxidant's potential to yield an electron by reducing the acceptor compound, which can be a metal. This property is quantified by the color change observed when the oxidant is reduced. These methods include the Ferric Reducing Antioxidant Power (FRAP) and the Folin–Ciocalteu methods. HAT-based methods measure the classical ability of an antioxidant to quench free radicals by hydrogen donation. The Crocin bleaching assay (CBA) is included among the HAT methods [27,28]. To evaluate the antioxidant properties of tomato samples, we used FRAP and Folin–Ciocalteu as SET methods and the Crocin bleaching assay as a HAT method. The analyses were performed on the methanolic extract obtained from fresh tomatoes (detailed procedures are reported in Supplementary Materials). The results were expressed as mean $\pm$ standard error of the mean (S.E.M.) of three replicates.

2.2. Animals

Male Wistar rats (4-week old) weighing 240–260 g were provided by Envigo S.r.l. They were housed two per cage and maintained on a 12 h on/off light cycle (8:00–20:00 h) at a constant temperature (22–24 °C) and humidity (50 $\pm$ 10%). During the acclimation period, animals were first fed with a standard chow diet providing 3.94 kcal/g and then divided into two homogenous groups with balanced weight. These groups were fed with standard laboratory food (NPD: normal pelletized diet, code PF1609, certificate EN 4RF25, Mucedola, Milan, Italy) or fed with HFD food with 60% of energy coming from fats (code PF4215-PELLET, Mucedola, Milan, Italy) in order to induce MetS, as assessed following criteria already established by previous literature [19,20]. Detailed description of the composition of the diet is included in Table 1, as in our previous study [20]. All rats had free access to food and water. Prior to starting the special diet, all animals were weighed. The experiment involved three stages—T0, T1, and T2—individuated according to procedures described in detail in Supplementary Materials. Animal care and handling throughout the experimental procedures were in accordance with the European Communities Council Directive (2010/63/EU). The experimental protocols were approved by the animal welfare committee of the University of Palermo, authorized by the Ministry of Health (Rome, Italy; Authorization Number 14/2022-PR), and conducted following the ARRIVE guidelines.

Table 1. Pellet composition is reported for both HFD and NPD. (SFA) Saturated Fat Acid, (MUFA) Monounsaturated Fatty Acid, (PUFA) Polyunsaturated Fatty Acid.

	Pellet HFD (PF4215)	NPDSND (PF1609)
Energy (Kcal/Kg)	5500–6000	3947
Fat Total (g/100 g)	60	3.50
SFA (g/60 g)	30	0.7
MUFA (g/60 g)	23	0.8
PUFA (g/60 g)	7	2
Crude protein (g/100 g)	23	22
Carbohydrates (Starch g/100 g)	-	35.18
Sugar (g/100 g)	-	5.66
Fiber (g/100 g)	5	4.5
Ash (g/100 g)	5.50	7.5
Vitamin A (IU)	8400	19.533
Vitamin D3 (IU)	2100	1260

Experimental Groups

Each experimental group consisted of $n = 6$ animals, except for one group (NPD, $n = 4$). At T0, animals were initially subdivided in NPD or HFD on the basis of the diet administered for 8 weeks until induction of MetS. At T1, once MetS induction was verified, animals were divided into 4 groups according to the type of diet (NPD and HFD) and treatment administered (golden tomato or red tomato) until T2, which was reached 4 weeks after T1. In particular, the normal control was fed normal diet feed (NPD) until T2 and the second one (HFD group), representing the MetS control, was fed with the HFD diet throughout the trial (from T0 to T2). The control NPD and HFD groups were subjected i.p. to the same stress conditions as the treated group, since they received during the last month of the experiment, from T1 to T2, a volume of vehicle equal to the tomato solution administered to the treated groups. Furthermore, one group (HFD/GT) was treated 1 mL daily with golden tomato (GT) in the last month of the trial (T1-T2). Lastly, a group of rats was orally treated with red tomato (HFD/RT) at full maturation, at the same dose and under the same experimental conditions as HFD/GT group for 1 month, in order to verify eventual specific effects of red tomato on the redox homeostasis biomarkers of MetS.

2.3. Preparation of the Orally Administered Tomato Solutions

The amount of tomato administered was 200 mg/kg body weight, corresponding to a daily intake of 300 g for a man with an average weight of 70 kg. The dose was established on the basis of valid toxicity tests for red tomatoes in the literature [29,30] and considered to be over the minimum dose exerting an eventual biological effect when translated from animal studies to humans [31].

The dose was obtained by dissolving 50 mg of freeze-dried fresh tomato in solution with 50 mL of water. The volume of golden tomato solutions orally-administered daily using a syringe was 1 mL. The groups not receiving the tomato solutions took the same volume (1 mL) of plain water. No animals showed signs of toxicity or intolerance during the treatments.

2.4. Biometric, Biochemical, and Oxidative Homeostasis Parameters Induced by MetS

At the T2 time point, the effect of treatments on the experimental groups was evaluated on the MetS-induced alterations in terms of biometric, biochemical, and oxidative homeostasis parameters. At the end of the experimental procedures, all animals were sacrificed using 2% isoflurane anesthesia followed by cervical dislocation in accordance with authorized procedures. Plasma samples were collected for subsequent analyses to evaluate lipid homeostasis, oxidative stress parameters, and plasma antioxidant status. Hepatic samples were also collected for malondialdehyde (MDA), RONS, and GSH determination as well as for histological evaluations.

2.4.1. Body Weight Gain

Body weight gain was evaluated at T2 for all animals after 4 weeks of nutritional treatments, calculating the Delta Body Weight (ΔBW) by subtracting the final rat weight from the initial weight recorded at T0.

2.4.2. Glucose and Lipid Homeostasis assays

Glucose Tolerance Test (GTT), a diagnostic tool for diabetes and an indicator of metabolic efficiency and insulin resistance, was conducted at T2 following established procedures [20,32] to evaluate the effect of nutritional treatments on glucose metabolism in MetS. To assess the effect of GT supplementation on lipid homeostasis in MetS, after sacrifice, blood samples of each animal were collected by cardiac puncture. Detailed procedures are described in our previous paper [20]. In the plasma samples, triglycerides (TG), total cholesterol (TC), low-density lipoprotein cholesterol (LDL), and high-density lipoprotein cholesterol (HDL) concentrations were quantified by commercial kits using the Free Carpe

Diem device (FREE® Carpe Diem; Diacron International, Grosseto, Italy). The data are expressed in mg/dL.

2.4.3. Oxidative Stress Parameters and Plasma Antioxidant Status

Plasma redox balance was assessed using Diacron kits following detailed procedures already published [20]. To assess the prooxidant status, the dROM (Reactive Oxygen Metabolites, primarily hydroperoxides) and the LP-CHOLOX test were carried out, the former assessing the levels of hydroperoxyl free radicals and the latter the levels of circulating lipid peroxides and, in particular, oxidized cholesterol. In the plasma samples, hydroperoxides, lipoperoxides, and oxidized cholesterol were measured by commercial kits using the Free Carpe Diem device (FREE® Carpe Diem; Diacron International, Grosseto, Italy). Data from dROM tests are expressed in arbitrary units, namely, Carratelli units (UCARR). The normal range of the test results was 250–300 U.CARR (Carratelli Units), where 1 U.CARR corresponds to 0.08 mg/dL of H_2O_2 [33]. In the LP-CHOLOX test, LP-CHOLOX (lipoperoxides and oxidized cholesterol) levels are detected based on peroxides' ability to facilitate the oxidation of Fe^{2+} to Fe^{3+}, which binds to an indicator mixture forming a colored complex detected by a spectrophotometer at 505 nm [34,35]. The results are expressed in mEq/L.

As for the plasma antioxidant status, the "BAP" test (Biological Antioxidant Potential) measures substances of an exogenous nature (ascorbate, tocopherols, carotenoids, and bioflavonoids) and substances of an endogenous nature (bilirubin, uric acid, and proteins) in plasma that have antioxidant potential and are capable of counteracting radical species. The analysis was performed using the Diacron kit by taking spectrophotometric readings at a wavelength of 505 nm as reported in a previous work and expressing the results as mmol/L [36]. Furthermore, the SHp test was used for evaluation of thiol groups to assess the reducing properties of tomato extracts that can counteract the oxidation of thiol groups and shift the balance in favor of reduced forms.

2.4.4. MDA Assay

Evaluation of MDA levels in liver homogenates was performed according to Ohkawa et al. [37]. Briefly, the reaction mixture contained 0.2 mL of whole homogenate, 0.2 mL of 8.1% sodium dodecyl sulphate (SDS), 1.5 mL of acetic acid solution adjusted at pH 3.5 with NaOH, and 1.5 mL of 1% thiobarbituric acid (TBA) aqueous solution. The mixture was finally made up to 4.0 mL with distilled water and heated at 95 °C for 60 min. After cooling with tap water, 1.0 mL of distilled water and 5.0 mL of a n-butanol/pyridine solution $(15/1, v/v)$ were added, and the mixture was shaken vigorously. After centrifugation at 4000 rpm for 10 min, the absorbance of the organic layer was measured at 532 nm. MDA levels were expressed as nmol MDA/g tissue, using 1,1,3,3,tetramethoxypropane as an external standard.

2.4.5. RONS Assay

RONS levels were detected in liver homogenates using 2′,7′dichlorodihydrofluorescein diacetate (H_2DCF-DA) as previously reported [38]. Briefly, the whole homogenate was centrifuged at 3500 rpm for 10 min at 4 °C and 100 µL of the supernatant was mixed with 5 µL of H_2DCF-DA (final concentration 10 µM). The mixture was incubated for 30 min at 37 °C protected from light and the fluorescence intensity was detected at 490 nm (excitation) and 540 nm (emission) by using a plate reader.

2.4.6. GSH Measurements

Hepatic GSH/GSSG levels were measured in the whole homogenate by employing a glutathione colorimetric assay kit according to the manufacturer's instructions (Invitrogen, Milan, Italy).

2.5. Histological Analyses

Hepatic samples were immediately stored in paraformaldehyde for 48 h, then were moved to a 20% PBS/sucrose solution, and, after 1 week, to 10% PBS/sucrose solution. At last, the solution was removed and samples were stored at $-80°$. Liver tissue sections (5 μm) were obtained from cryostat and stained with hematoxylin and eosin. Following staining, the slides were observed with an optical microscope (Microscope Axioscope 5/7 KMAT, Carl Zeiss, Oberkochen, Germany) connected to a digital camera (Microscopy Camera Axiocam 208 colour, Carl Zeiss). For the steatosis evaluation, a semiquantitative analysis was performed by two independent observers in a high-power field (HPF) (magnification $400\times$) and repeated for 10 HPFs.

2.6. Statistical Analyses

Statistical analysis was performed by using GraphPad Prism 9.02 (San Diego, CA, USA). Analyses of antioxidant composition between GT and RT samples were performed by an unpaired Student's *t*-test. Plasma glucose levels (GTT) were analyzed via a two-way repeated measures (RM) ANOVA, followed by Bonferroni post hoc test for significant differences for within- and between-subject comparisons, considering the effect of "time", "diet", and their interaction in the experimental groups. Values of ΔBW, GTT, TG, TC, AUC, MDA, and RONS levels and histological evaluations in liver were compared by a one-way ANOVA test followed by Bonferroni post hoc evaluations for differences between means and represented by scattered bar graphs, in which at least 4 animals per group were included for evaluation. Differences were considered significant when $p < 0.05$. The statistical power (g-power) was considered only if >0.75 and the effect size if >0.40. The results are presented as the mean $\pm$ standard error of the mean (S.E.M.), apart from GTT values at 0 and 120 min presented as box and whiskers plots.

3. Results

3.1. Analytical, Nutritional, and Antioxidant Composition of Tomato Food Matrices

The different ripening time determines the different phytonutrient composition and antioxidant properties of the tomato, making the red and golden tomato two different food matrices despite coming from the same cultivar and the same production conditions. In particular, the compounds identified in the two food matrices, together with the nutritional properties, micronutrient and organic acid composition are indicated in Supplementary Tables S1–S4. The chemical–physical characteristics used to distinguish the two groups of red (RT) and golden (GT) tomatoes are shown in Table 2.

Table 2. Analytical parameters of the two tomato samples expressed as percentages.

Analytical Parameters	Red Tomato	Golden Tomato
Moisture (% g)	94.2	91.5
Ash (%g)	0.8	0.9
Brix degree	5.8	5.3
Acidity (mg %)	0.5	0.6
pH	4.55	4.36

Based on the colorimetric analysis, four different colourings were considered with varying color gradations, evidenced by the C* and h° parameters calculated with the specific equations given in the text (Section 2.1.1), from the red fruits (C* = 25.11 and h° = 115.58) in a full ripening state to colorings of decreasing intensity up to the green fruits. Among these, only green (C* = 32.08 and h° = 264.56), pre-veraison (C* = 30.01 and h° = 264.72) and veraison tomatoes (C* = 30.07 and h° = 264.68) were used to produce "golden" sample GT.

As described in Table 2, tomatoes at veraison stage (GT) have a slightly lower water content in absolute value than ripe tomatoes (RT), making all macronutrients slightly more concentrated. Furthermore, as might be expected, GT has a higher degree of acidity and lower pH than RT. From a nutritional viewpoint, GT and RT show minor differences in macronutrients and energy value provided (data reported in supplementary material) even if the reduced degree of ripeness makes the tomato poorer in vitamin C by 45% and in pro-vitamin A, expressed as beta-carotene, by 69%, as shown in Table 3. Regarding the polyphenolic content and antioxidant properties of tomato samples, the different antioxidant properties of GT and RT samples were evaluated via HAT and SET methods. GT exhibits a superior reducing power compared to RT, and this can be observed from results obtained by FRAP and Folin–Ciocalteu methods. In detail, the FRAP assay revealed a significant increase in GT versus RT samples (t = 6.18, df = 13, $p < 0.0001$, Figure 1A). Conversely, the CBA assay revealed a significant increase in RT versus GT samples (t = 3.48, df = 12, $p = 0.0045$, Figure 1B). Lastly, the analysis of total polyphenolic content revealed that GT samples contained significantly higher levels of polyphenols as in Figure 1C (t = 4.923, df = 4, $p = 0.0079$). The antioxidant properties outlined in GT and RT samples could be ascribed to the differential composition in phytonutrients as listed in Table 2. In particular, we observed that, in the GT, the levels of naringenin and chlorogenic acid are 57% higher for the former and 81% higher for the latter, respectively, which stand out compared to RT. Meanwhile, RT is characterized by an increased content in lycopene and beta-carotene.

Table 3. Antioxidant components in tomato samples. The concentrations of golden and red (GT and RT) tomatoes are expressed as mg/100 g (Dried Weight, DW).

Antioxidant Components	Golden Tomato Concentration mg/100 gDW	Red Tomato Concentration mg/100 gDW
Vitamin C	170.4	311.2
β-carotene	39.08	126.22
Lycopene	333.0	1971.0
Phytoene	6.5	16.8
Lutein	7.78	6.37
Naringenin	38.31	16.5
Gallic acid	2.0	<0.2
Chlorogenic acid	9.8	1.9
Rutin	4.68	7.1

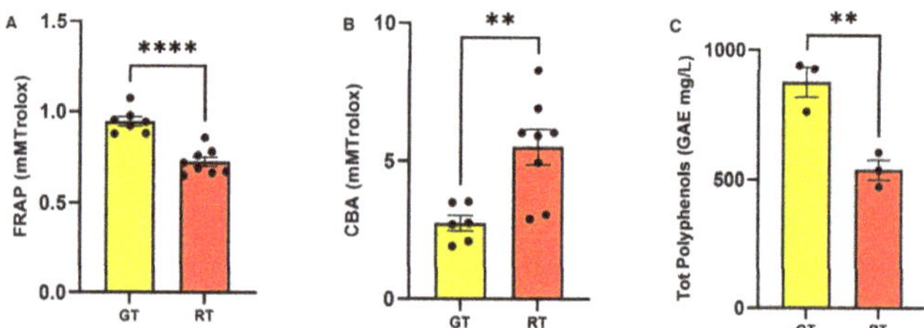

Figure 1. Analyses of antioxidant properties of golden tomato and red tomato (GT and RT) samples. Evaluations using (**A**) Ferric Reducing Antioxidant Power (FRAP, mM/trolox), (**B**) Crocin Bleaching Assay (CBA, mM/trolox), and (**C**) total polyphenolic content (GAE mg/L) were reported and significant differences following unpaired *t*-test between GT and RT are indicated as (****) $p < 0.0001$ and (**) $p < 0.01$.

3.2. Effects of GT Treatment on Body Weight in MetS

At T2, after 4-week treatment with GT and RT, ΔBW was compared via one-way ANOVA followed by Bonferroni post hoc test. Significant differences in body weight were found in HFD/GT that reduced their weight gain versus HFD and versus NPD in MetS ($F_{(2,13)}$ = 3.17, p = 0.0014, g-power: 1.307; effect size: 0.769; Figure 2). In detail, following treatment with GT, rats reached a mean % weight increase relative to initial weight of 78.75 ± 11.91 with respect to HFD (90.35 ± 10.74%g).

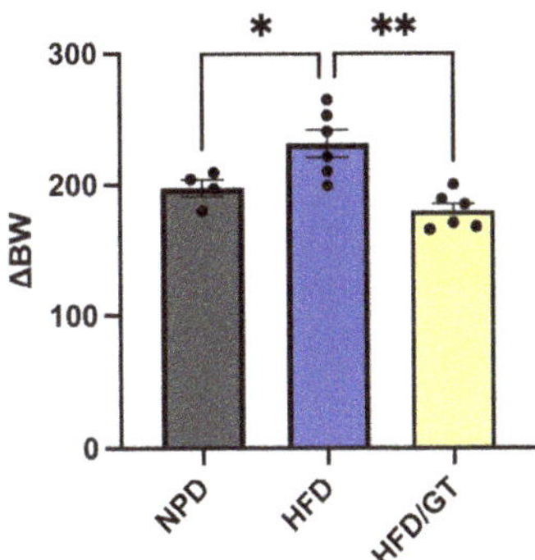

Figure 2. Variation of body weight gain (ΔBW) versus the initial weight in HFD/GT, HFD, and NPD experimental groups. Statistical significance by one-way ANOVA followed by post hoc Bonferroni is indicated as (**) for $p < 0.001$ versus HFD and (*) for $p < 0.005$ versus NPD.

3.3. Effects of GT Treatment on Glucose and Lipid Homeostasis in MetS

The effects of nutritional treatments in MetS were tested on glucose homeostasis. To this purpose, GTT test was performed at the end of nutritional treatments and outlined significant differences between experimental groups. To begin with, we evaluated the AUC, as described in the methods section, by one-way ANOVA followed by a Bonferroni post hoc test that highlighted a significant reduction in HFD/GT group versus HFD and a significant increase versus NPD ($F_{(2,10)}$ = 37.43, $p < 0.0001$, g-power: 0.819; effect size: 2.33; Figure 3A), suggesting an improvement in glucose tolerance tests following GT supplementation, though not returning to baseline. Besides this, a two-way RM ANOVA followed by a Bonferroni post hoc test performed on GTT at 0 and 120 min revealed marked differences in plasma glucose levels for time ($F_{(1,10)}$ = 119.8, $p < 0.0001$), diet ($F_{(2,10)}$ = 12.01, p = 0.0022), and their interaction ($F_{(2,10)}$ = 9.44, p = 0.0050) in HFD/GT compared to HFD and NPD groups (Figure 3B). In accordance with previous data [39], no significant differences are observed between groups in fasting glucose levels while there are different trends over time after i.p. glucose administration. In particular, a significant reduction in plasma glucose was evidenced by post hoc Bonferroni tests due to GT treatment, since the HFD/GT group at 120 min reached lower values versus HFD. However, glucose tolerance was ameliorated but not compensated, since the HFD/GT group was still significantly higher than NPD, not returning to baseline.

As for the evaluation of lipid homeostasis at T2, we assessed the plasma concentrations of TG, TC, LDL, and HDL in the experimental groups. On the one hand, the one-way ANOVA performed on the plasma levels of TG did not point out significant differences between groups, as shown in Table 4. On the other hand, the plasma levels of LDL analyzed by one-way ANOVA were remarkably decreased following GT treatment in the HFD/GT group with respect to HFD, and returned to basal values that were not statistically different than the NPD group ($F_{(2,13)}$ = 12.59, p = 0.0009; g-power: 0.769; effect size: 1.30; Table 4). Total cholesterol was markedly increased in HFD/GT rats versus HFD and NPD group ($F_{(2,11)}$ = 16.39, p = 0.0005; g-power: 0.937; effect size: 1.688; Table 4). Importantly, HDL cholesterol was increased by GT treatment in HFD rats reaching significantly higher levels with respect to HFD and NPD groups ($F_{(2,11)}$ = 20.84, *p* = 0.0002; g-power: 0.801; effect size: 1.358, Table 4).

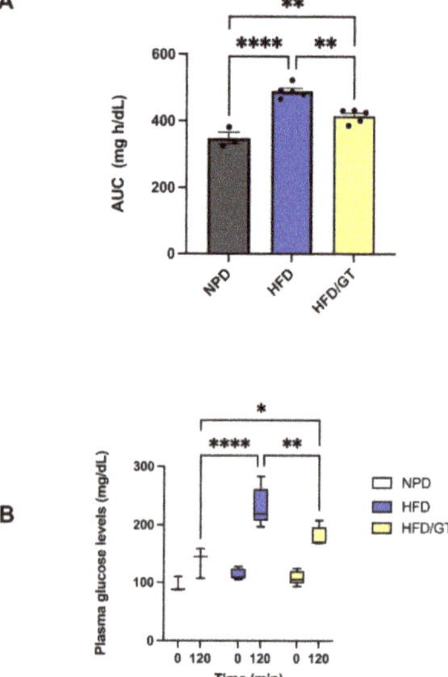

Figure 3. GTT test in experimental groups at the end of experimental procedures. (**A**) Area under the curve (AUC). Plasma glucose levels (mg/dL) per unit of time (h) difference between HFD/GT, HFD and NPD groups. (**B**) Glucose levels in GTT. Plasma glucose level (mg/dL) differences between groups after 0 and 120 min of GTT. Statistical significance by one-way ANOVA followed by post hoc Bonferroni is indicated as (****) *p* < 0.0001, (**) *p* < 0.01, and (*) *p* < 0.05.

Table 4. Biochemical parameters of lipid homeostasis triglycerides (TG), Total Cholesterol (TC), LDL Cholesterol, and HDL Cholesterol (mg/dL). Statistical significance for * *p* < 0.05 versus HFD and for # *p* < 0.05 versus NPD groups.

Experimental Groups	Triglycerides (mg/dL)	TC (mg/dL)	LDL Cholesterol (mg/dL)	HDL Cholesterol (mg/dL)
NPD	153.59 ± 61.49	60.90 ± 12.57	5.91 ± 3.15	34.35 ± 4.33
HFD	170.05 ± 57.44	86.57 ± 11.62 #	14.08 ± 3.13 #	30.02 ± 6.19
HFD/GT	183.58 ± 18.04	124.80 ± 23.12 *,#	9.98 ± 0.99 *,#	56.80 ± 9.01 *,#

3.4. Effects of GT on Plasma Redox Homeostasis Biomarkers in MetS

The antioxidant and prooxidant status was evaluated at time T2, after nutritional treatment, on plasma samples from all the experimental groups to explore the plasma redox balance in metabolic syndrome. Statistical analyses revealed that GT markedly modulated the antioxidant capacity of MetS animals.

In particular, a one-way ANOVA followed by post hoc test was conducted on mean values of SHp in HFD/GT that are higher than HFD, though still significantly reduced versus NPD group ($F_{(2,13)}$ = 27.09, *p* < 0.0001; g-power: 0.972; effect size: 1.869; Figure 4A). BAP values were reduced in both HFD/GT and HFD groups versus NPD ($F_{(2,13)}$ = 7.22, *p* = 0.0078; g-power: 0.905; effect size: 0.793; Figure 4B). Furthermore, statistical significance emerged from the analysis conducted on mean values of the prooxidant status, i.e., dROM and LP-CHOLOX levels. In detail, dROM levels were modified among the experimental groups, though post hoc analysis revealed a non-significant reduction induced by GT treatment versus HFD ($F_{(2,13)}$ = 6.15, *p* = 0.0132; g-power: 0.923; effect size: 0.809; Figure 4C).

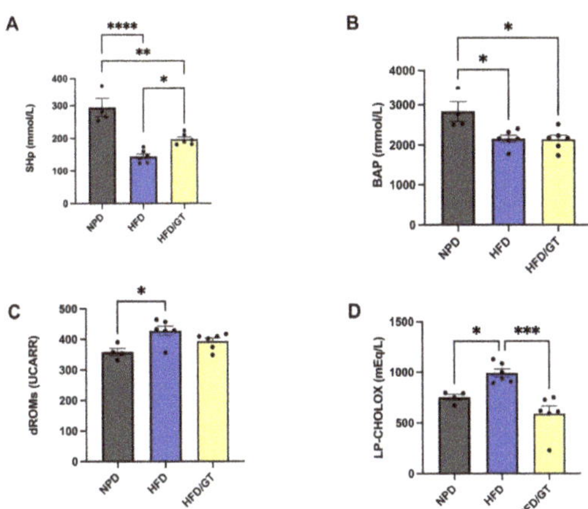

Figure 4. Plasma redox homeostasis biomarkers between HFD/GT, NPD, and HFD groups at the end of experimental procedures. Antioxidant status evaluated by (**A**) SHp test for thiol group levels (mmol/L) and (**B**) Biological Antioxidant Potential (BAP test) levels (mmol/L). Prooxidant status evaluated by (**C**) dROM test for differences in ROM (primarily hydroperoxide) levels (UCARR) and (**D**) LP-CHOLOX test for differences in LP-CHOLOX (lipoperoxides and oxidized cholesterol) levels (mEq/L). Statistical significance of Bonferroni post hoc tests are indicated for (*) $p < 0.05$, (**) $p < 0.01$, (***) $p < 0.001$ and (****) $p < 0.0001$, as represented in the graphs.

Lastly, LP-CHOLOX levels were significantly reduced in the HFD/GT group versus HFD, returning to baseline since non-significant differences emerged with the NPD group ($F_{(2,13)} = 13.63$, $p = 0.0006$; g-power: 0.776; effect size: 1.318; Figure 4D).

3.5. Effects of GT on Hepatic Steatosis

The histological evaluation performed on liver samples of the control NPD group showed an almost absent steatosis (average percentage of 3.2 ± 0.8) compared to the cases of the HFD group in which steatosis was found to be high (average percentage of 89.3 ± 1.7). In HFD liver tissue, macrovesicular steatosis with large lipid droplets was predominantly observed. In the liver samples of the GT group, the steatosis was microvesicular with an average percentage of 44.33 ± 14.5. Statistical evaluation by one-way ANOVA showed a significant decrease in the percentage of steatosis in HFD following GT supplementation versus HFD, though it was still significantly higher than NPD ($F_{(2,12)} = 6.59$, $p = 0.011$; g-power: 0.99; effect size: 5.07; Figures 5 and 6).

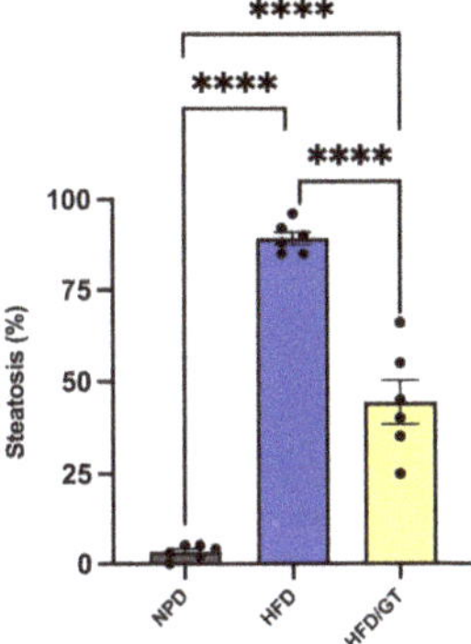

Figure 5. Histological evaluation of hepatic steatosis. Differences in hepatic steatosis (%) between NPD, HFD, and HFD/GT groups. **** for $p < 0.0001$.

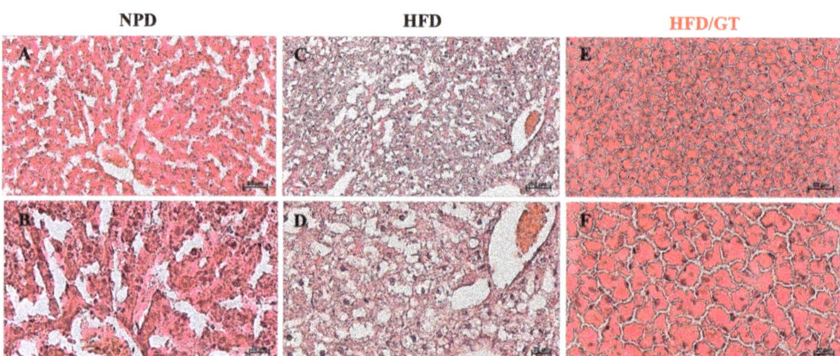

Figure 6. Histological features of liver tissue of experimental groups. Representative images of hematoxylin and eosin staining of liver tissue ((**A**,**B**): NPD; (**C**,**D**): HFD; (**E**,**F**): HFD/GT). (**A**,**C**,**E**): magnification 200×, scale bar 50 μm. (**B**,**D**,**F**): magnification 400×, scale bar 20 μm.

3.6. Effects of GT on MetS-Induced Hepatic Oxidative Stress

As shown in Figure 7A, the statistical analyses performed by one-way ANOVA on MDA levels in the liver showed significant differences between groups. In particular, GT significantly reduced the HFD-induced oxidative stress in the same tissue in comparison with the HFD group, though basal NPD values were not restored ($F_{(2,9)}$ = 124.2, $p < 0.0001$; g-power: 0.99; effect size: 5.19; Figure 7A).

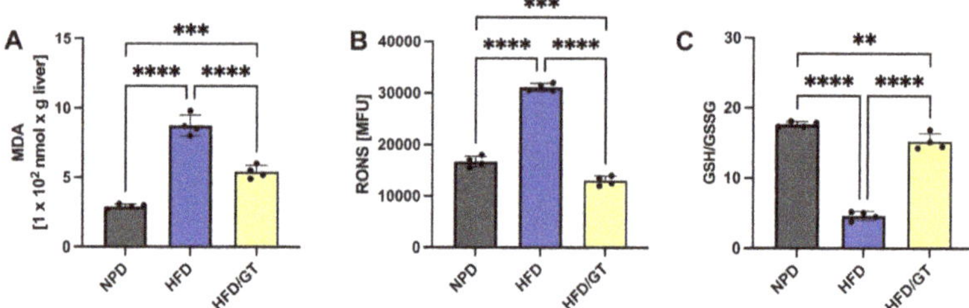

Figure 7. Levels of hepatic oxidative stress in HFD/GT, NPD, and HFD groups at T2. (**A**) Malondialdehyde (MDA) levels, (**B**) reactive oxygen and nitrogen species (RONS) levels, and (**C**) GSH/GSSG levels. Statistical significance by one-way ANOVA followed by post hoc Bonferroni is indicated as **** for $p < 0.0001$, *** for $p < 0.001$ and ** for $p < 0.01$.

The evaluation of RONS in the liver highlighted that the treatment with GT managed to recover the oxidative status induced by HFD by reducing hepatic RONS levels. In particular, GT was able to reduce RONS levels even in comparison with the NPD group as shown by post hoc significance in Figure 7B ($F_{(2,9)}$ = 436.3, $p < 0.0001$; g-power: 1; effect size: 8.61). Finally and in accordance with this evidence, GT treatment also ameliorated the ratio of GSH/GSSG in the liver by significantly increasing its levels, although not to the control levels (Figure 7C, $F_{(2,9)}$ = 289.2, $p < 0.0001$; g-power: 1; effect size: 7.561).

3.7. Effects of RT on MetS-Induced Systemic and Hepatic Oxidative Stress

The HFD/RT group of rats was employed in this study to unveil its specific effect on redox homeostasis in our experimental model. Therefore, we focused on the assessment of systemic and hepatic oxidative stress in MetS following nutritional treatment with RT at full maturation.

To begin with systemic biomarkers, we discovered that RT treatment does not manage to rescue SHp values versus HFD as analyzed by a one-way ANOVA followed by a Bonferroni post hoc analysis (Figure 8A). In contrast, RT treatment significantly increases BAP values versus HFD, returning to the basal values of NPD group ($F_{(2,13)}$ = 8.01, $p = 0.0054$;

g-power: 0.986; effect size: 0.968; Figure 8B). As for the evaluation of the prooxidant status, statistical significance emerged from the analysis conducted on mean values of dROM and LP-CHOLOX levels. In detail, dROM levels were markedly reduced by RT treatment versus HFD ($F_{(2,13)}$ = 21.02, $p < 0.0001$; g-power: 0.995; effect size: 1.72; Figure 8C) and LP-CHOLOX levels were significantly reduced in HFD/RT group versus HFD and also lower than the NPD group ($F_{(2,13)}$ = 32.79, $p < 0.0001$; g-power: 0.993; effect size: 2.15; Figure 8D).

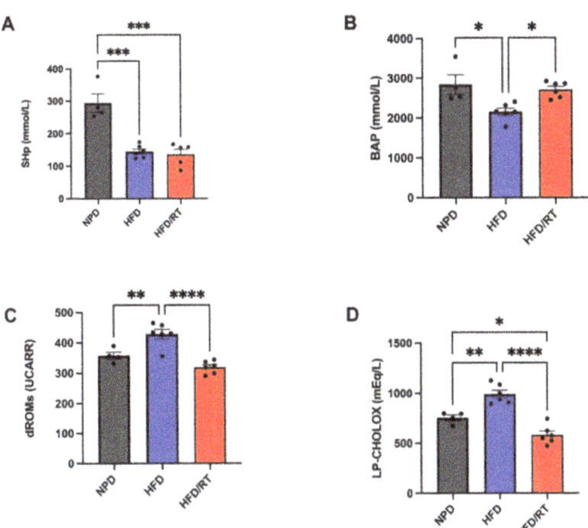

Figure 8. Plasma redox homeostasis biomarkers between HFD/RT, NPD, and HFD groups at the end of experimental procedures. Antioxidant status evaluated by (**A**) SHp test for thiol group levels (mmol/L) and (**B**) Biological Antioxidant Potential (BAP test) levels (mmol/L). Prooxidant status evaluated by (**C**) dROM test for differences in ROM (primarily hydroperoxides) levels (UCARR) and (**D**) LP-CHOLOX test for differences in LP-CHOLOX (lipoperoxides and oxidized cholesterol) levels (mEq/L). Statistical significance of Bonferroni post hoc tests are indicated for (*) $p < 0.05$, (**) $p < 0.01$, (***) for $p < 0.001$ and (****) $p < 0.0001$, as represented in the graphs.

Regarding hepatic biomarkers of oxidative stress, our outcomes revealed that RT at full maturation is able to significantly reduce the HFD-induced MDA levels in hepatic tissue in comparison with the HFD group, though not restoring to basal NPD values ($F_{(2,9)}$ = 163.4, $p < 0.0001$; g-power: 0.999; effect size: 6.26; Figure 9A). The evaluation of RONS in the liver highlighted that the treatment with RT managed to recover the oxidative status induced by HFD by reducing hepatic RONS levels and restoring basal NPD values ($F_{(2,9)}$ = 399.5, $p < 0.0001$, g-power: 1; effect size: 8.30; Figure 9B). Lastly, GSH levels in RT-treated rats were higher than the HFD group, but still significantly different from control NPD rats ($F_{(2,9)}$ = 434.8 $p < 0.001$; g-power: 1; effect size: 9.04; Figure 9C).

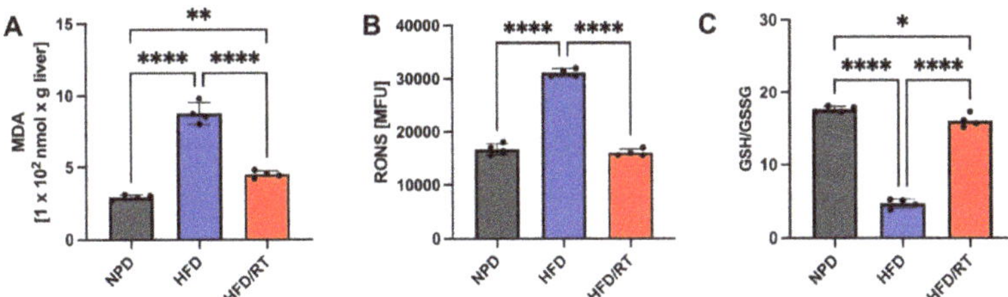

Figure 9. Levels of hepatic oxidative stress in HFD/RT, NPD, and HFD groups at T2. (**A**) Malondialdehyde (MDA) levels, (**B**) reactive oxygen and nitrogen species (RONS) levels, and (**C**) GSH/GSSG levels. Statistical significance by one-way ANOVA followed by post hoc Bonferroni is indicated as (*) for $p < 0.05$, (**) for $p < 0.01$ and (****) for $p < 0.0001$.

4. Discussion

GT is a food product harvested at an incomplete ripening stage and has a different nutritional and phytonutrient composition with respect to red tomato at full maturation. Not surprisingly, the two food matrices that we here analyzed for the first time differ to a great extent. To begin with, we revealed the higher acidity of GT samples compared to RT, which could account for a different bioavailability of phytonutrients [40] and for a different aggregation state of the polyphenolic compounds, which may explain the different antioxidant properties found in GT and RT food matrices. Indeed, we revealed the different phytonutrient composition that could be responsible for the higher antioxidant capacities of the GT in terms of reducing power compared to the RT. However, RT shows better radical scavenger activity than the GT evaluated by CBA. Among the phytonutrients, we demonstrated that GT has a higher naringenin and chlorogenic acid content than RT, which could be of striking importance since in vitro studies supported the beneficial effects of naringenin and of chlorogenic acid on MetS [23,41,42].

In the light of the intriguing antioxidant properties emerged by our evaluation of the GT food matrix, we orally administered GT extracts in an HFD rat model of MetS in order to further explore a putative protective role. Indeed, the HFD model was previously demonstrated to trigger an oxidative-dependent metabolic dysfunction that represents an undoubtedly valid model for the assessment of GT properties [20,43]. Our data revealed that one month of GT oral supplementation managed to reduce body weight gain in HFD rats. GT supplementation also counteracted the deleterious effect of the HFD on glucose tolerance. Nevertheless, the GT supplementation is able to ameliorate the glucose tolerance, but not to restore it to the basal values of the NPD group. Similarly to GT, some authors have shown that supplementation of naringenin or chlorogenic acid can improve but not normalize the manifestations of glucose tolerance [42,44,45]. Indeed, chlorogenic acid has anti-diabetic and anti-obesity properties by reducing glucose absorption in the small intestine through inhibition of the enzyme glucose-6-phosphate translocase, inhibiting the hepatic enzyme glucose-6-phosphatase, and increasing phosphorylation of AMP-activated protein kinase [46,47].

Noteworthy is that GT also reduced LDL cholesterol and increased HDL cholesterol versus both HFD and NPD rats. The same effect observed in our study was shown by administering 10 mg/Kg of chlorogenic acid to rats with hypercholesterolemia induced by the HFD diet [48]. This cholesterolemic-lowering effect could be due to the inhibitory action of naringenin and of chlorogenic acid on the enzyme HMG-CoA reductase already demonstrated [49–51]. On top of this, chlorogenic acid influences lipid metabolism by modulating the transcription of genes coding for lipogenic enzymes such as fatty acid synthase and acetyl-CoA carboxylase. In particular, it appears to induce down-regulation of LXRα and up-regulation of PPARα [52].

As for the effect of GT supplementation on altered plasma biomarkers of antioxidant defenses in MetS, we here revealed that GT markedly modulated the systemic antioxidant capacity in vivo in the HFD model. The different composition in phytochemicals of the two food matrices allows to justify the behavior on plasma antioxidant status observed after administration of GT. In particular, the oral administration of GT to HFD rats was able to enhance SHp levels. GT is, in fact, composed of chlorogenic acid and naringenin, molecules with high reducing power, hence explaining the effect observed on thiol groups. Furthermore, the influence of GT supplementation on SH groups is supported by the result obtained in terms of reducing power, measured by FRAP in GT. We previously uncovered that SHp values negatively correlate with lipid profile biomarkers in MetS [20], thus showing that GT supplementation resets—in favor of thiolic groups—the MetS-induced shifted balance towards disulfide compounds. This could be due to a GT-mediated compensation that increases antioxidant defenses to counteract the excessive free radicals. It could appear counterintuitive that GT treatment did not manage to rescue BAP levels in HFD rats, though it further supports the specific effect induced by the reducing power of GT. In accordance with an improved antioxidant status, the oxidative biomarkers, i.e., dROMs and

LP-CHOLOX, were powerfully decreased by GT supplementation in HFD rats, showing a better protection from plasmatic lipid peroxidation products. In detail, GT extracts reduced both LP-CHOLOX levels and, not significantly, dROMs values, though basal levels of normally fed rats were not restored. The protective effect of GT supplementation on plasma lipoperoxidation and the plasma antioxidant barrier could be ascribed to naringenin, whose ability to reduce lipid peroxidation and normalize antioxidant defenses by increasing the activity of antioxidant enzymes in the liver has been demonstrated in the literature [41]. Reduced plasmatic levels of hydroperoxides and lipoperoxides by GT supplementation are thought to well-correlate with the improved glucose profile revealed by our study.

Relevantly, not only does GT supplementation reduce systemic oxidative stress but also significantly counteracts the HFD-induced hepatic production of RONS. Interestingly, these species have been shown to enhance MDA levels in the liver [53]. This reactive aldehyde has been demonstrated to irreversibly form adducts with macromolecules, modifying cell function and contributing to MetS development [53–55]. Consistently with the reduction of hepatic RONS levels, treatment with GT also markedly reduced the HFD-induced increase in hepatic lipid peroxidation and increased the GSH/GSSG ratio in the liver. These results are in line with previously reported evidence showing how RT supplementation has been demonstrated to exert powerful, anti-oxidative effects able to counteract the onset and development of several pathological conditions in humans, including MetS [56]. Notwithstanding the fact that BAP was not modified by GT treatment, our data clearly demonstrate an improvement of the hepatic redox state. This evidence might well be associated with the observed amelioration of the hepatic metabolic functions that are strongly dependent on the endocellular redox state [57]. Relevantly, the current experimental evidence demonstrates, for the first time in an in vivo model of MetS, the ability for GT to also reduce the HFD-dependent hepatic oxidative stress and lipid peroxidation. In line with previously reported evidence [58], our data show that HFD-induced obesity is associated with the development of hepatic macrovesicular steatosis. Although tomato supplementation has already been reported to both ameliorate hepatic steatosis and reduce the risk of Nonalcoholic Fatty Liver Disease (NAFLD) development in rats [59], no data are available on the impact of GT treatment on hepatic dysfunction. Interestingly, our results show for the first time that GT consumption significantly improves hepatic steatosis. These effects on the HFD-induced liver structural damage are consistent and in agreement with the ability of GT to improve the plasma lipoprotein profile and to relieve hepatic oxidative stress.

In our study, we also tested the effects of red tomato at full maturation in vivo on the redox homeostasis biomarkers, considering that the effect of this food matrix has already been studied on dysmetabolism [30,60]. On this point, we uncovered that treating MetS rats with RT for one month is able to partially improve the systemic antioxidant endogenous barriers impaired by HFD, such as by increasing BAP levels and returning to basal control values, though not acting on the thiolic groups. Indeed, RT extracts appear to selectively target plasma antioxidant barriers in HFD rats, which could be due to the fact that RT is richer in carotenes, molecules capable of strongly influencing the BAP test. The effect exerted by RT on BAP test is in line with the antioxidant power measured by CBA in the RT food matrix. In contrast, the oral administration of GT to HFD rats was able to enhance SHp levels, but does not influence BAP levels in accordance with the outcomes of FRAP assay.

Even more evidently, RT reduced the prooxidant status in terms of lypoperoxides and hydroperoxides. Similarly, the hepatic oxidative stress biomarkers were positively modified by RT treatment in terms of MDA, RONS, and GSH levels. All these outcomes evaluated with crucial parameters of redox homeostasis confirmed the specific effects of RT in MetS, in accordance with previous literature [61], supporting the importance of tomato-based products in the prevention of oxidant-driven dysmetabolism.

A possible mechanism implicated in the GT-mediated protection exerted on HFD-induced liver damage could implicate adipokines (Figure 10). Indeed, alterations in adipokine levels are able to induce lipotoxicity and glucotoxicity in the liver with the development of steatosis through the involvement of TRP receptors [62]. Furthermore, according to data in the literature, naringenin—which we found to be specifically present in golden tomato—has been shown to act as a ligand for TRP receptors involved in many physiological processes that affect energy balance, inflammation, neuronal modulation, and oxidative stress [63–65]. It can be hypothesized that the reduction of free radicals observed in the liver and the regression of hepatic steatosis after treatment with the golden tomato could be associated with changes in the levels of the adipokines, leptin and adiponectin, as well as the involvement of TRP receptors, particularly TRPV1, as reported in the literature. This study opens up new possibilities for investigating the effect of golden tomato on the regression of NAFLD mediated by TRPV1 channels secondary to dietary activation of UCP2. Considering this, given the emerging clinical role of adipokines in cardiovascular disorders, the association of golden tomato with hypoadiponectinemia could suggest its role in the prevention of cardiovascular disorders.

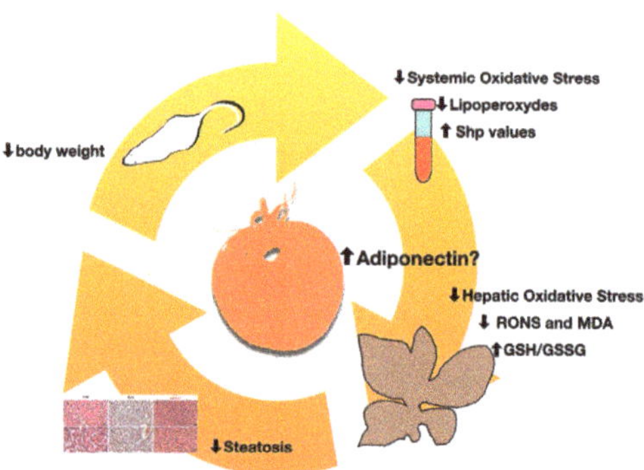

Figure 10. Schematic representation of possible pathophysiological processes implicated in the protection by GT in MetS. The protection exerted by GT on body weight gain, systemic, and hepatic oxidative stress levels (i.e., SHp, lipoperoxides, MDA, RONS and GSH/GSSG) could hypothetically be due to an increase in adiponectin.

Furthermore, it has been shown that there is a correlation between alterations in the gut microbiota and the development of NAFLD [66,67]. In this view, the intake of golden tomato, extremely rich in fiber, could lead to the production of acetic acid, butyric acid, and propionic acid (SCFAs) important in intestinal homeostasis. SCFAs produced by fiber fermentation are able to influence the regulation of intestinal hormones such as gut trypsin peptide, glucagon-like peptide-1 (GLP-1), leptin, and peptide tyrosine–tyrosine (PYY) involved in energy balance and to maintain the integrity of the intestinal barrier by reducing the entry of substances that can generate metabolic inflammation responsible for metabolic disorders [68–70]. The results of the present study provide also an opportunity to further investigate the effect of golden tomatoes on the gut microbiota, focusing also on pre/probiotic compounds that could play a role [71], and the possible preventive action of this food in metabolic syndrome.

5. Conclusions

Our research for the first time discloses the potential of "Golden Tomato" oral supplementation, via a comprehensive characterization of its biochemical, antioxidant, and disease-modifying properties, exploiting an in vivo rat model of MetS. The current investigation hints at GT as a powerful novel nutritional modulator and functional food, similarly to more widely known bioactive molecules that can also influence brain processes [72]. Indeed, in the context of nutritional supplementation and healthy diet, antioxidant-functional foods could underpin a robust strategy for the prevention and management of a wide plethora of physiological conditions, theoretically inasmuch as adaptive responses could be implicated in the homeostatic balance of complex networks [73]. This research hence substantiates the need for further investigations on tomato-based functional food, with a view to fully uncover GT nutritional application in food research.

Supplementary Materials: The following supporting information can be downloaded at: https://www.mdpi.com/article/10.3390/antiox12051121/s1, Table S1: Inclusion list of compounds sought and identified in the tomato samples analyzed; Table S2: Nutritional properties of tomato samples. The values are calculated on 100 g of edible part and are reported as the Mean ± SD of three repetitions; Table S3: Amount of micronutrients expressed in mg per 100 g of dry weight; Table S4: Quantity in mg of organic acids present in 100 g of dry product.

Author Contributions: Conceptualization: G.G. (Giuditta Gambino), G.G. (Giuseppe Giglia), G.F., P.S. and D.D.M.; methodology: G.G. (Giuditta Gambino), G.G. (Giuseppe Giglia), M.A., V.D.L., F.P.Z., F.R., I.R., F.V., F.S., E.P., G.A., P.S. and D.D.M.; validation, D.D.M. and G.G. (Giuseppe Giglia); formal analysis D.D.M., P.S., G.G. (Giuseppe Giglia), M.A., V.D.L., F.P.Z., F.R, I.R., F.V., F.S., E.P., G.A. and G.G. (Giuditta Gambino); investigation, D.D.M., P.S., G.G. (Giuseppe Giglia), M.A., F.P.Z., F.R., I.R., F.V., F.S., E.P., G.A., V.D.L. and G.G. (Giuditta Gambino); resources, D.D.M., G.F., P.S. and G.G. (Giuditta Gambino); data curation, D.D.M., P.S., G.G. (Giuseppe Giglia), M.A., F.P.Z., F.R. and G.G. (Giuditta Gambino); writing—original draft preparation, D.D.M., P.S., G.G. (Giuseppe Giglia), M.A., V.D.L., F.P.Z. and G.G. (Giuditta Gambino); writing—review and editing, all authors; supervision, D.D.M., P.S., G.G. (Giuseppe Giglia) and G.G. (Giuditta Gambino). All authors have read and agreed to the published version of the manuscript.

Funding: This work was partly supported by the project IN.PO.S.A-Grant Number: G66D20000170009, funded by PSR Sicilia (2014-2020)-16.1 and partly with FFR Funds provided by University of Palermo to Danila Di Majo and Giuditta Gambino.

Institutional Review Board Statement: The experimental protocols were approved by the animal welfare committee of the University of Palermo, authorized by the Ministry of Health (Rome, Italy; Authorization Number 14/2022-PR) and conducted following the ARRIVE guidelines.

Informed Consent Statement: Not applicable.

Data Availability Statement: The data presented in this study are available on reasonable request from the corresponding author.

Acknowledgments: We thank Aten Center, in particular Rosa Pitonzo, and Gaetano Caldara and Riccardo Messina from the University of Palermo. We also thank Daniele Gallo, Alessandro Strano, Miriana Scordino, and Antonino Vitale, studying at the University of Palermo.

Conflicts of Interest: The authors declare no conflict of interest.

References

1. Cruz-Carrión, Á.; Ruiz de Azua, M.J.; Bravo, F.I.; Aragonès, G.; Muguerza, B.; Suárez, M.; Arola-Arnal, A. Tomatoes consumed in-season prevent oxidative stress in Fischer 344 rats: Impact of geographical origin. *Food Funct.* **2021**, *12*, 8340–8350. [CrossRef] [PubMed]
2. Willcox, J.K.; Catignani, G.L.; Lazarus, S. Tomatoes and Cardiovascular Health. *Crit. Rev. Food Sci. Nutr.* **2003**, *43*, 1–18. [CrossRef] [PubMed]
3. Cheng, H.M.; Koutsidis, G.; Lodge, J.K.; Ashor, A.; Siervo, M.; Lara, J. Tomato and lycopene supplementation and cardiovascular risk factors: A systematic review and meta-analysis. *Atherosclerosis* **2017**, *257*, 100–108. [CrossRef] [PubMed]
4. Islam, N.; Akhter, N. Comparative Study of Protective Effect of Tomato Juice and N-Hexane Extract of Tomato on Blood Lipids and Oxidative Stress in Cholesterol-Fed Rats. *Anwer Khan Mod. Med. Coll. J.* **2017**, *8*, 30–37. [CrossRef]
5. Khayat Nouri, M.H.; Abad, A.N.A. Comparative Study of Tomato and Tomato Paste Supplementation on the Level of Serum Lipids and Lipoproteins Levels in Rats Fed With High Cholesterol. *Iran. Red Crescent Med. J.* **2013**, *15*, 287–291. [CrossRef]
6. Bose, K.S.C.; Agrawal, B.K. Effect of long term supplementation of tomatoes (cooked) on levels of antioxidant enzymes, lipid peroxidation rate, lipid profile and glycated haemoglobin in Type 2 diabetes mellitus. *West Indian Med. J.* **2006**, *55*, 274–278. [CrossRef]
7. Blum, A.; Merei, M.; Karem, A.; Blum, N.; Ben-Arzi, S.; Wirsansky, I.; Khazim, K. Effects of tomatoes on the lipid profile. *Clin. Investig. Med.* **2006**, *29*, 298–300.
8. Cuevas-Ramos, D.; Almeda-Valdés, P.; Chávez-Manzanera, E.; Meza-Arana, C.E.; Brito-Córdova, G.; Mehta, R.R.; Pérez-Méndez, O.O.; Gómez-Pérez, F.F.J. Effect of tomato consumption on high-density lipoprotein cholesterol level: A randomized, single-blinded, controlled clinical trial. *Diabetes Metab. Syndr. Obes. Targets Ther.* **2013**, *6*, 263–273. [CrossRef]

9. Thies, F.; Masson, L.F.; Rudd, A.; Vaughan, N.; Tsang, C.; Brittenden, J.; Simpson, W.G.; Duthie, S.; Horgan, G.W.; Duthie, G. Effect of a tomato-rich diet on markers of cardiovascular disease risk in moderately overweight, disease-free, middle-aged adults: A randomized controlled trial. *Am. J. Clin. Nutr.* **2012**, *95*, 1013–1022. [CrossRef]

10. Gambino, G.; Giglia, G.; Schiera, G.; Di Majo, D.; Epifanio, M.S.; La Grutta, S.; Baido, R.L.; Ferraro, G.; Sardo, P. Haptic Perception in Extreme Obesity: qEEG Study Focused on Predictive Coding and Body Schema. *Brain Sci.* **2020**, *10*, 908. [CrossRef]

11. Di Majo, D.; Cacciabaudo, F.; Accardi, G.; Gambino, G.; Giglia, G.; Ferraro, G.; Candore, G.; Sardo, P. Ketogenic and Modified Mediterranean Diet as a Tool to Counteract Neuroinflammation in Multiple Sclerosis: Nutritional Suggestions. *Nutrients* **2022**, *14*, 2384. [CrossRef]

12. Giammanco, M.; Lantieri, L.; Leto, G.; Plescia, F.; Di Majo, D. Nutrition, obesity and hormones. *J. Biol. Res.* **2018**, *91*. [CrossRef]

13. Spahis, S.; Borys, J.-M.; Levy, E. Metabolic Syndrome as a Multifaceted Risk Factor for Oxidative Stress. *Antioxid. Redox Signal.* **2017**, *26*, 445–461. [CrossRef] [PubMed]

14. Kim, B.; Feldman, E.L. Insulin resistance as a key link for the increased risk of cognitive impairment in the metabolic syndrome. *Exp. Mol. Med.* **2015**, *47*, e149. [CrossRef]

15. Huang, P.L. A comprehensive definition for metabolic syndrome. *Dis. Model. Mech.* **2009**, *2*, 231–237. [CrossRef]

16. Alberti, K.G.M.M.; Zimmet, P.; Shaw, J. Metabolic syndrome—A new world-wide definition. A Consensus Statement from the International Diabetes Federation. *Diabet. Med.* **2006**, *23*, 469–480. [CrossRef] [PubMed]

17. DeVallance, E.; Li, Y.; Jurczak, M.J.; Cifuentes-Pagano, E.; Pagano, P.J. The Role of NADPH Oxidases in the Etiology of Obesity and Metabolic Syndrome: Contribution of Individual Isoforms and Cell Biology. *Antioxid. Redox Signal.* **2019**, *31*, 687–709. [CrossRef]

18. Vatashchuk, M.V.; Bayliak, M.M.; Hurza, V.V.; Storey, K.B.; Lushchak, V.I. Metabolic Syndrome: Lessons from Rodent and Drosophila Models. *BioMed Res. Int.* **2022**, *2022*, 5850507. [CrossRef]

19. Rodríguez-Correa, E.; González-Pérez, I.; Clavel-Pérez, P.I.; Contreras-Vargas, Y.; Carvajal, K. Biochemical and nutritional overview of diet-induced metabolic syndrome models in rats: What is the best choice? *Nutr. Diabetes* **2020**, *10*, 24. [CrossRef]

20. Di Majo, D.; Sardo, P.; Giglia, G.; Di Liberto, V.; Zummo, F.P.; Zizzo, M.G.; Caldara, G.F.; Rappa, F.; Intili, G.; van Dijk, R.M.; et al. Correlation of Metabolic Syndrome with Redox Homeostasis Biomarkers: Evidence from High-Fat Diet Model in Wistar Rats. *Antioxidants* **2022**, *12*, 89. [CrossRef]

21. Li, P.; Lu, B.; Gong, J.; Li, L.; Chen, G.; Zhang, J.; Chen, Y.; Tian, X.; Han, B.; Guo, Y.; et al. Chickpea Extract Ameliorates Metabolic Syndrome Symptoms via Restoring Intestinal Ecology and Metabolic Profile in Type 2 Diabetic Rats. *Mol. Nutr. Food Res.* **2021**, *65*, e2100007. [CrossRef] [PubMed]

22. Zhao, M.; Cai, H.; Jiang, Z.; Li, Y.; Zhong, H.; Zhang, H.; Feng, F. Glycerol-Monolaurate-Mediated Attenuation of Metabolic Syndrome is Associated with the Modulation of Gut Microbiota in High-Fat-Diet-Fed Mice. *Mol. Nutr. Food Res.* **2019**, *63*, e1801417. [CrossRef] [PubMed]

23. Ali, M.Y.; Ibn Sina, A.A.I.; Khandker, S.S.; Neesa, L.; Tanvir, E.M.; Kabir, A.; Khalil, M.I.; Gan, S.H. Nutritional Composition and Bioactive Compounds in Tomatoes and Their Impact on Human Health and Disease: A Review. *Foods* **2020**, *10*, 45. [CrossRef] [PubMed]

24. Li, H.; Chen, A.; Zhao, L.; Bhagavathula, A.S.; Amirthalingam, P.; Rahmani, J.; Salehisahlabadi, A.; Abdulazeem, H.M.; Adebayo, O.; Yin, X. Effect of tomato consumption on fasting blood glucose and lipid profiles: A systematic review and meta-analysis of randomized controlled trials. *Phytother. Res.* **2020**, *34*, 1956–1965. [CrossRef]

25. Moncada, A.; Vetrano, F.; Esposito, A.; Miceli, A. Effects of NAA and *Ecklonia maxima* Extracts on Lettuce and Tomato Transplant Production. *Agronomy* **2022**, *12*, 329. [CrossRef]

26. McGuire, R.G. Reporting of Objective Color Measurements. *Hortscience* **1992**, *27*, 1254–1255. [CrossRef]

27. Gulcin, İ. Antioxidants and antioxidant methods: An updated overview. *Arch. Toxicol.* **2020**, *94*, 651–715. [CrossRef]

28. Benzie, I.F.F.; Strain, J.J. Ferric reducing/antioxidant power assay: Direct measure of total antioxidant activity of biological fluids and modified version for simultaneous measurement of total antioxidant power and ascorbic acid concentration. *Methods Enzymol.* **1999**, *299*, 15–27. [CrossRef]

29. Phachonpai, W.; Muchimapura, S.; Tong-Un, T.; Wattanathorn, J.; Thukhammee, W.; Thipkaew, C.; Sripanidkulchai, B.; Wannanon, P. Acute Toxicity Study of Tomato Pomace Extract in Rodent. *OnLine J. Biol. Sci.* **2013**, *13*, 28–34. [CrossRef]

30. Aborehab, N.M.; El Bishbishy, M.H.; Waly, N.E. Resistin mediates tomato and broccoli extract effects on glucose homeostasis in high fat diet-induced obesity in rats. *BMC Complement. Altern. Med.* **2016**, *16*, 1–10. [CrossRef]

31. Reagan-Shaw, S.; Nihal, M.; Ahmad, N. Dose translation from animal to human studies revisited. *FASEB J.* **2008**, *22*, 659–661. [CrossRef] [PubMed]

32. Bowe, J.E.; Franklin, Z.J.; Hauge-Evans, A.C.; King, A.J.; Persaud, S.J.; Jones, P.M. Metabolic Phenotyping Guidelines: Assessing Glucose Homeostasis in Rodent Models. *J. Endocrinol.* **2014**, *222*, G13–G25. [CrossRef] [PubMed]

33. Lee, S.; Hashimoto, J.; Suzuki, T.; Satoh, A. The effects of exercise load during development on oxidative stress levels and antioxidant potential in adulthood. *Free. Radic. Res.* **2017**, *51*, 179–186. [CrossRef] [PubMed]

34. Macri, A.; Scanarotti, C.; Bassi, A.M.; Giuffrida, S.; Sangalli, G.; Traverso, C.E.; Iester, M. Evaluation of oxidative stress levels in the conjunctival epithelium of patients with or without dry eye, and dry eye patients treated with preservative-free hyaluronic acid 0.15 % and vitamin B12 eye drops. *Graefe's Arch. Clin. Exp. Ophthalmol.* **2015**, *253*, 425–430. [CrossRef] [PubMed]

35. Mancini, S.; Mariani, F.; Sena, P.; Benincasa, M.; Roncucci, L. Myeloperoxidase expression in human colonic mucosa is related to systemic oxidative balance in healthy subjects. *Redox Rep.* **2017**, *22*, 399–407. [CrossRef]

36. Jansen, E.; Ruskovska, T. Serum Biomarkers of (Anti)Oxidant Status for Epidemiological Studies. *Int. J. Mol. Sci.* **2015**, *16*, 27378–27390. [CrossRef]

37. Ohkawa, H.; Ohishi, N.; Yagi, K. Assay for lipid peroxides in animal tissues by thiobarbituric acid reaction. *Anal. Biochem.* **1979**, *95*, 351–358. [CrossRef]

38. Terzo, S.; Attanzio, A.; Calvi, P.; Mulè, F.; Tesoriere, L.; Allegra, M.; Amato, A. Indicaxanthin from *Opuntia ficus-indica* Fruit Ameliorates Glucose Dysmetabolism and Counteracts Insulin Resistance in High-Fat-Diet-Fed Mice. *Antioxidants* **2021**, *11*, 80. [CrossRef]

39. Auberval, N.; Dal, S.; Bietiger, W.; Pinget, M.; Jeandidier, N.; Maillard-Pedracini, E.; Schini-Kerth, V.; Sigrist, S. Metabolic and oxidative stress markers in Wistar rats after 2 months on a high-fat diet. *Diabetol. Metab. Syndr.* **2014**, *6*, 130. [CrossRef]

40. Di Majo, D.; La Neve, L.; La Guardia, M.; Casuccio, A.; Giammanco, M. The influence of two different pH levels on the antioxidant properties of flavonols, flavan-3-ols, phenolic acids and aldehyde compounds analysed in synthetic wine and in a phosphate buffer. *J. Food Compos. Anal.* **2011**, *24*, 265–269. [CrossRef]

41. Alam, M.A.; Subhan, N.; Rahman, M.M.; Uddin, S.J.; Reza, H.M.; Sarker, S.D. Effect of Citrus Flavonoids, Naringin and Naringenin, on Metabolic Syndrome and Their Mechanisms of Action. *Adv. Nutr.* **2014**, *5*, 404–417. [CrossRef]

42. Santana-Gálvez, J.; Cisneros-Zevallos, L.; Jacobo-Velázquez, D.A. Chlorogenic Acid: Recent Advances on Its Dual Role as a Food Additive and a Nutraceutical against Metabolic Syndrome. *Molecules* **2017**, *22*, 358. [CrossRef]

43. Giammanco, M.; Aiello, S.; Casuccio, A.; La Guardia, M.; Cicero, L.; Puleio, R.; Vazzana, I.; Tomasello, G.; Cassata, G.; Leto, G.; et al. Effects of 3,5-Diiodo-L-Thyronine on the Liver of High Fat Diet Fed Rats. *J. Biol. Res.* **2016**, 89. [CrossRef]

44. Li, S.; Zhang, Y.; Sun, Y.; Zhang, G.; Bai, J.; Guo, J.; Su, X.; Du, H.; Cao, X.; Yang, J.; et al. Naringenin improves insulin sensitivity in gestational diabetes mellitus mice through AMPK. *Nutr. Diabetes* **2019**, *9*, 28. [CrossRef] [PubMed]

45. Zygmunt, K.; Faubert, B.; MacNeil, J.; Tsiani, E. Naringenin, a citrus flavonoid, increases muscle cell glucose uptake via AMPK. *Biochem. Biophys. Res. Commun.* **2010**, *398*, 178–183. [CrossRef] [PubMed]

46. Ma, Y.; Gao, M.; Liu, D. Chlorogenic Acid Improves High Fat Diet-Induced Hepatic Steatosis and Insulin Resistance in Mice. *Pharm. Res.* **2015**, *32*, 1200–1209. [CrossRef]

47. Jin, S.; Chang, C.; Zhang, L.; Liu, Y.; Huang, X.; Chen, Z. Chlorogenic Acid Improves Late Diabetes through Adiponectin Receptor Signaling Pathways in db/db Mice. *PLoS ONE* **2015**, *10*, e0120842. [CrossRef]

48. Wan, C.-W.; Wong, C.N.-Y.; Pin, W.-K.; Wong, M.H.-Y.; Kwok, C.-Y.; Chan, R.Y.-K.; Yu, P.H.-F.; Chan, S.-W. Chlorogenic Acid Exhibits Cholesterol Lowering and Fatty Liver Attenuating Properties by Up-regulating the Gene Expression of PPAR-α in Hypercholesterolemic Rats Induced with a High-Cholesterol Diet. *Phytother. Res.* **2013**, *27*, 545–551. [CrossRef]

49. Lee, M.-K.; Moon, S.-S.; Lee, S.-E.; Bok, S.-H.; Jeong, T.-S.; Park, Y.B.; Choi, M.-S. Naringenin 7-O-cetyl ether as inhibitor of HMG-CoA reductase and modulator of plasma and hepatic lipids in high cholesterol-fed rats. *Bioorg. Med. Chem.* **2003**, *11*, 393–398. [CrossRef]

50. Andrade-Pavón, D.; Gómez-García, O.; Villa-Tanaca, L. Molecular Recognition of Citroflavonoids Naringin and Naringenin at the Active Site of the HMG-CoA Reductase and DNA Topoisomerase Type II Enzymes of *Candida* spp. and *Ustilago maydis*. *Indian J. Microbiol.* **2022**, *62*, 79–87. [CrossRef]

51. Cho, A.-S.; Jeon, S.-M.; Kim, M.-J.; Yeo, J.; Seo, K.-I.; Choi, M.-S.; Lee, M.-K. Chlorogenic acid exhibits anti-obesity property and improves lipid metabolism in high-fat diet-induced-obese mice. *Food Chem. Toxicol.* **2010**, *48*, 937–943. [CrossRef] [PubMed]

52. Huang, K.; Liang, X.-C.; Zhong, Y.-L.; He, W.-Y.; Wang, Z. 5-Caffeoylquinic acid decreases diet-induced obesity in rats by modulating PPARα and LXRα transcription. *J. Sci. Food Agric.* **2015**, *95*, 1903–1910. [CrossRef] [PubMed]

53. Paolisso, G.; Gambardella, A.; Tagliamonte, M.R.; Saccomanno, F.; Salvatore, T.; Gualdiero, P.; D'Onofrio, M.V.; Howard, B.V. Does free fatty acid infusion impair insulin action also through an increase in oxidative stress? *J. Clin. Endocrinol. Metab.* **1996**, *81*, 4244–4248. [CrossRef]

54. McKeegan, K.; Mason, S.A.; Trewin, A.J.; Keske, M.A.; Wadley, G.D.; Della Gatta, P.A.; Nikolaidis, M.G.; Parker, L. Reactive oxygen species in exercise and insulin resistance: Working towards personalized antioxidant treatment. *Redox Biol.* **2021**, *44*, 102005. [CrossRef]

55. Allegra, M.; Gentile, C.; Tesoriere, L.; Livrea, M.A. Protective effect of melatonin against cytotoxic actions of malondialdehyde: An in vitro study on human erythrocytes. *J. Pineal Res.* **2002**, *32*, 187–193. [CrossRef]

56. Li, N.; Wu, X.; Zhuang, W.; Xia, L.; Chen, Y.; Wu, C.; Rao, Z.; Du, L.; Zhao, R.; Yi, M.; et al. Tomato and lycopene and multiple health outcomes: Umbrella review. *Food Chem.* **2021**, *343*, 128396. [CrossRef]

57. Kakimoto, P.A.; Kowaltowski, A.J. Effects of high fat diets on rodent liver bioenergetics and oxidative imbalance. *Redox Biol.* **2016**, *8*, 216–225. [CrossRef] [PubMed]

58. Li, J.; Wang, T.; Liu, P.; Yang, F.; Wang, X.; Zheng, W.; Sun, W. Hesperetin ameliorates hepatic oxidative stress and inflammation via the PI3K/AKT-Nrf2-ARE pathway in oleic acid-induced HepG2 cells and a rat model of high-fat diet-induced NAFLD. *Food Funct.* **2021**, *12*, 3898–3918. [CrossRef]

59. Elvira-Torales, L.I.; Navarro-González, I.; González-Barrio, R.; Martín-Pozuelo, G.; Doménech, G.; Seva, J.; García-Alonso, J.; Periago-Castón, M.J. Tomato Juice Supplementation Influences the Gene Expression Related to Steatosis in Rats. *Nutrients* **2018**, *10*, 1215. [CrossRef]

60. Alam, P.; Raka, M.A.; Khan, S.; Sarker, J.; Ahmed, N.; Nath, P.D.; Hasan, N.; Mohib, M.M.; Tisha, A.; Taher Sagor, M.A. A clinical review of the effectiveness of tomato (*Solanum lycopersicum*) against cardiovascular dysfunction and related metabolic syndrome. *J. Herb. Med.* **2019**, *16*, 100235. [CrossRef]

61. Ferron, A.J.T.; Aldini, G.; Francisqueti-Ferron, F.V.; Silva, C.C.V.d.A.; Bazan, S.G.Z.; Garcia, J.L.; de Campos, D.H.S.; Ghiraldeli, L.; Kitawara, K.A.H.; Altomare, A.; et al. Protective Effect of Tomato-Oleoresin Supplementation on Oxidative Injury Recoveries Cardiac Function by Improving β-Adrenergic Response in a Diet-Obesity Induced Model. *Antioxidants* **2019**, *8*, 368. [CrossRef] [PubMed]

62. Araújo, M.C.; Soczek, S.H.S.; Pontes, J.P.; Marques, L.A.C.; Santos, G.S.; Simão, G.; Bueno, L.R.; Maria-Ferreira, D.; Muscará, M.N.; Fernandes, E.S. An Overview of the TRP-Oxidative Stress Axis in Metabolic Syndrome: Insights for Novel Therapeutic Approaches. *Cells* **2022**, *11*, 1292. [CrossRef] [PubMed]

63. Holzer, P.; Izzo, A.A. The pharmacology of TRP channels. *Br. J. Pharmacol.* **2014**, *171*, 2469–2473. [CrossRef] [PubMed]

64. Gambino, G.; Gallo, D.; Covelo, A.; Ferraro, G.; Sardo, P.; Giglia, G. TRPV1 channels in nitric oxide-mediated signalling: Insight on excitatory transmission in rat CA1 pyramidal neurons. *Free. Radic. Biol. Med.* **2022**, *191*, 128–136. [CrossRef]

65. Gambino, G.; Rizzo, V.; Giglia, G.; Ferraro, G.; Sardo, P. Cannabinoids, TRPV and Nitric Oxide: The Three Ring Circus of Neuronal Excitability. *Brain Struct. Funct.* **2020**, *225*, 1–15. [CrossRef]

66. De Vadder, F.; Kovatcheva-Datchary, P.; Goncalves, D.; Vinera, J.; Zitoun, C.; Duchampt, A.; Bäckhed, F.; Mithieux, G. Microbiota-Generated Metabolites Promote Metabolic Benefits via Gut-Brain Neural Circuits. *Cell* **2014**, *156*, 84–96. [CrossRef]

67. Wang, P.-X.; Deng, X.-R.; Zhang, C.-H.; Yuan, H.-J. Gut microbiota and metabolic syndrome. *Chin. Med. J.* **2020**, *133*, 808–816. [CrossRef]

68. Dabke, K.; Hendrick, G.; Devkota, S. The gut microbiome and metabolic syndrome. *J. Clin. Investig.* **2019**, *129*, 4050–4057. [CrossRef]

69. Tralongo, P.; Tomasello, G.; Sinagra, E.; Damiani, P.; Leone, A.; Palumbo, V.; Giammanco, M.; Di Majo, D.; Abruzzo, A.; Bruno, A.; et al. The Role of Butyric Acid as a Oprotective Agent against Inflammatory Bowel Disease. *Euromediterranean Biomed. J.* **2014**, *9*, 24–35. [CrossRef]

70. Tomasello, G.; Zeenny, M.; Giammanco, M.; Di Majo, D.; Traina, G.; Sinagra, E.; Damiani, P.; Zein, R.; Jurjus, A.; Leone, A. Intestinal Microbiota Mutualism and Gastrointestinal Diseases. *Euromediterr. Biomed. J.* **2015**, *10*, 65–75. [CrossRef]

71. Spooner, H.C.; Derrick, S.A.; Maj, M.; Manjarín, R.; Hernandez, G.V.; Tailor, D.S.; Bastani, P.S.; Fanter, R.K.; Fiorotto, M.L.; Burrin, D.G.; et al. High-Fructose, High-Fat Diet Alters Muscle Composition and Fuel Utilization in a Juvenile Iberian Pig Model of Non-Alcoholic Fatty Liver Disease. *Nutrients* **2021**, *13*, 4195. [CrossRef] [PubMed]

72. Gambino, G.; Brighina, F.; Allegra, M.; Marrale, M.; Collura, G.; Gagliardo, C.; Attanzio, A.; Tesoriere, L.; Di Majo, D.; Ferraro, G.; et al. Modulation of Human Motor Cortical Excitability and Plasticity by Opuntia Ficus Indica Fruit Consumption: Evidence from a Preliminary Study through Non-Invasive Brain Stimulation. *Nutrients* **2022**, *14*, 4915. [CrossRef] [PubMed]

73. Palermo, A.; Giglia, G.; Vigneri, S.; Cosentino, G.; Fierro, B.; Brighina, F. Does habituation depend on cortical inhibition? Results of a rTMS study in healthy subjects. *Exp. Brain Res.* **2011**, *212*, 101–107. [CrossRef] [PubMed]

Article

Ethnopharmacological Effects of *Urtica dioica*, *Matricaria chamomilla*, and *Murraya koenigii* on Rotenone-Exposed *D. melanogaster*: An Attenuation of Cellular, Biochemical, and Organismal Markers

Shabnam Shabir [1], Sumaira Yousuf [1], Sandeep Kumar Singh [2,*], Emanuel Vamanu [3,*] and Mahendra P. Singh [1,*]

[1] School of Bioengineering and Biosciences, Lovely Professional University, Phagwara 144411, India
[2] Indian Scientific Education and Technology Foundation, Lucknow 226002, India
[3] Faculty of Biotechnology, University of Agricultural Sciences and Veterinary Medicine, 011464 Bucharest, Romania
* Correspondence: sandeeps.bhu@gmai.com (S.K.S.); email@emanuelvamanu.ro (E.V.); mahendra.19817@lpu.co.in (M.P.S.)

Abstract: Natural antioxidants derived from plants have been proven to have significant inhibitory effects on the free radicals of living organisms during actively metabolization. Excessive production of free radicals increases the risk of neurodegenerative diseases, such as Alzheimer's disease, Parkinson's disease, and motor sclerosis. This study aimed to compare the ethnopharmacological effects of *Urtica dioica* (UD), *Matricaria chamomilla* (MC), and *Murraya koenigii* (MK) on the amelioration of rotenone-induced toxicity in wild-type *Drosophila melanogaster* (Oregon R$^+$) at biochemical, cellular, and behavioral levels. Phytoextracts were prepared from all three plants, i.e., UD, MC, and MK (aqueous and ethanolic fractions), and their bioactive compounds were evaluated using in vitro biochemical parameters (DPPH, ABTS, TPC, and TFC), UV-Vis, followed by FT-IR and HPLC. Third instar larvae and freshly eclosed flies were treated with 500 µM rotenone alone or in combination with UD, MC, and MK for 24 to 120 h. Following exposure, cytotoxicity (dye exclusion test), biochemical (protein estimation and acetylcholinesterase inhibition assays), and behavioral assays (climbing and jumping assays) were performed. Among all three plant extracts, MK exhibited the highest antioxidant properties due to the highest TPC, TFC, DPPH, and ABTS, followed by UD, then MC. The overall trend was MK > UD > MC. In this context, ethnopharmacological properties mimic the same effect in *Drosophila*, exhibiting significantly ($p < 0.05$) reduced cytotoxicity (trypan blue), improved biochemical parameters (proteotoxicity and AChE activity), and better behavioral parameters in the organisms cotreated with phyto extracts compared with rotenone. Conclusively, UV-Vis, FTIR, and HPLC analyses differentiated the plant extracts. The findings of this research may be beneficial in the use of select herbs as viable sources of phyto-ingredients that could be of interest in nutraceutical development and various clinical applications.

Keywords: antioxidants; acetylcholinesterase; 1,1-diphenyl-2-picrylhydrazyl; HPLC; medicinal plants; oxidative stress

Citation: Shabir, S.; Yousuf, S.; Singh, S.K.; Vamanu, E.; Singh, M.P. Ethnopharmacological Effects of *Urtica dioica, Matricaria chamomilla,* and *Murraya koenigii* on Rotenone-Exposed *D. melanogaster*: An Attenuation of Cellular, Biochemical, and Organismal Markers. *Antioxidants* **2022**, *11*, 1623. https://doi.org/10.3390/antiox11081623

Academic Editors: Antonella D'Anneo and Marianna Lauricella

Received: 27 July 2022
Accepted: 17 August 2022
Published: 21 August 2022

Publisher's Note: MDPI stays neutral with regard to jurisdictional claims in published maps and institutional affiliations.

1. Introduction

The cellular redox status is determined by the balance between antioxidants and oxidants. The oxidative state of the cell is defined by an imbalance between these two, which can result in apoptosis or necrosis [1]. Reactive oxygen species (ROS) are primarily responsible for the high susceptibility of brain cells to oxidative stress [2]. Although oxygen is a relatively nonreactive substance, it can be metabolized in the body to create highly reactive free radicals, such as hydroxyl radicals (OH$^-$), superoxide anions (O$_2$$^-$), and many other reactive species [3]. These free radical species contribute to the pathophysiology of many neurodegenerative diseases, including amyloidosis, α-synucleinopathies, aging, and TDP-43 proteinopathies, which are caused by the inhibition of acetylcholinesterase [4]. Detoxification is aided by cellular defense mechanisms that involve endogenous antioxidants and antioxidant enzymes such as superoxide dismutase, glutathione reductase, lipid peroxidase, glutathione, and catalase [5]. The deterioration of these defense mechanisms damages significant cell biomolecules (lipids, DNA, and proteins) and ultimately leads to cell death [6].

Currently, conventional medicines or therapeutic drugs (levodopa or dopaminergic agonists) have only been used to treat motor symptoms by restoring neurotransmitters [7]. The extended use of these drugs, however, can have negative effects, such as fatigue and other motor difficulties [8]. Considering these limitations, there is a need for novel natural neuroprotective agents to halt or slow the progression of various neurodegenerative disorders [9]. According to cumulative evidence, nutraceuticals and other phytochemicals have been shown to have neuroprotective effects and alleviate the symptoms of neurodegenerative diseases by activating the PI3K/Akt/Nrf2 pathway by scavenging free radicals [10]. Herbal biomolecules have been used as phytomedicines in healthcare industries since the emergence of civilization; according to the World Health Organization (WHO), phytomedicine is used as a primary source of therapy by ~79% of the world's population [11]. *Urtica dioica* (UD), *Matricaria chamomilla* (MC), and *Murraya koenigii* (MK) have all been used as well-known cognition enhancers and nerve relaxants in the Ayurvedic medical system [12]. The phytochemicals of these three herbs, such as quercetin, polyphenols, alkaloids, reducing sugars, and vitamins, are known for their protective effects. UD has been extensively studied and has shown promising results in the treatment of prostate enlargement [13], colon carcinogenesis in rats [14], and protecting against hyperglycemia [15], hypertension, [16] and hypercholesterolemia [17]. Several studies have demonstrated the beneficial effects of MC against diabetes by regulating GLP-1, which is crucial in stimulating insulin gene transcription [18]. In addition, chamomile oil significantly decreased osteoarthritis [19], and may also help treat lung cancer [20]. Mondal et al. (2022) found that MK modulated various cellular programs and signaled cascades to intervene as an antioxidant in normal cells and as a pro-oxidant in lung carcinoma cells [21], protecting against liver damage caused by TPA [22]. Moreover, previous studies on the MK leaf fraction have observed its efficacy in treating hyperglycemia [23]. Although UD, MC, and MK have been used in several in vivo and in vitro experiments, there is still little evidence supporting of the therapeutic effect of these three herbs on cellular and neurological complications.

Considering the general protective and cognitive effects of UD, MC, and MK that have been reported in the literature. This study aimed to determine the bioactive compounds of these botanicals (UD, MC, and MK) through biochemical and analytical (UV-Vis, FTIR, and HPLC) methods. Furthermore, we attempted to investigate the cellular and neurological toxicities in a nontarget in vivo model of *D. melanogaster* through a widely used neurotoxic natural pesticide, rotenone (ROT). In addition, we also investigated its organismal effect on *Drosophila*, as it has been demonstrated to be a model framework of neurodegenerative diseases, particularly Parkinson's disease.

2. Materials and Methods

2.1. Chemicals and Reagents

Ellman's reagent (DTNB), 2,2′-azinobis (3-ethylbenzothiazoline-6-sulfonic acid) (ABTS), gallic acid ($C_7H_6O_5$), 1,-1-diphenyl-2-picrylhydrazyl (DPPH), ascorbic acid ($C_6H_8O_6$), sodium carbonate (Na_2CO_3), trichloroacetic acid ($C_2HCl_3O_2$), sulfuric acid (H_2SO_4), ferric chloride ($FeCl_3$), ethanol, gallic acid ($C_7H_6O_5$), and aluminum chloride hexahydrate ($AlCl_3$ $6H_2O$) were obtained from Hi-Media (Mumbai, India). The solvent utilized for HPLC analysis was of HPLC grade, whereas all other organic solvents and chemicals were of analytical grade. Rutin, quercetin, rotenone, and acetylcholinesterase were obtained from

Sigma (Roedermark, Germany). Millipore grade water was used. Calorimetric analysis was performed using a Shimadzu UV-1601 spectrophotometer (Tokyo, Japan).

2.2. Plant Materials Used in the Study

The young leaves of *Urtica dioica* (UD) were collected from the local kitchen gardens and apple orchards of Sopore (Sopore, India) before the plants started developing seeds. *Matricaria chamomilla* (MC) flowers were procured from the Mediaroma Agro Producer Company Limited Kaskanj (Kasganj, India), and young leaves of *Murraya koenigii* (MK) were obtained from the garden maintained by Lovely Professional University (Phagwara, India). The identification and authentication of the medicinal plants were confirmed by a taxonomist from the Plant Sciences Division of the CSIR-Indian Institute of Integrative Medicine (IIIM) (Jammu, India).

2.2.1. Drying, Processing, and Extraction of Samples

The collected plant leaves and flowers were washed under tap water and dried at room temperature for a week. After drying, all dry plant components were crushed to a fine powder with a mechanical grinder and then sieved through a 40-micron sieve to obtain fine particles. A 10% aqueous extract was prepared using five grams (± 0.05) of the powdered sample mixed with 50 mL deionized water and steeping at 95–100 °C for 10–15 min. Then, the extracts were filtered by using Whatman No. 1 filter paper, and a 10% ethanolic extract was prepared using Soxhlet. Twenty grams (± 0.05) of powdered plant samples were inserted into a Whatman 25×100 mm celluloid thimble by adding 200 mL of ethanol as a solvent at boiling temperature (70 °C). A dark green extract was obtained from UD and MK leaves, whereas the extract obtained from MC flowers was a pale yellow. The extract was then evaporated using a vacuum rotatory evaporator at 70 °C or less to remove the solvents. The crude extracts were weighed and stored at 4 °C in an airtight dark box for future assessment [24].

2.2.2. Determination of Plant Yield

The percentage yields of UD, MC, and MK samples were calculated using the following formula [25]:

$$W_2 - W_1 \times 100 \tag{1}$$

where W_2 signifies the weight of the extract including the container, W_1 signifies the weight of the container itself, and W_0 signifies the weight of the initial dried sample.

2.2.3. Free Radical Scavenging Activity Using the DPPH Radical Assay

Antioxidant potential was measured using the 1,-1-diphenyl-2-picrylhydrazyl (DPPH) test [26]. A freshly prepared solution of DPPH (0.011 gm) was taken in 50 mL methanol for spectrophotometric measurements. The DPPH solution was further diluted with methanol, and the optical density (OD) was set between 0.8–1. Different concentrations of plant fractions were added to every 2 mL of DPPH mixture. Absorbance was measured at 517 nm using a Shimadzu UV-1601 spectrophotometer (Tokyo, Japan) after 30 min of incubation. Methanol was used as a blank, and DPPH was used as a control. Triplicate experiments were performed. Using the following equation, the radical scavenging activity was calculated as percent inhibition (1%) of the DPPH radical:

$$\text{DPPH inhibition (\%)} = [(A_{control} - A_{sample})/A_{control}] \times 100 \tag{2}$$

where A_{sample} represents absorbance of the plant extract sample and $A_{control}$ represents absorbance of the DPPH solution as a control.

2.2.4. ABTS Radical Cation Decolorization Assay

The 2,2′-azinobis (3-ethylbenzothiazoline-6-sulfonic acid) radical cation decolorization analyte was also used to assess the efficacy of plant fractions to scavenge free radicals, which is based on the reduction of ABTS radicals by antioxidants in plant extracts [27]. The radical cation formed when the ABTS stock solution (0.036 g in 10 mL methanol) was mixed with potassium persulfate (0.057 g in 10 mL methanol) at a 1:1 ratio. Then, the mixture was incubated in darkness for 16 h at ambient temperature. To attain an optical density (OD) of 0.8–1, the ABTS solution was further diluted with methanol. Every 2 mL of ABTS solution had extracts of various concentrations added to it. All samples were measured at 745 nm after 30 min of incubation. The percentage of scavenging activity was determined using the following equation:

$$\text{ABTS inhibition } (\%) = [(A_{control} - A_{sample})/A_{control}] \times 100 \qquad (3)$$

where A_{sample} signifies absorbance of the plant extract sample and $A_{control}$ signifies absorbance of the ABTS solution as a control.

2.2.5. EC$_{50}$ (Dose-Response Relationship)

EC_{50} is the half-maximal effective concentration of an antibody, drug, or toxicant that elicits a reaction halfway between the baseline and maximum, following a specific duration of exposure. Using CompuSyn software, data analysis for free radical scavenging activity and dose-response studies were performed to assess the potency of the selected herbs. A lower EC_{50} indicates greater radical scavenging activity [28]. SPSS software was used for statistical analysis.

2.2.6. Determination of Total Phenolic Content (TPC)

The TPC of the plant fractions was measured by using Folin–Ciocalteu's colorimetric method [29]. Each plant fraction was coupled with 2.5 mL of the FC reagent (1:10 v/v) and vortexed. After 5 min, 2 ml of Na_2CO_3 (7.5%) were incorporated. Then, the solution was placed for approximately 90 min at room temperature before taking the OD at 760 nm by using a UV-Vis spectrophotometer. The results are given in mg GAE/g of dry weight. Triplicates of each sample were analyzed.

$$C = \frac{c \times V}{m} \qquad (4)$$

where 'C' indicates the total phenolic component content in (mg g^{-1}) plant extract in GAE, 'c' indicates the gallic acid concentration (mg mL^{-1}), 'V' indicates the volume of extract in microliters (μL), and 'm' indicates the weight of crude plant extract in grams.

The correlation coefficient (R^2) value was determined using the mean of three absorbance determinations for each concentration. The equation is shown below:

$$Y = mx + c \qquad (5)$$

where 'Y' signifies extract absorbance, 'm' signifies the slope of the calibration curve, 'x' signifies extract concentration, and 'c' is the intercept. Concentrations of extracts were calculated using this regression equation. The phenolic content was estimated using the calculated value for each extract concentration.

2.2.7. Determination of Total Flavonoid Content (TFC)

Total flavonoids were quantified using the aluminum chloride colorimetric technique [30]. The plant extracts were combined with methanol (1.5 mL), 100 μL of (10%) aluminum chloride followed by 0.1 mL of potassium acetate (1 M), and finally 2.8 mL of deionized water. The reaction mixture was placed for 40 min at ambient temperature, and the absorbance of the solution was obtained at 415 nm. Quercetin was used to create a calibration curve. The total flavonoid content was calculated in terms of quercetin equivalents (mg QE/g dry weight). Triplicate readings were taken for each plant sample.

$$C = \frac{c \times V}{m} \tag{6}$$

where 'C' is the total phenolic content in mg g^{-1} plant extract in E, "c" reflects the quantity of quercetin determined by the calibration graph (mg/mL), 'V' shows the volume of plant extract in μL and 'm' is the weight of crude plant extract in grams. The absorbance of each concentration of the extract was measured using the procedure described above. Then, using the calculations above, the total flavonoid content was determined.

2.2.8. Preliminary Qualitative Screening Analysis of Plant Extracts

Bioactive compounds such as polyphenols, reducing sugars, alkaloids, terpenoids, glycosides, flavonoids, saponins, and amino acids are mostly responsible for curative capabilities such as menstruation problems, muscular spasms, anemia, ulcers, hemorrhoids, inflammation, and wound healing [31]. The presence of phytochemicals is determined using conventional qualitative test methods, which include the following:

(a) *Detection of phenols (ferric chloride test):* An amount of 2 mL of plant extract was combined in a test tube with 2 mL of 5% ferric chloride aqueous solution. The presence of phenols was indicated by a deep blue-green solution [32].

(b) *Detection of flavonoids (Alkaline reagent test):* A few drops of NaOH (20%) solution were added to 2 mL of plant extract, which displayed a yellowish red color within a second and turned transparent with the addition of diluted HCl, displaying a positive result [33].

(c) *Detection of alkaloids (Wagner's test):* To 4 mL of plant extract, 3 drops of Wagner's reagent were added. The appearance of a reddish-brown precipitate indicated a positive outcome [33].

(d) *Detection of tannins (FeCl3 solution test):* An alcoholic ferric chloride (10%) solution was added to 2 mL of plant extract. The appearance of the blue/green color suggested a positive outcome [33].

(e) *Detection of carbohydrates (Molisch's test):* An amount of 3 mL of extract and 3 mL of H_2SO_4 were placed in a test tube; a few drops of Molisch's reagent were added (conc.). Allowing it to stand 3 min, the appearance of a red/dull violet tone at the interphase of the two layers showed a successful outcome [32].

(f) *Detection of saponins (saponin foam test):* Five milliliters of distilled water was combined with 500 μL of plant fractions. The presence of saponins is indicated by foaming (formation of creamy tiny bubbles) [34].

(g) *Detection of terpenoids (Salkowski test):* An amount of 2 mL of extract and a few drops of conc. H_2SO_4 was mixed with 1 mL of chloroform. The appearance of a reddish-brown precipitate indicated a positive outcome [32].

(h) *Detection of steroids (Liebermann-Burchard test):* An amount of 3 mL of acetic anhydride was mixed with 5 mL of plant extract. Then, 2 mL of H_2SO_4 was added to it. The presence of steroids was indicated by a shift in color from violet to bluish green [34].

(i) *Detection of glycosides (Kellar-Kiliani test):* An amount of 1 mL of glacial acetic acid was mixed with 2 mL of plant extract. Then, 1 mL of $FeCl_3$ and 1 mL of (conc.) H_2SO_4 was mixed into it. The appearance of glycosides was confirmed by the solution's greenish-blue color [34].

2.2.9. UV-Visible Spectroscopic Analysis

The ultraviolet spectral data were obtained by a Shimadzu UV-1601 spectrophotometer (Tokyo, Japan). UV-Vis spectroscopy is concerned with the absorption of radiation in the ultraviolet and visible spectra and mostly used for quantitative analysis [35]. This radiation permits electrons in atoms or molecules to shift from lower to higher energy levels. The level of radiation absorbed is proportional to the number of molecules in the solution under specified conditions. Spectral data were used to demonstrate a link between absorption concentration and intensity. Quality control might thus be evaluated with a UV-Vis spectrometer without the need for expensive markers by establishing a library of spectrum data from actual raw samples [36]. Extracts of UD, MC, and MK were used for UV-Visible analysis. The samples were tested with a spectral range of 200–800 nm at 1 nm intervals at room temperature.

2.2.10. FT-IR Spectroscopic Analysis

Fourier transform infrared spectrophotometry (FTIR) is one of the most potent instruments for detecting types of chemical bonds, molecular structures, and functional groups in substances. The absorbed wavelength of light is indicative of the chemical bond, as shown in the annotated spectrum. The chemical bonds of a molecule can be identified by interpreting the infrared absorption spectrum [37].

Ethanolic extracts of UD, MC, and MK were used for FTIR analysis. To make translucent sample discs, 10 mg of the crude plant extract sample was enclosed in 100 mg of KBr pellet. Then, the crude extract of each plant material was subjected to FT-IR spectroscopy (Perkin-Elmer Spectrum 2 with ATR and Pellet accessories). The samples were tested in the infrared band with a spectral range of 400–4000 cm^{-1} and a resolution of ± 4 cm^{-1} at room temperature.

2.2.11. HPLC Chromatographic Analysis

A Shimadzu Prominence I LC2030 Plus HPLC system (Kyoto, Japan) equipped with a Shimadzu LC 2030 UV-Vis detector was used to separate natural compounds from crude extracts of UD, MC, and MK. The standard external technique was used to perform HPLC analysis under isocratic conditions. Before running in the column, the mobile phase was degassed and filtered through a membrane using methanol and 0.5% acetic acid in water (90:10 v/v). Column C-18 (4.6 $\times$ 250 mm) with a 5 μm particle size was used and maintained at 25 °C temperature. Each injection volume was prepared at 20 μL and then injected into the HPLC. Samples were filtered using a 0.45 mm membrane filter (Millipore) before being put in vials, employing a 1.0 mL/min flow rate.

Spectral information was analyzed in the 200–400 nm region, and chromatograms were detected at a wavelength of 280 nm. Based on previous findings, the quantitative quantification of each bioactive component contained in the plant extracts was determined [38–40]. Peak identification was carried out by comparing the retention times of specific standards with those of the extract. The retention times of specific standards were compared with those of the extract to identify the peak.

2.3. Fly Strain and Maintenance of Culture

A wild-type *Drosophila melanogaster* strain (Oregon R$^+$) that was kindly gifted by Dr. Anurag Sharma, Senior Assistant Professor, NITTE (Deemed to be University), Mangalore, India, and maintained on a standard *Drosophila* diet (containing maize powder, agar-agar, sugar, yeast, sodium benzoate, and propionic acid) was used for rearing. The flies were retained in a 12-h dark/light cycle at 24 $\pm$ 1 °C and 65–70% humidity levels. [41].

2.3.1. Plant Concentration and Rotenone Exposure

To determine whether the treatment has any impact on the survival of the flies during the experimental period, preliminary studies were conducted with small numbers of flies, testing several concentrations (0.01, 0.025, 0.05, and 0.1%) of UD, MC, and MK. However, only one concentration, i.e., 0.1% per unit of the medium, was selected as the optimum concentration from the conclusive studies. Nevertheless, to evaluate the cellular and neurological protective properties of UD, MC, and MK, the concentration of rotenone (500 μM) used was selected based on our findings in *Drosophila* and those of other published studies [42–44].

2.3.2. Treatment Schedule

For the experimental setup, flies/third instar larvae of *Drosophila* were divided into six groups. There were two groups of controls: Group I was fed the larval standard *Drosophila* food as a control, whereas Group II was fed food mixed with 0.1% DMSO as a vehicle control. Group III represented ROT (500 µM) treatment alone; Group IV comprised ROT with UD extract (0.1%); Group V consisted of ROT and MC extract (0.1%); Group VI was ROT cotreated with MK extract (0.1%). Larvae were permitted to feed either normally or with food that had been exposed to ROT or ROT+ fractions for 24 and 48 h. The flies were exposed for 120 h (5 days) and were assessed for jumping and climbing. We determined the modulatory effect of UD, MC, and MK fractions on rotenone-induced lethality, locomotor dysfunctions, inhibition of acetylcholinesterase, cellular toxicity, and proteotoxicity.

2.3.3. Trypan Blue Dye Exclusion Assay

Dye exclusion was employed as described by Krebs and Feder (1997) with slight modifications [45]. This is a simple and quick method that distinguishes living and nonliving cells in tissue. It is used to detect cell death in the whole larvae and larval gut. At the end point of treatment, 5–10 larvae were thrice washed with 0.1 M phosphate buffer saline (pH 7.4), then whole or dissected midguts of larvae were incubated in trypan blue solution (0.2 mg/mL in 50.0 mM PBS, pH 7.4) for 15 min followed by three washes with 0.1 M phosphate buffer saline (pH 7.4). The larvae were analyzed by using a stereomicroscope; the images were acquired for trypan blue scoring, and were analyzed thoroughly.

2.3.4. Homogenate Preparation

To obtain 10% homogenate/cytosol, the midgut of the third instar larvae was dissected from control (normal/untreated), DMSO, ROT, and ROT with phyto extract groups and homogenized in ice-cold 0.1 M phosphate buffer at a pH of 7.4, containing 0.15 M KCl. Following homogenization, the samples underwent centrifugation at $10,000 \times g$ at 4 °C for 15 min. After that, a nylon mesh sieve with a pore size of 10 mm was used to filter the supernatant, which was then used for various assays. [46].

2.3.5. Protein Estimation

The method of Lowry et al. (1951) was used to measure the protein concentrations in the whole-body homogenates using bovine serum albumin as the standard [41].

2.3.6. Acetylcholinesterase (AChE) Enzymatic Assay

AChE activity was measured as previously mentioned [47]. Briefly, the reaction was started by adding acetylthiocholine iodide (78 mM) to 0.1 M phosphate buffer with a pH of 8.0, which also contained 5,5′-dithio-bis-(2-nitrobenzoic acid) (DTNB 10 mM) and sample (cytosolic) 0.01 mg protein. The change in absorbance was observed at 412 nm for three minutes. The amount of substrate hydrolyzed/min/mg protein was used to express enzyme activity.

2.3.7. Measurement of Locomotor Deficits: Climbing Assay

The climbing assay was performed with some modifications, as previously described [48]. A vertical plastic tube measuring 18 cm in length and 2 cm in diameter was filled with twenty adult flies. During a 30 s time frame, flies followed tapping sounds until they reached the bottom of the vials; flies were scored if they crossed the 15 cm line. The average percentage of flies that cross the 15 cm line is represented by the climbing scores. The average number of flies above and below 15 cm (n_{top} and n_{bot}), expressed as a percentage of the total number of flies, determines the scores (n_{tot}). The results are displayed in $\pm$SD of the results from three independent observations. A performance index (PI) was calculated for each experiment and was given as follows: $1/2[(n_{tot} + n_{top} - n_{bot})/n_{tot}]$.

2.3.8. Jumping Assay

The jumping activity was performed to examine neuromuscular activity [49]. The speed of locomotor activity seems to have an impact on the threshold for the jumping reaction. One at a time, newly emerged flies were put into a labelled vial 1–10 cm, and the height the fly jumped from the bottom of the vial was recorded. The jumping activity was determined to be the mean number of jumps across five replicates. Five replicates of each group of 100 flies each were used.

2.4. Statistical Analysis

To analyze the UV-Vis, FT-IR, and HPLC profiles, professional assessment software was used (Digitized Quantitative Evaluation System of Herbal Medicine Chromatographic Fingerprints 4.0). EC_{50} analysis was performed using CompuSyn software (version 1.0). Chemometric data were expressed as the mean ± SEM, and using SPSS software (version 18), a two-way ANOVA and Tukey's test were used to compare n = 3 for significant differences. Significance was ascribed to a *p* value < 0.05.

3. Results

The results of the assays and analyses used to determine the feasibility of UD, MC, and MK extracts against rotenone-induced lethality, locomotor dysfunctions, inhibition of acetylcholinesterase, cellular toxicity, and proteotoxicity are presented in the subsections below.

3.1. Analytical Assays

3.1.1. Percentage Yield of Bioactive Compounds

The results showed that MK leaf extraction yield was higher than that of UD and MC when distilled water was used as the extracting solvent (Table 1). The yield of ethanolic extraction was again higher in MK, followed by MC and UD. The findings also revealed that variation in the yield depends on the extraction solvent used. The variability in extract quantities from the plant materials used in this study could be due to the varying accessibility of extractable components caused by plant position.

Table 1. The percentage yield of plant extracts using aqueous and ethanolic extraction methods.

S. No.	Plant Species	Code	Plant Part Used	Solvents (%) Yield (*w/w*)	
				Aqueous Extracts	Ethanolic Extracts
1.	*U. dioica*	UD	Leaves	14.34	10.45
2.	*M. chamomilla*	MC	Flower	8.56	14.68
3.	*M. koenigii*	MK	Leaves	21.62	15.27

3.1.2. Antioxidant Potential of UD, MC, and MK

DPPH is widely used to evaluate the antioxidant and antiradical potential of plant fractions. The ability of antioxidants to scavenge DPPH radicals is assumed to be attributed to their hydrogen donating capabilities. The antioxidant potential of both aqueous and ethanolic extracts of UD, MC, and MK was measured using a free radical DPPH assay. As shown in Figures 1a and 2c, MK had the highest DPPH radical scavenging activity in both aqueous and ethanolic extracts, followed by UD and MC. Compared with aqueous extracts, ethanolic extracts of all three plants had the strongest scavenging efficacy.

ABTS is an unstable colored free radical that is used to investigate the antioxidant properties of both hydrophobic and hydrophilic antioxidants found in food extracts. The ABTS radical scavenging activity of the different extracts with different extraction solvents (aqueous and ethanol, Figures 1b and 2d) was found to be higher in MK, followed by UD and MC. These results suggest that MK has a higher efficacy in scavenging free radicals along with greater antiradical and antioxidant activity.

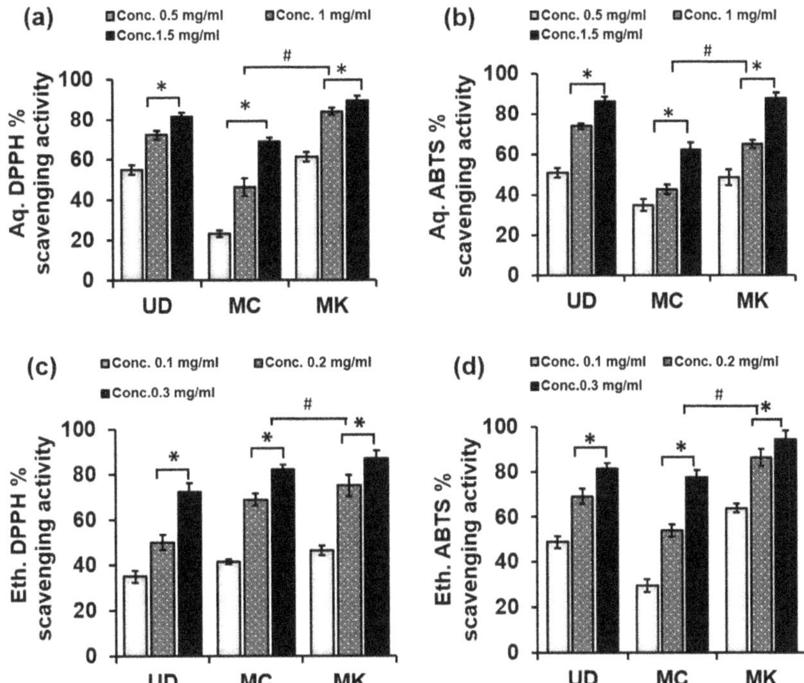

Figure 1. The effect of different concentrations of *U. dioica* (UD), *M. chamomilla* (MC), and *M. koenigii* (MK) in (**a,b**) aqueous and (**c,d**) ethanolic extractions on the (**a,c**) DPPH and (**b,d**) ABTS free radical scavenging assay. Data represent mean ± SD for n = 3. Statistically significance ascribed as * $p < 0.05$ (intragroup) and # $p < 0.05$ (intergroup) compared with 0.5 mg/mL and 0.1 mg/mL of the respective groups.

3.1.3. EC$_{50}$ Prediction Using Statistical Models

EC$_{50}$ is a significant parameter for assessing antioxidant activity and can be used to compare the antioxidant capacities of different materials. The EC$_{50}$ can be calculated, using various models, by interpolating data from a suitable curve or performing a nonlinear regression of the data using different components. The leaves of MK were found to have a higher antioxidant capacity due to its greater total phenolic and flavonoid content, with EC$_{50}$ values of 0.33 and 0.10 mg/mL for aqueous and ethanolic extractions on DPPH scavenging and 0.51 and 0.07 mg/mL for aqueous and ethanolic extractions on ABTS radicals, respectively (Figure 2; Table 2).

Table 2. Estimated EC$_{50}$ (mg/mL) of UD, MC, and MK were obtained by different models using DPPH and ABTS assays.

Assays	Plant Species	Plant Part Used	EC$_{50}$ (mg/mL) of Aqueous Extracts	EC$_{50}$ (mg/mL) of Ethanolic Extracts
DPPH	UD	Leaves	0.42	0.16
	MC	Flowers	1.00	0.12
	MK	Leaves	0.33	0.10
ABTS	UD	Leaves	0.55	0.10
	MC	Flowers	1.04	0.16
	MK	Leaves	0.51	0.07

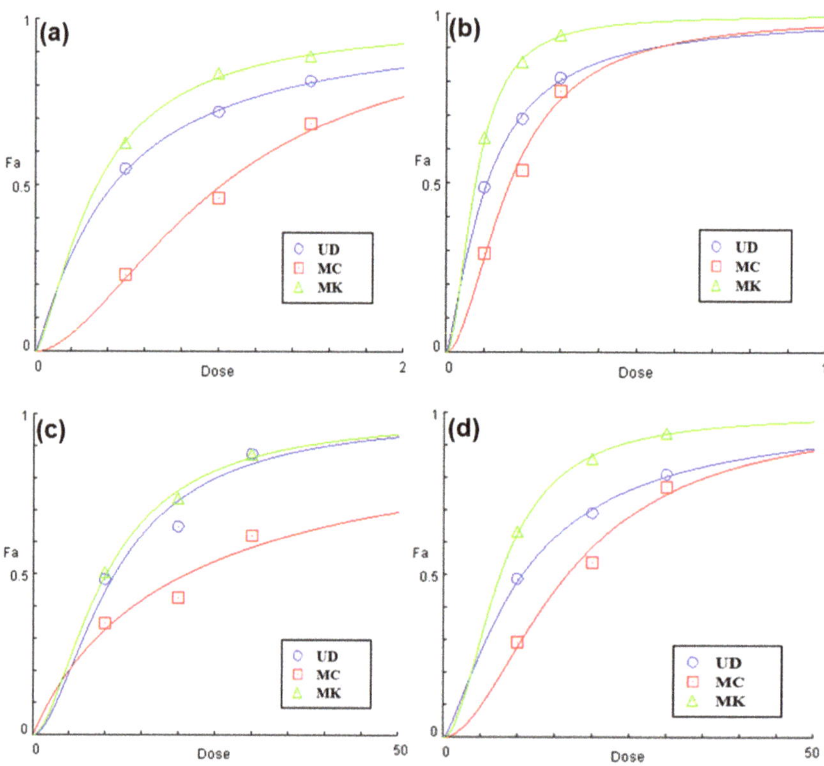

Figure 2. Dose-response profiles of the estimated EC_{50} (mg/mL) of (**a,c**) aqueous and (**b,d**) ethanolic extracts of *U. dioica* (UD), *M. chamomilla* (MC), and *M. koenigii* (MK) on the (**a,b**) DPPH and (**c,d**) ABTS assays.

3.1.4. Phenolic and Flavonoid Potential of UD, MC, and MK

Individual aqueous and ethanolic extracts of UD, MC, and MK were quantified for total phenolic and flavonoid content. The total phenolic content (TPC) was calculated based on the gallic acid standard curve. In terms of aqueous extracts, MK (35.14 mg (GAE)/g) was found to have the highest TPC, followed by UD (26.08 mg (GAE)/g), then MC (24.01 mg (GAE)/g). The trend was the same in the case of ethanolic extracts: MK (48.93 mg (GAE)/g) was followed by UD (42.93 mg (GAE)/g) and was lowest in MC (40.5 mg (GAE)/g), as shown in Figure 3 and Table 3. The total flavonoid content (TFC) was calculated based on the quercetin standard curve. It was highest in MK leaves (9.64 mg (QE)/g), followed by MC flowers (5.46 mg (QE)/g), and lowest in UD leaves (5.45 mg (QE)/g) in the case of aqueous extracts, and the trend was the same for ethanolic extracts: MK (22.88 mg (GAE)/g), MC (12.64 mg (GAE)/g), and UD (12.48 mg (GAE)/g), as shown in Figure 3 and Table 3. The total amounts of phenolic and flavonoid content in the aqueous fractions of UD, MC, and MK were lower than in the ethanolic extracts. This may have been due to the aqueous solvent and extraction process employed. It has been shown that the solvent employed for extraction may be to blame for the quantity of phenolic content in the extract [24]. MK had the highest TPC and TFC levels compared with UD and MC in both aqueous and ethanolic extracts. The presence of hydroxyl groups in phenols allows them to scavenge free radicals, proving that they are the most essential phytochemicals.

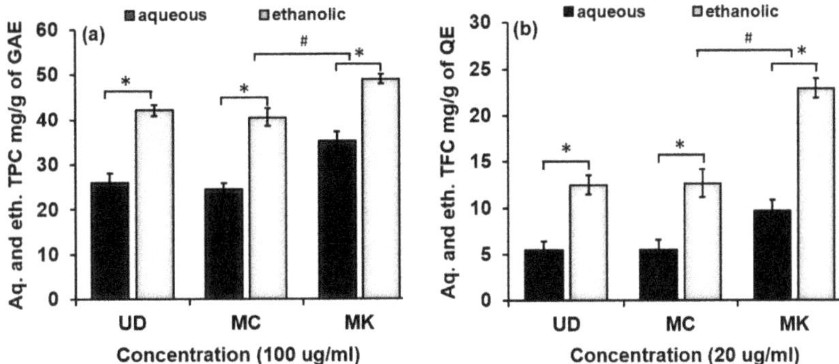

Figure 3. Total (**a**) phenolic and (**b**) flavonoid contents of aqueous and ethanolic fractions of *U. dioica*, *M. chamomilla*, and *M. koenigii*. Data are shown as the mean $\pm$ SD for n = 3. Statistical significance is ascribed as * $p < 0.05$ (intragroup) and # $p < 0.05$ (intergroup) of the respective groups.

Table 3. TPC and TFC results of UD, MC, and MK are expressed in mean $\pm$ SD.

Plant Species	Plant Part Used	TPC mg (GAE)/g (Aqueous)	TPC mg (GAE)/g (Ethanolic)	TFC mg (QE)/g (Aqueous)	TFC mg (QE)/g (Ethanolic)
U. dioica	Leaves	26.08 $\pm$ 2.02	42.16 $\pm$ 2.06	5.458 $\pm$ 2.3	12.48 $\pm$ 1.04
M. chamomilla	Flowers	24.01 $\pm$ 1.50	40.5 $\pm$ 4.04	5.465 $\pm$ 1.06	12.64 $\pm$ 2.3
M. koenigii	Leaves	35.14 $\pm$ 3.0	48.93 $\pm$ 2.03	9.641 $\pm$ 2.5	22.88 $\pm$ 1.05

3.1.5. Preliminary Qualitative Screening Analysis of Plant Extracts

Phytochemicals are chemical molecules produced by plants due to their normal metabolic activities. These substances are referred to as secondary metabolites. There is still limited understanding of the benefits of plants because of a lack of raw data and experiments following proper scientific standards. Phytoconstituents such as amino acids, polyphenols, reducing sugars, alkaloids, terpenoids, glycosides, carbohydrates, and saponins are mostly responsible for plant curative capabilities, such as menstruation problems, muscular spasms, anemia, ulcers, hemorrhoids, inflammation, and wound healing. To confirm the existence of phytoconstituent substances, a phytochemical screening study was performed on the crude extracts of UD, MC, and MK, along with appropriate chemical tests. The presence of bioactive phytochemical substances such as phenols, flavonoids, alkaloids, tannins, carbohydrates, saponins, terpenoids, steroids, and glycosides are represented in Table 4.

Table 4. Preliminary qualitative screening of secondary metabolites of crude extracts of UD, MC, and MK.

S. No.	Phytoconstituents	Tests	Positive Results	UD	MC	MK
1.	Phenols	Ferric chloride test	Bluish-green	+	+	+
2.	Flavonoids	Alkaline reagent test	Orange-red	+	+	+
3.	Alkaloids	Wagner's test	Red precipitate	+	+	+
4.	Tannins	FeCl$_3$ test	Black blue	+	+	+
5.	Carbohydrates	Molisch's test	Red or dull violet	+	−	+
6.	Saponins	Foam test	White precipitate	+	+	+
7.	Terpenoids	Salkowski test	Change from pink to violet	+	+	+
8.	Steroids	Liebermann's test	violet to blue or green color	+	+	+
9.	Glycosides	Keller-Killiani test	Brick red	−	+	+

Note: The presence of phytoconstituents is indicated by a '+' sign, whereas the lack of phytoconstituents is indicated by '−'.

3.1.6. UV-Visible Analysis

Spectral data were observed in the range of 200–800 nm at intervals of 1 nm and showed maximum absorption at 277 and 321 nm in the case of UD, 266 and 323 nm for MC, and 286 and 316 nm for MK (Figure 4). With increasing concentration, absorption increased. This technique is a quantitative analysis of UD, MC, and MK and their absorption of radiation in the ultraviolet and visible spectra. This radiation enables electrons in atoms or molecules to shift from lower to higher energy levels. Under regulated conditions, the amount of radiation absorbed is proportional to the intensity of chemicals in the plant extracts. Spectral analysis showed peaks in crude extracts of UD, MC, and MK, indicating the presence of a variety of chemicals, particularly providing information about unsaturated bonds in conjugated or aromatic components.

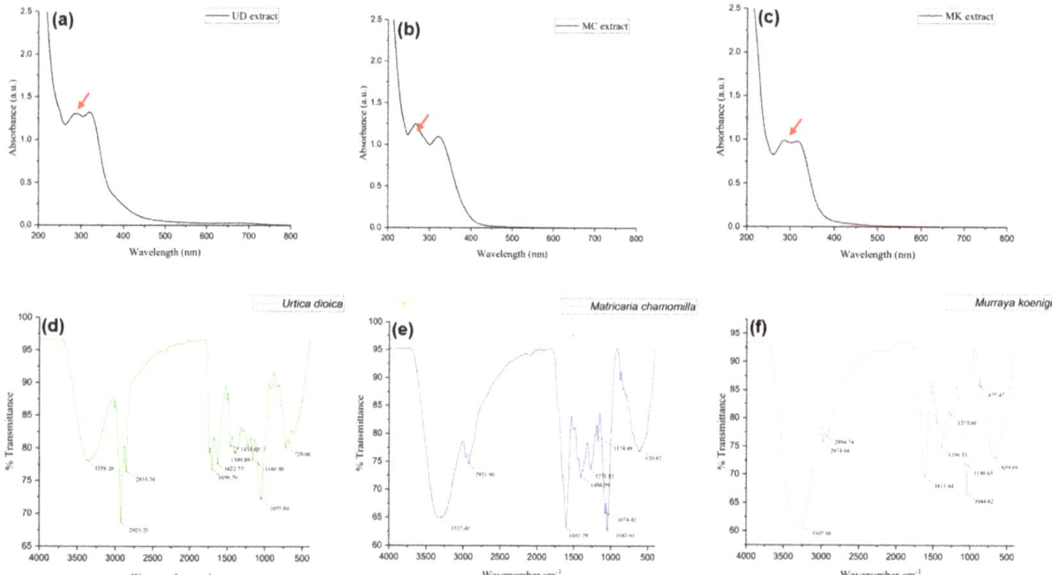

Figure 4. UV-Visible spectra of (**a**) UD, (**b**) MC, and (**c**) MK fractions. Red arrows represent the peak of extract (conjugated or chemical bonds). The FT-IR absorption spectrum of (**d**) UD, (**e**) MC, and (**f**) MK with a scan range of 400–4000 cm^{-1}.

3.1.7. Fourier Transform Infrared Spectrophotometer (FTIR)

The functional groups of bioactive components present in plant extracts of UD, MC, and MK were identified using the FT-IR spectra depending on peak values in the IR radiation area. When the extract was run through FTIR, the functional groups of the constituents were segregated based on the ratio of their peaks. The existence of alkanes, amino acids, aldehydes, phenols, secondary alcohols, aromatic amines, ketones, and halogen compounds was verified by FTIR analysis (Figure 4).

In the ethanolic extracts of UD, MC, and MK, the band between 3500 and 3200 cm^{-1} was assigned to an O–H stretching, indicating the presence of hydroxyl and phenolic groups. These groups are found in cellulose, hemicellulose, and lignin structures and could be related to the hydroxylated compounds (polyphenols) and moisture content of the extract. A minor intensity peak in the region of 2950–2960 cm^{-1} showed an O–H (stretch), suggesting the existence of amide groups and alcohols, as well as C–H vibrations of the CH$_3$ group. In the area of 1650–1600 cm^{-1}, an acceptable N=O (stretch) vibration was recorded. The weak peaks in this area suggest the existence of C=C (stretch). Absorption in this region may indicate the existence of carbonyl groups. A doublet band was seen in the fingerprint area between 1455 and 1370 cm^{-1}, indicating the C-H (bending) vibration of the methyl group (–CH$_3$) molecule. Furthermore, a broad peak was found between 1100 and 1000 cm^{-1}, suggesting the presence of alcohol groups within the compound structure. This is most likely due to an alkoxy C–O (stretching) vibration. In addition, the UD, MC, and MK fractions showed aromatic C–H bonds between 720 and 620 cm^{-1} in the infrared spectrum (Table 5).

Table 5. FT-IR frequency range and functional groups are present in the extracts of UD, MC, and MK.

S. No.	Frequency Range (cm^{-1})			Functional Groups	Phytocompounds Identified
	UD	**MC**	**MK**		
1.	3358.28	3327.40	3307.88	H-bonded, OH stretching	Hydroxyl compounds
2.	2923.25	2921.90	2974.04	Asymmetric stretching –CH(CH$_2$) vibration	Saturated aliphatic Compounds (Lipids)
3.	2853.54	2802.86	2804.74	Symmetric stretching –CH$_2$(CH$_2$) vibration	Proteins, lipids
4.	1698.70, 1622.77	1601.79	1611.44	C=O stretching vibration	Ketone compound
5.	1455.85	1404.99	-	C=C-C aromatic ring stretching	Aromatic compound
6.	1399.89	-	1396.51	O-H, alcoholic group	Phenol or tertiary alcohol
7.	-	1271.82	1275.04	CN stretching	Aromatic primary amine
8.	1160.88	1174.09	1140.63	Polymeric OH, C-O stretching	Cyclic ether
9.	1055.84	1074.42, 1043.61	1044.62	Phosphate ion	Phosphate compound
10.	-	-	877.47	*p*-O-C stretching	Aromatic phosphate
11.	720.06	620.82	659.68	C-Cl stretching	Aliphatic chloro compound

Finally, the results revealed that the chemical structures of UD, MC, and MK are extremely polar (lignin (+)-neoolivil, 3,4-divanillyltetrahydrofuran, isolariciresinol, (−)-secoisolariciresinol, and pinoresinol), which stimulates the proliferation of human lymphocytes and has anti-inflammatory effects, as evidenced by the presence of wide peaks within the spectra.

3.1.8. High-Performance Liquid Chromatography (HPLC)

The external standard technique was used to perform HPLC experiments under isocratic conditions. Ethanolic extracts of UD, MC, and MK were analyzed directly using the total extracts without any manipulation. The retention time of the chromatographic peaks of plant extracts was compared with reference standards (rutin and quercetin), and DAD spectra (200–400 nm) of existing literature were analyzed (Figure 5). Our findings revealed the presence of 24 compounds (Table 6) in the ethanolic extracts of UD, MC, and MK, including quercetin, coumaric acid, chlorogenic acid, gallic acid, apigenin, myricetin, ferulic acid, fumaric acid, rutin, isorhamnetin, kaempferol, etc.

In the ethanolic extract of UD, three classes of phenols were characterized: anthocyanin compounds (rosinidin 3-*O*-rutinoside; peonidin 3-*O*-rutinoside; and peonidin 3-O-(6′-O-coumaroyl glucoside), hydroxycinnamic acid derivatives (p-coumaric acid; chlorogenic acid; caffeoylquinic acid; and 2-O-caffeoylmalic acid), and flavonoids (rutin; isorhamnetin 3-O-rutinoside; quercetin; p-coumaroyl glucoside; kaempferol 3-O-rutinoside; kaempferol 3-O-and glucoside; and quercetin 3-O-glucoside).

The polyphenolic compounds found in the ethanolic fractions of MC flowers were identified as essential constituents, such as quercetin (quercetin-7-O-β-glucoside; quercetin-3-O-β-rutinoside; and quercetin-3-O-β-galactoside), apigenin (apigenin-7-O-7-glucoside; apigenin-7-O-apiosyl-glucoside; and apigenin-7-O-glucosyl-6′-acetate), luteolin (luteolin-7-O-β-glucoside; luteolin-4′-O-7-β-glucoside; and luteolin-7-O-β-rutinoside), isorhamnetin (isorhamnetin-7-O-β-glucoside), patuletin (patuletin-7-O-β-glucoside), eupatoletin, astragalin, chrysosplenol, and spinacetin. The MK ethanolic fraction was examined by HPLC-DAD, which permitted the identification of important components such as chlorogenic acid, quercitrin, citric acid, piperine, 7 p-coumaric acid, hesperidin, rutin, gallic acid, β-terpineol, ferulic acid, catechin, naringenin, D-α-pinene, di-α-phellandrene, dipentene, D-sabinene, caryophyllene, nicotinic acid, koenigine-quinone A, and koenigine-quinone B. All these secondary metabolites have been shown to have cerebrovascular protective, neuroprotective, and cardiovascular protective properties. In addition, it also acts as an anti-carcinogenic, anti-tumor, anti-inflammatory, antimicrobial, antiviral, and antibacterial agent and protects against oxidative stress-related diseases.

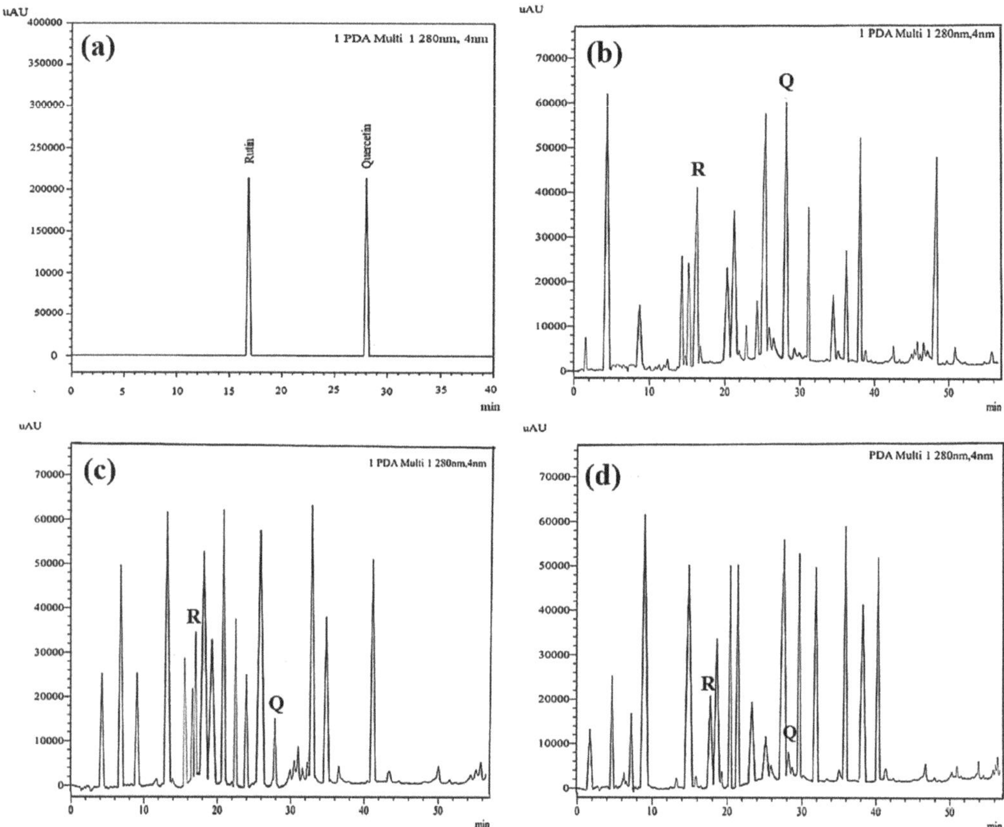

Figure 5. HPLC profiles acquired at 280 nm of (**a**) standard rutin and quercetin and ethanolic extracts obtained from (**b**) *U. dioica* leaves, (**c**) *M. chamomilla* flowers, and (**d**) *M. koenigii* leaves, showing different bioactive compounds. R stands for rutin, and Q stands for quercetin.

3.2. Cellular Assays

Cytotoxicity of Rotenone and Amelioration of Cytotoxicity through Bioactive Compounds UD, MC, and MK Determined by a Dye Exclusion Test (Trypan Blue) in Whole Larvae and Tissues of ROT-Exposed Organisms

To determine if exposure to ROT causes any tissue damage, we analyzed trypan blue staining in whole larvae and tissues of D. melanogaster (Figure 6). Of the larvae exposed to ROT, 95% showed blue staining in the whole larvae and their tissues (brain ganglia, salivary gland, midgut, and gastric caeca). ROT coexposed with MK exhibited significantly less blue staining than in the ROT + UD and ROT + MC groups in the whole larvae and the abovementioned tissues, respectively.

Table 6. Major phytochemical compounds identified in ethanolic extracts *of U. dioica, M. chamomilla,* and *M. koenigii.*

S. No	Retention Time (R_t min.)			Compound	Molecular Formulae	Chemical Structure	Molecular Weight (g/mol)	Pharmacological Actions
	UD	MC	MK					
1.	1.931	-	1.645	Fumaric acid	$C_4H_4O_4$		116.07	Reduces gallstone formation, used for the treatment of multiple sclerosis and psoriasis.
2.	4.240	4.235	4.276	Gallic acid	$C_7H_6O_5$		170.12	Expectorant, cytotoxic steroid, memory enhancer, anti-inflammatory, anti-neoplastic, and antioxidant properties.
3.	-	7.523	7.473	Protocatechuic acid	$C_7H_6O_4$		154.12	Neuroprotective, antioxidant, anticancer, antibacterial, anti-aging, and anti-asthma properties.
4.	9.211	-	9.865	Catechins	$C_{15}H_{14}O_6$		290.27	Used to prevent and treat various diseases, high antioxidant activity, and used in cosmetics.
5.	-	11.387	-	4-O-Caffeoylquinic acid	$C_{16}H_{18}O_9$		354.31	Cytoprotective, neuroprotective, and hepatoprotective effects.
6.	15.982	14.683	15.554	Caffeic acid derivative	$C_9H_8O_4$		180.16	Prevents DNA damage and oxidative stress induced by free radicals.
7.	16.683	16.299	16.693	Epicatechin	$C_{15}H_{14}O_6$		290.27	Reduces blood glucose levels in diabetic patients and stimulates mitochondrial respiration.

Table 6. *Cont.*

S. No	Retention Time (R_t min.)			Compound	Molecular Formulae	Chemical Structure	Molecular Weight (g/mol)	Pharmacological Actions
	UD	MC	MK					
8.	17.248	17.558	17.280	Rutin	$C_{27}H_{30}O_{16}$	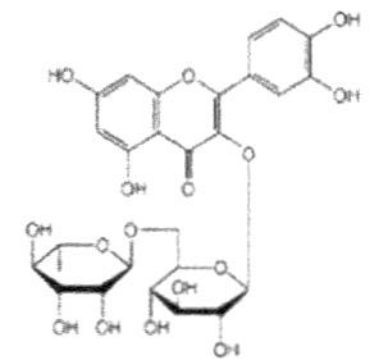	610.5	Hypolipidemic, anti-protozoal, vasoactive, cytoprotective, anti-allergic, anti-platelet, anti-hypertensive, and anti-spasmodic properties.
9.	17.515	17.428	17.625	Syringic acid	$C_9H_{10}O_5$	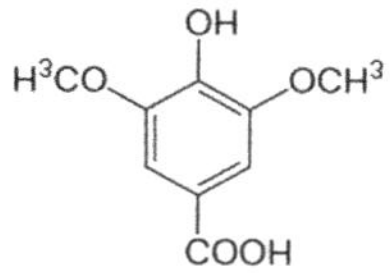	198.17	Used in the prevention of CVDs, cancer, diabetes, and possesses antioxidant activities.
10.	18.094	18.763	18.651	Isorhamnetin-3-O-glucoside	$C_{22}H_{22}O_{12}$	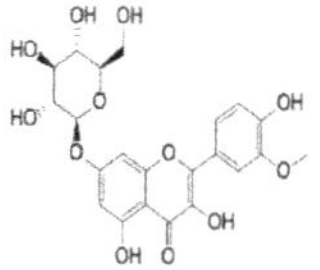	478.4	Anti-viral, antioxidant, anticancer, anti-tumor, anti-inflammatory, and antimicrobial properties.
11.	-	19.243	-	Apigenin-7-O-glucoside	$C_{21}H_{20}O_{10}$	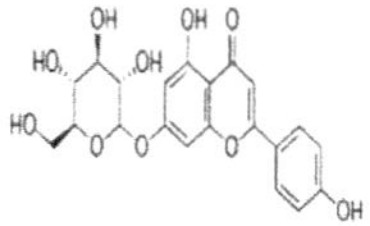	432.4	Prominent chemopreventive, anti-candidal effect, antifungal potential, and strengthens the failing heart.
12.	20.728	20.152	20.835	p-coumaric acid	$C_9H_8O_3$	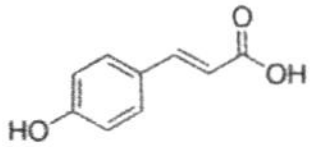	164.16	Anti-inflammatory, antimicrobial, anti-viral, and antibacterial properties.
13.	-	21.739	-	4,5-O-dicaffeoylquinic acid	$C_{25}H_{24}O_{12}$	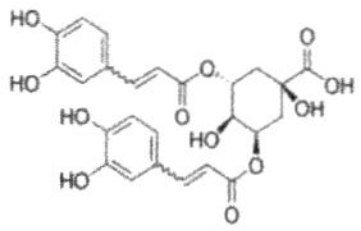	516.4	In melanocytes, significantly reduces tyrosinase activity and melanin synthesis in a dose-dependent manner.

Table 6. *Cont.*

S. No	Retention Time (R_t min.)			Compound	Molecular Formulae	Chemical Structure	Molecular Weight (g/mol)	Pharmacological Actions
	UD	MC	MK					
14.	22.564	22.571	22.677	2-O-Caffeoylmalic acid	$C_{13}H_{12}O_8$		296.230	Prevents ROS production and possesses high antioxidant activity.
15.	23.232	23.924	23.232	Ferulic acid	$C_{10}H_{10}O_4$		194.18	Wide range of therapeutic uses against various diseases including cancer, arthritis, etc.
16.	-	24.579	-	Naringin	$C_{27}H_{32}O_{14}$		580.5	Anti-carcinogenic and acts as inhibitor of selected cytochrome P_{450} enzymes.
17.	25.247	25.460	25.814	Quercetin (quercetin-3-O-rhamonoside)	$C_{21}H_{20}O_{11}$		448.4	Used in the treatment of inflammatory, allergic, and metabolic disorders and act as anti-protozoal.
18.	27.723	27.787	27.379	Myricetin	$C_{15}H_{10}O_8$		318.23	Acts as an anti-epileptic, anti-amyloidogenic, anti-diabetic, antioxidant, antibacterial, anti-ulcer, antiviral, anticancer, and anti-inflammatory agent.

Table 6. *Cont.*

S. No	Retention Time (R_t min.)			Compound	Molecular Formulae	Chemical Structure	Molecular Weight (g/mol)	Pharmacological Actions
	UD	MC	MK					
19.	28.231	28.789	29.031	Quercetin	$C_{21}H_{20}O_{11}$	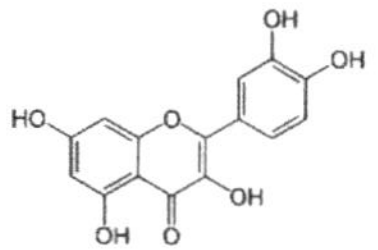	448.4	Decreases tumor necrosis factor (TNF-α) production in macrophages and LPS-driven IL-8 synthesis in lung A549 cells generated by lipopolysaccharide (LPS).
20.	32.156	32.956	31.456	Kaempferol	$C_{15}H_{10}O_6$	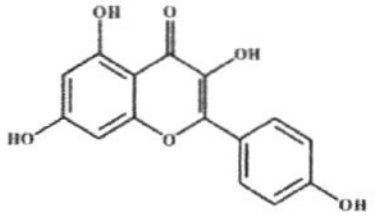	286.24	Anxiolytic, anti-diabetic, anti-estrogenic, anti-osteoporotic, cardioprotective, and neuroprotective properties.
21.	-	34.320	-	Luteolin	$C_{15}H_{10}O_6$	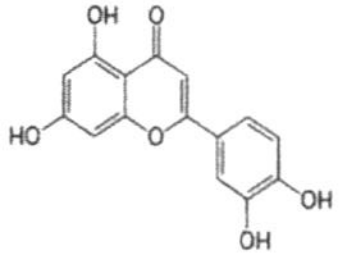	286.24	Exhibits anti-inflammatory properties due to ability to regulate transcription factors like NF-B, AP-1, and STAT3.
22.	-	34.745	-	Cirsiliol	$C_{17}H_{14}O_7$	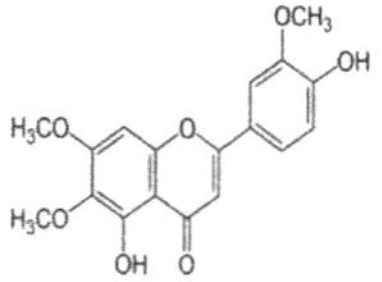	330.29	Act as inhibitor of arachidonate 5-lipoxygenase and has anticancer, hypnotic, sedative, and anti-inflammatory properties.
23.	35.339	35.445	36.021	Isorhamnetin	$C_{16}H_{12}O_7$	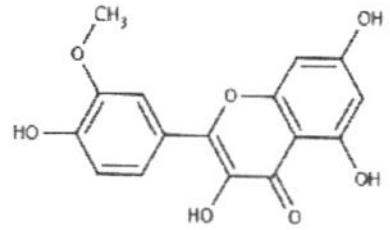	316.26	Cerebrovascular and cardiovascular protective properties; in addition, has antioxidant, anti-tumor, anti-inflammatory, organ protection, and obesity prevention properties.

Table 6. *Cont.*

S. No	Retention Time (R_t min.)			Compound	Molecular Formulae	Chemical Structure	Molecular Weight (g/mol)	Pharmacological Actions
	UD	MC	MK					
24.	37.277	-	37.552	Kaempferol 3-O-glucoside	$C_{21}H_{14}O_{11}$	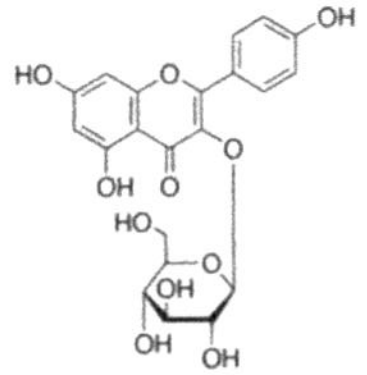	448.38	Lowers the risk of chronic diseases, particularly cancer, and boosts the body's antioxidant defenses against free radicals.
25.	39.901	41.026	40.516	Isorhamnetin 3-O-rutinoside	$C_{28}H_{32}O_{16}$	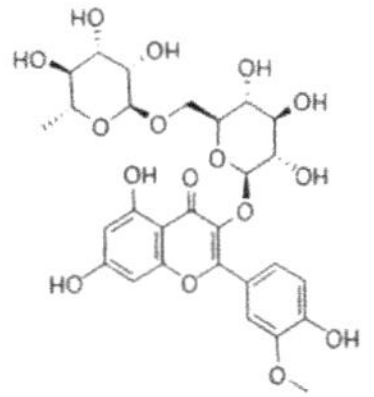	624.5	Inhibits membrane proteins and has anti-apoptosis, antioxidation, anti-tumor, anti-inflammation, antiviral, antibacterial, anti-amyloidogenic, and anti-diabetic properties.

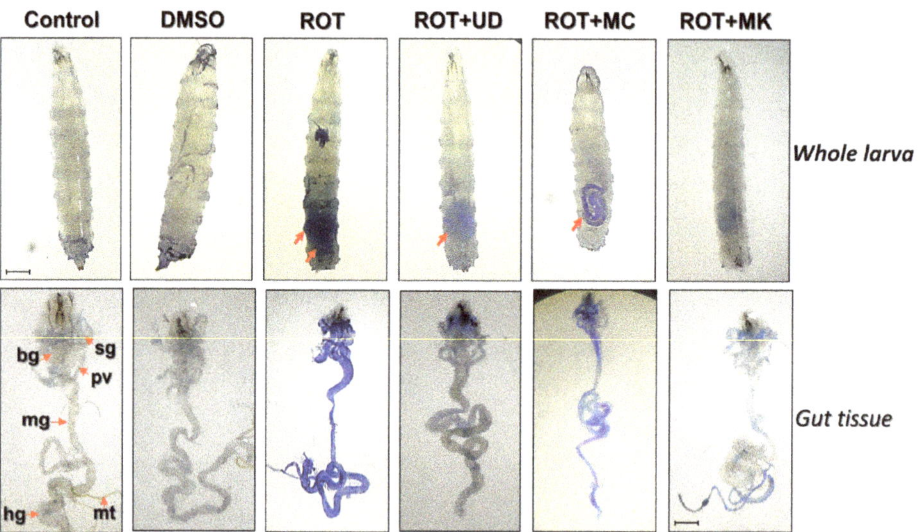

Figure 6. Dye exclusion test through trypan blue staining in third instar larvae exposed to rotenone and cotreated with UD, MC, and MK, as shown in the upper panels. The lower panels show dissected third instar larvae stained with trypan blue. Seventy-two hour ($\pm$2 h) old larvae (early third instar) of D. melanogaster (Oregon R^{+}) were exposed to ROT 500 µM alone or in combination with UD, MC, and MK for 48 h. Arrows of the upper panel show cytotoxicities in the whole larvae through trypan blue staining. Note: bg= brain ganglia, sg= salivary glands, pv= proventriculus, mg= midgut, mt= malpighian tubules, and hg = hind gut. The bar represents 100 µm. ROT= rotenone; UD = *Urtica dioica*, MC = *Matricaria chamomilla* and MK = *Murraya koenigii*.

3.3. Biochemical Assays

3.3.1. Decreased Protein Content in *D. melanogaster* Treated with Rotenone after 24 and 48 h

Third instar larvae of *D. melanogaster* exposed to 500 µM ROT exhibited a significant reduction ($p < 0.05$) in the total protein content of their tissues. After 24 h, the protein content in the larvae was reduced in the ROT group (9.34 $\pm$ 0.150 mg/mL) compared with the control group (12.88 $\pm$ 0.313 mg/mL). ROT coexposed with MK (11.63 $\pm$ 0.225 mg/mL) showed highest protein levels, followed by ROT + UD (10.05 $\pm$ 0.381 mg/mL), and ROT + MC (9.46 $\pm$ 0.174 mg/mL). After 48 h, in comparison with the control group (11.47 $\pm$ 0.328 mg/mL), the ROT treatment decreased the protein content in the larvae (7.37 $\pm$ 0.225 mg/mL). Increased protein levels were observed in ROT coexposed with MK (11.71 $\pm$ 0.263 mg/mL), followed by ROT + UD (9.07 $\pm$ 0.196 mg/mL), and ROT + MC (7.72 $\pm$ 0.213 mg/mL) (Figure 7).

3.3.2. Rotenone Inhibits AChE Activity in *D. melanogaster*, and This Effect Is Reversed by Phytoextraction

In this study, it was found that when the larvae were exposed to ROT for 24 h, they exhibited statistically significant ($p < 0.001$) inhibition of AChE activity compared with the control or DMSO, and ~60% reduced AChE levels were observed in this group. When ROT was coexposed with MK, AChE levels were improved, and only 9.7% inhibition was evident compared with the control. These elevated levels of AChE were significant when compared with the ROT-treated groups. The AChE levels in the ROT + UD and ROT + MC groups were also significantly improved (40.5% and 52.0% inhibition, respectively). Maximum inhibition of AChE levels was present in ROT-exposed organisms after 48 h (69.13% compared with control larvae), and the greatest improvement from ROT-induced toxicity was observed in the ROT + MK group (5.63% compared with control groups). Significantly higher AChE levels were also observed in the ROT + UD and ROT + MC groups (38.15 and 47.45% inhibition, respectively) than in the ROT-treated group (Figure 8).

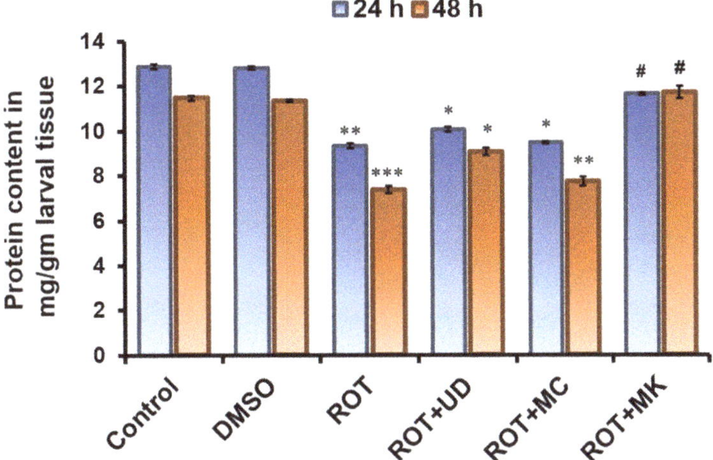

Figure 7. Total protein content in third instar larvae of *D. melanogaster* (Oregon R$^+$) exposed to 500 μM rotenone for 24 and 48 h. Data represent the mean ± SD of three identical experiments made in three replicates. Significance is ascribed as * $p < 0.05$, ** $p < 0.01$, *** $p < 0.001$ vs. control or DMSO control. $^\#$ = significance at * $p < 0.05$ as compared with 500 μM rotenone. UD = *Urtica dioica*, MC = *Matricaria chamomilla* and MK = *Murraya koenigii*.

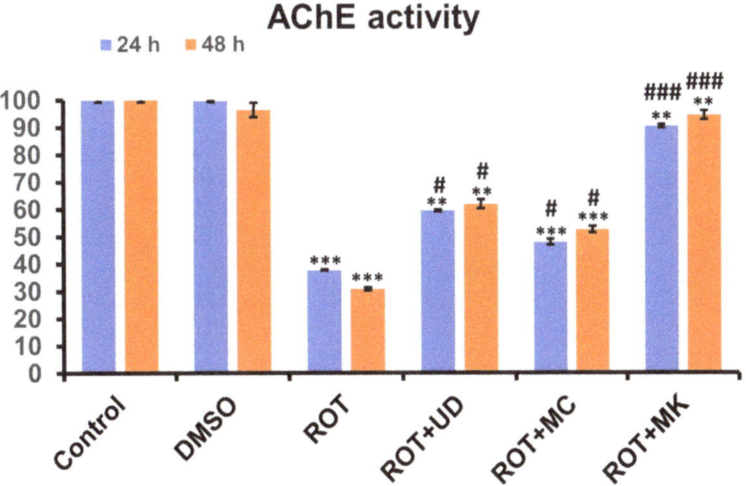

Figure 8. Acetylcholinesterase activity in the third instar larvae of *D. melanogaster* (Oregon R$^+$) exposed to 500 μM ROT alone or in combination with UD, MC, and MK for 24 and 48 h. Data represent mean ± SD (n = 3); significance ascribed as ** $p < 0.01$, *** $p < 0.001$ vs. control or DMSO control. $^\#$ is ascribed as significance at $p < 0.05$, $^{\#\#\#}$ $p < 0.001$ as compared with 500 μM rotenone. UD = *Urtica dioica*, MC = *Matricaria chamomilla* and MK = *Murraya koenigii*.

3.4. Behavioral Assays

3.4.1. Rotenone Affects Locomotor Behavior in *D. melanogaster*

After 30 s, the control and DMSO-treated flies demonstrated maximum climbing ability (only 11% and 12% reduction, respectively). The greatest reduction in climbing ability was observed in ROT-treated *Drosophila* (50.5%), and flies found it difficult to climb the plastic tube walls. The groups receiving ROT + phytoextracts exhibited varying levels of improvement in their climbing skills. All nutraceuticals improved the climbing ability of flies. Among the nutraceutical groups, ROT + MK (15%) exhibited the greatest improvement, followed by ROT + UD (26%), then ROT + MC (37%). To identify any significant differences, the mean ± SEM was compared using an unpaired Student's *t*-test. Significance was ascribed at $p < 0.001$ (Figure 9A).

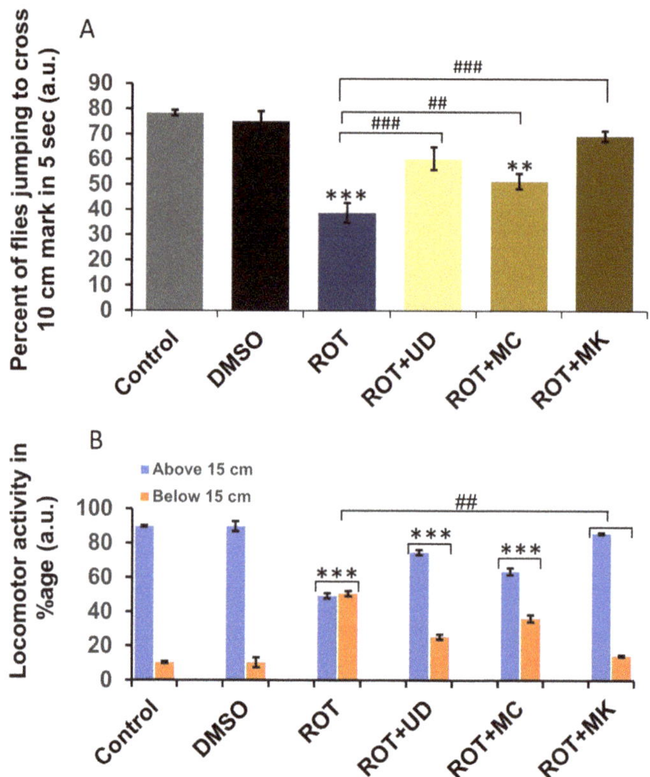

Figure 9. (**A**) Jumping and (**B**) climbing activity of *D. melanogaster* (Oregon R⁺) flies exposed to ROT 500 μM alone or in combination with UD, MC, and MK for 120 h; significance is ascribed as ** $p < 0.01$, *** $p < 0.001$ vs. control or DMSO control. ## ascribed as significance at $p < 0.01$, ### $p < 0.001$ compared with 500 μM rotenone. UD = *Urtica dioica*, MC = *Matricaria chamomilla* and MK = *Murraya koenigii*.

3.4.2. Significant Changes in the Jumping Activity of ROT-Exposed Flies

We observed significantly decreased jumping behavior in the flies treated with ROT (62%) compared with the control and DMSO (22 and 25%, respectively). The groups receiving ROT + phytoextracts exhibited varying levels of improvement in their jumping skills. All nutraceuticals improved the jumping ability of flies. ROT + MK (31%) exhibited the greatest improvement, followed by ROT + UD (40%), and ROT + MC (49%). An unpaired Student's t-test was used to compare the mean $\pm$ SEM. Significance was ascribed at $p < 0.001$ (Figure 9A).

4. Discussion

The present study demonstrates the protective efficacy of UD, MC, and MK against rotenone-induced cellular, neurological, and organismal toxicity in a nontarget organism, *Drosophila melanogaster*. Unlike synthetic drugs, herbal medicines have a comprehensive structure of chemical elements. As a result, the methods of choice for identifying a 'botanical medicine' are primarily designed to obtain a unique fingerprint of certain plants that indicates the existence of quality-defining active chemical elements [50]. It is noteworthy that the Ayurvedic medical system has well-established the benefits of UD, MC, and MK for improving cognition, memory, and learning. To our knowledge, the current study is the first to report the comparative efficacy of UD, MC, and MK against various ROT-induced toxicities in *Drosophila*.

In this study, we examine how different aqueous and ethanolic fractions of UD, MC, and MK react to different radicals. Our results demonstrate that the ABTS activity in MK extracts did not differ significantly from the free radical scavenging potential measured by the DPPH assay because both assays use the same mechanism (single-electron transfer). Among the three extracts, MK displayed the greatest scavenging activity in both the DDPH and ABTS radical scavenging assay, as shown in Figure 1. Previous research has shown that MK has great antioxidant potential [51]. MK leaves had greater antioxidant capacities than UD leaves and MC flowers, owing to greater total phenolic and flavonoid content, with EC_{50} values for aqueous and ethanolic extracts in DPPH scavenging of 0.33 and 0.10 mg/mL, respectively, and EC_{50} values for aqueous and ethanolic extracts in ABTS scavenging of 0.51 and 0.07 mg/mL, respectively. Aqueous extracts of UD, MC, and MK had lower total phenolic and flavonoid concentrations their ethanolic counterparts. This could be due to the aqueous solvent and extraction method used. It was also discovered that the extraction solvent could be at fault for the amount of phenolic content in the extracted samples [52]. In both the aqueous and ethanolic fractions, the TPC and TFC were highest in MK, followed by UD, and MC had the lowest TPC values. This could be due to the extraction solvent used, as ethanolic extraction solvent was previously regarded to be the best for extracting total flavonoids [53]. Overall, MK exhibited the highest antioxidant properties due to the highest TPC, TFC, DPPH, and ABTS, followed by UD and MC. The aromatic leaves of MK have 11–21 pinnate leaflets that are each 2–4 cm (0.80–1.59 in) long and 1–2 cm (0.38–0.80 in) wide. It is known that MK leaves are a major source of phyto-carbazole alkaloids, comprising mahanimbine, koenine, girinimbine, murrayacine, koenidine, mahanine, and 8,8'-biskoenigine, with promising pharmacological activities [51]. Bioactive phytochemical substances of these three herbs, such as phenols, flavonoids, alkaloids, tannins, saponins, terpenoids, steroids, and glycosides, are primarily responsible for curative properties such as menstrual problems, muscular spasms, anemia, ulcers, hemorrhoids, inflammation, and wound healing [54].

UV-Vis spectral analysis was recorded in the range of 200–800 nm with intervals of 1 nm and showed maximum absorption at 277 and 321 nm for UD, 266 and 323 nm for MC, and 286 and 316 nm for MK. Beer's rule asserts that the amount of light absorbed at a specific frequency is proportional to the sample's composition absorption coefficient [55]. As a result, all spectral fluctuations caused by spectrophotometer and sample inaccuracies must be rectified before information processing. Peaks in crude extracts indicated the presence of a variety of ingredients or chemicals, particularly unsaturated bonds in conjugated or aromatic components [56].

A Fourier transform infrared spectrometer was used to determine the functional groups of the bioactive components in the plant fractions based on peak intensity in the infrared radiation (IR) region. The specific wavenumbers and intensities were determined in the range of 4000–400 cm^{-1} [57]. Both stretching and bending vibration assignments were compared with data from the literature. Figure 5 represents the FTIR spectrum of UD,

MC, and MK extracts in the form of KBr pellets and shows the presence of phenols, alcohols, ketones, nitro compounds, esters, carboxylic acids, ethers, aliphatic fluoro, alkenes, and aromatic rings. The broad absorption bands observed at 3358.28, 3327.40, and 3307.88 cm^{-1} were attributed to the stretching of hydroxyl groups and H–bonding in alcohol or phenol groups [58]. The weak absorption peaks in alkanes were detected at 2923.25, 2921.90, and 2974.04 cm^{-1}, which correspond to C–H stretching. N–H bends in primary amines were indicated by the high absorption peaks at 1698.70, 1622.77, 1601.79, and 1611.44 cm^{-1}. C–C stretching in aromatic groups was assigned to the medium peaks at 1455.85 and 1404.99 cm^{-1}. The rocking of the methyl group was assigned to the vibrational absorption bands at 1399.89 and 1396.51 cm^{-1}. C-O stretching was represented by distinct bands at 1271.82 and 1275.04 cm^{-1}. The C–N stretch in aliphatic amines was assigned to the thin peaks at 1055.84, 1074.42, 1043.61, and 1044.62 cm^{-1}. The aromatic H out-of-plane bending had bands at 877.47, 720.06, 659.68, and 620.82 cm^{-1} [59].

The HPLC technique is repeatable, sensitive, and reliable. The existence of 24 chemicals in the ethanolic fractions of UD, MC, and MK was identified using HPLC. In the UD ethanolic extract, three classes of phenols were characterized: anthocyanin compounds (rosinidin 3-O-rutinoside; peonidin 3-O-rutinoside; and peonidin 3-O-6′-O-coumaroyl glucoside), hydroxycinnamic acid derivatives (p-coumaric acid; chlorogenic acid; caffeoylquinic acid; and 2-O-caffeoylmalic acid), and flavonoids (rutin; isorhamnetin 3-O-rutinoside; quercetin; p-coumaroyl glucoside; kaempferol 3-O-rutinoside; and quercetin 3-O-glucoside), as described earlier [38]. The polyphenolic compounds found in the ethanolic fractions of MC flowers were identified by comparing with a previous study [60]. Our findings showed the existence of important constituents, such as quercetin (quercetin-7-O-β-glucoside; quercetin-3-O-β-rutinoside; and quercetin-3-O-β-galactoside), apigenin (apigenin-7-O-7-glucoside; apigenin-7-O-apiosyl-glucoside; and apigenin-7-O-glucosyl-6′-acetate), luteolin (luteolin-7-O-β-glucoside; luteolin-4′-O-7-β-glucoside; and luteolin-7-O-β-rutinoside), isorhamnetin (isorhamnetin-7-O-β-glucoside), patuletin (patuletin-7-O-β-glucoside), eupatoletin, astragalin, and spinacetin [61]. The MK ethanolic fraction was examined by HPLC-DAD, which permitted the identification of important components such as chlorogenic acid, quercitrin, citric acid, piperine, 7 p-coumaric acid, hesperidin, rutin, gallic acid, β-terpineol, ferulic acid, catechin, naringenin, D-α-pinene, di-α-phellandrene, dipentene, D-sabinene, caryophyllene, nicotinic acid, koenigine-quinone A, and koenigine-quinone B [62]. The various phytochemicals in the UD, MC, and MK extracts that are responsible for their antioxidant and protective potential have been well identified. UD has been extensively studied and has shown prominent results in the treatment of prostate enlargement [13], preventing colon carcinogenesis in rats [14], and providing a protective effect against hyperglycemia [15], hypertension [16], and hypercholesterolemia [17]. Numerous studies have shown that MC counteracts diabetes by controlling GLP-1, which is essential for promoting insulin gene transcription [18]. Chamomile oil has also been shown to significantly reduce osteoarthritis [19], and it may be useful in the treatment of lung cancer [20]. Mondal et al., 2022 found that MK modulates various cellular programs and signaling cascades to intervene as an antioxidant in normal cells, as a pro-oxidant in lung carcinoma cells [21], and protects against liver damage caused by TPA [22]. Additionally, earlier research on the MK leaf fraction reported its efficacy in the management of hyperglycemia [23]. Although UD, MC, and MK have been studied in some in vivo and in vitro experiments, there is little evidence for the effect of these three herbs on cellular and neurological complications.

Considering the general protective, organismal, or cognitive effects of UD, MC, and MK that have been reported in the literature, the present study treated third instar larvae and freshly eclosed flies with 500 µM ROT alone or in combination with UD, MC, and MK for 24 to 120 h. Following exposure, cytotoxicity assays (dye exclusion test), biochemical assays (protein estimation and acetylcholinesterase inhibition assays), and behavioral assays (climbing and jumping assays) were performed. ROT is a well-known generator of reactive oxygen species (ROS), which cause cellular damage and eventually lead to necrosis or programmed cell death. Lacking an effective antioxidant system, cells are unable to prevent the harm caused by ROS. $_L$-DOPA appears to simply act as a dopamine precursor to restore endogenous dopamine deficits, as previous studies have shown that feeding the drug to flies did not reduce cell loss [63].

The cytotoxicity of rotenone and amelioration of cytotoxicity through the use of bioactive compounds (UD, MC, and MK) were determined through a dye exclusion test (trypan blue) in whole larvae and tissues of ROT-exposed organisms. Of the larvae exposed to ROT, 95% showed blue staining in the whole larvae and their tissues (brain ganglia, salivary gland, midgut, and gastric caeca). ROT combined with MK exhibited significantly less blue staining than the ROT + UD and ROT + MC groups in the whole larvae and the abovementioned tissues, respectively. This observation is supported by a previous study on ROT, which found that nutraceuticals significantly improved cell viability [64]. This protective property of UD, MC, and MK may be attributed to the presence of bioactive components responsible for quenching free radicals or due to the upregulation of antioxidative defense mechanisms.

To comprehend the unfavorable effects of ROT that increase cellular oxidant levels and cause proteins to undergo oxidative post-translational modifications, biochemical studies were carried out after giving treatment for 24 and 48 h. Third instar larvae of *D. melanogaster* exposed to 500 µM ROT exhibited a statistically significant decrease ($p < 0.001$) in the total protein content of their tissues. After 24 h, the protein content in the larvae was reduced in the ROT treated group compared with the control and DMSO control group. ROT coexposed with MK showed improved protein concentrations, followed by ROT + UD and ROT + MC. After 48 h, the trend remained the same in comparison with the control group, and the ROT treatment decreased the protein content of the larvae. Increased protein was observed in ROT coexposed with MK, followed by ROT + UD, and lowest in ROT + MC. This finding is consistent with previous research that found various pesticides led to reduced protein content in organisms [65].

Acetylcholinesterase (AChE) is a vital enzyme of the cholinergic system that modulates physiological processes, including memory and locomotor activities. It hydrolyzes acetylcholine to choline and acetate, thereby terminating cholinergic neurotransmission between synapses. In this study, it was found that when the larvae were exposed to ROT for 24 h, they exhibited a statistically significant ($p < 0.001$) inhibition of AChE activity compared with the control or DMSO and had ~60% reduced AChE levels. When ROT was coexposed with MK, AChE levels were improved, and only 9.7% inhibition was evident. AChE levels were improved to a lesser extent in the ROT + UD and ROT + MC groups compared with the control. These elevated levels of AChE were significant when compared with the ROT-treated groups. The maximum inhibition of AChE levels was present in ROT-exposed organisms after 48 h compared with control larvae, and the highest rescue from ROT-induced toxicity was observed in the ROT + MK group, followed by the ROT + UD and ROT + MC groups. We observed that inhibition of AChE in ROT-exposed organisms and nutraceuticals helps to rescue AChE levels, which is supported by previous observations [66,67].

An organism's behavior reflects its typical physiological activity. Climbing and jumping activities in this context reflect the physiological condition of the organism. Therefore, a high rate of locomotor deficits as evaluated by the climbing assay may indicate rotenone-induced neurotoxicity. Due to their propensity to remain at the base of the plastic tube, flies with locomotor deficits do not appear to have normal leg coordination. This phenotypic expression has previously been attributed to the high energy needs of the muscles used for walking and flying, which are packed with mitochondria. Although speculative, uncoupled mitochondrial machinery may likely be to blame for the same underlying conditions of severe complex I inhibition. Surprisingly, MK > UD > MC were able to significantly rescue flies from worsening locomotor dysfunctions, showing that they may be able to protect by restoring the dopamine pool at the mitochondrial level. This finding supports previous research that showed a strong link between dopamine deficiency and locomotor dysfunction [68]. The adverse effect of the pesticide on the organism was shown by the significant decrease in jumping behavior in exposed organisms, which was followed by an inhibition of AChE activity. Inhibition of AChE activity has previously been described as a sign of poor locomotor activity [69]. All nutraceuticals improved the jumping ability of flies. ROT + MK exhibited the highest rescue, followed by ROT + UD, and ROT + MC. In this context, the isolation of naturally occurring antioxidants have raised interest in plant biomass, which has proven to be rich in compounds. Plants possess the ability to biosynthesize a variety of non-enzymatic antioxidants that can reduce ROS-induced oxidative damage [70–73].

Taken together, the current study suggests that UV-Vis, FTIR, and HPLC analysis differentiates the extracts of UD, MC, and MK. The comparative account of these medicinal herbs revealed significant variation, which can be used to identify plants that have the most phytoconstituents to be used as phyto remedies for a variety of diseases. Based on our biochemical evidence, we conclude that short-term dietary feeding of UD, MC, and MK to *Drosophila melanogaster* has the propensity to attenuate ROT-induced oxidative stress because of its antioxidative properties and capacity to regulate antioxidant defenses. Additionally, their neuroprotective properties were demonstrated by their potential to significantly alleviate rotenone-induced oxidative stress, enhance locomotion, and restore AChE levels. Moreover, these findings reveal that these plants are medicinally important and should be studied further to locate bioactive compounds and determine their significance in pharmaceutical industries. In the future, we will work with advanced spectroscopic and nanotechnology-based investigations for the identification and structural elucidation of compounds present in UD, MC, and MK against ROT-induced toxicities.

5. Conclusions

Current research provides evidence that the antioxidant and antiradical activities of three ethnomedicinal plants collected from different geographical origins and varieties have statistically significant ($p < 0.001$) varied complex chemical mixtures. This investigation has provided preliminary information to determine the chemical composition of *U. dioica*, *M. chamomilla*, and *M. koenigii* using UV-Vis, FT-IR, and HPLC techniques. From the above investigations, it can be concluded that MK has higher anti-radical activity in aqueous as well as ethanolic extracts in DPPH, ABTS, TPC, and TFC assays among the screened plants. The presence of the O-H, C-H, N-H, C-O, C=O, C-C, C-N, N=C, S=O, and C=N groups were predicted by UV-Vis, FT-IR, and HPLC. Additionally, we conclude that short-term nutritional feeding of UD, MC, and MK to *Drosophila* has the potential to reduce ROT-induced oxidative stress due to its antioxidative properties and capacity to regulate antioxidant defense mechanisms. They are promising plants for future research due to their antioxidant capabilities and might help slow or prevent the process of oxidative stress-related diseases. The findings of this study will be useful in the quality control of raw herbaceous material to verify their potential for phytopharmaceutical applications and health-promoting properties that could be used in drug discovery. However, further research is needed for a better understanding of their bioactivity, toxicity profile, and impact on the ecosystem and agricultural commodities.

Author Contributions: Conceptualization, M.P.S. and S.S.; methodology, M.P.S. and S.S.; software, S.S. and S.Y.; validation, M.P.S. and S.S.; formal analysis, M.P.S. and S.S.; resources, M.P.S.; writing—original draft preparation, S.S., S.Y., M.P.S., E.V. and S.K.S.; supervision, M.P.S. All authors have read and agreed to the published version of the manuscript.

Funding: This research received no external funding.

Institutional Review Board Statement: Not applicable.

Informed Consent Statement: Not applicable.

Data Availability Statement: Data are contained within the article.

Acknowledgments: We acknowledge Lovely Professional University for providing the infrastructure and reagents required for our research. The authors are also grateful to Anurag Sharma, NITTE (Deemed to be University), Mangalore, India, for kindly providing a gift of the *Drosophila* stock (Oregon R$^+$).

Conflicts of Interest: The authors declare no conflict of interest.

References

1. Miliaraki, M.; Briassoulis, P.; Ilia, S.; Michalakakou, K.; Karakonstantakis, T.; Polonifi, A.; Bastaki, K.; Briassouli, E.; Vardas, K.; Pistiki, A.; et al. Oxidant/Antioxidant Status Is Impaired in *Sepsis* and Is Related to Anti-Apoptotic, Inflammatory, and Innate Immunity Alterations. *Antioxidants* **2022**, *11*, 231. [CrossRef]

2. Sies, H.; Belousov, V.V.; Chandel, N.S.; Davies, M.J.; Jones, D.P.; Mann, G.E.; Murphy, M.P.; Yamamoto, M.; Winterbourn, C. Defining roles of specific reactive oxygen species (ROS) in cell biology and physiology. *Nat. Rev. Mol. Cell Boil.* **2022**, *23*, 499–515. [CrossRef] [PubMed]

3. Sahoo, B.M.; Banik, B.K.; Borah, P.; Jain, A. Reactive Oxygen Species (ROS): Key Components in Cancer Therapies. *Anticancer Agents Med. Chem.* **2022**, *22*, 215–222. [CrossRef] [PubMed]

4. Abramov, A.Y. Redox biology in neurodegenerative disorders. *Free Rad. Boil. Med.* **2022**, *188*, 24–25. [CrossRef]

5. Deepashree, S.; Shivanandappa, T.; Ramesh, S.R. Genetic repression of the antioxidant enzymes reduces the lifespan in *Drosophila melanogaster*. *J. Comp. Physiol. B Biochem. Syst. Environ. Physiol.* **2022**, *192*, 1–13. [CrossRef] [PubMed]

6. Hajam, Y.A.; Rani, R.; Ganie, S.Y.; Sheikh, T.A.; Javaid, D.; Qadri, S.S.; Pramodh, S.; Alsulimani, A.; Alkhanani, M.F.; Harakeh, S.; et al. Oxidative Stress in Human Pathology and Aging: Molecular Mechanisms and Perspectives. *Cells* **2022**, *11*, 552. [CrossRef] [PubMed]

7. Beckers, M.; Bloem, B.R.; Verbeek, M.M. Mechanisms of peripheral levodopa resistance in Parkinson's disease. *NPJ Park. Dis.* **2022**, *8*, 56. [CrossRef] [PubMed]

8. Bandopadhyay, R.; Mishra, N.; Rana, R.; Kaur, G.; Ghoneim, M.M.; Alshehri, S.; Mustafa, G.; Ahmad, J.; Alhakamy, N.A.; Mishra, A. Molecular Mechanisms and Therapeutic Strategies for Levodopa-Induced Dyskinesia in Parkinson's Disease: A Perspective Through Preclinical and Clinical Evidence. *Front. Pharmacol.* **2022**, *13*, 805388. [CrossRef] [PubMed]

9. Gulcan, H.O. Selected Natural and Synthetic Agents Effective against Parkinson's Disease with Diverse Mechanisms. *Curr. Top. Med. Chem.* **2022**, *22*, 199–208. [CrossRef]

10. Tan, M.A.; Sharma, N.; An, S. Phyto-Carbazole Alkaloids from the Rutaceae Family as Potential Protective Agents against Neurodegenerative Diseases. *Antioxidants* **2022**, *11*, 493. [CrossRef]

11. Nath, M.; Debnath, P. Therapeutic role of traditionally used Indian medicinal plants and spices in combating COVID-19 pandemic situation. *J. Biomol. Str. Dynam.* **2022**, *40*, 1–20. [CrossRef] [PubMed]

12. Anand, U.; Tudu, C.K.; Nandy, S.; Sunita, K.; Tripathi, V.; Loake, G.J.; Dey, A.; Proćków, J. Ethnodermatological use of medicinal plants in India: From ayurvedic formulations to clinical perspectives—A review. *J. Ethnopharmacol.* **2022**, *284*, 114744. [CrossRef] [PubMed]

13. Bhusal, K.K.; Magar, S.K.; Thapa, R.; Lamsal, A.; Bhandari, S.; Shrestha, J. Nutritional and pharmacological importance of stinging nettle (*Urtica dioica* L.): A review. *Heliyon* **2022**, e09717. [CrossRef] [PubMed]

14. Uyar, A.; Doğan, A.; Yaman, T.; Keleş, Ö.F.; Yener, Z.; Çelik, İ.; Alkan, E.E. The Protective Role of *Urtica dioica* Seed Extract Against Azoxymethane-Induced Colon Carcinogenesis in Rats. *Nutr. Cancer* **2022**, *74*, 306–319. [CrossRef]

15. Chehri, A.; Yarani, R.; Yousefi, Z.; Novin Bahador, T.; Shakouri, S.K.; Ostadrahimi, A.; Mobasseri, M.; Pociot, F.; Araj-Khodaei, M. Anti-diabetic potential of Urtica Dioica: Current knowledge and future direction. *J. Diabetes Metab. Disord.* **2022**, *21*, 931–940. [CrossRef]

16. Ahmadipour, B.; Khajali, F. Expression of antioxidant genes in broiler chickens fed nettle (*Urtica dioica*) and its link with pulmonary hypertension. *Anim. Nutr.* **2019**, *5*, 264–269. [CrossRef]

17. Samakar, B.; Mehri, S.; Hosseinzadeh, H. A review of the effects of *Urtica dioica* (nettle) in metabolic syndrome. *Iranian J. Basic Med. Sci.* **2022**, *25*, 543. [CrossRef]

18. Perestrelo, B.O.; Carvalho, P.M.; Souza, D.N.; Carneiro, M.J.; Cirino, J.; Carvalho, P.O.; Sawaya, A.; Oyama, L.M.; Nogueira, F.N. Antioxidant effect of chamomile tea on the salivary glands of streptozotocin-induced diabetic rats. *Braz. Oral Res.* **2022**, *36*, e034. [CrossRef]

19. Shoara, R.; Hashempur, M.H.; Ashraf, A.; Salehi, A.; Dehshahri, S.; Habibagahi, Z. Efficacy and safety of topical *Matricaria chamomilla* L. (chamomile) oil for knee osteoarthritis: A randomized controlled clinical trial. *Complement. Ther. Clin. Pr.* **2022**, *21*, 181–187. [CrossRef]

20. Mojibi, R.; Morad Jodaki, H.; Mehrzad, J.; Khosravi, A.R.; Sharifzadeh, A.; Nikaein, D. Apoptotic Effects of Caffeic Acid Phenethyl Ester and *Matricaria chamomilla* essential oil on A549 Non-small Cell Lung Cancer Cells. *Iranian J. Vet. Med.* **2022**. [CrossRef]

21. Mondal, P.; Natesh, J.; Penta, D.; Meeran, S.M. Extract of *Murraya koenigii* selectively causes genomic instability by altering redox-status by targeting PI3K/AKT/Nrf2/caspase-3 signaling pathway in human non-small cell lung cancer. *Phytomedicine* **2022**, *104*, 154272. [CrossRef] [PubMed]

22. Aniqa, A.; Kaur, S.; Negi, A.; Sadwal, S.; Bharati, S. Phytomodulatory effects of *Murraya koenigii* in DMBA/TPA induced angiogenesis, hepatotoxicity and renal toxicity during skin carcinogenesis in mice. *J. Adv. Sci. Res.* **2022**, *13*, 153–165. [CrossRef]

23. Sanmugarajah, V.; Rajkumar, G. A Review of Anti-hyperglycemic Effects of Curry Leaf Tree (*Murraya koenigii*). *Borneo J. Pharm.* **2022**, *5*, 104–114. [CrossRef]

24. Markom, M.; Hasan, M.; Daud, W.R.W.; Singh, H.; Jahim, J.M. Extraction of hydrolyzable tannins from Phylla nthusniruri Linn: Effects of solvents and extraction methods. *Sep. Purify. Technol.* **2007**, *52*, 487–496. [CrossRef]

25. Adams, C.; Thapa, S.; Kimura, E. Determination of a plant population density threshold for optimizing cotton lint yield: A synthesis. *Field Crops Res.* **2019**, *230*, 11–16. [CrossRef]

26. Baliyan, S.; Mukherjee, R.; Priyadarshini, A.; Vibhuti, A.; Gupta, A.; Pandey, R.P.; Chang, C.M. Determination of Antioxidants by DPPH Radical Scavenging Activity and Quantitative Phytochemical Analysis of *Ficus religiosa*. *Molecules* **2022**, *27*, 1326. [CrossRef]

27. Wołosiak, R.; Drużyńska, B.; Derewiaka, D.; Piecyk, M.; Majewska, E.; Ciecierska, M.; Worobiej, E.; Pakosz, P. Verification of the Conditions for Determination of Antioxidant Activity by ABTS and DPPH Assays-A Practical Approach. *Molecules* **2021**, *27*, 50. [CrossRef]

28. Gudimella, K.K.; Gedda, G.; Kumar, P.S.; Babu, B.K.; Yamajala, B.; Rao, B.V.; Singh, P.P.; Kumar, D.; Sharma, A. Novel synthesis of fluorescent carbon dots from biobased Carica Papaya Leaves: Optical and structural properties with antioxidant and anti-inflammatory activities. *Environ. Res.* **2022**, *204*, 111854. [CrossRef]

29. Clarke, G.; Ting, K.N.; Wiart, C.; Fry, J. High correlation of 2,2-diphenyl-1-picrylhydrazyl (DPPH) radical scavenging, ferric reducing activity potential, and total phenolic content indicates redundancy in the use of all three assays to screen for antioxidant activity of extracts of plants from the Malaysian rainforest. *Antioxidants* **2013**, *2*, 1–10. [CrossRef]

30. Do, Q.D.; Angkawijaya, A.E.; Tran-Nguyen, P.L.; Huynh, L.H.; Soetaredjo, F.E.; Ismadji, S.; Ju, Y.H. Effect of extraction solvent on total phenol content, total flavonoid content, and antioxidant activity of *Limnophila aromatic*. *J. Food Drug Anal.* **2014**, *22*, 296–302. [CrossRef]

31. Sreenivasulu, N.; Fernie, A.R. Diversity: Current and prospective secondary metabolites for nutrition and medicine. *Curr. Opin. Biotechnol.* **2022**, *74*, 164–170. [CrossRef] [PubMed]

32. Sofowora, A. *Medicinal Plants and Traditional Medicine in Africa*; Spectrum Books Ltd.: Ibadan, Nigeria, 1993; pp. 191–289. [CrossRef]

33. Kumar, G.S.; Jayaveera, K.N.; Kumar, C.K.; Sanjay, U.P.; Swamy, B.M.; Kumar, D.V. Antimicrobial effects of Indian medicinal plants against acne-inducing bacteria. *Trop. J. Pharm. Res.* **2007**, *6*, 717–723. [CrossRef]

34. Olugbenga, O.O.; Adebola, S.S.; Friday, A.D.; Mercy, A.T.; Keniokpo, O.S. Effect of dietary tomato powder on growth performance and blood characteristics of heat-stressed broiler chickens. *Trop. Anim. Heal. Prod.* **2022**, *54*, 1–7. [CrossRef] [PubMed]

35. Mannu, A.; Poddighe, M.; Garroni, S.; Malfatti, L. Application of IR and UV-Vis spectroscopies and multivariate analysis for the classification of waste vegetable oils. *Resour. Conserv. Recycl.* **2022**, *178*, 106088. [CrossRef]

36. Tian, W.; Chen, G.; Gui, Y.; Zhang, G.; Li, Y. Rapid quantification of total phenolics and ferulic acid in whole wheat using UV-Vis spectrophotometry. *Food Control* **2021**, *123*, 107691. [CrossRef]

37. Ismail, M.M.; Morsy, G.M.; Mohamed, H.M.; El-Mansy, M.A.M.; Abd-Alrazk, M.M.A. FT-IR spectroscopic analyses of 4-hydroxy-1-methyl-3-2-nitro-2-oxoacetyl-2 (1H) quinolinone (HMNOQ). *Spectrochim. Acta Part A Mol. Biomol. Spectrosc.* **2013**, *113*, 191195. [CrossRef] [PubMed]

38. Pinelli, P.; Ieri, F.; Vignolini, P.; Bacci, L.; Baronti, S.; Romani, A. Extraction and HPLC analysis of phenolic compounds in leaves, stalks, and textile fibers of *Urtica dioica* L. *J. Agric. Food Chem.* **2008**, *56*, 9127–9132. [CrossRef]

39. Miguel, F.G.; Cavalheiro, A.H.; Spinola, N.F.; Ribeiro, D.L.; Barcelos, G.R.M.; Antunes, L.M.G.; Hori, J.I.; Marquele-Oliveira, F.; Rocha, B.A.; Berretta, A.A. Validation of an RP-HPLC-DAD method for chamomile (*Matricaria recutita*) preparations and assessment of the marker, apigenin-7-glucoside, safety, and anti-inflammatory effect. *Evid. Based Complement. Altern. Med.* **2015**, *2015*, 1–9. [CrossRef]

40. Pandit, S.; Kumar, M.; Ponnusankar, S.; Pal, B.C.; Mukherjee, P.K. RP-HPLC-DAD for simultaneous estimation of mahanine and mahanimbine in *Murraya koenigii*. *Biomed. Chromatogr.* **2011**, *25*, 959–962. [CrossRef]

41. Singh, M.P.; Reddy, M.M.; Mathur, N.; Saxena, D.K.; Chowdhuri, D.K. Induction of hsp70, hsp60, hsp83 and hsp26 and oxidative stress markers in benzene, toluene and xylene exposed *Drosophila melanogaster*: Role of ROS generation. *Toxicol. Appl. Pharmacol.* **2009**, *235*, 226–243. [CrossRef]

42. Hosamani, R.; Muralidhara. Neuroprotective efficacy of Bacopa monnieri against rotenone induced oxidative stress and neurotoxicity in *Drosophila melanogaster*. *NeuroToxicology* **2009**, *30*, 977–985. [CrossRef] [PubMed]

43. Akinade, T.C.; Babatunde, O.O.; Adedara, A.O.; Adeyemi, O.E.; Otenaike, T.A.; Ashaolu, O.P.; Johnson, T.O.; Terriente-Felix, A.; Whitworth, A.J.; Abolaji, A.O. Protective capacity of carotenoid trans-astaxanthin in rotenone-induced toxicity in *Drosophila melanogaster*. *Sci. Rep.* **2022**, *12*, 4594. [CrossRef] [PubMed]

44. Kumar, P.P.; Bawani, S.S.; Anandhi, D.U.; Prashanth, K. Rotenone mediated developmental toxicity *in Drosophila melanogaster*. *Environ. Toxicol. Pharmacol.* **2022**, *93*, 103892. [CrossRef] [PubMed]

45. Singh, M.P.; Ram, K.R.; Mishra, M.; Shrivastava, M.; Saxena, D.K.; Chowdhuri, D.K. Effects of coexposure of benzene, toluene, and xylene to *Drosophila melanogaster*: Alteration in hsp70, hsp60, hsp83, hsp26, ROS generation and oxidative stress markers. *Chemosphere* **2010**, *79*, 577–587. [CrossRef]

46. Singh, M.P.; Mishra, M.; Sharma, A.; Shukla, A.K.; Mudiam, M.K.; Patel, D.K.; Ram, K.R.; Chowdhuri, D.K. Genotoxicity and apoptosis in *Drosophila melanogaster* exposed to benzene, toluene and xylene: Attenuation by quercetin and curcumin. *Toxicol. Appl. Pharmacol.* **2011**, *253*, 14–30. [CrossRef]

47. Ellman, G.L.; Courtney, K.D.; Andres, V.; Feather-stone, R.M. A new and rapid colorimetric determination of acetylcholinesterase activity. *Biochem. Pharmacol.* **1961**, *7*, 88–95. [CrossRef]

48. Nitta, Y.; Sugie, A. Studies of neurodegenerative diseases using *Drosophila* and the development of novel approaches for their analysis. *Fly* **2022**, *16*, 275–298. [CrossRef]

49. Sharma, A.; Mishra, M.; Shukla, A.K.; Kumar, R.; Abdin, M.Z.; Chowdhuri, D.K. Organochlorine pesticide, endosulfan induced cellular and organismal response in *Drosophila melanogaster*. *J. Hazard. Mater.* **2012**, *221–222*, 275–287. [CrossRef]

50. Chaitanya, M.V.N.L.; Ali, H.S.; Usamo, F.B. Regulatory considerations of herbal biomolecules. In *Herbal Biomolecules in Healthcare Applications*; Academic Press: Cambridge, MA, USA, 2022; pp. 669–676. [CrossRef]

51. Mitra, E.; Ghosh, A.K.; Ghosh, D.; Mukherjee, D.; Chattopadhyay, A.; Dutta, S.; Bandyopadhyay, D. Protective effect of aqueous Curry leaf (*Murraya koenigii*) extracts against cadmium-induced oxidative stress in rat heart. *Food Chem. Toxicol.* **2012**, *50*, 1340–1353. [CrossRef]

52. Biswas, A.K.; Chatli, M.K.; Sahoo, J. Antioxidant potential of curry (*Murraya koenigii* L.) and mint (*Mentha spicata*) leaf extracts and their effect on color and oxidative stability of raw ground pork meat during refrigeration storage. *Food Chem.* **2012**, *133*, 467–472. [CrossRef]

53. Stalikas, C.D. Extraction, separation, and detection methods for phenolic acids and flavonoids. *J. Sep. Sci.* **2007**, *30*, 3268–3295. [CrossRef] [PubMed]

54. Schreiner, M.; Mewis, I.; Huyskens-Keil, S.; Jansen, M.A.K.; Zrenner, R.; Winkler, J.B.; Krumbein, A. UV-B-induced secondary plant metabolites-potential benefits for plant and human health. *Crit. Rev. Plant Sci.* **2012**, *31*, 229–240. [CrossRef]

55. Jain, P.K.; Soni, A.; Jain, P.; Bhawsar, J. Phytochemical analysis of *Mentha spicata* plant extract using UV-VIS, FTIR, and GC/MS technique. *J. Chem. Pharm. Res.* **2016**, *8*, 16. [CrossRef]

56. Wyrostek, J.; Kowalski, R.; Pankiewicz, U.; Solarska, E. Estimation of the content of selected active substances in primary and secondary herbal brews by UV-VIS and GC-MS spectroscopic analyses. *J. anal. Met. Chem.* **2020**, *2020*, 1–11. [CrossRef]

57. Maobe, M.A.; Nyarango, R.M.; Box, P.O. Fourier transformer infrared spectrophotometer analysis of *Urtica dioica* medicinal herb used for the treatment of diabetes, malaria, and pneumonia in Kisii region, Southwest Kenya. *World Appl. Sci. J.* **2013**, *21*, 1128–1135. [CrossRef]

58. Qais, F.A.; Shafiq, A.; Khan, H.M.; Husain, F.M.; Khan, R.A.; Alenazi, B.; Ahmad, I. Antibacterial effect of silver nanoparticles synthesized using *Murraya koenigii* (L.) against multidrug-resistant pathogens. *Bioinorg. Chem. Appl.* **2019**, *2019*, 1–11. [CrossRef]

59. Jiang, X.; Li, S.; Xiang, G.; Li, Q.; Fan, L.; He, L.; Gu, K. Determination of the acid values of edible oils via FTIR spectroscopy based on the OH stretching band. *Food Chem.* **2016**, *212*, 585–589. [CrossRef]

60. Tsivelika, N.; Irakli, M.; Mavromatis, A.; Chatzopoulou, P.; Karioti, A. Phenolic Profile by HPLC-PDA-MS of Greek Chamomile Populations and Commercial Varieties and Their Antioxidant Activity. *Foods* **2021**, *10*, 2345. [CrossRef]

61. Nagappan, T.; Ramasamy, P.; Wahid, M.E.A.; Segaran, T.C.; Vairappan, C.S. Biological activity of carbazole alkaloids and essential oil of *Murraya koenigii* against antibiotic-resistant microbes and cancer cell lines. *Molecules* **2011**, *16*, 9651–9664. [CrossRef]

62. Ningappa, M.B.; Srinivas, L. Purification and characterization of~ 35 kDa antioxidant protein from curry leaves (*Murraya koenigii* L.). *Toxicol. Vitr.* **2008**, *22*, 699–709. [CrossRef]

63. Javed, H.; Meeran, M.; Azimullah, S.; Bader Eddin, L.; Dwivedi, V.D.; Jha, N.K.; Ojha, S. α-Bisabolol, a Dietary Bioactive Phytochemical Attenuates Dopaminergic Neurodegeneration through Modulation of Oxidative Stress, Neuroinflammation, and Apoptosis in Rotenone-Induced Rat Model of Parkinson's disease. *Biomolecules* **2020**, *10*, 1421. [CrossRef] [PubMed]

64. Pandareesh, M.D.; Shrivash, M.K.; Naveen Kumar, H.N.; Misra, K.; Srinivas Bharath, M.M. Curcumin Monoglucoside Shows Improved Bioavailability and Mitigates Rotenone Induced Neurotoxicity in Cell and *Drosophila* Models of Parkinson's Disease. *Neurochem. Res.* **2016**, *41*, 3113–3128. [CrossRef] [PubMed]

65. Gupta, S.C.; Siddique, H.R.; Mathur, N.; Mishra, R.K.; Mitra, K.; Saxena, D.K.; Chowdhuri, D.K. Adverse effect of organophosphate compounds, dichlorvos and chlorpyrifos in the reproductive tissues of transgenic *Drosophila melanogaster*: 70 kDa heat shock protein as a marker of cellular damage. *Toxicology* **2007**, *238*, 1–14. [CrossRef] [PubMed]

66. Rao, S.V.; Yenisetti, S.C.; Rajini, P.S. Evidence of neuroprotective effects of saffron and crocin in a *Drosophila* model of parkinsonism. *Neurotoxicology* **2016**, *52*, 230–242. [CrossRef] [PubMed]

67. Krishna, G.; Muralidhara. Aqueous extract of tomato seeds attenuates rotenone-induced oxidative stress and neurotoxicity in *Drosophila melanogaster*. *J. Sci. Food Agric.* **2016**, *96*, 1745–1755. [CrossRef] [PubMed]

68. Riemensperger, T.; Issa, A.R.; Pech, U.; Coulom, H.; Nguyễn, M.V.; Cassar, M.; Jacquet, M.; Fiala, A.; Birman, S. A single dopamine pathway underlies progressive locomotor deficits in a *Drosophila* model of Parkinson disease. *Cell Rep.* **2013**, *5*, 952–960. [CrossRef] [PubMed]

69. Casanova, Y.; Negro, S.; Barcia, E. Application of neurotoxin-and pesticide-induced animal models of Parkinson's disease in the evaluation of new drug delivery systems. *Acta Pharm.* **2022**, *72*, 35–58. [CrossRef]

70. Zheng, L.; Yu, P.; Zhang, Y.; Wang, P.; Yan, W.; Guo, B.; Huang, C.; Jiang, Q. Evaluating the bio-application of biomacromolecule of lignin-carbohydrate complexes (LCC) from wheat straw in bone metabolism via ROS scavenging. *Int. J. Biol. Macromol.* **2021**, *176*, 13–25. [CrossRef]

71. Dong, H.; Zheng, L.; Yu, P.; Jiang, Q.; Wu, Y.; Huang, C.; Yin, B. Characterization and application of lignin–carbohydrate complexes from lignocellulosic materials as antioxidants for scavenging *in vitro* and *in vivo* reactive oxygen species. *ACS Sustain. Chem. Eng.* **2019**, *8*, 256–266. [CrossRef]
72. Gu, J.; Guo, M.; Zheng, L.; Yin, X.; Zhou, L.; Fan, D.; Shi, L.; Huang, C.; Ji, G. Protective Effects of Lignin-Carbohydrate Complexes from Wheat Stalk against Bisphenol a Neurotoxicity in Zebrafish via Oxidative Stress. *Antioxidants* **2021**, *10*, 1640. [CrossRef]
73. Vasile, D.; Enescu, C.M.; Dincă, L. Which Are the Main Medicinal Plants That Could Be Harvested from Eastern Romania? *Sci. Papers. Ser. Manag. Econ. Eng. Agric. Rural Dev.* **2018**, *18*, 523–528. Available online: https://managementjournal.usamv.ro/index.php/scientific-papers/1623-which-are-the-main-medicinal-plants-that-could-be-harvested-from-eastern-romania-1623#spucontentCitation67 (accessed on 12 August 2022).

Article

Assessment of *Gnaphalium viscosum* (Kunth) Valorization Prospects: Sustainable Recovery of Antioxidants by Different Techniques

Stanislava Boyadzhieva [1], Jose A. P. Coelho [2,3], Massimiliano Errico [4,*], H. Elizabeth Reynel-Avilla [5,6], Dragomir S. Yankov [1], Adrian Bonilla-Petriciolet [5] and Roumiana P. Stateva [1]

1 Institute of Chemical Engineering, Bulgarian Academy of Sciences, 1113 Sofia, Bulgaria
2 Instituto Superior de Engenharia de Lisboa, Instituto Politécnico de Lisboa, Rua Conselheiro Emídio Navarro 1, 1959-007 Lisboa, Portugal
3 Centro de Química Estrutural, Institute of Molecular Sciences, Instituto Superior Técnico, Universidade de Lisboa, Av. Rovisco Pais, 1049-001 Lisboa, Portugal
4 Department of Green Technology, Faculty of Engineering, University of Southern Denmark, Campusvej 55, 5230 Odense, Denmark
5 Instituto Tecnológico de Aguascalientes, Aguascalientes 20256, Mexico
6 CONACYT, Ciudad de México 03940, Mexico
* Correspondence: maer@igt.sdu.dk

Citation: Boyadzhieva, S.; Coelho, J.A.P.; Errico, M.; Reynel-Avilla, H.E.; Yankov, D.S.; Bonilla-Petriciolet, A.; Stateva, R.P. Assessment of *Gnaphalium viscosum* (Kunth) Valorization Prospects: Sustainable Recovery of Antioxidants by Different Techniques. *Antioxidants* **2022**, *11*, 2495. https://doi.org/10.3390/antiox11122495

Academic Editors: Antonella D'Anneo and Marianna Lauricella

Received: 9 November 2022
Accepted: 14 December 2022
Published: 19 December 2022

Publisher's Note: MDPI stays neutral with regard to jurisdictional claims in published maps and institutional affiliations.

Abstract: This work investigates the prospects for exploitation of *Gnaphalium viscosum* (Kunth) abundant but with limited applications till present biomass. The feasibility of traditional techniques (two-phase solvent, and the benchmark Soxhlet extraction) and supercritical extraction without/with a cosolvent at $T = 40–60\ °C$ and $p = 30–50\ MPa$ was examined to explore the possibility of recovering phytochemicals from *G. viscosum* leaves, flowers and stems. The efficiency of the techniques was assessed and compared based on yield, influence of solvents used, total phenolic content and antioxidant activity of the extracts. Phenolics of different complexities were identified and quantified by applying LC (LC–MS/MS, and LC–HRAM), while the fatty acid profile was determined by GC–FID. The results of this extensive study demonstrated the huge valorization potential and prospects of *G. viscosum*, since highly potent antioxidants such as kaempferol, kaempferol-3-O-β-d-glucoside (astragalin), and chlorogenic acid were ascertained in considerable amounts. Furthermore, for the first time, the presence of leontopodic acid, a greatly substituted derivative of glucaric acid, was detected in the species.

Keywords: Mexican Gordolobo; supercritical CO_2 extraction; phytochemicals; antioxidants; fatty acids; biomass valorization

1. Introduction

Gnaphalium L. is a genus of flowering plants, commonly called cudweeds, which includes approximately 200 species of the Compositae (Asteraceae) family. It is widespread in temperate and subtropical regions of the world [1,2]. Of the *Gnaphalium* genus, at least 26 species are referred to as "Gordolobo", also known in English as Mexican Mullein. However, the latter should not be confused with the plant Great Mullein (species *Verbascum Thapsus* L. *Scrophulariaceae* family), which is native to Europe, Africa, and Asia, and is not commonly found in Mexico.

A detailed review, spanning over recent decades, of the *Gnaphalium* genus phytochemical and biological characteristics was published by Zheng et al. [1]. The authors reported that approximately 125 metabolites were identified in the genus comprising, among others, flavonoids, sesquiterpenes, diterpenes, triterpenoids, phytosterols, anthraquinones, acetylenic compounds, carotenoids and some long-chain unsaturated fatty acids. Furthermore, it was shown that extracts of the flowers and leaves of *Gnaphalium* species possess antioxidant, antibacterial, antifungal, anticomplement, antitussive, expectorant, insect antifeedant, cytotoxic, anti-inflammatory, antidiabetic, and antihypouricemic activity.

In another recent review on the application of Mexican plants, Quinones-Bastidas and Navarrete [2] outlined the applications of tea infusion of the inflorescences of the *Gnaphalium* genus in the treatment of asthma, flu, cough, fever and bronchial infections. Attention was also drawn to the fact that at least 10 species of Mexican Gordolobo (MG) have been used for centuries in folk medicine in Mexico and other Latin American counties to treat respiratory ailments and digestive disorders. Mata et al. [3] corroborated and commented that, obviously, the ancient Aztecs were well aware of the medicinal uses of some *Gnaphalium* species.

Some preliminary results about the antimycobacterial activity of MG species on Mycobacterium tuberculosis were reported by Hernández [4]. Another study investigated the antioxidant activity and cytotoxic effect of the flowers and leaf extracts of a particular *Gnaphalium* species (*G. viscosum*) recovered by methanol, *n*-hexane and ethyl acetate, on malignant human cell lines of the cervix and breast [5]. It was speculated that the effect could be a result of the presence of the flavonoid 5-hydroxy-3,7dimetoxi flavone, a compound that helps in the prevention and treatment of various diseases, including cancer.

From the analyses of results published till present, it is evident that extracts of the areal parts of *Gnaphalium* species were mainly obtained by organic solvents (e.g., *n*-hexane, methanol, ethanol, and ethyl acetate) and water. The extract composition was identified by applying different analytical methods. For example, Villagomez-Ibarra et al. [6] examined successive hexane, ethyl acetate and methanol extracts obtained by maceration of air-dried and powdered plant material (flowers, leaves and stems) of three Mexican *Gnaphalium* species. They confirmed the presence of previously isolated constituents such as diterpenoids, flavonoids, and acetylenic compounds, as well as some that are new, and have not been detected before, such as carotenoids, ent-Kaur-16-en-19-oic acid (kaurenoic acid), 13-epi-sclareol, beta-sitosterol, and stigmasterol.

Ontiveros-Rodríguez et al. [7] argued that specialized metabolites of medicinal plants can be considered as their chemical fingerprinting, which is important in order to introduce quality control on medicinal plant species and guarantee the safety of consumption. In view of this, they advocated an NMR-based protocol that can be applied to determine the chemical profiling of commercial samples of MG acquired from different vendors in Mexico City. In order to establish the compositional differences between these samples, special emphasis was placed on the flavones that characterize MG. Extracts from 17 retail samples of MG flowers recovered with a gradient of water:chloroform (1:4, 1:2 and 0:1) in an ultrasonic bath were prepared. The organic phase of the three extracts was analyzed by ^{1}H NMR. Formic, malic, gallic, fumaric and malonic acids, the amino acids alanine, asparagine and valine, as well as a complex mixture of sugars were reported. Flavones (e.g., gnaphaliin A, gnaphaliin B, and araneol) were identified by their corresponding ^{1}D ^{1}H NMR spectra.

From the retrospective analysis of the literature, it can be concluded that plants of the *Gnaphalium* genus, and in particular MG, can be classified as a sustainable resource of vital specialized secondary metabolites with biological activities and prospects for potential applications in modern food, pharmaceutical and other industries, targeting human health and well-being.

As known, the implementation and use of feasible methods and/or intensified technologies towards valorization of renewable biomass is a key factor in the development of sustainable processes. There is a considerable number of conventional and non-conventional methods to isolate secondary metabolites from plants. Still, to the best of our knowledge, till present, as discussed briefly above, the bioactive constituents of different *Gnaphalium*, and MG in particular, species were usually recovered by traditional techniques applying organic solvents. Furthermore, a systematic and in-depth comparison of the influence of the techniques operating parameters on the secondary metabolites content in the extracts obtained is rare.

Hence, there is a niche, unexplored till present, regarding acquiring insight into the viability, potential, and possible limitations of different methods/techniques—from traditional to sustainable, with low environmental impact, applying green solvents—for extraction of high-quality bioactive compounds from MG species.

Of the different MG species, *Gnaphalium viscosum* (Kunth), an annual or biannual herb, 30–100 cm high, with hairy or downy leaves and inconspicuous flowers, was selected as an object of investigation in the present study. It was not a random choice but one motivated by several important facts—*G. Viscosum* is abundantly present in 27 of Mexico 32 states [7], and can also be found from Canada to Honduras, as well as in Northern South America; is among the species mainly considered as MG by the Mexican Herbal Pharmacopoeia [8]; has demonstrated great potential as an antibacterial agent [2,6]; its cultivation is not limited by any specific requirements or inclusions in protected species lists, and obtaining its biomass would require minimum investment. In view of the above, the main aim of our research was to explore the potential of three techniques to recover secondary metabolites—antioxidants and fatty acids—from the leaves, flowers and stems of *G. viscosum*. To achieve that goal, the efficiency and sustainability of the techniques were compared on the basis of extraction yield, composition and quality of extracts recuperated.

The first two methods are conventional extractions applying organic solvents. The third one is extraction with supercritical CO_2 ($scCO_2$), either neat or with a co-solvent. Supercritical extraction (SCE) is considered to be among the most sustainable green alternatives to the conventional ones. Its particular benefits include no waste production, shorter extraction time, automation, lower solvent consumption, and no presence of organic solvents in the extracts. Of particular importance for the recovery of heat-sensitive compounds are the low-temperature operative conditions of SCE, as prolonged heating during the removal of solvent could lead to degradation of secondary metabolites [9].

To the best of our knowledge, the application of neat $scCO_2$ and $scCO_2$ with a co-solvent to the recovery of bioactives from *G. viscosum* is unique. Moreover, in the open literature, we only found one article devoted to the SCE of Mullein species, namely extraction of dried areal parts of Common mullein (*Verbascum thapsus* L.) by $scCO_2$ at 30 MPa and 313.2 K with the view to determine the antibacterial activity of the extracts [10].

The composition of phenolics in representative extracts was identified and quantified by LC (liquid chromatography high-resolution accurate mass—LC–HRAM, and liquid chromatography with tandem mass spectrometry—LC–MS/MS) analyses, while the fatty acid profile was determined by GC–FID. Furthermore, total phenolic content was measured, and the antioxidant activity of the extracts was determined by ABTS and DPPH.

On the basis of the results obtained, it was possible to critically analyze the influence of the techniques' specifics (operational parameters and application of given solvents/co-solvents) on the quality of the extracts recovered. Subsequently, that info can be used to reveal *G. viscosum* valorization prospects and potential for its efficient utilization in pharmaceutical, food, cosmetic, and other industries.

As far as we are aware, the present research, which combined the concerted efforts of scientists from four countries, is the first to report the application of conventional methods and a supercritical fluid extraction technique to the recovery and analyses of secondary metabolites derived from the MG species *G. viscosum* (Kunth).

2. Materials and Methods

2.1. Plant Material

The species *Gnaphalium Viscosum* (Kunth) biomass was purchased from a local herb pharmacy in Aguascalientes, Mexico. Gordolobo material was separated by hand into leaves, flowers, and stems. Each fraction was grinded in a household blender (Heinner, Bucharest, Romania). The average particle diameter of the material, subjected to further extractions, was determined to be less than 1 mm.

2.2. Chemicals and Reagents

The main standards used in the analyses of phenolic compounds by LC–MS/MS, and all other relevant data were presented in detail in a previous work [9]. The additional standards were leontopodic acid A, cat. N 6026S, and leontopodic acid B, cat. N 6032S.

Chemicals applied for TPC, antioxidant activity assays (ATBS and DPPH): ethanol HPLC grade (Panreac, Barcelona, Spain), gallic acid, Folin–Ciocalteu reagent 2 N, Trolox (6-hydroxy-2,5,7,8-tetramethylchroman-2-carboxylic acid), DPPH• (2,2-diphenyl-1-picrylhydrazyl) (Sigma-Aldrich, St. Louis, MO, USA), sodium carbonate (Merck, Darmstadt, Germany), potassium persulfate and absolute ethanol (Neon, Suzano, SP, Brazil), and quercetin dihydrate (Sigma-Aldrich Chemie GmbH., Steinheim, Germany).

Chemicals used for the GC–FID analyses: Supelco 37 Component FAME Mix (CRM47885), toluene (pure, VWR International, France), sulfuric acid (98%, MerckKGaA, Germany), sodium chloride (pure, MerkKGaA, Darmstadt, Germany), potassium bicarbonate (pure, VWR International, Paris, France), chloroform (99.8% VWR International, Paris, France), sodium sulfate (pure, Sigma-Aldrich Chemie GmbH, Taufkirchen, Germany), and helium (99.9999%, Air Liquide A/S).

The rest of the reagents applied were of the highest purity: methanol $\geq$ 99.9%, ethanol $\geq$ 99.8% and *n*-hexane $\geq$ 99% were purchased from Honeywell Riedel-de-Haen (Seelze, Germany), ethyl acetate $\geq$ 99.5% from JLS-Chemie Handel GmbH (Hannover, Germany), methyl tert-butyl ether $\geq$ 99.8%, and acetonitrile $\geq$ 99.9% from Sigma-Aldrich (Darmstadt, Germany) and bone dry grade CO_2 (99.99% pure; No water, Messer, Sofia, Bulgaria).

2.3. Two-Phase Solvent Extraction

From each of the three samples, 1.00 g of material was weighed by an analytical balance, and to each of them 25 mL of solution A (water/methanol (3/1, (*v/v*)) and 15 mL of solution B (methyl tert-butyl ether/methanol (3/1, vol.)) were added. Each mixture was stirred in a mechanical homogenizer for 1 min at 36000 rpm and at room temperature. Consequently, the mixture was stirred in a rotator for 24 h, after which the samples were centrifuged for 20 min at 6500 rpm and at 4 °C.

The upper (top) layer containing non-polar substances and chlorophyll was removed and subsequently evaporated to dryness in a rotary vacuum evaporator at a temperature below 35 °C. All samples produced an oil-resembling product.

The bottom water–alcohol layer was filtered on a paper membrane; the solid mass was washed three times with 10 mL of solution A. Then, the combined filtrate was evaporated to dryness on a rotary evaporator at a temperature below 40 °C. The resulting solid residue was dissolved in 10% acetonitrile, and lyophilized.

The extracts were subsequently analyzed by LC–HRAM and LC–MS/MS.

2.4. Soxhlet Extraction

All experiments were performed by means of the Soxhlet apparatus ISOLAB NS29/32+ 34/35 (Merck KGaA, Darmstadt, Germany). Three solvents with different polarities (*n*-hexane, ethyl acetate and ethanol) were used (Table 1). The ratio between the liquid: solid phases was 30:1.

In all experiments, the extraction cartridge was filled with 7.0 $\pm$ 0.1 g dry material (*G. viscosum* leaves, flowers or stems). The extraction time was different and depended on discoloration of the solvent. After each extraction, the solvent from the liquid extract was evaporated under vacuum using a Hei-VAP Rotary Evaporator (Heidolph Instruments GmbH&Co. KG, Schwabach, Germany).

Table 1. Soxhlet extraction yields of *G. viscosum* leaves, flowers and stems using different solvents.

Solvent	Extraction Yield (wt%)
Leaves	
Ethanol	17.18 ± 0.84
Ethyl Acetate	7.27 ± 0.36
n-Hexane	3.99 ± 0.21
Flowers	
Ethanol	12.23 ± 0.56
Ethyl Acetate	3.96 ± 0.19
n-Hexane	3.48 ± 0.17
Stems	
Ethanol	8.20 ± 0.38
Ethyl Acetate	2.88 ± 0.14
n-Hexane	1.81 ± 0.09

Extraction yield expressed in wt% (mean ± standard deviation).

The resulting dry extract was further dried to a constant weight in an air circulation oven, at 333.15 ± 2.0 K, and the yield was evaluated according to Equation (1):

$$\text{Yield}(\%) = \frac{\text{mass of extract (g)}}{\text{mass of sample (g)}} * 100 \tag{1}$$

The extracts obtained were placed in glass vials and kept at 4 °C until analysis by LC–MS/MS and GC–FID. Experiments were performed in triplicates and total extraction yield was expressed as the mean ± standard deviation.

2.5. Supercritical Fluid Extraction (SFE)

In our study, the SFE experiments were performed in a flow apparatus (SFT-110-XW, Supercritical Fluid Technologies Inc., Newark, DE, USA), equipped with two parallel 50 cm^3 internal volume extractors made from stainless steel tubing (7 cm long, internal diameter 3.02 cm) and temperature controllers for extraction vessels and restrictor valves, which can be adjusted up to 393.2 K.

The required pressure of the CO_2 from the tank (room temperature) is ensured by a SFT Nex10 SCF pump actuated from a compressor model HYAC50-25, Hyundai, Seoul, Republic of Korea. The maximum pressure is 60 MPa.

The CO_2 flow rate at the outlet of the extraction cell is measured by a flow meter and a totalizer from Alicat Scientific (Tucson, AZ, USA), model M-5SLPM-D/5M. In experiments with a co-solvent, an additional pump (LL-Class, State College, PA, USA) is used.

The extraction with neat scCO$_2$ of *G. viscosum* biomass was performed at $T = (313.2, 323.2$ and $333.2)$ K and $p = (30, 40$ and $50)$ MPa. The scCO$_2$ flow rate was 1.9×10^{-3} kg·min^{-1}. For the extractions carried out using CO_2 with a co-solvent, ethanol, the two pumps (for CO_2 and the co-solvent) were adjusted so that the final value of the CO_2 flow rate was 1.9×10^{-3} kg·min^{-1}. The values in percentage of the co-solvent were (5% and 10%) mol fractions, accordingly.

In all experiments, approximately 5 g dry sample of the respective matrix section of the *G. viscosum* plant biomass was placed in the processing vessel. The bottom and top of the extractor contain two metal frits (2 μm), and its lower part was filled with propylene wool, which was also placed at the top to avoid the entrainment of any material. The uncertainties of the temperature and pressure measurements were 1 °C and 0.1 MPa, respectively. The gravimetric measurements were performed on an analytical balance with an uncertainty of 0.1 mg and with a coverage factor of 2.

Once the system has equilibrated at the selected pressure and temperature, the static/dynamic valve on the oven is opened following the restrictor valve opening to achieve the required flow rate of liquid carbon dioxide through the system and the dynamic extraction takes place.

The extract fractions, without replicates, were collected at ambient pressure into glass vials, placed in an ice bath. The vials were changed every two minutes until no extract was collected in two consecutive vials. For the cases of SFE with a co-solvent, the solvent was evaporated in an air circulation oven at 338.2 K until constant weight.

The samples were kept at 277.2 K in the dark until analysis with GC–FID (neat scCO$_2$) and LC–MS/MS (with ethanol as a co-solvent).

2.6. Characterization and Quantification of the Extracts

2.6.1. LC–High-Resolution Accurate Mass analysis (LC–HRAM)

LC–HRAM analysis is a powerful tool that allows detection of complex analytes at low concentrations and accurate identification of components. In our case, the analyses were carried out on a Q Exactive® hybrid quadrupole-Orbitrap® mass spectrometer (Thermo Scientific Co., Waltham, MA, USA) equipped with a HESI® (heated electrospray ionization) module, a TurboFlow® Ultra High-Performance Liquid Chromatography (UHPLC) system (Thermo Scientific Co., Waltham, MA, USA) and a HTC PAL® autosampler (CTC Analytics, Zwingen, Switzerland).

Chromatographic Conditions

The chromatographic separations of the analyzed compounds was achieved on a Nucleo shell C18 (100 × 2.1 mm, 2.7 µm) analytical column (Macherey-Nagel, Germany), using gradient elution at a 300 µL/min flow rate. The eluents used were: A—0.1% formic acid in water; B—0.1% formic acid in ACN. The following binary gradient was used: Start at 0% B, hold for 2 min; 0–40% B—26 min, 40–90% B—3 min; 90% B—1 min; 90–0% B for 2 min and 0% B for 3 min.

Mass Spectrometry Conditions

Full-scan mass spectra over the m/z range 100–1200 were acquired in the negative ion mode (NIM) at resolution settings of 70,000. Parallel reaction monitoring (PRM) mode at resolution settings of 17,500 and 0.5 amu isolation window of precursor ions was used for quantitative analysis. Qualitative analyses were carried out using AIF (all ion fragmentation), top N (5) and PRM scans of operation of mass analyzer in the negative mode.

The mass spectrometer operating parameters used in the negative ionization mode were as those described in detail in a previous work [9], with the following changes: capillary temperature—320 °C; probe heater temperature—300 °C; auxiliary gas flow 12 units; sweep gas 2 units (units refer to arbitrary values set by the Q Exactive Tune software) and S-Lens RF level of 50.00. All derivatives were quantified using 5 ppm mass tolerance filters to their theoretical calculated m/z values. Data acquisition and processing were carried out applying the software package (Thermo Scientific Co., Waltham, MA, USA), as reported in [9].

Quantitative Analysis

Standards available at the lab were used. The results obtained are based on external calibration and use of PRM in the negative mode of operation of the mass analyzer, and represent the mean values based on 3 replicates (independently processed samples).

2.6.2. Liquid Chromatography with Tandem Mass Spectrometry (LC–MS/MS) Analysis

The methodology of quali-quantification of phenolics was explained in detail in an earlier work [9]. Here, just the main steps involved are summarized.

Standard and sample preparations were analogous to those reported in [9].

The LC–MS/MS analyses were carried out on a Q Exactive® hybrid quadrupole-Orbitrap® mass spectrometer (Thermo Scientific Co., Waltham, MA, USA) equipped with a HESI® (heated electrospray ionization) module, a TurboFlow® Ultra High-Performance Liquid Chromatography (UHPLC) system (Thermo Scientific Co., Waltham, MA, USA) and a HTC PAL® autosampler (CTC Analytics, Switzerland).

Chromatographic Conditions

The chromatographic separations of the compounds were performed on a Nucleodur C18 Gravity (100 × 2.1 mm, 1.8 μm) analytical column (Macherey-Nagel, Düren, Germany) using gradient elution at a 0.3 mL·min^{-1} flow rate, and eluents A—0.1% formic acid in water; B—0.1% formic acid in ACN (please see above).

Mass spectrometry conditions: Full-scan mass spectra over the m/z range 100–1200 were obtained in NIM at resolution settings of 70,000. The PRM mode was analogous to that of LC–HRAM analyses.

The operating parameters of the mass spectrometer were those reported in [9]. The quantification of the compounds identified was performed as also described in [9], and the results represent the mean values based on 3 replicates (independently processed samples).

2.6.3. Gas Chromatography (GC) Analysis

The nature and profile of the fatty acids in certain extracts of *G. viscosum* recovered by Soxhlet *n*-hexane and neat scCO$_2$ were determined by Gas Chromatography—Flame Ionization Detection (GC–FID) of methyl esters (FAME). The methodology is described in detail in [11]. In brief:

Sample Preparation

In a test tube, approximately 5 mg of the sample was dissolved in 1 mL toluene. A volume of 2 mL of 1% sulfuric acid in methanol was added and left overnight at 50 °C. A volume of 5 mL of sodium chloride (5%) solution was added before the esters were extracted with hexane (2 × 5 mL) and layers separated. The hexane layer was transferred to a clean test tube and washed with 4 mL of potassium bicarbonate (2%) solution. The hexane layer was then dried over anhydrous sodium sulfate. The solution was filtered through a 0.45 μm syringe filter and the solvent removed with a rotary evaporator under vacuum. The dried extract was redissolved in 1 mL hexane and 0.250 mL of chloroform before analysis.

Chromatographic Conditions

Samples were run on a GC Agilent 7890B equipped with an FID detector and a Agilent J&W DB-FFAP column (30 mm × 0.032 mm × 0.25 μm), with an injection volume of 1 μL. Split/splitless injection mode, 280 °C, split ratio 50:1, carrier gas—helium, 42 cm/s, constant flow mode.

The temperature gradient of the oven was set to: 120 °C (2 min), 5 °C/min to 140 °C (3 min); 20 °C/min to 250 °C (10 min). The FID was operating at 280 °C, hydrogen: 40 mL/min; air: 400 mL/min; make-up gas: 25 mL/min

Identification was performed based on the different retention times of the analytes. Analyses were performed in triplicate and the results are presented as the relative percent of each fatty acid in each of the samples analyzed.

2.6.4. Measurement of Total Phenolic Content

Total phenolic content (TPC) of all analyzed samples was determined using the Folin–Ciocalteu reagent [12]. In brief, a 20 μL aliquot of each sample was mixed with 100 μL Folin–Ciocalteu phenol reagent and 300 μL freshly prepared Na$_2$CO$_3$ (15% (w/v)) and was incubated for 2 min at room temperature. The reaction mixtures were completed to a final volume of 2 mL with deionized water, vortexed and incubated further for 2 h at room temperature. A 200 μL aliquot from each sample was transferred to a 96-well plate and the absorbance was measured at 765 nm using Varioskan multiplate reader (Thermo Electron Corporation, Vantaa, Finland). A calibration curve was obtained using the standard quercetin (0–700 mg/L) and the results were expressed as mg quercetin equivalents/L (mg QE/L) of sample. All determinations were carried out in triplicates and the results are expressed as the mean values.

2.6.5. Antioxidant Activity

DPPH Assay

The free radical scavenging activity of the samples was determined by the DPPH assay as previously described [13]. The DPPH solution (0.1 mM) was freshly prepared in methanol and 980 μL of the solution were mixed with a 20 μL aliquot from each analyzed sample. Methanol was used as the negative control. The reaction mixture was incubated for 1 h at room temperature in the dark and the absorbance was measured at 518 nm using a Varioskan multiplate reader. A calibration curve was obtained using Trolox (0–1.0 mM) and the results were expressed as mM Trolox equivalents (mM TE). All determinations were carried out in triplicates and the results are expressed as the mean values.

ABTS Assay

ABTS radical scavenging activity was determined by direct absorbance measurement of the radical (ABTS+) [14]. The radical was generated by mixing 2.5 mL of ABTS (7 mM solution in water) with 44 μL of 140 mM potassium persulfate. The solution was kept in the dark for 12–16 h till the development of a blue-green color, and diluted with 70% methanol to final absorbance of 0.700 ± 0.020 at 734 nm. The solution was used on the same day by mixing 200 μL of the diluted ABTS+ with 5 μL of fresh standard (0–1.0 mM Trolox) or sample. After 5 min, the absorbance was measured at 734 nm, using methanol as blank. The results were expressed as mM Trolox equivalents (mM TE). Determinations were performed in triplicates and expressed as the mean values.

3. Results and Discussion

This section is organized in the following way: firstly, the extraction techniques' efficiency assessed based on the yields achieved is analyzed and compared. The impact of operating parameters (solvents, temperature and pressure) is also discussed. Then, the chemical composition, total phenolic content and antioxidant activity of chosen flowers, leaves, and stem extracts recovered by the three extraction methods and identified and quantified by the analytical methods employed are presented and compared.

3.1. Extraction Yield

For the two-phase solvent extraction, the yields of the lyophilized dry matter were as follows: leaves—134.5 mg, flowers—110. 3 mg, stems—86.7 mg or (13.45, 11.03, 8.67)%, respectively.

Extraction yields obtained by Soxhlet with the three solvents, with polarity relative to water of 0.654, 0.228, 0.09 (ethanol, ethyl acetate and *n*-hexane), respectively [15], are displayed in Table 1.

The extractions with neat scCO$_2$ were performed on *G. viscosum* flowers and leaves only. The reason behind that was that the yield of the *n*-hexane Soxhlet extraction of the stems fraction was quite low, and typicallyscCO$_2$ yield is lower than the former. The SCE temperature was varied in the range 313.2–333.2 K, and pressure in the range 30–50 MPa. Representative cumulative experimental kinetic extraction curves were plotted to assess the effect of operating conditions (at a constant flow rate) on the yield, as well as a comparison with the Soxhlet *n*-hexane extraction yield (represented by the continuous line, parallel to the abscissa), are shown on Figures 1 and 2 for the flowers and leaves, respectively.

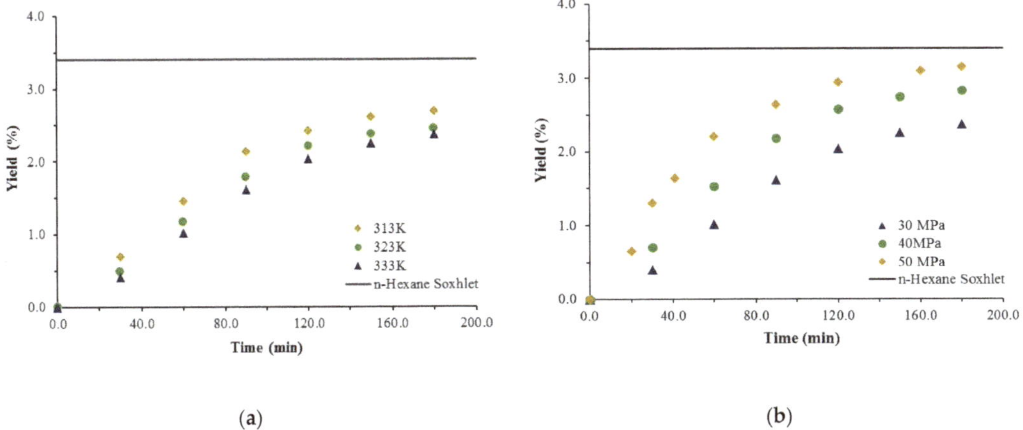

(**a**)

(**b**)

Figure 1. Cumulative experimental kinetic extraction curves plotted vs. the extraction time, at a scCO$_2$ flow rate of 1.9×10^{-3} kg/min. Influence of temperature at a constant pressure of 30 MPa (**a**) and pressure at a constant temperature of 333 K (**b**) on the yield for *G. viscosum* flowers.

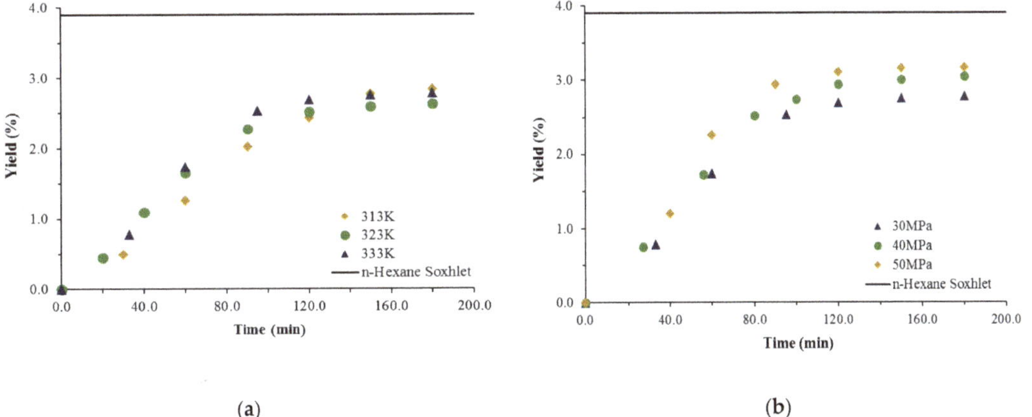

(**a**)

(**b**)

Figure 2. Cumulative experimental kinetic extraction curves plotted vs. the extraction time, at a scCO$_2$ flow rate of 1.9×10^{-3} kg/min. Influence of temperature at a constant pressure of 30 MPa (**a**) and pressure at a constant temperature of 333 K (**b**) on the yield for *G. viscosum* leaves.

The impact of the temperature on the extraction yield (Figures 1a and 2a) is not definitive in the sense that at lower temperatures, despite an increase in the yield, it was quite insignificant. Yield decrease with the rise in temperature was discussed by Coelho et al. [16], who explained that in part with the balance between two opposite effects: increasing the temperature decreases the density of the scCO$_2$ and thus its solubility capacity; but at the same time, it increases the vapor pressure of the compounds, consequently enhancing their solubility in the supercritical fluid. The influence of pressure, on the other hand, is in all cases positive, clearly demonstrated and straightforward—at a given temperature increasing the pressure improves the yield (Figures 1b and 2b). Thus, the highest yields for the flowers and leaves—3.1% and 3.17%, respectively—are achieved at the highest pressure applied and, though still lower, they are almost commensurable with that of *n*-hexane Soxhlet (3.48%). Moreover, it should be noted that 80% of the extract in both cases is recovered for a much shorter time than that required by the Soxhlet *n*-hexane.

With regard to the influence of solvents/co-solvents, the highest yields for the Soxhlet extractions were achieved by the solvent with the highest polarity among those examined—

ethanol, followed by ethyl acetate, and *n*-hexane. For the SCE, experiments were performed with 10% ethanol on the three areal parts of *G. viscosum*. In addition, scCO$_2$ with 5% ethanol was applied to the leaves only, as their yield with Soxhlet ethanol was the highest among all measured.

The cumulative experimental extraction curves plotted to assess the effect of the addition of a co-solvent to scCO$_2$ at the previously determined most favorable towards the yields values of the operation parameters temperature and pressure are shown in Figure 3.

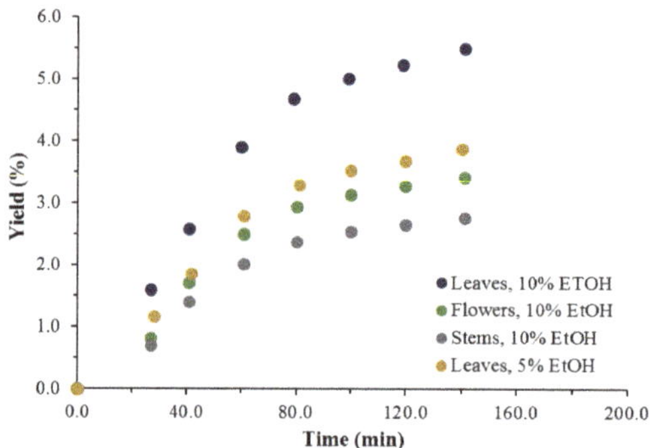

Figure 3. Cumulative experimental kinetic extraction curves plotted vs. the extraction time, at a scCO$_2$ flow rate of 1.9×10^{-3} kg/min, T = 333 K, p = 50 MPa. Influence of the co-solvent ethanol on the extraction process.

The scCO$_2$ + ethanol extraction process was realized for approximately 30% less time than that required with neat CO$_2$, at the same flowrate. Nevertheless, the yields achieved are still considerably lower than those of Soxhlet ethanol. For example, the highest yield for the leaves is approximately 3-fold lower than that of the latter but is obtained with a much lower amount of ethanol (only 10% in the composition of the solvent), and at a much shorter operation time. The scCO$_2$ + ethanol yields are commensurable with Soxhlet ethyl acetate yields, e.g., 2.76 vs. 2.88% for the stems.

The two-phase solvent extraction yields for the leaves and flowers, respectively, are lower than the corresponding ones of Soxhlet ethanol. However, for the stems, this technique renders a slightly higher yield than that of Soxhlet ethanol (8.6 vs. 8.2%), respectively, and for all three *G. viscosum* fractions its yields are higher than those of scCO$_2$ +ethanol. Still, the latter surmounts important limitations of conventional extractions as it uses less organic solvent and does not cause any harm to the environment.

3.2. Phytochemical Analysis

3.2.1. Analysis and Quantification of Antioxidants

Firstly, extracts of the two-phase solvent technique were analyzed by LC–HRAM. Prompted by a previous work [17], the analyses were concentrated on demonstrating whether presence of derivatives of caffeoylquinic, and in particular caffeoyl-D-glucaric acids would be detected. All samples were analyzed in identical concentrations and at identical analytical conditions.

The results obtained are illustrated by a mass chromatogram for the leaf extracts only. Thus, Figure 4 displays an exemplary mass chromatogram of compounds containing MS/MS fragment ions specific to substances comprising caffeoyl-D-glucaric acids. For the flower and stem extracts, the picture is identical with some variations.

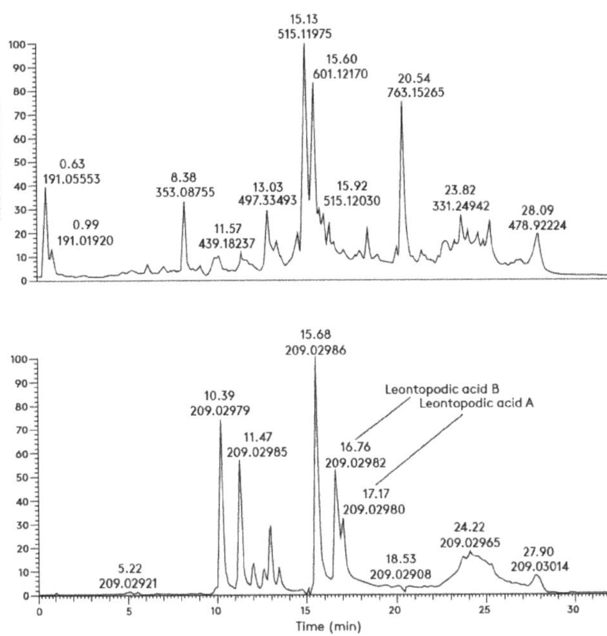

Figure 4. Mass chromatogram of compounds containing MS/MS fragment ion [M − H]− = 209.023 specific to substances comprising caffeoyl-D-glucaric acids.

Quantitative analyses were performed for 3-O-caffeoylquinic (chlorogenic) and 5-O-caffeoylquinic (neo-chlorogenic) acids, and for the highly substituted glucaric acid derivatives leontopodic acids A and B, using standards available at the laboratory. The results obtained are based on external calibration and use of PRM in the negative mode of operation of the mass analyzer and are displayed in Table 2.

Table 2. LC–HRAM analyses of *G. viscosum* flowers, leaves and stems extracts obtained by the two-phase solvent extraction.

	3-O-Caffeoylquinic (Chlorogenic) Acid	5-O-Caffeoylquinic (Neo-Chlorogenic) Acid	Glucaric Acid Derivatives	
			Leontopodic Acid A	Leontopodic Acid B
		ng/mg		
Flowers	4710.0	286.3	12.9	167.2
Leaves	3366.9	198.8	8.1	141.1
Stems	2752.8	98.7	14.1	1132.7

Relative standard deviation (RSD) = ±3.4%.

The highest quantity of chlorogenic acid is registered in the flowers, and the lowest in the stems. The amounts of the neo-chlorogenic acid are much lower and follow the same trend. The quantities of leontopodic acids A and B are the highest in the stems and lowest in the leaves. However, the amounts of leontopodic acid B are much higher, e.g., in the stems, its quantity is over 80-fold higher than that of leontopodic acid A.

Subsequently, in order to obtain a deeper knowledge and improved comprehension about the extract composition, particularly regarding the presence of phenolics of different chemical complexities, certain extracts of *G. viscosum* flowers, leaves and stems recovered by the three techniques applied were analyzed by LC–MS/MS.

The results obtained are displayed in Tables 3 and 4, respectively.

Table 3. LC–MS/MS analysis of phenolic compounds in selected *G. viscosum* flowers, leaves and stems extracts obtained by the two-phase solvent and Soxhlet ethanol extractions.

Compound	Two-Phase Solvent Technique			Soxhlet EtOH		
	Flowers	Leaves	Stems	Flowers	Leaves	Stems
	ng/mg					
Phenolic Acids						
Hydroxycinnamic and caffeoylquinic acid derivatives						
caffeic acid	3.15	4.28	1.95	5.99	17.21	9.24
o-coumaric acid	2.21	5.03	4.35	10.61	16.09	4.65
p-coumaric acid	0.14	0.16	0.19	0.43	0.31	0.25
m-coumaric acid	0.05	0.81	0.18	0.34	0.39	0.89
ferulic acid	5.04	4.77	4.62	14.01	19.23	6.11
cinnamic acid	2.63	11.14	3.48	3.99	12.53	4.56
3-O-caffeoylquinic (chlorogenic) acid	3618.95	2116.35	1925.44	5668.11	2421.34	2041.15
Hydroxybenzoic acid derivatives						
gallic acid	6.24	1.71	5.94	5.94	1.48	0.57
vanillic acid	69.66	64.21	46.35	10.23	21.97	1.89
ellagic acid	2.79	11.59	1.75	9.78	2.98	7.96
gentisic acid	351.82	181.58	52.26	866.72	465.26	284.73
protocatechinic acid	55.46	53.48	14.09	35.58	33.44	0.23
o-hydroxybenzoic acid	22.95	19.38	7.22	0.33	2.35	0.93
m-hydroxybenzoic acid	4.03	4.93	0.96	7.25	6.73	4.26
syringic acid	5.46	9.11	6.73	12.92	16.12	3.02
3-OH-4-methoxybenzoic acid	46.41	52.70	263.47	8.58	46.51	24.04
Flavonoids						
Flavonols						
quercetin	331.41	39.93	17.41	450.27	92.48	30,96
myrecitrin	790.80	200.78	56.72	1288.26	398.57	439.65
myrecitin	n.d.	n.d.	n.d.	n.d.	n.d.	n.d.
rutin	4.80	1.46	0.98	5.91	38.25	90.06
resveratrol	n.d.	n.d.	n.d.	n.d.	n.d.	n.d.
kaempferol	798.86	82.59	13.79	1106.91	72.32	39.68
kaempferol-3-O-glycoside	45999.92	6589.52	1207.40	76129.67	7993.67	9827.97
kaempferitin	n.d.	n.d.	n.d.	n.d.	n.d.	n.d.
fisetin	7.93	2.17	1.01	21.12	2.37	2.32
Flavones						
luteolin	6.24	1.71	0.44	5.94	1.48	0.57
apigenin	77.97	7.22	1.02	81.78	4.47	2.39
Flavan-3-ols						
catechin	n.d.	n.d.	n.d.	n.d.	n.d.	n.d.
epicatechin	0.21	0.30	0.06	1.31	0.29	0.13
epigallocatechin	3.74	n.d.	n.d.	n.d.	n.d.	n.d.
epigallocatechin gallate	0.00	n.d.	0.01	0.01	0.02	0.02
epicatechin gallate	0.19	0.22	1.91	0.22	0.12	0.16
Flavanones						
hisperidin	2.43	0.028	0.03	0.71	0.29	0.42
naringenin	11.02	2.08	0.58	19.91	2.52	1.95
Proanthocyanidins						
procyanidin B1	7.14	7.15	7.14	7.17	7.13	7.14
procyanidin B3	n.d.	n.d.	3.19	n.d.	4.86	n.d.
Caffeoyl-D-glucaric acid derivatives						
Leontopodic acid A	29.55	211.28	119.13	6.62	16.80	21.93
Leontopodic acid B	134.94	95.02	1011.05	2.23	71.36	4.37

Relative standard deviation (RSD) = ±2.3%. n.d—Not detected.

Table 4. LC–MS/MS analysis of phenolic compounds in selected *G. viscosum* flowers, leaves and stems extracts obtained by scCO$_2$ + 10% ethanol for the flowers, leaves and stems, and 5% ethanol for the leaves.

Compound	Flowers	Leaves	Stems	Leaves (5% EtOH)
		[ng/mg]		
		Phenolic Acids		
	Hydroxycinnamic and caffeoylquinic acid derivatives			
caffeic acid	1999.37	604.53	113.46	39.30
o-coumaric acid	173.29	5.59	36.91	4.95
p-coumaric acid	1.38	0.38	0.36	0.42
m-coumaric acid	10.32	2.30	1.74	1.04
ferulic acid	206.24	39.26	22.39	12.88
cinnamic acid	9.94	21.07	4.85	15.31
3-O-caffeoylquinic (chlorogenic) acid	62.49	60.89	16.64	2.88
		Hydroxybenzoic acid derivatives		
gallic acid	6.73	4.07	10.46	0.21
vanillic acid	772.21	162.03	51.54	46.25
ellagic acid	24.70	3.02	1.69	1.19
gentisic acid	1091.63	184.44	49.51	14.93
protocatechinic acid	32.55	24.56	1.21	1.33
o-hydroxybenzoic acid	174.88	41.38	17.79	8.04
m-hydroxybenzoic acid	47.90	19.33	5.39	3.99
syringic acid	160.85	50.37	33.33	17.08
3-OH-4-methoxybenzoic acid	726.99	138.40	48.99	45.90
		Flavonoids		
		Flavonols		
quercetin	3215.37	206.14	123.49	21.76
myrecitrin	307.14	32.38	5.62	0.95
myrecitin	n.d	n.d	n.d	n.d
rutin	17.33	11.49	5.13	1.98
resveratrol	n.d.	n.d.	n.d.	n.d.
kaempferol	10731.82	383.51	99.97	33.20
kaempferol-3-O-glycoside	2574.92	221.52	45.81	8.49
kaempferitin	n.d	n.d	0.01	n.d
fisetin	30.46	3.55	0.36	0.14
		Flavones		
luteolin	434.97	14.99	4.89	1.54
apigenin	15.01	0.19	0.19	0.17
		Flavan-3-ols		
catechin	n.d.	n.d.	n.d.	n.d.
epicatechin	1.70	0.01	n.d.	0.00
epigallocatechin	1.35	0.11	0.01	0.02
epigallocatechin gallate	0.10	n.d.	n.d.	n.d.
epicatechin gallate	0.12	0.02	0.01	0.01
		Flavanones		
hisperidin	1.19	0.71	0.89	0.76
naringenin	3294.13	83.91	13.42	23.06
		Proanthocyanidins		
procyanidin B1	n.d.	n.d.	n.d.	n.d.
procyanidin B3	n.d.	n.d.	n.d.	n.d.
		Glucaric acid derivatives		
Leontopodic acid A	n.d.	n.d.	n.d.	n.d.
Leontopodic acid B	n.d.	34.93	0.39	n.d.

Relative standard deviation (RSD) = ±2.3%. n.d—Not detected.

The quali-quantification of the extracts demonstrate that the three areal parts of *G. viscosum* are rich in important multifunctional ingredients, belonging to the hydroxycinnamic, caffeoylquinic, and hydroxybenzoic acids, respectively, and to several flavonoid subgroups, etc. In the recuperated by the different techniques *G. viscosum* extracts though, the quantities of the bioactives vary sometimes by orders of magnitude, which demonstrates the influence of the recovery methods operating conditions and solvents applied.

Chlorogenic (3-O-caffeoylquinic) acid is by far the most abundant among all acids in the extracts recovered by the two-phase solvent extraction and Soxhlet ethanol. In the flower extracts of the latter, chlorogenic acid quantity is approximately 1.5 higher than that in the corresponding extract of the first technique, while the amounts in the leaves and stems are lower and commensurable for both techniques. A dramatic change in the quantities of chlorogenic acid is observed for the $scCO_2$ + ethanol extracts. Though the highest amount is still found in the flowers, it is approximately 90-fold lower than that recovered by Soxhlet ethanol and over 57-fold lower than the one of the two-phase solvent extraction. Obviously, the change in the chlorogenic acid amounts registered in the $scCO_2$ + ethanol extracts can be an indication of the fact that the quantity of the co-solvent ethanol applied is far from sufficient to recover chlorogenic acid in full. That is further supported by the results obtained for the leaf extracted with 5% ethanol—the quantity of chlorogenic acid is over 21-fold lower than in the leaves but recuperated by 10% ethanol.

The demonstrated richness in chlorogenic acid of the flower, leaf and stem extracts obtained by the two-phase solvent and Soxhlet ethanol provides valuable information that can lead to new prospects for their use for health benefits, since chlorogenic acid has proven antidiabetic, anticarcinogenic, anti-inflammatory and antiobesity impacts [18].

Another interesting characteristic of the $scCO_2$ + ethanol flower, leaf and stem extracts is that caffeic acid is now dominant among all in the hydroxycinnamic and caffeoylquinic acid derivatives group. Moreover, its quantity in the flower extract is the highest among all acids of the two groups—and for comparison, is approximately 334-fold higher than that recovered by Soxhlet ethanol (1999.37 vs. 5.99) ng/mg. Ferulic and *o*-coumaric acids are the second and third most abundant acids in the flowers—a trend similar to that exhibited by the Soxhlet ethanol and the two-phase solvent extracts but their quantities are by an order of magnitude lower, e.g., 206.24 vs. 14.01 ng/mg ferulic acid in the $scCO_2$ + ethanol vs. Soxhlet ethanol flower extracts. In the leaves and stems, the quantities of ferulic and *o*-coumaric acids for the three techniques are in the same range of magnitude.

An intriguing and a completely different picture is obtained for the hydroxybenzoic acids recuperated by the three techniques. Though, in all extracts, gentisic acid is in the highest amounts, its quantities in the $scCO_2$ + ethanol flower extract are approximately 1.5-fold higher than Soxhlet ethanol and over 3-fold higher than that of two-phase solvent extraction. For the leaves and stems, however, the trend is reversed; the quantities of gentisic acid in the Soxhlet ethanol extracts are higher than those of $scCO_2$+ ethanol, the latter being commensurable with those of the two-phase solvent.

$scCO_2$ + ethanol extraction demonstrates a steady trend in the recovery of vanillic, *o*-hydroxybenzoic, syringic and 3-OH-4-methoxybenzoic acids. Their quantities in the flower, leaf and stem extracts are much higher than those recuperated by the two-phase solvent and Soxhlet ethanol, respectively. The only exception is the amount of 3-OH-4-methoxybenzoic acid detected in the stems extract of the former. Moreover, if compared to Soxhlet ethanol, it shows higher selectivity regarding the above acids.

The results obtained reveal that for the flowers, $scCO_2$ + ethanol, at the operating conditions tested, exhibits much higher selectivity towards certain hydroxybenzoic acids and promotes their recovery in higher amounts, when compared to the two-phase solvent and Soxhlet + ethanol. For the leaves, and stems, however, there is not a steady clear trend demonstrated, gentisic acid being the most prominent example—its quantities in the extracts recovered by Soxhlet ethanol are higher than those registered in $scCO_2$ leaf and stem extracts, respectively.

A closer examination supports the assumption that the amounts of the acids obtained depend not only on the matrix (flowers vs. leaves vs. stems), but on the extraction method, and in particular solvents applied.

In addition to the acids groups, several subgroups of the complex secondary metabolites belonging to the Flavonoids family were detected. Among those, the most remarkable by far is the flavonols subgroup.

A notable feature of *G. viscosum* is the presence of kaempherol-3-O-β-d-glucoside, known as astragalin, in very high quantities in all extracts recovered by the two-phase solvent and Soxhlet ethanol. The highest amount was identified in the flowers, followed by the leaves and stems for both techniques. Kaempherol-3-O-β-d-glucoside quantities in the extracts of the two-phase solvent and Soxhlet ethanol are not only the highest in the flavonols subgroup, but the highest among all secondary metabolites identified and quantified in the extracts obtained by the three techniques. Moreover, the amount of 76,129.67 ng/mg detected for kaempherol-3-O-β-d-glucoside in the Soxhlet ethanol flower extract is the absolute maximum among all phenolics identified and quantified by the three techniques.

While the quantities of that compound in the extracts recovered by the two-phase solvent are lower, still they do not differ by orders of magnitude—76,129.67, 7993.67, and 9827.97 ng/mg vs. 45,999.92, 6589.52, and 1207.40 ng/mg—when compared to Soxhlet ethanol, which is in a striking contradiction with the amounts registered in the $scCO_2$ + ethanol extracts—2574.92, 221.52, and 45.81 ng/mg. It should also be noted that kaempherol-3-O-β-d-glucoside quantities in the Soxhlet ethanol stem extracts are higher than those in the leaves, a tendency not observed for the extracts of the two-phase solvent and $scCO_2$ + ethanol. Moreover, the amount of astragalin in the Soxhlet ethanol stems extract is in the second place among all flavonoids and is over 8-fold higher than that in the extract of the two-phase solvent.

The other bioactives in relatively high amounts in the Soxhlet ethanol extracts are myricitrin, kaempferol, and quercetin (in diminishing amounts in that order). The trend observed for the two-phase solvent extraction is almost analogous but the quantities of the first two compounds in the line are commensurable.

For the extracts of $scCO_2$ + ethanol, as mentioned previously, the picture is completely different: now kaempferol, and not kaempherol-3-O-β-d-glucoside whose quantities have diminished to the fourth place following those of kaempferol, naringenin and quercetin, is the most abundant in the flavonols group. In addition, kaempferol amount is by far the highest among all phenolics identified and quantified in the $scCO_2$ + ethanol extracts, and is higher than that detected in the two-phase solvent and Soxhlet ethanol extracts (e.g., in the flowers is approximately 10-fold higher than that recovered by Soxhlet ethanol). The quantity of quercetin is the second highest. It is over 7.5-fold higher than the corresponding that in Soxhlet ethanol, which performs better than the two-phase solvent regarding that compound.

Kaempferol and kaempferol-3-O-glucoside are very important antioxidants. Many studies have described the beneficial effects of dietary kaempferol in reducing the risk of chronic diseases, especially cancer, as it inhibits cancer cell growth and angiogenesis and acts as a powerful promoter of apoptosis, while preserving normal cell viability [19]. Furthermore, kaempferol has a role as an antibacterial agent, a human xenobiotic, and blood serum metabolite, a human urinary metabolite, and a geroprotector.

Astragalin is a multifaceted phytochemical with broad and diversified pharmacological applications such as anticancer, anti-inflammatory, antioxidant, neuroprotective, antidiabetic, cardioprotective, antiulcer, antifibrotic, and antiosteoporotic properties [20]. Astragalin exhibits high antioxidant activity by scavenging radicals as well as inhibiting pro-oxidant enzymes and activating antioxidant enzymes. It was demonstrated that kaempherol-3-O-β-d-glucoside has a very good curative effect on cancer, and shows a superior pharmacological effect compared with quercetin.

In the rest of the flavonoid subgroups, the quantities of the metabolites detected are not high, with the only exception being flavanone naringenin recovered by $scCO_2$ + ethanol in *G. viscosum* flowers. Its amount is the second among all flavonoids recuperated by that technique, being lower only than kaempferol, and slightly higher but commensurable with quercetin. On the other hand, its amount is over 172-fold higher than the quantity registered in the flower extract of Soxhlet ethanol, and the two-phase solvent. The other compound present in relatively high amount is luteolin, its quantities in the flowers being much higher than those of the two conventional techniques, following the pattern of naringenin.

Myricitrin, naringenin and luteolin are also powerful antioxidants. Myricitrin exhibits antitumor and hepatoprotective properties, while shows strong anti-inflammatory and antioxidant activities, and is beneficial for the treatment of obesity, diabetes, hypertension, and metabolic syndrome. Luteolin possesses antioxidative, antitumor, and anti-inflammatory properties, and exhibits cardiac protective effects.

Finally, the occurrence of two unique caffeoyl-D-glucaric acid derivatives, namely leontopodic acids A and B, is yet another very important characteristic of *G. viscosum*. Though the presence of leontopodic acids A and B in some *Gnaphalium* species, and in particular in five European members of the genus habitual to the Alps, was earlier registered [17], still, as far as we are aware, this is the first work which reports the presence of the two acids in *G. viscosum*. That is a very important finding as, until recently, those acids were predominantly associated only with the protected and difficult to cultivate alpine species *Leontopodium alpinum* (edelweiss) [21,22]. Discovering new sources for those highly potent antioxidants is of considerable importance; moreover, when those resources, such as *G. viscosum*, are abundant and in practice not limited by specific cultivation requirements, or inclusion in protected species lists, etc.

The largest amount of leontopodic acid B was detected in the stem extracts recovered by the two-phase solvent extraction—the quantity being approximately 8- to 10-fold higher than that registered in the flowers and leaves, respectively. The amounts of acid B recovered by Soxhlet ethanol and $scCO_2$ + ethanol are orders of magnitude lower than those recuperated in all three extracts of the two-phase solvent. Additionally, the highest amounts of the acid are detected not in the stems but in the leaves for both techniques. With regard to leontopodic acid A, again, the two-phase solvent is the best performer, with the amount in the leaves being the highest, a tendency not observed for Soxhlet ethanol, for which the largest quantity is found in the stem extracts. The acid is not detected in any of the three *G. viscosum* extracts recovered by $scCO_2$ + ethanol. Obviously, water/methanol and methyl tert-butyl ether/methanol, which are applied as solvents in the two-phase extraction, exhibit much higher selectivity to leontopodic acids A and B. Those findings are corroborated by the LC–HRAM data, which confirmed that *G. viscosum* stems are a rich source of leontopodic acid B when compared to the flowers and leaves. Soxhlet ethanol does not exhibit good selectivity towards glucaric acid derivatives, which applies even more strongly to $scCO_2$ + ethanol extraction.

Leontopodic acids A and B are known for their strong antioxidant potential, anti-inflammatory, antiaging, memory improving, etc., effects. They also possess DNA-protecting properties, help resisting hepatitis virus, protect the liver, and have greater antioxidant potential than alpha-tocopherol (a form of vitamin E). Moreover, because of their beneficial effects on stimulating several key genes and proteins responsible for epidermal protection, the acids are essential in cosmetic formulations.

When analyzing the results obtained, some general observations can be made: it is the usual assumption that because of the presence of unsaturated bonds, polyphenols are heat sensitive and can be oxidized at higher temperatures, particularly when the recovery process is over extended period. In view of this, it would have been expected that $scCO_2$ + ethanol would perform better when compared for example to Soxhlet ethanol, taking into consideration the lower temperatures and much shorter extraction times applied. While that might be true for some of the compounds (kaempferol being one such example), for others it is far from so—see chlorogenic acid, and of course astragalin. A possible explanation might be that for certain secondary metabolites, the combined effect of lower temperature, shorter time and a low amount of co-solvent used is negative. As mentioned briefly previously, that assumption is corroborated when the respective quantities in the leaves recovered with either $scCO_2$ + 10 or 5% ethanol are compared.

Consequently, it is only fair to conclude that temperature influences a phenolic compound sensitivity in a complex and not straightforward way, and depends on numerous factors, among those physicochemical and biochemical characteristics and type of the particular metabolite, and lastly, but very importantly, on the nature of the plant matrix.

In summary, it should be underlined that the analyses performed have brought to light *G. viscosum* capabilities as a master chemist able to synthesize and store important and highly effective secondary phytochemicals, some of which are unique, with considerable potential for broad and diversified applications in the manufacturing of pharmaceuticals, food additives, cosmetics formulations, etc., targeted at human benefit.

3.2.2. Total Phenolic Content (TPC) and Antioxidant Activity (AA)

Table 5 shows that the flower $scCO_2$ + ethanol extract exhibits the highest phenolic content, followed by those of the two-phase solvent and Soxhlet ethanol. Flower extracts, are, however, an exception, since the TPC of the leaves and stems of the two-phase solvent extracts are higher than both Soxhlet ethanol and $scCO_2$ + ethanol, the latter being the lowest of the three.

Table 5. Total phenolic content, antioxidant activity and IC_{50} by DPPH and ABTS of *G. viscosum* flowers, leaves and stems extracts.

Extraction Method	TPC	DPPH		ABTS	
Two-Phase Solvent	**Quercetin eq. [µg/mg]**	**Trolox eq. [mM]**	**IC_{50} mg Extract**	**Trolox eq. [mM]**	**IC_{50} mg Extract**
flowers	130.87	12.72	1.34	9.48	2.27
leaves	95.70	9.25	2.72	6.07	4.11
stems	76.83	4.93	2.46	4.01	7.01
Soxhlet ethanol					
flowers	113.28	6.85	0.68	5.44	1.24
leaves	69.54	3.76	0.96	3.27	2.02
stems	58.81	4.08	1.96	2.15	3.22
$scCO_2$ + 10% ethanol					
flowers	162.11	1.64	3.15	2.71	36.08
leaves	36.96	0.65	18.45	1.14	10.98
stems	6.39	0.25	90.10	0.35	48.20
Leaves 5% EtOH	8.05	0.22	126.42	0.55	25.93

Relative standard deviation (RSD): $RSD_{DPPH} = \pm3.01\%$; $RSD_{ATBS} = \pm3.95\%$; $RSD_{IC50} = \pm1.6\%$.

As known, TPC only measures total phenols in the extracts without any identification of the compounds. Additionally, since phenolics show higher affinity towards polar solvents, obviously, the viable explanation of the TPC trends observed is that the solvents applied by the two-phase solvent (water/methanol, and methyl tert-butyl ether/methanol) are more powerful and perform better than ethanol. The latter holds particularly for the $scCO_2$+ ethanol extraction, for which the low amount of ethanol used when compared to Soxhlet was not enough to recuperate all phenolic compounds available. Hence, the TPC values calculated for the $scCO_2$ + ethanol leaf and stem extracts are 3–9-fold lower than those for the two-phase solvent and Soxhlet ethanol.

The fact that the TPC calculated for the flower extracts is the highest for the three techniques is expected, as the flowers are richer in certain polyphenolic acids (e.g., chlorogenic and gentisic acids mentioned above), as well as in some prominent members of the flavonoids (astragalin, etc.), when compared to the leaves and stems of *G. viscosum*.

With regard to the DPPH free radical scavenging activity and ABTS radical scavenging assay, the same trend as for the TPC is observed; however, this time, there are no exceptions, and the pattern is very clear—the two-phase solvent extraction performs the best, followed by Soxhlet ethanol and $scCO_2$ + ethanol. Thus, the activity towards the DPPH radical decreased in the order flowers > leaves > stems for the two-phase solvent extraction and $scCO_2$ + ethanol, and flowers > stems > leaves for Soxhlet ethanol (the activity of leaf and stem extracts; however, this is almost commensurable), while for the AA by ATBS, the first order is confirmed with no exceptions.

It can be speculated that certain components extracted were stronger radical scavengers than the rest. Consequently, though they in general present in lower amounts in the extracts of the two-phase solvent when compared to those of Soxhlet ethanol (leontopodic acid B being a remarkable exception), such components can influence and exert a positive influence on the AA.

Moreover, the extracts examined were more effective in DPPH than ABTS radical scavenging (as clearly demonstrated by the IC50 values calculated). That can be a result of the complexity, polarity and chemical properties which could lead to diverging bioactivity. Consequently, the observation reported by some authors that certain compounds might exhibit high scavenging activity in one assay while concurrently lower activity in the other assay [23,24] is supported by our results.

3.2.3. GC–FID

GC–FID analyses of *G. viscosum* leaves and flower extracts obtained by Soxhlet *n*-hexane and neat $scCO_2$ were also performed. In what follows, the results for selected leaf extracts are presented in Table 6. The fatty acid profiles determined for the flower extracts are similar, with some not very substantial deviations.

Table 6. Fatty acid composition from FAME GC–FID analysis of selected *G. viscosum* leaves extracts recovered by different extraction methods, expressed as the relative percentage of total fatty acids identified.

Fatty Acid	Soxhlet *n*-Hexane		$scCO_2$, $T = 40\,°C$, $p = 40$ MPa		$scCO_2$, $T = 60\,°C$, $p = 40$ MPa		$scCO_2$, $T = 60\,°C$, $p = 50$ MPa	
	%	Str. Dev.	%	Str. Dev.	%	Str. Dev.	%	Str. Dev.
Lauric acid, 12:0	n.d.		n.d.		0.40	0.05	n.d	
Myristic acid, 14:0	4.68	0.48	5.67	0.97	3.04	0.40	3.19	1.2
Palmitic acid, 16:0	22.77	1.13	49.42	9.59	31.57	5.53	19.36	6.35
Palmitoleic acid, 16:1	n.d.		n.d.	-	n.d.		4.46	2.62
Stearic acid, 18:0	13.67	1.35	n.d.		n.d		n.d.	
Linoleic acid, 18:2	20.42	3.44	31.34	7.44	22.25	11.57	27.35	4.83
Arachidic acid, 20:0	n.d		n.d.	-	11.12	0.81	33.32	15.8
γ-Linolenic acid, 18:3	25.48	8.38	n.d.		22.05	6.17	n.d.	
Behenic acid, 22:0	5.42	2.20	7.15	1.79	4.83	0.94	6.52	0.7
Lignoceric acid, 24:0	7.55	3.60	6.41	1.56	4.74	1.59	5.81	1.28
SFA	54.1		68.66		55.7		68.18	
MUFA	n.d.		n.d.		n.d.		4.46	
DUFA	20.42		31.34		22.25		27.35	
PUFA	25.48		n.d.		22.05		n.d.	
PUFA:SFA	0.471		0.0		0.386		0.0	

n.d—Not detected.

The FAME analysis results for the Soxhlet *n*-hexane and neat $scCO_2$ exhibit qualitatively similar fatty acid profiles with the predominant presence of saturated fatty acids (SFAs), with chain lengths of 12–24 carbon atoms (Table 6). Furthermore, a limited number of certain mono-, di- and polyunsaturated fatty acids (MUFAs, DUFAs, and PUFAs, respectively) were also detected, namely palmitoleic, linoleic and γ linolenic acids. Still, it should be noted that neither of the extracts examined contained the three acids—for example, only one of them contained palmitoleic acid in a relatively small amount, linoleic acid was found in all four extracts, while PUFA γ linoleic acid was found in only two of them.

The results also demonstrate the influence of the two techniques and, in further detail, reveal the effect of $scCO_2$ extraction process operating conditions on the FAME composition. Thus, as mentioned above, SFAs are the dominant ones, with the lowest percentage being registered in the extract recovered by Soxhlet *n*-hexane, while the highest—is in the extract of $scCO_2$ at $T = 40\,°C$, $p = 40$ MPa. The latter is practically commensurable with that of the extract recuperated at the highest temperature and pressure applied, in which the over 2.5-fold lower percentage of palmitic acid is compensated by the relatively high percentage of arachidic acid, which is not detected in the extract recovered at the lower temperature and pressure.

With regard to unsaturated fatty acids, the best performer is Soxhlet + n-hexane, the second best is scCO$_2$ at $T = 60\,^{\circ}$C and $p = 40$ MP. In both extracts, linoleic and γ linolenic acids are detected, with the respective percentages in the two extracts deviating by not very large margins. Hence, a viable assumption can be that in the case of scCO$_2$ extraction higher temperature and lower pressure promote the recovery of unsaturated fatty acids.

Additional useful information can be obtained if the ratio (PUFA:SFA) is calculated for each extract presented in Table 6.

On the one hand, for a plant extract to be considered as a suitable source for biofuels production, in addition to other important requirements, its (PUFA:SFA) should be low. On the other hand, however, the PUFA:SFA ratio is among the important parameters currently used to assess nutritional quality of foods. According to the World Health Organization (WHO), the PUFA:SFA ratio should be above 0.4 in the human diet in order to reduce the risk of developing cardiovascular and other chronic diseases [25]. The PUFA:SFA ratios, which were calculated only for the two *G. viscosum* leaf extracts containing γ linolenic acids, show that the Soxhlet n-hexane one is the higher of the two but not by an order of magnitude, and complies with the WHO requirements.

A different FAME profile was exhibited by the flower extracts at the lowest pressure applied in the scCO$_2$ extraction ($p = 30$ bar) and $T = 40\,^{\circ}$C. SFAs are the dominant ones, with the palmitic acid relative presence being the highest (41. 6%). Still, it is lower than the relative presence of the only unsaturated fatty acid detected—the DUFA linoleic acid with 43.99%, respectively.

If the FAME profile of the flower extract is compared with that of the leaves at the same operating conditions ($T = 60\,^{\circ}$C, $p = 50$ MPa), again the SFAs are the dominant ones, with capric and lauric acids also present, though with low relative percentages. Two unsaturated fatty acids are detected, palmitoleic and linoleic acid. The latter is the most abundant, with a percentage higher than that registered in the leaves (34.42%). In complete analogy with the leaves, the PUFA γ linolenic acid is not found.

4. Conclusions

G. viscosum is a widely distributed plant with enormous potential. In this study, three different extraction techniques were applied to recover highly potent metabolites from the plant's flowers, leaves, and stems. Advanced analysis techniques such as LC–MS/MS, and LC–HRAM were used to characterize the extracts and important representatives of phenolic acids (hydroxycinnamic and caffeoylquinic acid derivatives), hydroxybenzoic acid derivatives, and flavonoids, flavones, flavan-3-ols, flavanones and proanthocyanidins were determined. For the first time, in the present study, the powerful antioxidants leontopodic acids A and B were identified and quantified in the species. Additionally, the fatty acid profile of the extracts obtained was determined by applying GC–FID. This extensive experimental work for extraction, characterization, and evaluation of the antioxidant capacity of *G. viscosum* extracts represents a significant and important contribution in pushing forward the knowledge boundaries about the potential of this biomass and the prospects for its efficient valorization.

Author Contributions: Conceptualization, J.A.P.C., R.P.S., A.B.-P., M.E., D.S.Y. and H.E.R.-A.; methodology J.A.P.C., S.B., A.B.-P., R.P.S. and M.E.; validation S.B., J.A.P.C. and M.E.; formal analysis, S.B., J.A.P.C., M.E., H.E.R.-A. and R.P.S.; investigation S.B., J.A.P.C., R.P.S. and D.S.Y.; writing—original draft preparation, R.P.S., J.A.P.C. and M.E.; writing—review and editing, R.P.S., J.A.P.C., M.E. and A.B.-P.; resources A.B.-P., R.P.S., D.S.Y. and M.E.; project administration R.P.S., M.E. and J.A.P.C.; funding acquisition, R.P.S., M.E., J.A.P.C. and A.B.-P. All authors have read and agreed to the published version of the manuscript.

Funding: This project has received funding from the European Union's Horizon 2020 research and innovation program under the Marie Sklodowska-Curie grant agreement No 778168. Centro de Química Estrutural is a Research Unit funded by Fundação para a Ciência e Tecnologia through projects UIDB/00100/2020 and UIDP/00100/2020. Institute of Molecular Sciences is an Associate Laboratory funded by FCT through project LA/P/0056/2020.

Institutional Review Board Statement: Not applicable.

Informed Consent Statement: Not applicable.

Data Availability Statement: Not applicable.

Conflicts of Interest: The authors declare no conflict of interest.

References

1. Zheng, X.; Wang, W.; Piao, H.; Xu, W.; Shi, H.; Zhao, C. The genus *Gnaphalium* L. (Compositae): Phytochemical and pharmacological characteristics. *Molecules* **2013**, *18*, 8298–8318. [CrossRef] [PubMed]
2. Quiñonez-Bastidas, G.N.; Navarrete, A. Mexican Plants and Derivates Compounds as Alternative for Inflammatory and Neuropathic Pain Treatment—A Review. *Plants* **2021**, *10*, 865. [CrossRef]
3. Mata, R.; Figueroa, M.; Navarrete, A.; Rivero-Cruz, I. Progress in the Chemistry of Organic Natural Products. In *Chemistry and Biology of Selected Mexican Medicinal Plants*, 1st ed.; Kinghorn, A.D., Falk, H., Gibbons, S., Kobayashi, J., Asakawa, Y., Liu, J.-K., Eds.; Springer Nature: Cham, Switzerland, 2019; Volume 108, p. 301. [CrossRef]
4. Hernández, B.R. Actividad Antimicobacteriana Sobre Mycobacterium Tuberculosis y/o Activadora de Macrófagos de Extractos de Plantas Mexicanas Conocidas. Master's Thesis, Universidad Autónoma De Nuevo León Facultad De Ciencias Biológicas, San Nicolás de los Garza, Mexico, 2004. Available online: http://eprints.uanl.mx/1587/1/1020150313.PDF (accessed on 3 August 2022).
5. Hernández Gómez, K.A. Análisis Fitoquímico y Citotóxico de Extractos de Gnaphalium Viscosum (Kunth) Sobre Líneas Celulares Humanas Malignas de Cérvix (SiHa) y Mama (Mda). Master's Thesis, Autonomous University of the State of Hidalgo, Pachuca, Mexico, 2018. Available online: http://dgsa.uaeh.edu.mx:8080/bibliotecadigital/handle/231104/2117 (accessed on 3 August 2022).
6. Villagomez-Ibarra, J.R.; Sanchez, M.; Espejo, O.; Zuniga-Estrada, A.; Torres-Valencia, J.M.; Joseph-Nathan, P. Discussed the antimicrobial activity of three Mexican Gnaphalium species. *Fitoterapia* **2001**, *72*, 692–694. [CrossRef] [PubMed]
7. Villaseñor, J.L. Check list of the native vascular plants of Mexico (Catálogo de las plantas vasculares nativas de México). *Rev. Mex. De Biodivers.* **2016**, *87*, 559–902. [CrossRef]
8. Ontiveros-Rodríguez, J.C.; Serrano-Contreras, J.I.; Villagomez-Ibarra, J.R.; García-Gutierrez, H.A.; Zepeda-Vallejo, L.G. A semi-targeted NMR-based chemical profiling of retail samples of Mexican gordolobo. *J. Pharmaceut. Biomed.* **2022**, *212*, 114651. [CrossRef] [PubMed]
9. Stefanov, S.M.; Fetzer, D.E.L.; Custodio de Souza, A.R.; Corazza, M.L.; Hamerski, F.; Yankov, D.S.; Stateva, R.P. Valorization by compressed fluids of Arctium lappa seeds and roots as a sustainable source of valuable compounds. *J. CO_2 Util.* **2022**, *56*, 101821. [CrossRef]
10. Mišić, D.; Ašanin, R.; Ivanović, J.; Zizovic, I. Investigation of antibacterial activity of supercritical extracts of plants, as well as of extracts obtained by other technological processes on some bacteria isolated from animals. *Acta Vet.-Beogr.* **2009**, *59*, 557–568.
11. Christie, W.W.; Han, X. Preparation of derivatives of fatty acids. In *Lipid Analysis*, 4th ed.; Christie, W.W., Han, X., Eds.; Woodhed Publishing: Sawston, UK; Elsevier: Cambridge, UK, 2012; pp. 145–158, ISBN 9780955251245. [CrossRef]
12. Singleton, V.; Rossi, J. Colorimetry of total phenolic with phosphomolybdic-phosphotungstic acid reagents. *Am. J. Enol. Viticult.* **1965**, *16*, 144–158.
13. Soler-Rivas, C.; Espín, J.C.; Wichers, H.J. An easy and fast test to compare total free radical scavenger capacity of foodstuffs. *Phytochem. Anal.* **2000**, *11*, 330–338. [CrossRef]
14. Nenadis, N.; Wang, L.F.; Tsimidou, M.; Zhang, H.Y. Estimation of scavenging activity of phenolic compounds using the ABTS (•+) assay. *J. Agric. Food Chem.* **2004**, *52*, 4669–4674. [CrossRef] [PubMed]
15. Reichardt, C. *Solvents and Solvent Effects in Organic Chemistry*, 3rd ed.; Wiley-VCH Verlag GmbH & Co. KGaA: Weinheim, Germany, 2003.
16. Coelho, J.P.; Filipe, R.M.; Robalo, M.P.; Boyadzhieva, S.S.; Cholakov, G.S.; Stateva, R.P. Supercritical CO_2 extraction of spent coffee grounds. Influence of co-solvents and characterization of the extracts. *J. Supercrit. Fluid* **2020**, *161*, 104825. [CrossRef]
17. Cicek, S.S.; Untersulzner, C.; Schwaiger, S.; Zidorn, C. Caffeoyl-D-Glucaric Acid Derivatives in the Genus Gnaphalium (Asteraceae: Gnaphalieae). *Rec. Nat. Prod.* **2012**, *6*, 311–315.
18. Tajik, N.; Tajik, M.; Mack, I.; Enck, P. The potential effects of chlorogenic acid, the main phenolic components in coffee, on health: A comprehensive review of the literature Review. *Eur. J. Nutr.* **2017**, *56*, 2215–2244. [CrossRef] [PubMed]
19. Allen, Y.; Chen, Y.; Chen, C. A review of the dietary flavonoid, kaempferol on human health and cancer chemoprevention. *Food Chem.* **2013**, *138*, 2099–2107. [CrossRef]
20. Riaz, A.; Rasul, A.; Hussain, G.; Zahoor, M.K.; Jabeen, F.; Subhani, Z.; Younis, T.; Ali, M.; Sarfraz, I.; Selamoglu, Z. Astragalin: A Bioactive Phytochemical with Potential Therapeutic Activities. *Adv. Pharmacol. Sci.* **2018**, *2018*, 9794625. [CrossRef] [PubMed]

21. Schwaiger, S.; Cervellati, R.; Seger, C.; Ellmerer, E.P.; About, N.; Renimel, I.; Godenir, C.; André, P.; Gafner, F.; Stuppner, H. *Leontopodic acid*—A novel highly substituted glucaric acid derivative from edelweiss (*Leontopodium alpinum* Cass.) and its antioxidative and DNA protecting properties. *Tetrahedron* **2005**, *61*, 4621–4630. [CrossRef]
22. Cho, W.K.; Kim, H.-I.; Kim, S.-Y.; Seo, H.H.; Song, J.; Kim, J.; Shin, D.S.; Jo, Y.; Choi, H.; Lee, J.H.; et al. Anti-Aging Effects of *Leontopodium alpinum* (Edelweiss) Callus Culture Extract through Transcriptome Profiling. *Genes* **2020**, *11*, 230. [CrossRef] [PubMed]
23. Guerrini, A.; Sacchetti, G.; Rossi, D.; Paganetto, G.; Muzzoli, M.; Andreotti, E. Bioactivities of *Piper aduncum* L. and Piper obliquum Ruiz & Pavon (Piperaceae) essential oils from Eastern Ecuador. *Environ. Toxicol. Pharmacol.* **2009**, *27*, 39–48. [PubMed]
24. Wang, W.; Wu, N.; Zu, Y.G.; Fu, Y. Antioxidative activity of *Rosmarinus officinalis* L. essential oil compared to its main components. *Food Chem.* **2008**, *108*, 1019–1022. [CrossRef] [PubMed]
25. World Health Organization. *Diet, Nutrition and the Prevention of Chronic Diseases*; Technical Report Series, No. 916 (TRS 916); WHO: Geneva, Switzerland, 2003; pp. 87–88. Available online: http://whqlibdoc.who.int/trs/WHO_TRS_916.pdf (accessed on 3 August 2022).

antioxidants

MDPI

Article

Antioxidant, Anti-Inflammatory, and Antibacterial Properties of an *Achillea millefolium* L. Extract and Its Fractions Obtained by Supercritical Anti-Solvent Fractionation against *Helicobacter pylori*

Marisol Villalva [1], Jose Manuel Silvan [1], Teresa Alarcón-Cavero [2,3], David Villanueva-Bermejo [4], Laura Jaime [4], Susana Santoyo [4] and Adolfo J. Martinez-Rodriguez [1,*]

[1] Microbiology and Food Biocatalysis Group (MICROBIO), Department of Biotechnology and Food Microbiology, Institute of Food Science Research (CIAL, CSIC-UAM), C/Nicolás Cabrera, 9. Cantoblanco Campus, Autonomous University of Madrid, 28049 Madrid, Spain
[2] Microbiology Department, Hospital Universitario de La Princesa, 28006 Madrid, Spain
[3] Department of Preventive Medicine, Public Health and Microbiology, School of Medicine, Autonomous University of Madrid, 28029 Madrid, Spain
[4] Department of Production and Characterization of Novel Foods, Institute of Food Science Research (CIAL, CSIC-UAM), C/Nicolas Cabrera, 9. Cantoblanco Campus, Autonomous University of Madrid, 28049 Madrid, Spain
* Correspondence: adolfo.martinez@csic.es; Tel.: +34-91-001-79-64

Citation: Villalva, M.; Silvan, J.M.; Alarcón-Cavero, T.; Villanueva-Bermejo, D.; Jaime, L.; Santoyo, S.; Martinez-Rodriguez, A.J. Antioxidant, Anti-Inflammatory, and Antibacterial Properties of an *Achillea millefolium* L. Extract and Its Fractions Obtained by Supercritical Anti-Solvent Fractionation against *Helicobacter pylori*. *Antioxidants* **2022**, *11*, 1849. https://doi.org/10.3390/antiox11101849

Academic Editors: Antonella D'Anneo and Marianna Lauricella

Received: 17 August 2022
Accepted: 16 September 2022
Published: 20 September 2022

Publisher's Note: MDPI stays neutral with regard to jurisdictional claims in published maps and institutional affiliations.

Abstract: The main objective of this work is to evaluate the potential utility of an *Achillea millefolium* extract (yarrow extract, YE) in the control of *H. pylori* infection. The supercritical anti-solvent fractionation (SAF) process of YE allowed the obtaining of two different fractions: yarrow's precipitated fraction (YPF), enriched in most polar phenolic compounds (luteolin-7-*O*-glucoside, luteolin, and 3,5-dicaffeoylquinic acid), and yarrow's separator fraction (YSF), enriched in monoterpenes and sesquiterpenes, mainly containing camphor, artemisia ketone, and borneol. YE was effective in reducing reactive oxygen species (ROS) production in human gastric AGS cells by 16% to 29%, depending on the *H. pylori* strain. YPF had the highest inhibitory activity (38–40%) for ROS production. YE modulated the inflammatory response in AGS gastric cells, decreasing IL-8 production by 53% to 64%. This IL-8 inhibition also showed a strain-dependent character. YPF and YSF exhibited similar behavior, reducing IL-8 production, suggesting that both phenolic compounds and essential oils could contribute to IL-8 inhibition. YSF showed the highest antibacterial activity against *H. pylori* (6.3–7.1 log CFU reduction, depending on the strain) and lower MIC (0.08 mg/mL). Results obtained have shown that YE and SAF fractions (YPF and YSF) were effective as antioxidant, anti-inflammatory, and antibacterial agents regardless of the *H. pylori* strain characteristics.

Keywords: *Achillea millefolium*; yarrow extract; *H. pylori*; supercritical anti-solvent fractionation; anti-inflammatory activity; antioxidant activity; antibacterial activity

1. Introduction

Helicobacter pylori (*H. pylori*) is one of the most prevalent human pathogens, as over half of the world's population is colonized with this Gram-negative bacterium [1]. The gastric colonization by *H. pylori* occurs asymptomatically in most individuals, although most people infected with *H. pylori* usually have histological changes in gastric mucosa consistent with the presence of gastritis. However, long-term infection with the pathogen can cause a wide range of clinical manifestations associated with several diseases, including gastric inflammation, peptic ulcer, gastric cancer, gastric mucosa-associated lymphoid-tissue lymphoma, and other extra-gastric pathologies [2]. Due to the high correlation between *H. pylori* infection and gastric cancer, most therapeutic guidelines aim to eradicate this pathogen using a combination of antibiotics with a proton pump inhibitor in triple or quadruple therapy [3]. However, there are a number of concerns related to the use of eradicative therapies, especially in asymptomatic individuals. First, the global increase in antibiotic resistance [4] and the significant distress that antibiotic therapy causes in the microbiota [5]; and second, the relationship that has been found between the use of eradicative therapies and the emergence or worsening of other pathologies, such as esophageal reflux [6]. This situation has led to increased interest in bioactive compounds obtained from natural sources for the treatment of *H. pylori* infection [7]. Natural extracts not only with antibacterial activity against *H. pylori* but also with anti-inflammatory and antioxidant properties could be potentially interesting in *H. pylori* treatment [8–11]. This is because the immune response to *H. pylori* is a combination of events involving both protective and damaging responses to the host. In fact, it has been described that much of the pathological evidence related to *H. pylori* infection may be due more to the effects of the host's immune system than to the bacterial infection itself [12].

Achillea millefolium L.—traditionally known as yarrow—is a flowering plant commonly used in folk medicine not only in Europe but also in Asia, Africa, and America [13]. Due to the widely known benefits of this plant, the study of its composition and biological properties has awakened a constant interest in developing pharmaceutical, nutraceutical, and food products [14–16]. Dried and fresh upper parts from yarrow have been used to prepare aqueous and alcoholic extracts for the treatment of several health problems, such as diabetes and cardiovascular, respiratory, hepatobiliary, spasmodic, and gastrointestinal disorders [17]. In addition, yarrow has been used externally for the treatment of skin and mucous membrane inflammation [18]. The main bioactive compounds present in different yarrow extracts have been associated to health benefits. The presence of phenolic compounds, specifically chlorogenic and dicaffeoylquinic acids, luteolin, apigenin, and quercetin, as well as volatile fraction constituents, predominantly terpenes, such as borneol, camphor, 1,8-cineole, and chamazulene, have been related to antioxidant, anti-inflammatory, antibacterial, antitumor, and antidiabetic properties [19–22]. The antioxidant effect of yarrow extracts has been extensively studied in both in vitro and in vivo models; likewise, the radical scavenging capacity, intracellular oxidative damage, and reduction in lipid peroxidation in rats have been reported [13,23–26]. The anti-inflammatory properties of yarrow ethanolic extracts have shown their role in the suppression of pro-inflammatory cytokines [26,27]. Regarding antibacterial activity, aqueous and ethanolic extracts of yarrow have been effective against different microorganisms, including those causing skin infections, such as *Staphylococcus aureus*, *Staphylococcus epidermidis*, and *Pseudomonas aeruginosa*, and others related with gastrointestinal diseases, such as *Salmonella thypi* and *Escherichia coli* [19,28,29]. However, the effect of yarrow on *H. pylori* is scarcely known despite it being one of the main human pathogens. Only two previous studies screening different extracts obtained from plants used in traditional medicine have shown an antibacterial effect of yarrow against *H. pylori* [30,31]. There are no previous reports on the antioxidant and anti-inflammatory effect of yarrow on human gastric cells infected with *H. pylori*.

Since continuously increasing research on bioactive components and rising interest in high quality ingredients is evident, manufacturers are motivated to use enriched extracts, fractions, or purified components instead of crude extracts. Furthermore, the use of clean and sustainable extract processes is an essential requirement nowadays. For that purpose, different approaches have been explored to obtain fractions enriched in bioactive molecules from plant extracts, mainly phenolic compounds, such as the use of membrane technology [32], solid-phase extraction with reusable macroporous resins [33], and supercritical anti-solvent fractionation (SAF) with CO_2 at supercritical conditions as a solvent [34,35]. The SAF technique has gained interest as a fractionation or purification process with the potential to reduce the number of steps, since as well as separation of compound(s) occurring in the precipitate, a dried enriched-precipitate is produced [36]. Another advantage of SAF is the low use of chemicals and the reduction in waste that is due to CO_2 being recycled for further extractions. Recently, we have demonstrated that SAF resulted in an adequate method to improve the antioxidant and anti-inflammatory properties of a yarrow ethanolic extract [37], although its impact against *H. pylori* is unknown. For this reason, in this study, we have evaluated the antioxidant, anti-inflammatory, and antibacterial properties of a yarrow extract and its fractions obtained by SAF against three different *H. pylori* strains.

2. Materials and Methods

2.1. Sample Material and Ultrasound-Assisted Extraction of Yarrow

Inflorescences and upper dried leaves from yarrow (*Achillea millefollium* L.) were purchased from a local herbalist (Murciana Herbolisteria, Murcia, Spain). The sun-dried plant from a Bulgarian variety was ground in a hammer mill (Premill 250, Lleal S.A., Granollers, Spain) and sieved to reduce its particle size (<500 µm). Then, the UAE extraction was carried out by using an ultrasonic device (Branson Digital Sonifier 250, Danbury, CT, USA) with a power of 200 W and frequencies of 60 kHz. For this purpose, 40 g of ground and sieved yarrow plant were added to 400 mL of pure ethanol (Panreac Madrid, Spain) for 30 min at 40 °C. An output of 70% with respect to the nominal amplitude was applied during extraction. Finally, the obtained yarrow extract (YE) was concentrated to a final concentration of 17.9 mg/mL by rotary evaporation at 35 °C (IKA RV-10 control, VWR, Madrid, Spain) and stored at −20 °C.

2.2. Supercritical Anti-Solvent Fractionation (SAF) of Yarrow Extract

Fractionation of YE was performed by means of a piece of supercritical technology equipment (Thar SF2000, Thar Technology, Pittsburgh, PA, USA) with two pumps for the separate supply of supercritical CO_2 (SC-CO_2) and YE solution, and a precipitation vessel and two separators' vessels (0.5 L each), with independent control of temperature and pressure as described by Villanueva-Bermejo et al. [35]. Briefly, SC-CO_2 was pumped into the precipitation vessel until 15 MPa of pressure and 40 °C were attained. Then, the solution of YE (17.9 mg/mL concentration) was pumped into the precipitator while maintaining the SC-CO_2 flow. A CO_2/extract flow ratio of 31.3 g/g (50 g/min for CO_2 and 1.6 g/min for YE) was employed. During the process, both separators' vessels were kept at ambient pressure. After system depressurization, two fractions were collected, one corresponding to the YE components that were not soluble in the SC-CO_2+ethanol mixture and precipitated in the precipitation vessel (yarrow's precipitated fraction, YPF). The second fraction corresponded to the YE components soluble in the SC-CO_2+ethanol recovered in the separators (yarrow's separator fraction, YSF) with an oleoresin appearance. To obtain a dried YSF fraction, the samples of both separator vessels were recovered with ethanol and combined in a single fraction to finally remove the solvent by rotary evaporation under vacuum. The YPF and YSF fractions were kept at −20 °C in darkness until analysis.

2.3. Chemical Characterization of YE and Its Fractions by HPLC-PAD-ESI-QTOF-MS and GC-MS Analyses

The phenolic composition was determined by HPLC using an Agilent HPLC 1260 Infinity series system (Agilent Technologies Inc., Santa Clara, CA, USA) according to the Villalva et al. [37] methodology. Chromatographic separation was carried out by using a reverse phase ACE Excell 3 Super C18 column (150 mm × 4.6 mm, 3 µm particle size) from Advanced Chromatography Technologies (Aberdeen, Scotland), thermostated at 35 °C and protected by an ACE 3 C18-AR (10 mm, ×3 mm) guard column. Dry samples were dissolved in DMSO (HPLC grade, ≥99.7%) (Merck, Madrid, Spain) to allow a final concentration of 5 mg/mL and filtered by a PVDF filter (0.45 µm) before injection (20 µL). For identification purposes, the retention time (Rt) and UV–Vis spectrum of each chromatographic peak were compared with the analytical standards (Phytolab, Madrid, Spain); additionally, the accurate mass from HPLC-ESI-QTOF-MS in negative mode analysis was used for compounds assignment, as previously described in Villalva et al. [37]. For quantification, standard calibration curves were built for each pure compound, namely, caffeic acid, caftaric acid, chlorogenic acid, cryptochlorogenic acid, 1,5-dicaffeoylquinic acid (DCQA), 3,4-DCQA, 3,5-DCQA, 4,5-DCQA, ferulic acid, neochlorogenic acid, rosmarinic acid, apigenin, apigenin-7-*O*-glucoside, diosmetin, homoorientin, luteolin-*β*-7-*O*-glucuronide, luteolin-7-*O*-glucoside, schaftoside, vicenin 2, casticin, quercetin, rutin, and vitexin. Moreover, luteolin-6,8-di-*C*-glucoside and 6-hydroxyluteolin-7-*O*-glucoside were quantified by the calibration curve of orientin and luteolin-7-*O*-glucoside. In addition, vicenin 2 and schaftoside calibration curves were used for apigenin-*C*-hexoside-*C*-pentoside and schaftoside isomer quantification; as well, quercetin and casticin were used for methoxyquercetin isomer and centaureidin, respectively.

Volatile compounds from yarrow extracts were characterized by GC-MS using an Agilent 7890A system (Agilent Technologies, Santa Clara, CA, USA) equipped with a split/splitless auto-injector (G4513A), a flame ionization detector, a triple-axis mass spectrometer detector (5975C), and GC/MS Solution software. Extracts were dissolved in ethanol (5 mg/mL final concentration), filtered (0.45 μm), and injected (1 μL) in splitless mode. Then, the chromatographic analysis was carried out as described by Villalva et al. [37]. Briefly, the mass spectrometer operated under electron impact mode (70 eV) and it was used in total ion current (TIC) mode (mass range from 40 to 500 m/z). The analysis was performed using an Agilent HP-5MS capillary column (30 m × 0.25 mm i.d., 0.25 μm phase thickness) and the following chromatographic method: 40 °C initial temperature, from 40 °C to 150 °C at 3 °C min^{-1}, isothermal at 150 °C for 10 min, then increased from 150 to 300 °C at 6 °C min^{-1}, and finally isothermal at 300 °C for 1 min. Helium (99.99%) was employed as the carrier gas (1 mL/min flow rate). The temperature used for the injector was 250 °C. For the identification of volatile compounds, the obtained mass spectral fragmentation patterns were contrasted with those of the Wiley 229 mass spectral library. In addition, their corresponding retention indices were calculated and compared to the information reported in the literature [38–41] and contained in the NIST database.

2.4. Helicobacter pylori, Growth Media, and Culture Conditions

H. pylori strains (Hp48, Hp53, and Hp59) were isolated from gastric mucosal biopsies obtained from symptomatic patients from the Microbiology Department, Hospital Universitario La Princesa (Madrid, Spain). Biopsies were cultured in selective (Pylori agar, BioMerieux, Madrid, Spain) and non-selective media (blood-supplemented Columbia Agar, BioMerieux, Madrid, Spain). Hp48 and Hp59 strains are resistant to metronidazole, while Hp53 is a multi-resistant strain with resistance to amoxicillin, clarithromycin, and rifampicin. *H. pylori* strains were stored at −80 °C in Brucella broth (BB) (Becton, Dickinson, & Co., Madrid, Spain) with 20% glycerol. The agar-plating medium consisted of Müeller–Hinton agar supplemented with 5% defibrinated sheep blood (MHB) (Becton, Dickinson, & Co.), and the liquid growth medium consisted of BB supplemented with 10% horse serum (HS) (Biowest, Barcelona, Spain). *H. pylori* inoculum strains were prepared as follows: the frozen stored strains were reactivated by inoculation (200 μL) in a MHB plate and incubation in a microaerophilic atmosphere using a variable atmosphere incubator (VAIN) (85% N_2, 10% CO_2, 5% O_2) (MACS-VA500, Don Whitley Scientific, Bingley, UK) at 37 °C for 72 h. Bacterial biomass grown in one MHB plate was collected with a sterile cotton swab and suspended in 2 mL of BB supplemented with 10% HS (BB-HS) or a culture medium cell (~1 × 10^8 colony forming units/mL (CFU/mL)), and was used as an experimental bacterial inoculum in the different experimental assays.

2.5. Human Gastric Epithelial Cell Cultures

The human gastric epithelial cell line AGS was obtained from the American Type Culture Collection (ATCC, Barcelona, Spain). Cells were grown in Dulbecco's Modified Eagle's Medium/F12 (DMEM/F12) (Lonza, Madrid, Spain) supplemented with 10% fetal bovine serum (FBS) of South American origin (Hyclone, GE Healthcare, Logan, UK) and 1% penicillin/streptomycin (5000 U/mL) (Lonza). Cells were plated at densities of ~1 × 10^6 cells in 75 cm^2 culture flasks (Sarstedt, Barcelona, Spain) and incubated at 37 °C under 5% CO_2 in a humidified incubator until 90% confluence was reached. The culture cell medium was changed every two days. Before a confluent monolayer appeared, a cell sub-culturing process was carried out. All experiments were performed between passage 5 and 15 to ensure cell uniformity and reproducibility.

2.6. Cell Viability

Before conducting experiments on antioxidant and anti-inflammatory activity, it was necessary to evaluate the potential cytotoxicity of YE and its fractions (YPF and YSF) against the AGS cell line at different concentrations. For this purpose, cell viability was determined by the MTT (3-(4,5-dimethylthiazol-2-yl)-2,5-diphenyltetrazolium bromide) (Merck) reduction assay, as was previously described by Silvan et al. [10]. Confluent cell cultures (~90%) were trypsinized (Trypsin/EDTA solution 170,000 U/L) (Lonza) and cells were seeded (~5 $\times$ 10^4 cells per well) in 96-well plates (Sarstedt) and incubated in cell culture medium at 37 °C under 5% CO_2 in a humidity incubator for 24 h. Cell culture medium was replaced with a serum-free cell culture medium containing YE and its fractions at 0.4, 0.2, and 0.08 mg/mL (final concentration), and cells were incubated at 37 °C under 5% CO_2 in a humidity incubator for 24 h. Viability control cells (non-treated) were incubated in a serum-free cell culture medium without samples. Thereafter, cells were washed twice with phosphate-buffered saline (PBS) (Lonza), and the medium was replaced with 200 μL of serum-free cell culture medium plus 20 μL of MTT solution in PBS (5 mg/mL) that were added to each well for the quantification of the living, metabolically active cells after 1 h incubation at 37 °C under 5% CO_2 in a humidity incubator. MTT is reduced to purple formazan in the mitochondria of living cells. Formazan crystals in the wells were solubilized in 200 μL of DMSO. After incubation, cell concentration was estimated as ranging from ~5 $\times$ 10^4 to 5.5 $\times$ 10^4 cells per well. Finally, absorbance was measured at 570 nm wavelengths, employing a microplate reader Synergy HT (BioTek Instruments Inc., Winooski, VT, USA). Cell viability was calculated considering controls containing the serum-free medium as 100% viable cells, and using the following formula:

$$\text{Cell viability (\%) = (absorbance of sample)/(absorbance of control)} \times 100$$

Data represent the mean and standard deviation (SD) of triplicates of three independent experiments (n = 9).

2.7. Antioxidant Activity of YE and Its Fractions against Intracellular Reactive Oxygen Species (ROS) Production on H. pylori-Infected Gastric Cells

Intracellular ROS were measured by the DCFH-DA (carboxy-2′,7′-dichloro-dihydro-fluorescein diacetate) (Merck) assay, as previously reported by Silvan et al. [10]. Cells were seeded (~5 $\times$ 10^4 cells per well in 500 μL) in 24-well plates (Sarstedt) and incubated at 37 °C under 5% CO_2 in a humidity incubator until a monolayer was formed. Cells were incubated with YE and its fractions (YPF and YSF) (0.08 mg/mL) dissolved in a serum-free cell culture medium for 24 h. After that, cells were washed twice with PBS and incubated with 20 mM DCFH-DA (Merck) at 37 °C for 30 min. Next, cells were washed twice with PBS to remove the unabsorbed probe and were then infected with *H. pylori* inoculum strains (500 μL) suspended in a serum/antibiotics-free cell culture medium (~1 $\times$ 10^8 CFU/mL). ROS production was immediately monitored for 180 min in a Synergy HT (BioTek Instruments Inc.) fluorescent microplate reader using λ_{ex} 485 nm and λ_{em} 530 nm. After incubation, cell concentration was estimated as ranging from ~5 $\times$ 10^5 to 5.5 $\times$ 10^5 cells per well. After being oxidized by intracellular oxidants, DCFH-DA changes to dichloro-fluorescein (DCF) and emits fluorescence. Cells incubated only with the *H. pylori* inoculum were used as an oxidation control (100% of intracellular ROS production). All samples were analyzed in triplicate in three independent experiments (n = 9).

2.8. Anti-Inflammatory Activity of YE and Its Fractions on H. pylori-Infected Gastric Cells

The inflammatory response was evaluated as IL-8 production in AGS cells after being infected with different *H. pylori* strains following the procedure described by Silvan et al. [9]. Briefly, human gastric AGS cells were seeded (~5 × 10^4 cells/well) in 24-well plates (Sarstedt) and incubated in a cell culture medium at 37 °C under 5% CO_2 in a humidity incubator until a monolayer was formed. Cells were incubated with YE and its fractions (YPF and YSF) (0.08 mg/mL) at 37 °C in a 5% CO_2 humidified atmosphere for 2 h. Cells were washed twice with PBS and infected with 0.5 mL of *H. pylori* inoculum prepared in a serum/antibiotics-free cell culture medium (~1 × 10^8 CFU/mL for all tested strains). The infected cells were incubated at 37 °C under 5% CO_2 for 24 h to allow the bacteria to adhere and invade the cells. Uninfected and nontreated cells were included in the experiment as a negative and positive control of IL-8 production, respectively. At the end of incubation, cell supernatants were collected, particulate material was removed by centrifugation (10 min at 12,000 rpm), and samples were stored at −20 °C until analyses were performed. The amounts of secreted interleukin IL-8 in the collected supernatant from gastric epithelial cell samples were determined by an ELISA assay. A commercially available ELISA kit (Diaclone, Besancon, France) for the quantitation of IL-8 cytokine was used as described by the manufacturer's instructions. Absorbance was measured at 450 nm using a microplate reader Synergy HT (BioTek Instruments Inc.). Since, in the absence of bacteria, gastric AGS cells release small amounts of IL-8 [42], titers of cytokine released by AGS cells (pg/mL) were determined experimentally. The data represent the mean and SD of triplicates of three independent experiments ($n = 9$).

2.9. Antibacterial Activity of YE and Its Fractions against H. pylori Strains

The antibacterial activity of YE and its fractions (YPF and YSF) against the *H. pylori* strains was tested following the procedure described by Silvan et al. [10]. Briefly, 1 mL of the sample at 0.4, 0.2, 0.14, and 0.08 mg/mL (final concentration) was transferred into different flasks containing 4 mL of BB-HS. Bacterial inoculum (100 µL of ~1 × 10^8 CFU/mL) was then inoculated into the flasks under aseptic conditions. The culture was incubated in the VAIN in the conditions described above. *H. pylori* growth controls were prepared by transferring 100 µL of bacterial inoculum (~1 × 10^8 CFU/mL) to 5 mL of BB-HS. After 24 h incubation, serial decimal dilutions of cultures were prepared in 0.9% saline solution (NaCl). Then, they were plated onto fresh MHB agar and incubated at 37 °C under microareophillic conditions in the VAIN. After 72 h of incubation, the CFU were assessed. Results were expressed as CFU/mL.

2.10. Statistical Analysis

Results were reported as means ± SD. Significant differences among the data were estimated by applying analysis of variance (ANOVA). Tukey's least significant differences (LSD) test was used to evaluate the significance of the analysis. Differences were considered significant at $p < 0.05$. All statistical tests were performed with IBM SPSS Statistics for Windows, Version 27.0 (IBM Corp., Armonk, NY, USA).

3. Results

3.1. Characterization of YE and Its Fractions

Phenolic composition of YE and its fractions and the details of HPLC-ESI-QTOF-MS of the identified phenolic compounds are shown in Table 1 and Table S1 (supplementary material). Phenolic compounds from two different families (phenolic acids and flavonoids) were identified in the extract and fractions. Flavonoids were the major family within YE (2924.4 mg/100 g), constituting 77% of the total phenolic compounds. Among the flavonoids, flavones were the prevalent group (2018.1 mg/100 g), representing 69% of flavonoids. Luteolin-7-O-glucoside (768.7 mg/100 g; 38% of flavones) and luteolin (447.4 mg/100 g; 22.2% of flavones) were the main compounds identified within YE and in the flavones group.

Table 1. Phenolic composition and quantification of YE and its fractions (mg/100 g dry sample).

Phenolic Compounds	YE	YPF	YSF
Phenolic acids			
Hydroxycinnamic acids			
Caffeic acid [1]	17.4 ± 0.1 [a]	-	18.4 ± 0.1 [b]
Caftaric acid [1]	<L.Q.	22.5 ± 0.3	-
Chlorogenic acid [1]	61.7 ± 0.2 [a]	190.7 ± 4.1 [b]	-
Cryptochlorogenic acid [1]	1.1 ± 0.1 [a]	4.4 ± 1.4 [a]	-
1,5- DCQA [1]	68.7 ± 0.7 [a]	179.7 ± 8.7 [b]	-
3,4- DCQA [1]	38.3 ± 5.1 [a]	69.1 ± 0.5 [b]	-
3,5- DCQA [1]	361.7 ± 1.8 [b]	1163.4 ± 10.2 [c]	10.4 ± 0.1 [a]
4,5- DCQA [1]	96.6 ± 0.9 [a]	318.7 ± 0.2 [b]	-
Ferulic acid [1]	7.9 ± 1.7 [b]	4.3 ± 0.1 [a]	-
Neochlorogenic acid [1]	5.8 ± 0.1 [a]	13.8 ± 0.14 [b]	-
Rosmarinic acid [1]	185.0 ± 1.2	-	-
Σ *Total Phenolic acids*	844.2 [b]	1966.6 [c]	28.8 [a]
Flavonoids			
Flavones			
Amentoflavone	41.9 ± 0.1 [a]	41.2 ± 0.1 [a]	62.2 ± 0.1 [b]
Apigenin [1]	195.7 ± 0.4 [b]	474.2 ± 0.4 [c]	92.5 ± 0.1 [a]
Apigenin-*C*-hexoside-*C*-pentoside	30.3 ± 0.1 [a]	84.7 ± 0.2 [b]	-
Apigenin-7-*O*-glucoside [1]	179.5 ± 0.8 [a]	587.7 ± 1.3 [b]	-
Diosmetin [1]	50.1 ± 0.1 [a]	72.8 ± 0.1 [b]	50.2 ± 0.1 [a]
Homoorientin [1]	2.5 ± 0.7 [a]	15.5 ± 0.1 [b]	-
6-Hydroxyluteolin-7-*O*-glucoside	145.2 ± 0.7 [a]	466.2 ± 0.4 [b]	-
Luteolin [1]	447.4 ± 1.2 [b]	1304.0 ± 10.0 [c]	95.5 ± 1.2 [a]
Luteolin-6,8-di-*C*-glucoside	46.9 ± 0.2 [a]	151.1 ± 0.1 [b]	-
Luteolin-7-*β*-glucuronide [1]	19.9 ± 1.1 [a]	59.3 ± 2.8 [b]	-
Luteolin-7-*O*-glucoside [1]	768.7 ± 8.0 [a]	2385.3 ± 97.5 [b]	-
Schaftoside [1]	27.2 ± 0.1 [a]	88.8 ± 0.4 [b]	-
Schaftoside isomer	26.2 ± 0.3 [a]	89.6 ± 0.2 [b]	-
Vicenin 2 [1]	36.6 ± 0.5 [a]	111.7 ± 0.6 [b]	-
Σ *Total Flavones*	2018.1 [b]	5932.1 [c]	300.4 [a]
Flavonols			
Casticin [1]	28.6 ± 0.1 [a]	-	61.8 ± 0.6 [b]
Centaureidin	391.3 ± 0.4 [a]	107.8 ± 0.1 [b]	669.6 ± 0.4 [c]
Methoxyquercetin isomer	376.0 ± 0.8 [b]	751.3 ± 0.8 [c]	265.0 ± 0.2 [a]
Quercetin [1]	47.0 ± 0.1 [a]	143.6 ± 0.8 [b]	-
Rutin [1]	50.6 ± 1.0 [a]	133.7 ± 2.2 [b]	-
Vitexin [1]	12.8 ± 0.8 [a]	24.8 ± 0.7 [b]	-
Σ *Total Flavonols*	906.3 [a]	1161.2 [c]	996.4 [b]
Σ *Total Flavonoids*	2924.4 [b]	7093.3 [c]	1296.8 [a]
Σ *Total phenolic compounds*	3768.6 [b]	9060.0 [c]	1325.7 [a]

YE: yarrow extract. YPF: yarrow's precipitator fraction. YSF: yarrow's separator fraction. <L.Q.: below limit of quantification. [1] Comparison with authentic standard. [a,b,c] Values in the same row marked with different superscript letters indicates statistical differences ($p < 0.05$).

In the flavonols group (906.3 mg/100 g; 31% of flavonoids), centaureidin and methoxyquercetin isomer were the major compounds accounting for 43% and 41% of total flavonols, respectively. With regards to phenolic acids (22% of total phenolic compounds), chlorogenic acid and its derivatives (1,5-DCQA, 3,4-DCQA, 3,5-DCQA, and 4,5-DCQA) were predominant (75% of total phenolic acids), 3,5-DCQA (361.7 mg/100 g) being the most abundant phenolic acid in YE (43% of total phenolic acids). Rosmarinic acid content was also relevant in YE (22% of total phenolic acids).

Concerning SAF fractions, YPF showed a similar phenolic composition to YE, but it was enriched 2.4 times in total phenolic compounds (9060 mg/100 g) in comparison with YE (3768 mg/100 g). Flavonoid content increases up to 7093.3 (2.4 times more than YE), representing 78% of total phenolic compounds in YPF, similar to that obtained in YE (77%). Within the flavonoids compounds, and as was observed in YE, flavones were the prevalent group (5932.1 mg/100 g), increasing its content up to 83% of total flavonoids, luteolin-7-*O*-glucoside (40% of total flavones) and luteolin (22% of total flavones) being the major compounds in this class of compounds. Phenolic acids concentration increases 2.3 times in YPF compared to YE, chlorogenic acid derivatives being the major compounds in this fraction (98% of total phenolic acids), outstanding the 3,5-DCQA as the most abundant phenolic acid (1163.4 mg/100 g).

On the other hand, only some low-polarity phenolic compounds were recovered as part of YSF (1325.7 mg/100 g), mainly flavonoid compounds (1296.8 mg/100 g) representing 97% of total phenolic compounds identified. Mostly, aglycones of flavonoids, the lesser polar compounds originally described in YE, were found in this fraction. Among them, the biflavonoid amentoflavone (62.2 mg/100 g) and methoxylated flavonols casticin (61.8 mg/100 g) and centaureidin (669.6 mg/100 g) were in significantly ($p < 0.05$) higher concentrations in YSF than in YE. Due to the oleoresin appearance of YSF, it was expected that it could contain volatile oil components. That hypothesis was confirmed with a GC-MS analysis and the results are presented in Table 2.

Table 2. Volatile compounds identified by GC-MS in YE and YSF represented as peak area contributions (normalized percentage of area).

Rt	Compound	YE	YSF
11.9	Yomogi alcohol	2.5	2.7
13.1	Eucalyptol	5.2	4.2
13.5	γ-Vinyl-γ-valerolactone	2.1	1.8
14.6	Artemisia ketone	13.3	11.5
18.4	Camphor	16.7	15.0
19.5	Borneol	10.5	10.5
20.7	3,7-dimethyl-1,5-Octadiene-3,7-diol	7.1	7.5
22.8	trans-Chrysanthenyl acetate	2.9	3.0
24.3	(5E)-5,9-dimethyl-5,8-decadien-2-one	1.8	1.6
24.7	2,6-dimethyl-1,7-octadiene-3,6-diol	10.5	11.3
26.0	N.i.	8.7	10.4
30.3	Jasmone	1.9	1.8
31.2	β-Caryophyllene	1.7	1.9
38.2	β-Caryophyllene oxide	6.0	7.0
41.7	β-Eudesmol	1.5	1.9
51.8	Saussurea lactone	4.1	3.9
52.8	Hexahydrofarnesyl acetone	3.6	4.2
	ΣAUC	23.6×10^6	43.8×10^6

Rt: retention time. YE: yarrow extract. YSF: yarrow's separator fraction. N.i.: non-identified compound. AUC: area under curve.

As shown, a great abundance of monoterpenes and sesquiterpernes was found for both YE and YSF. In particular, four monoterpenes, camphor, artemisia ketone, borneol, and 2,6-dimethyl-1,7-octadiene-3,6-diol, were the most abundant compounds in both extracts. When comparing the total peak area contribution, it can be observed that YSF (43.8×10^6 AUC) represented a double richness of volatile compounds with respect to YE (23.6×10^6 AUC). The fraction obtained in the precipitator vessel (YPF) was also analyzed; however, as expected, it lacks volatile components (data not shown).

3.2. Antioxidant Activity of YE and Its Fractions against Intracellular ROS Production in H. pylori-Infected AGS Cells

Before the antioxidant activity experiments, the viability of the AGS cells was evaluated in the presence of YE and its corresponding fractions (YPF and YSF). For this purpose, AGS cells were placed in contact with variable concentrations of YE and its fractions (0.08 to 0.40 mg/mL), and the MTT assay was performed. The data obtained demonstrated that the maximum concentration that was non-cytotoxic resulted in 0.08 mg/mL for all tested samples (>95% cell viability) (data not shown). Higher concentrations of YE and its fractions significantly reduced cell viability compared to the control of untreated cells (viability lower than 80%). Infection of gastric cells with *H. pylori* strains (Hp48, Hp53, and Hp59) induced ROS production in AGS cells (data not shown), as has been demonstrated in previous works using this cell model [10]. As shown in Figure 1, in all cases, YE and YPF significantly ($p < 0.05$) reduced intracellular ROS production in AGS-infected cells in comparison with the control group (untreated infected cells). However, YSF had a strain-dependent behavior and only significantly inhibited ($p < 0.05$) ROS production when AGS cells were infected with the Hp48 strain.

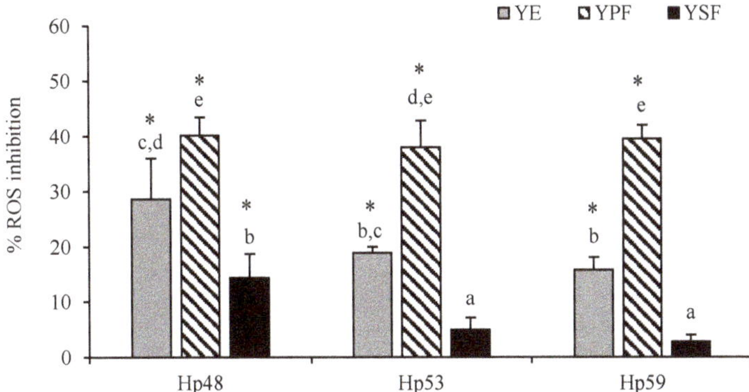

Figure 1. Inhibition effect of yarrow extract (YE) and its fractions (YPF and YSF) (0.08 mg/mL) on ROS production by human gastric epithelial AGS cells after *H. pylori* strains infection. Values are the mean $\pm$ SD ($n = 3$). * Asterisk indicates significant differences compared to the untreated infected control (no inhibition) ($p < 0.05$). [a,b,c,d,e] Different letters indicate statistical difference between samples and *H. pylori* strains ($p < 0.05$).

The inhibition effect of YE on ROS production ranged from 16% to 29% depending on the *H. pylori* strain. YPF, the fraction enriched in phenolic compounds, was the most active fraction regardless of the strain used. It provoked the inhibition of intracellular ROS production of about 40%. YSF, which contained only the most non-polar phenolic compounds, showed a lower antioxidant activity (3–14%) than the YE, which is also coherent with the presence of lower amounts of phenolic compounds in this fraction.

3.3. Effect of the YE and Its Fractions on the Inflammatory Response Induced by H. pylori in AGS Cells

Previously, we evaluated in vitro the secretion of different pro-inflammatory cytokines produced in *H. pylori*-infected AGS cells, IL-8 being the most secreted cytokine, similarly to that described by others [2]. For this reason, we selected IL-8 as a biomarker to evaluate the anti-inflammatory effect of YE and its fractions on AGS cells infected by *H. pylori* strains. As can be observed in Figure 2, the background level of IL-8 production in uninfected AGS cells was 105.0 ± 12.0 pg/mL (Ctrl. AGS; untreated and uninfected cells). Infection with *H. pylori* strains effectively stimulated the secretion of IL-8 pro-inflammatory cytokine (Ctrl. Hp; untreated infected control) in AGS cells (413 to 521 pg/mL). Furthermore, IL-8 production showed a strain-dependent character, since statistical differences between strains were found ($p < 0.05$).

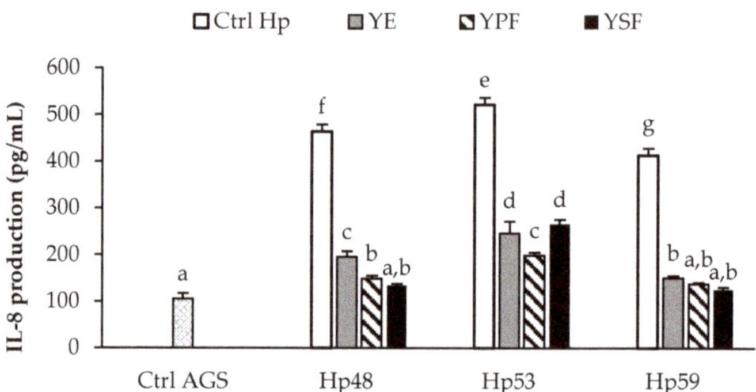

Figure 2. Inhibition effect of yarrow extract (YE) and its SAF fractions (YPF, yarrow's precipitator fraction and YSF, and yarrow's separator fraction) (0.08 mg/mL) on pro-inflammatory cytokine IL-8 production (pg/mL) by human gastric epithelial AGS cells infected by *H. pylori* strains. Control Hp (Ctrl Hp) represents the values obtained from untreated cells infected with *H. pylori* strains. Control AGS (Ctrl AGS) represents the values obtained from untreated and uninfected AGS cells. Values are the mean $\pm$ SD ($n = 3$). [a–g] Different letters indicate statistical differences between treatments for each *H. pylori* strain ($p < 0.05$).

For all strains, YE significantly ($p < 0.05$) decreased IL-8 production by 53% to 64% when compared to its respective control Hp. Unlike the antioxidant activity, it was more difficult in this case to evaluate the impact of each fraction on the observed behavior. For two of the strains (Hp48 and Hp59), both the fraction enriched in phenolic compounds (YPF) and the fraction containing essential oils (YSF) showed similar behaviors, reducing IL-8 production, suggesting that the two types of compounds could contribute to IL-8 inhibition. In contrast, for the Hp53 strain, YPF presented a greater contribution to IL-8 inhibition than YSF ($p < 0.05$).

3.4. Antibacterial Activity of YE and Its Fractions against H. pylori Strains

The antibacterial effect of YE, YPF, and YSF against *H. pylori* growth is presented in Table 3. YE was significantly ($p < 0.05$) effective as an antibacterial agent against all *H. pylori* strains tested, although the effect was greater or lesser depending on the strain and varied in a range of CFU reduction between 4.8 and 7.1 log. However, MIC was the same for all strains (0.14 mg/mL). Analysis of the contribution of each fraction to the antibacterial effect showed that YSF, the fraction enriched in volatile compounds, had a significantly ($p < 0.05$) greater antibacterial effect (6.3–7.1 log CFU reduction, depending on the strain) and lower MIC (0.08 mg/mL) than YE. On the other hand, phenolic-enriched YPF also significantly ($p < 0.05$) reduced bacterial growth of all strains and this reduction was independent of the strain used.

Table 3. Antibacterial activity of YE and its fractions (YPF and YSF) at 0.4 mg/mL against *H. pylori* strains. Results represent the mean $\pm$ standard deviation of colony forming units (CFU)/mL ($n = 3$).

Strains	Control Growth	YE CFU/mL	YE Log CFU Reduction	YE MIC (mg/mL)	YPF CFU/mL	YPF Log CFU Reduction	YPF MIC (mg/mL)	YSF CFU/mL	YSF Log CFU Reduction	YSF MIC (mg/mL)
Hp48	$3.6 \pm 0.3 \times 10^8$ [c,A]	$<1.0 \times 10^2$ [a,A]	>6.3	0.14	$4.0 \pm 0.7 \times 10^5$ [b,A]	2.4	0.08	$<1.0 \times 10^2$ [A,a]	>6.3	0.08
Hp53	$3.4 \pm 1.6 \times 10^8$ [d,A]	$5.0 \pm 1.5 \times 10^3$ [b,B]	4.8	0.14	$6.5 \pm 1.3 \times 10^5$ [c,A]	2.2	0.14	$<1.0 \times 10^2$ [a,A]	>6.8	0.08
Hp59	$9.9 \pm 0.9 \times 10^8$ [d,A]	$1.4 \pm 0.2 \times 10^2$ [b,A]	7.1	0.14	$4.8 \pm 1.6 \times 10^6$ [c,B]	2.4	0.14	$<1.0 \times 10^2$ [a,A]	>7.1	0.08

CFU detection limit was 1.00×10^2 CFU/mL. MIC: minimal inhibitory concentration (mg/mL). YE: yarrow extract. YPF: yarrow's precipitator fraction. YSF: yarrow´s separator fraction. [a,b,c,d] Different lowercase letters denote significant differences within a line ($p < 0.05$). [A,B] Different uppercase letters denote significant differences within a column ($p < 0.05$).

4. Discussion

The phenolic composition of the YE obtained by ethanolic extraction was similar to that reported in previous works for this same yarrow variety [26,35]. The use of ethanol or ethanol mixtures as extraction solvents has been described as a useful method to obtain extracts rich in bioactive phenolic compounds and volatile essential oils from yarrow [26,43]. Because of the well-known bioactivities of phenolic compounds contained in yarrow, the SAF technique was employed to selectively obtain enriched fractions from YE, according to its greater or lesser affinity to the SC-CO$_2$ and ethanol mixture performing as solvents. YPF was enriched in phenolic compounds, while YSF was enriched in monoterpenes and sesquiterpenes, which are very abundant compounds in yarrow's essential oil [13,19]. It has been described that these fractions represent an advantage in the recovering of the extract with high purity and free of solvent, contributing to producing high-quality products [36]. YE and its fractions (YPF and YSF) demonstrated their potential utility for use in both the control of *H. pylori* growth and the modulation of the oxidative and inflammatory response of the human gastric cells associated with *H. pylori* infection. Modulation of the oxidative and inflammatory response in the gastric epithelium has been shown to be particularly relevant in preventing tissue damage and the progression of pathologies associated with *H. pylori* infection [2]. YPF, which presents phenolic compounds 2.4 times more concentrated than YE, had the highest inhibitory activity for ROS production. This behavior seems consistent with the potent antioxidant activity described for many of the major phenolic compounds identified in this fraction. For example, the flavones luteolin-7-*O*-glucoside and luteolin, the most predominant phenolic compounds in the YPF fraction, have been described as potent antioxidant agents, since their molecular structure, formed by a 2–3 carbon double bond of C ring (C2=C3) conjugated with a carbonyl group in C4, confers them with the capacity to react and neutralize ROS, behaving as scavengers in the cellular processes that generate this type of molecules [44]. Other major compounds in YE concentrated in the YPF, such as 3,5-DCQA, have also been shown to have a relevant capacity to scavenge intracellular ROS [45]. In general, since in YPF most of the phenolic compounds present are in a higher concentration than in YE, it is expected that many of them, whose antioxidant properties have been described [46–48], may contribute to a higher inhibition of ROS production found for YPF. On the other hand, the scarce presence of phenolic compounds in the YSF fraction was correlated with low antioxidant activity. The high antioxidant capacity of phenolic-compound-enriched YPF also led to a decrease in IL-8 production. In the case of YE and YSF, not only phenolic compounds but also some essential oils seemed to be involved in their anti-inflammatory capacity. Similar results have recently been obtained evaluating the effect of a yarrow extract and its fractions on differentiated human macrophages, observing that the inhibition in the secretion of some pro-inflammatory cytokines (IL-6, IL-1β, and TNF-α) could be related to the presence of essential oils such as camphor, borneol, or artemisia ketone, which constituted approximately 30% of the fraction studied [37]. Likewise, luteolin-7-*O*-glucoside and luteolin, predominant phenolic compounds in the YE and YPF, have been shown to be able to downregulate IL-1β, IL-6, and TNF-α production acting on NF-κB, MAPK, and JAK/STAT inflammatory pathways by reducing inflammation in cellular models [44,49]. It has also been reported in experiments carried out with a hydro-alcoholic extract of thistle that 3,5-dicaffeoylquinic acid, another major phenolic compound in YE and YPF, was primarily responsible for inhibiting the secretion of IL-8 and NF-κB pathways in human gastric epithelial AGS cells [50].

Although YE and its two fractions were effective as inhibitors of *H. pylori* growth, the contribution of YSF was higher in the antibacterial activity of the extracts. Numerous essential oils are known to have significant antibacterial activity against *H. pylori* [51]. Particularly in yarrow, the major volatile compounds identified (camphor, borneol, and artemisia ketone) have also been shown to be effective as inhibitors of *H. pylori* growth [48–54]. Although the *H. pylori* strain may influence the intensity of the bioactive response obtained, the present work showed that YE and its fractions were effective as antioxidant, anti-inflammatory, and antibacterial agents regardless of the characteristics of the used strain.

5. Conclusions

Among other uses, yarrow has been widely utilized as a part of folk medicine to alleviate symptoms related to gastrointestinal discomfort, many of them similar to those associated with *H. pylori* infection. The historical background of its efficacy in the treatment of these pathologies is complemented in this work by more scientifically based evidence to support the pharmacological effects of various compounds present in YE against *H. pylori*. YE may be potentially effective in combating oxidative stress and modulating the inflammatory response associated with gastric *H. pylori* infection. In addition, YE exhibits strong antibacterial activity against *H. pylori*. Both the phenolic compounds and essential oils present in the extract appear to contribute to the bioactive properties of the extract, although the degree of contribution varies depending on each property (antioxidant, anti-inflammatory, or antibacterial). The SAF technique allows the obtaining of YE fractions enriched in phenolic compounds or essential oils, on the basis of the concept of green extraction, and may be useful in the design of bioactive extracts against *H. pylori* in which it is desirable to enhance specific bioactivity. This approach is attractive in terms of cost, tolerability, and cultural acceptability and can be especially useful in those countries where modern health facilities and access to certain pharmacological substances are not always adequate or available.

Supplementary Materials: The following supporting information can be downloaded at: https://www.mdpi.com/article/10.3390/antiox11101849/s1, Table S1: Phenolic compounds identified in yarrow samples by using HPLC-ESI-QTOF-MS.

Author Contributions: Conceptualization, A.J.M.-R., L.J., S.S. and M.V.; methodology, A.J.M.-R. and J.M.S.; validation, A.J.M.-R., J.M.S. and M.V.; formal analysis, D.V.-B., J.M.S., M.V.; investigation, J.M.S., D.V.-B. and M.V.; resources, A.J.M.-R., T.A.-C., L.J. and S.S.; data curation, A.J.M.-R., J.M.S. and M.V.; writing—original draft preparation, A.J.M.-R., J.M.S. and M.V.; writing—review and editing, A.J.M.-R., J.M.S. and M.V.; visualization, A.J.M.-R. and J.M.S.; supervision, J.M.S. and M.V.; project administration, A.J.M.-R.; funding acquisition, A.J.M.-R. All authors have read and agreed to the published version of the manuscript.

Funding: This research was funded by Projects AGL2017-89566-R (HELIFOOD) (MCIN/AEI/10.13039/501100011033/ (Spanish Ministry of Science and Innovation) and Fondo Europeo de Desarrollo Regional (FEDER) "Una manera de hacer Europa") and ALIBIRD-CM2020 P2018/BAA-4343 (Comunidad de Madrid, Spain).

Institutional Review Board Statement: Not applicable.

Informed Consent Statement: Not applicable.

Data Availability Statement: The data presented in this study are available in this manuscript.

Acknowledgments: The authors thank Soledad Diaz Palero for experimental support (Garantía Juvenil CAM 2020, Ref. 37722).

Conflicts of Interest: The authors declare no conflict of interest.

References

1. Kotilea, K.; Bontems, P.; Touati, E. Epidemiology, diagnosis and risk factors of *Helicobacter pylori* infection. In *Advances in Experimental Medicine and Biology*; Kamiya, S., Backert, S., Eds.; Springer: New York, NY, USA, 2019; Volume 1149, pp. 17–33.
2. Silvan, J.M.; Martinez-Rodriguez, A.J. Modulation of inflammation and oxidative stress in *Helicobacter pylori* infection by bioactive compounds from food components. In *Current Advances for Development of Functional Foods Modulating Inflammation and Oxidative Stress*; Hernandez-Ledesma, B., Martinez-Villaluenga, C., Eds.; Academic Press: London, UK, 2021; pp. 499–516.
3. Alarcon, T.; Urruzuno, P.; Martinez, M.J.; Domingo, D.; Llorca, L.; Correa, A.; Lopez-Brea, M. Antimicrobial susceptibility of 6 antimicrobial agents in *Helicobacter pylori* clinical isolates by using EUCAST breakpoints compared with previously used breakpoints. *Enferm. Infecc. Microbiol. Clin.* **2017**, *35*, 278–282. [CrossRef] [PubMed]

4. Alba, C.; Blanco, A.; Alarcón, T. Antibiotic resistance in *Helicobacter pylori*. *Curr. Opin. Infect. Dis.* **2017**, *35*, 489–497. [CrossRef] [PubMed]

5. Chen, C.C.; Liou, J.M.; Lee, Y.C.; Hong, T.C.; El-Omar, E.M.; Wu, M.S. The interplay between *Helicobacter pylori* and gastrointestinal microbiota. *Gut Microbes.* **2021**, *13*, 1909459. [CrossRef]

6. Zhao, Y.; Li, Y.; Hu, J.; Wang, X.; Ren, M.; Lu, G.; Lu, X.; Zhang, D.; He, S. The effect of *Helicobacter pylori* eradication in patients with gastroesophageal reflux disease: A meta-analysis of randomized controlled studies. *Dig. Dis.* **2020**, *38*, 261–268. [CrossRef]

7. Abou Baker, D. Plants against *Helicobacter pylori* to combat resistance: An ethnopharmacological review. *Biotechnol. Rep.* **2020**, *26*, e00470. [CrossRef] [PubMed]

8. Silvan, J.M.; Gutiérrez-Docio, A.; Moreno-Fernandez, S.; Alarcón-Cavero, T.; Prodanov, M.; Martinez-Rodriguez, A.J. Procyanidin-Rich Extract from Grape Seeds as a Putative Tool against *Helicobacter pylori*. *Foods* **2020**, *9*, 1370. [CrossRef] [PubMed]

9. Silvan, J.M.; Guerrero-Hurtado, E.; Gutiérrez-Docio, A.; Alarcón-Cavero, T.; Prodanov, M.; Martinez-Rodriguez, A.J. Olive-Leaf Extracts Modulate Inflammation and Oxidative Stress Associated with Human *H. pylori* Infection. *Antioxidants* **2021**, *10*, 2030. [CrossRef] [PubMed]

10. Silvan, J.M.; Gutierrez-Docio, A.; Guerrero-Hurtado, E.; Domingo-Serrano, L.; Blanco-Suarez, A.; Prodanov, M.; Alarcon-Cavero, T.; Martinez-Rodriguez, A.J. Pre-Treatment with Grape Seed Extract Reduces Inflammatory Response and Oxidative Stress Induced by *Helicobacter pylori* Infection in Human Gastric Epithelial Cells. *Antioxidants* **2021**, *10*, 943. [CrossRef]

11. Villalva, M.; Silvan, J.M.; Guerrero-Hurtado, E.; Gutierrez-Docio, A.; Navarro del Hierro, J.; Alarcón-Cavero, T.; Prodanov, M.; Martin, D.; Martinez-Rodriguez, A.J. Influence of In Vitro Gastric Digestion of Olive Leaf Extracts on Their Bioactive Properties against *H. pylori*. *Foods* **2022**, *11*, 1832. [CrossRef]

12. Kusters, J.G.; van Vliet, A.H.; Kuipers, E.J. Pathogenesis of *Helicobacter pylori* infection. *Clin. Microbiol. Rev.* **2006**, *19*, 449–490. [CrossRef]

13. Garcia-Oliveira, P.; Barral, M.; Carpena, M.; Gullón, P.; Fraga-Corral, M.; Otero, P.; Prieto, M.A.; Simal-Gandara, J. Traditional plants from *Asteraceae* family as potential candidates for functional food industry. *Food Funct.* **2021**, *12*, 2850–2873. [CrossRef]

14. Gaweł-Bęben, K.; Strzępek-Gomółka, M.; Czop, M.; Sakipova, Z.; Głowniak, K.; Kukula-Koch, W. Achillea millefolium L. and *Achillea biebersteinii* Afan. hydroglycolic extracts–bioactive ingredients for cosmetic use. *Molecules* **2020**, *25*, 3368. [CrossRef] [PubMed]

15. Rakmai, J.; Cheirsilp, B.; Torrado-Agrasar, A.; Simal-Gándara, J.; Mejuto, J.C. Encapsulation of yarrow essential oil in hydroxypropyl-beta-cyclodextrin: Physiochemical characterization and evaluation of bio-efficacies. *CyTA-J. Food* **2017**, *15*, 409–417. [CrossRef]

16. Vitas, J.S.; Cvetanović, A.D.; Mašković, P.Z.; Švarc-Gajić, J.V.; Malbaša, R.V. Chemical composition and biological activity of novel types of kombucha beverages with yarrow. *J. Funct. Foods* **2018**, *44*, 95–102. [CrossRef]

17. Barda, C.; Grafakou, M.E.; Tomou, E.M.; Skaltsa, H. Phytochemistry and Evidence-Based Traditional Uses of the Genus *Achillea* L.: An Update (2011–2021). *Sci. Pharm.* **2021**, *89*, 50. [CrossRef]

18. Mainka, M.; Czerwińska, M.E.; Osińska, E.; Ziaja, M.; Bazylko, A. Screening of antioxidative properties and inhibition of inflammation-linked enzymes by aqueous and ethanolic extracts of plants traditionally used in wound healing in Poland. *Antioxidants* **2021**, *10*, 698. [CrossRef] [PubMed]

19. Ali, S.I.; Gopalakrishnan, B.; Venkatesalu, V. Pharmacognosy, phytochemistry and pharmacological properties of *Achillea millefolium* L.: A review. *Phytother. Res.* **2017**, *31*, 1140–1161. [CrossRef] [PubMed]

20. Mouhid, L.; Gómez de Cedrón, M.; Quijada-Freire, A.; Fernández-Marcos, P.J.; Reglero, G.; Fornari, T.; Ramírez de Molina, A. Yarrow supercritical extract ameliorates the metabolic stress in a model of obesity induced by high-fat diet. *Nutrients* **2019**, *12*, 72. [CrossRef]

21. Pereira, J.M.; Peixoto, V.; Teixeira, A.; Sousa, D.; Barros, L.; Ferreira, I.C.F.R.; Vasconcelos, M.H. *Achillea millefolium* L. hydroethanolic extract inhibits growth of human tumor cells lines by interfering with cell cycle and inducing apoptosis. *Food Chem. Toxicol.* **2018**, *118*, 635–644. [CrossRef]

22. Vidović, S.; Vasić, A.; Vladić, J.; Jokić, S.; Aladić, K.; Gavarić, A.; Nastić, N. Carbon dioxide supercritical fluid extracts from yarrow and rose hip herbal dust as valuable source of aromatic and lipophilic compounds. *Sustain. Chem. Pharm.* **2021**, *22*, 100494. [CrossRef]

23. Farhadi, N.; Babaei, K.; Farsaraei, S.; Moghaddam, M.; Pirbalouti, A.G. Changes in essential oil compositions, total phenol, flavonoids and antioxidant capacity of *Achillea millefolium* at different growth stages. *Ind. Crops. Prod.* **2020**, *152*, 112570. [CrossRef]

24. Salehi, B.; Selamoglu, Z.; Sevindik, M.; Fahmy, N.M.; Al-Sayed, E.; El-Shazly, M.; Büsselberg, D. *Achillea* spp.: A comprehensive review on its ethnobotany, phytochemistry, phytopharmacology and industrial applications. *Cell. Mol. Biol.* **2020**, *66*, 78–103. [CrossRef] [PubMed]

25. Salomon, L.; Lorenz, P.; Bunse, M.; Spring, O.; Stintzing, F.C.; Kammerer, D.R. Comparison of the Phenolic Compound Profile and Antioxidant Potential of *Achillea atrata* L. and *Achillea millefolium* L. *Molecules* **2021**, *26*, 1530. [CrossRef] [PubMed]

26. Villalva, M.; Santoyo, S.; Salas-Pérez, L.; Siles-Sánchez, M.D.L.N.; Rodríguez García-Risco, M.; Fornari, T.; Reglero, G.; Jaime, L. Sustainable extraction techniques for obtaining antioxidant and anti-inflammatory compounds from the *Lamiaceae* and *Asteraceae* species. *Foods* **2021**, *10*, 2067. [CrossRef]

27. Strzępek-Gomółka, M.; Gaweł-Bęben, K.; Kukula-Koch, W. *Achillea* species as sources of active phytochemicals for dermatological and cosmetic applications. *Oxid. Med. Cell. Long.* **2021**, *2021*, 6643827. [CrossRef]

28. Abdossi, V.; Kazemi, M. Bioactivities of *Achillea millefolium* essential oil and its main terpenes from Iran. *Int. J. Food Prop.* **2016**, *19*, 1798–1808. [CrossRef]
29. Verma, R.S.; Joshi, N.; Padalia, R.C.; Goswami, P.; Singh, V.R.; Chauhan, A.; Darokar, M.P. Chemical composition and allelopathic, antibacterial, antifungal and in vitro acetylcholinesterase inhibitory activities of yarrow (*Achillea millefolium* L.) native to India. *Ind. Crops. Prod.* **2017**, *104*, 144–155. [CrossRef]
30. Mahady, G.B.; Pendland, S.L.; Stoia, A.; Hamill, F.A.; Fabricant, D.; Dietz, B.M.; Chadwick, L.R. In vitro susceptibility of *Helicobacter pylori* to botanical extracts used traditionally for the treatment of gastrointestinal disorders. *Phytother. Res.* **2005**, *19*, 988–991. [CrossRef] [PubMed]
31. Zaidi, S.F.H.; Yamada, K.; Kadowaki, M.; Usmanghani, K.; Sugiyama, T. Bactericidal activity of medicinal plants, employed for the treatment of gastrointestinal ailments, against *Helicobacter pylori*. *J. Ethnopharmacol.* **2009**, *121*, 286–291. [CrossRef]
32. Gutierrez-Docio, A.; Almodóvar, P.; Moreno-Fernandez, S.; Silvan, J.M.; Martinez-Rodriguez, A.J.; Alonso, G.L.; Prodanov, M. Evaluation of an integrated ultrafiltration/solid phase extraction process for purification of oligomeric grape seed procyanidins. *Membranes* **2020**, *10*, 147. [CrossRef] [PubMed]
33. Villalva, M.; Jaime, L.; Aguado, E.; Nieto, J.A.; Reglero, G.; Santoyo, S. Anti-inflammatory and antioxidant activities from the basolateral fraction of Caco-2 cells exposed to a rosmarinic acid enriched extract. *J. Agric. Food Chem.* **2018**, *66*, 1167–1174. [CrossRef] [PubMed]
34. Quintana, S.E.; Hernández, D.M.; Villanueva-Bermejo, D.; García-Risco, M.R.; Fornari, T. Fractionation and precipitation of licorice (*Glycyrrhiza glabra* L.) phytochemicals by supercritical antisolvent (SAS) technique. *LWT* **2020**, *126*, 109315. [CrossRef]
35. Villanueva-Bermejo, D.; Zahran, F.; García-Risco, M.R.; Reglero, G.; Fornari, T. Supercritical fluid extraction of Bulgarian *Achillea millefolium*. *J. Supercrit. Fluids* **2017**, *119*, 283–288. [CrossRef]
36. Gil-Ramírez, A.; Rodriguez-Meizoso, I. Purification of Natural Products by Selective Precipitation Using Supercritical/Gas Antisolvent Techniques (SAS/GAS). *Sep. Purif. Rev.* **2019**, *50*, 32–52. [CrossRef]
37. Villalva, M.; Jaime, L.; Villanueva-Bermejo, D.; Lara, B.; Fornari, T.; Reglero, G.; Santoyo, S. Supercritical anti-solvent fractionation for improving antioxidant and anti-inflammatory activities of an *Achillea millefolium* L. extract. *Food Res. Int.* **2019**, *115*, 128–134. [CrossRef] [PubMed]
38. Judzentiene, A.; Mockutė, D. Chemical composition of essential oils produced by pink flower inflorescences of wild *Achillea millefolium* L. *Chemija* **2004**, *15*, 28–32.
39. Bimbiraitė, K.; Ragažinskienė, O.; Maruška, A.; Kornyšova, O. Comparison of the chemical composition of four yarrow (*Achillea millefolium* L.) morphotypes. *Biologija* **2008**, *54*, 208–212. [CrossRef]
40. Falconieri, D.; Piras, A.; Porcedda, S.; Marongiu, B.; Gonçalves, M.J.; Cabral, C.; Cavaleiro, C.; Salgueiro, L. Chemical composition and biological activity of the volatile extracts of *Achillea millefolium*. *Nat. Prod. Commun.* **2011**, *6*, 1527–1530. [CrossRef] [PubMed]
41. Marzouki, H.; Piras, A.; Porcedda, S.; Falconieri, D.; Bagdonaite, E. Influence of extraction methods on the composition of essential oils of *Achillea millefolium* L. from Lithuania. *J. Biodivers. Manag. For.* **2014**, *3*, 1–4.
42. Silvan, J.M.; Mingo, E.; Martinez-Rodriguez, A.J. Grape seed extract (GSE) modulates *Campylobacter* pro-inflammatory response in human intestinal epithelial cell lines. *Food Agric. Immunol.* **2017**, *28*, 739–753. [CrossRef]
43. Ivanović, M.; Grujić, D.; Cerar, J.; Islamčević Razboršek, M.; Topalić-Trivunović, L.; Savić, A.; Kočar, D.; Kolar, M. Extraction of Bioactive Metabolites from *Achillea millefolium* L. with Choline Chloride Based Natural Deep Eutectic Solvents: A Study of the Antioxidant and Antimicrobial Activity. *Antioxidants* **2022**, *11*, 724. [CrossRef] [PubMed]
44. Caporali, S.; De Stefano, A.; Calabrese, C.; Giovannelli, A.; Pieri, M.; Savini, I.; Tesauro, M.; Bernardini, S.; Minieri, M.; Terrinoni, A. Anti-Inflammatory and Active Biological Properties of the Plant-Derived Bioactive Compounds Luteolin and Luteolin 7-Glucoside. *Nutrients* **2022**, *14*, 1155. [CrossRef] [PubMed]
45. Zha, R.P.; Xu, W.; Wang, W.Y.; Dong, L.; Wang, Y.P. Prevention of lipopolysaccharide-induced injury by 3,5-dicaffeoylquinic acid in endothelial cells. *Acta Pharmacol. Sin.* **2007**, *28*, 1143–1148. [CrossRef]
46. Pérez-Torres, I.; Castrejón-Téllez, V.; Soto, M.E.; Rubio-Ruiz, M.E.; Manzano-Pech, L.; Guarner-Lans, V. Oxidative Stress, Plant Natural Antioxidants, and Obesity. *Int. J. Mol. Sci.* **2021**, *22*, 1786. [CrossRef] [PubMed]
47. Khan, J.; Deb, P.K.; Priya, S.; Medina, K.D.; Devi, R.; Walode, S.G.; Rudrapal, M. Dietary Flavonoids: Cardioprotective Potential with Antioxidant Effects and Their Pharmacokinetic, Toxicological and Therapeutic Concerns. *Molecules* **2021**, *26*, 4021. [CrossRef]
48. Speisky, H.; Shahidi, F.; Costa de Camargo, A.; Fuentes, J. Revisiting the Oxidation of Flavonoids: Loss, Conservation or Enhancement of Their Antioxidant Properties. *Antioxidants* **2022**, *11*, 133. [CrossRef] [PubMed]
49. De Stefano, A.; Caporali, S.; Di Daniele, N.; Rovella, V.; Cardillo, C.; Schinzari, F.; Minieri, M.; Pieri, M.; Candi, E.; Bernardini, S. Anti- Inflammatory and Proliferative Properties of Luteolin-7-O-Glucoside. *Int. J. Mol. Sci.* **2021**, *22*, 1321. [CrossRef] [PubMed]
50. Marengo, A.; Fumagalli, M.; Sanna, C.; Maxia, A.; Piazza, S.; Cagliero, C.; Rubiolo, P.; Sangiovanni, E.; Dell'Agli, M. The hydro-alcoholic extracts of Sardinian wild thistles (*Onopordum* spp.) inhibit TNF alpha-induced IL-8 secretion and NF-kappa B pathway in human gastric epithelial AGS cells. *J. Ethnopharmacol.* **2018**, *210*, 469–476. [CrossRef]
51. El-Sherbiny, G.M.; Elbestawy, M.K. A review–plant essential oils active against *Helicobacter pylori*. *J. Essent. Oil Res.* **2022**, *34*, 203–215. [CrossRef]
52. Bergonzelli, G.E.; Donnicola, D.; Porta, N.; Corthésy-Theulaz, I.E. Essential oils as components of a diet-based approach to management of *Helicobacter* infection. *Antimicrob. Agents Chemother.* **2003**, *47*, 3240–3246. [CrossRef] [PubMed]

53. Eftekhar, F.; Nariman, F.; Yousefzadi, M.; Hadian, J.; Ebrahimi, S.N. Anti-*Helicobacter pylori* activity and Essential Oil Composition of *Thymus caramanicus* from Iran. *Nat. Prod. Commun.* **2009**, *4*, 1139–1142. [CrossRef] [PubMed]
54. Jeong, M.; Park, J.M.; Han, Y.M.; Kangwan, N.; Kwon, S.O.; Kim, B.N.; Kim, W.H.; Hahm, K.B. Dietary intervention of Artemisia and Green Tea extracts to rejuvenate *Helicobacter pylori*-associated chronic atrophic gastritis and to prevent tumorigenesis. *Helicobacter* **2016**, *21*, 40–59. [CrossRef] [PubMed]

antioxidants

MDPI

Comment

Comment on Villalva et al. Antioxidant, Anti-Inflammatory, and Antibacterial Properties of an *Achillea millefolium* L. Extract and Its Fractions Obtained by Supercritical Anti-Solvent Fractionation against *Helicobacter pylori*. Antioxidants 2022, *11*, 1849

Rafał Frański [1,*] and Monika Beszterda-Buszczak [2]

[1] Faculty of Chemistry, Adam Mickiewicz University, Uniwersytetu Poznańskiego 8, 61-614 Poznań, Poland
[2] Department of Food Biochemistry and Analysis, Poznań University of Life Sciences, Mazowiecka 48, 60-623 Poznań, Poland
* Correspondence: franski@amu.edu.pl

Abstract: Villalva et al. evaluated the potential utility of an *Achillea millefolium* (yarrow) extract in the control of *H. pylori* infection. The agar-well diffusions bioassay was applied to determine the antimicrobial activity of yarrow extracts. The supercritical anti-solvent fractionation process of yarrow extract was made to give two different fractions with polar phenolic compounds and monoterpenes and sesquiterpenes, respectively. Phenolic compounds were identified by HPLC-ESIMS by using the accurate masses of $[M-H]^-$ ions and the characteristic product ions. However, some of the reported product ions seem to be disputable, as described below.

Keywords: yarrow; flavonoids; fragmentation pathway; mass spectrometry; electrospray ionization

Citation: Frański, R.; Beszterda-Buszczak, M. Comment on Villalva et al. Antioxidant, Anti-Inflammatory, and Antibacterial Properties of an *Achillea millefolium* L. Extract and Its Fractions Obtained by Supercritical Anti-Solvent Fractionation against *Helicobacter pylori*. Antioxidants 2022, *11*, 1849. *Antioxidants* **2023**, *12*, 1226. https://doi.org/10.3390/antiox12061226

Academic Editors: Antonella D'Anneo and Marianna Lauricella

Received: 8 November 2022
Revised: 21 April 2023
Accepted: 1 June 2023
Published: 7 June 2023

Eradication of *Helicobacter pylori* has become a serious challenge due to increasing antimicrobial resistance. That Gram-negative, microaerophilic bacterium that is known to affect over 50% of the worldwide population occurs initially during childhood but, if left untreated, could persist for life, resulting in a broad spectrum of gastropathies. It is particularly important because *H. pylori* is involved in the development of 80% of gastric cancers and 5.5% of all malignant conditions worldwide [1–3].

The most effective conventional therapies for *H. pylori* infection require a minimum of two antibiotics (commonly, amoxicillin and clarithromycin) in combination with a gastric acid inhibitor or bismuth to guarantee high eradication rates. That is why nowadays, much research work is concentrated on the search for new potential anti-*H. pylori* candidates, among others, vegetables and plant extracts showing antibacterial properties [4,5]. They act by inhibiting bacterial enzymes, suppressing nuclear factor-κB, adhesions with gastric mucosa, and by inhibiting oxidative stress [6,7]. Organic extracts (ethanol, methanol, acetone, chloroform, petroleum ether, and mixtures of mentioned) of *Acacia nilotica, Alchornea triplinervia, Arrabidaea chica, Bridelia micrantha, Calotropis procera, Camellia sinensis, Carum carvi, Cocculus hirsutus, Derris trifoliate, Geranium wilfordii, Hydrastis canadensis, Myristica fragrans, Xanthium brasilicum,* and *Trachyspermum copticum* and many others have demonstrated antibacterial activity against clinical isolates of *H. pylori* [8–15]. In the study by Villalva and coworkers, the potential utility of an *Achillea millefolium* (yarrow) extract in the control of *H. pylori* infection was evaluated [16]. The supercritical anti-solvent fractionation process of yarrow extract was made to give two different fractions with polar phenolic compounds and monoterpenes and sesquiterpenes.

Yarrow (*Achillea millefolium* L.) is one of the most commonly used medicinal herbs, in both folk and conventional medicine, for over 3000 years, growing wild and as a cultivated plant in the region of Eurasia and North America [17]. According to the conducted studies,

Achillea millefolium L. is a biologically active plant that demonstrates multiple beneficial effects, including antioxidant, anti-inflammatory, spasmolytic, diaphoretic, hepatoprotective, choleretic, antipyretic, analgesic, antimicrobial and anticancer properties [18–20]. *Achillea millefolium* L. helps eliminate toxins from the body, controls bleeding, lowers blood pressure, relieves menstrual pain, and is used in the treatment of various diseases such as diabetes, tuberculosis, Alzheimer's, and Parkinson's disease [21–23]. Moreover, as indicated by Tilwani and coworkers, the yarrow-treated SARS-nCoV-2 cell exhibits the disintegration of the virus membrane [24].

Different extracts (hexane, petroleum ether, and methanol) of *A. millefolium* aerial parts were found to be active toward the following pathogens: *Bacillus cereus, Staphylococcus aureus, Escherichia coli, Klebsiella pneumoniae, Pseudomonas aeruginosa, Salmonella enteritidis, Yersinia enterocolitica, Aspergillus niger*, and *Candida albicans* [17,25,26]. As the literature indicates, the Gram-positive bacteria were more susceptible than the Gram-negative ones, whereas *S. aureus* and *B. cereus* were the most susceptible Gram-positive bacteria [25,27].

In the study by Villalva et al., the agar-well diffusions bioassay was applied to determine the antimicrobial activity of yarrow extracts (YE). YE was significantly ($p < 0.05$) effective as an antibacterial agent against all *H. pylori* strains tested (Hp48, Hp53, and Hp59) in the range of CFU reduction between 4.8 and 7.1 log [16]. Moreover, even better results were obtained for the fraction enriched in volatile compounds. In turn, in the experiment with the inflammatory response induced by *H. pylori* in AGS cells, interleukin 8 (IL-8) has been used as a biomarker. It is a well-documented fact that *H. pylori* infection is associated with an increase in IL-8 concentration, both in vitro and in vivo; it was among the first cytokines described to be produced by infected gastric epithelium; and leads to the recruitment of leukocytes in the gastric mucosa, representing a major step in the regulation of immune-inflammatory responses [2]. In the study by Villalva et al., the application of YE decreased IL-8 synthesis by 53% to 64% in human gastric epithelial cells, with a suggestion that the two types of compounds could contribute to this inhibition—not only phenolic compounds but also some essential oils. In addition, the authors analyzed the antioxidant activity of YE and its fractions against intracellular reactive oxygen species (ROS) synthesis in *H. pylori*-infected AGS cells. The inhibition effect of YE on ROS production depended on the examined *H. pylori* strain and ranged from 16% to 29%, while the fraction enriched in phenolic compounds was more active, regardless of the strain used, with about 40% intracellular ROS inhibition.

In the study by Villalva et al., the volatile compounds have been identified by GC-MS, and only retention times have been provided for them (Table 2 of [16]). Phenolic compounds have been identified by HPLC-ESIMS in the negative ion mode, and aside from the retention times, the accurate masses of [M−H]⁻ ions and the characteristic product ions (MS/MS ions) have also been provided (Supplementary Material, Table S1 of [16]). However, some of the product ions seem to be disputable, as described below.

The most disputable is the product ion at m/z 112 reported for apigenin and diosmetin. This is the only product ion reported for these two compounds and has 100% ri (Supplementary Material, Table S1 of [16]). According to the published data, apigenin should yield characteristic product ions at m/z 225, 151, 149, 117, and 107, although the relative abundances of these ions may be different, depending on the instrumental conditions used [28–31]. Diosmetin should yield an abundant product ion at m/z 284 as a result of methyl radical loss—a characteristic feature of methoxylated flavonoids [32,33]. The other characteristic diosmetin product ions should be at m/z 256, 227, 151, and 107 [34]. Besides diosmetin, Villalva et al. have found one other methoxylated flavonoid, named 'methoxyquercetin isomer' (the third most abundant flavonoid in yarrow's precipitated fraction, Table 1 of [16]); however, taking into account the m/z of [M−H]⁻ at 315, it should be *O*-methyl quercetin. Villalva et al. claim only the detection of a product ion at m/z 301, thus the loss of mass 14 (elimination of a CH_2 moiety). As mentioned above, the characteristic feature of the fragmentation of [M−H]⁻ ions of methoxylated flavonoids is the loss of mass 15 (loss of a methyl radical). Furthermore, other product ions should be observed as well, enabling at least tentative identification of this compound (isorhamnetin glycosides have already been found in the *Achillea millefolium* L. [35,36]). It has to be stressed that the two other methoxylated flavonoids, centaureidin and casticin, have been observed to have methyl radical losses by Villalva et al. [16]. The authors claim the detection of one biflavonoid, namely amentoflavone ([M−H]⁻ at m/z 537), for which they have reported two product ions at m/z 519 and 495. However, the characteristic amentoflavone product ions are at m/z 443, 417, 375 (the most abundant), and 331 [37,38].

Villalva et al. have also detected a number of flavone C-glycosides. Among them are three isomers, namely apigenin-C-hexoside-C-pentoside, schaftoside, and schaftoside isomer ([M−H]⁻ at m/z 563). For all these three compounds, Villalva et al. have obtained two identical product ions with identical relative abundances (m/z 473 and 443, 10 and 20% ri, respectively). The loss of mass 90 and 120 is a characteristic feature of flavone C-glycosides fragmentation [39]; thus, the product ions at m/z 473 and 443 are typical of these compounds. However, other product ions should also be detected, e.g., at m/z 383 and 353, and at least minor differences in relative ion abundances should be observed for these three isomers detected [40,41]. Villalva et al. claim the detection of luteolin-6,8-di-C-glucoside ([M−H]⁻ at m/z 609), for which they have detected two product ions at m/z 489 and 325 [16]. The first one (loss of mass 120) is a characteristic product ion of luteolin-6,8-di-C-glucoside, often having 100% ri; however, at m/z 325 it is not the characteristic one. Other characteristic product ions which should be observed for this compound are at m/z 591, 399, 369, and 327 [41,42]. Villalva et al. claim the detection of vicenin 2 (apigenin 6,8-di-C-glucoside); however, no product ions have been reported for this compound [16]. For the two last flavone C-glycosides, namely homoorientin (luteolin 6-C-glucoside) and vitexin (apigenin 8-C-glucoside), the reported by Villalva et al. product ions match perfectly with those, the most abundant ones, reported elsewhere (although vitexin cannot be classified as flavonols) [43,44].

The product ions detected by Villalva et al. for other flavonoids are in agreement with those reported in the literature, at least as concerns the most characteristic ones. For example, for quercetin, the authors have detected only the product ion at m/z 151 [16]. Although deprotonated quercetin molecule yields a few other product ions (Figure 1), that at m/z 151 (formally $^{1,3}A^-$ product ion) is the most abundant, formed through the retro-Diels–Alder reaction, and can be regarded as a diagnostic ion for 5,7-dihydroxyflavonoids [28,45,46].

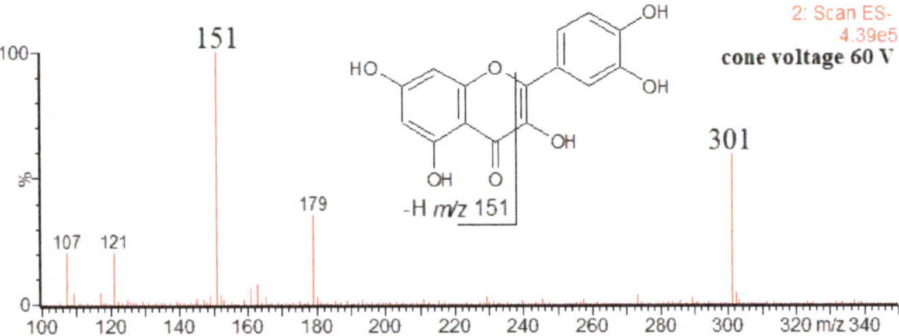

Figure 1. ESI mass spectrum (CID in-source) of quercetin obtained upon HPLC-MS(−) analysis of *Prunus persica* bark extract [46].

Villalva et al. have also detected a number of hydroxycinnamic acids and their conjugates, and the detection of most of them does not raise any doubts. The only disputable ones are the product ions reported for chlorogenic acid (*trans*-5-O-caffeoylquinic acid) and its isomer, cryptochlorogenic acid (*trans*-4-O-caffeoylquinic acid). For chlorogenic acid and cryptochlorogenic acid, Villalva et al. have obtained two identical product ions of almost identical relative abundances, namely at m/z 191 (100% ri) and 161 (10–11% ri) [16]. However, the fragmentation patterns of these compounds should be significantly different. Chlorogenic acid yields abundant product ions at m/z 191, and other product ions have very low abundances (practically are not detectable), whereas cryptochlorogenic acid, except the product ion at m/z 191, yields abundant product ions at m/z 179, 173, and 135 [47–50]. On the other hand, it has to be stressed that the product ions detected by Villalva et al. for neochlorogenic acid (*trans*-3-*O*-caffeoylquinic acid) perfectly match those reported elsewhere with respect to their m/z and ri values [47–49]. The detection of other hydroxycinnamic acids and their conjugates also does not raise any doubts, although the accurate m/z of ferulic acid $[M−H]^-$ ion should be 193.0504, not 103.0504 (minor typos error). It is also worth adding that the elution order of the mono-*O*-caffeoylquinic acid isomers ($[M−H]^-$ at m/z 353) and di-*O*-caffeoylquinic acid isomers ($[M−H]^-$ at m/z 515) reported by Villalva et al. perfectly matches that obtained elsewhere during reversed-phase liquid chromatographic analysis [51].

It should be emphasized that our very specific comments concerning the product ions do not affect the paper by Villalva et al., the paper is characterized by a high scientific level, and the authors finding may imply the next practical application of *Achillea millefolium* L. extracts. On the other hand, the correction of Table S1 is desirable, maybe as a reply to our comment or as a corrigendum to their paper since it surely would improve the quality of the paper.

Author Contributions: Conceptualization, R.F. and M.B.-B.; writing—original draft preparation, R.F. and M.B.-B.; writing—review and editing, R.F. and M.B.-B. All authors have read and agreed to the published version of the manuscript.

Conflicts of Interest: The authors declare no conflict of interest.

References

1. Cardos, I.A.; Zaha, D.C.; Sindhu, R.K.; Cavalu, S. Revisiting therapeutic strategies for *H. pylori* treatment in the context of antibiotic resistance: Focus on alternative and complementary therapies. *Molecules* **2021**, *26*, 6078. [CrossRef] [PubMed]
2. Dincă, A.L.; Melit, L.E.; Mărginean, C.O. Old and new aspects of *H. pylori*-associated Inflammation and gastric cancer. *Children* **2022**, *9*, 1083. [CrossRef] [PubMed]
3. Georgopoulos, S.; Papastergiou, V. An update on current and advancing pharmacotherapy options for the treatment of *H. pylori* infection. *Expert Opin. Pharmacother.* **2021**, *22*, 729–741. [CrossRef] [PubMed]
4. Ouyang, Y.; Wang, M.; Xu, Y.L.; Zhu, Y.; Lu, N.H.; Hu, Y. Amoxicillin-vonoprazan dual therapy for *Helicobacter pylori* eradication: A systematic review and meta-analysis. *J. Gastroenterol. Hepatol.* **2022**, *37*, 1666–1672. [CrossRef]
5. Safavi, M.; Shams-Ardakani, M.; Foroumadi, A. Medicinal plants in the treatment of *Helicobacter pylori* infections. *Pharmac. Biol.* **2015**, *53*, 939–960. [CrossRef]
6. Salehi, B.; Sharopov, F.; Martorell, M.; Rajkovic, J.; Ademiluyi, A.O.; Sharifi-Rad, M.; Fokou, P.V.T.; Martins, N.; Iriti, M.; Sharifi-Rad, J. Phytochemicals in *Helicobacter pylori* infections: What are we doing now? *Int. J. Mol. Sci.* **2018**, *19*, 2361. [CrossRef]
7. Sathianarayanan, S.; Aparna, V.; Biswas, R.; Anita, B.; Sukumaran, S.; Venkidasamy, B. A new approach against *Helicobacter pylori* using plants and its constituents: A review study. *Microb. Pathog.* **2022**, *168*, 105594. [CrossRef]
8. Amin, M.; Anwar, F.; Naz, F.; Mehmood, T.; Saariet, N. Anti-*Helicobacter pylori* and urease inhibition activities of some traditional medicinal plants. *Molecules* **2013**, *18*, 2135–2149. [CrossRef]
9. He, C.; Chen, J.; Liu, J.; Li, Y.; Zhou, Y.; Mao, T.; Li, Z.; Qin, X.; Jin, S. *Geranium wilfordii* maxim.: A review of its traditional uses, phytochemistry, pharmacology, quality control and toxicology. *J. Ethnopharmacol.* **2022**, *285*, 114907. [CrossRef]
10. Kaewkod, T.; Wangroongsarb, P.; Promputtha, I.; Tragoolpua, Y. Inhibitory efficacy of *Camellia sinensis* leaf and medicinal plant extracts on *Helicobacter pylori* standard and isolate strains growth, urease enzyme production and epithelial cell adhesion. *Chiang Mai J. Sci.* **2021**, *48*, 56–73.
11. Mafioleti, L.; da Silva, I.F., Jr.; Colodel, E.M.; Flach, A.; de Oliveira Martins, D.T. Evaluation of the toxicity and antimicrobial activity of hydroethanolic extract of *Arrabidaea chica* (Humb. & Bonpl.) B. Verl. *J. Ethnopharmacol.* **2013**, *150*, 576–582. [PubMed]
12. Okeleye, B.I.; Bessong, P.O.; Ndip, R.N. Preliminary phytochemical screening and in vitro anti-*Helicobacter pylori* activity of extracts of the stem bark of *Bridelia micrantha* (Hochst., Baill., Euphorbiaceae). *Molecules* **2011**, *16*, 6193–6205. [CrossRef]
13. Poovendran, P.; Kalaigandhi, V.; Poongunran, E. Antimicrobial activity of the leaves of *Cocculus hirsutus* against gastric ulcer producing *Helicobacter pylori*. *J. Pharm. Res.* **2011**, *4*, 4294–4295.
14. Suthisamphat, N.; Dechayont, B.; Phuaklee, P.; Prajuabjinda, O.; Vilaichone, R.K.; Itharat, A.; Prommee, N. Anti-*helicobacter pylori*, anti-inflammatory, cytotoxic, and antioxidant activities of mace extracts from *Myristica fragrans*. *Evid. Based Complement. Alternat. Med.* **2020**, *2020*, 7576818. [CrossRef] [PubMed]
15. Uyub, A.M.; Nwachukwu, I.N.; Azlan, A.A.; Fariza, S.S. In-vitro antibacterial activity and cytotoxicity of selected medicinal plant extracts from Penang Island Malaysia on metronidazole-resistant-*Helicobacter pylori* and some pathogenic bacteria. *Ethnobot. Res. Appl.* **2010**, *8*, 95–106. [CrossRef]

16. Villalva, M.; Silvan, J.M.; Alarcón-Cavero, T.; Villanueva-Bermejo, D.; Jaime, L.; Santoyo, S.; Martinez-Rodriguez, A.J. Antioxidant, anti-inflammatory, and antibacterial properties of an *Achillea millefolium* L. extract and its fractions obtained by supercritical anti-solvent fractionation against *Helicobacter pylori*. *Antioxidants* **2022**, *11*, 1849. [CrossRef]

17. Grigore, A.; Colceru-Mihul, S.; Bazdoaca, C.; Yuksel, R.; Ionita, C.; Glava, L. Antimicrobial activity of an *Achillea millefolium* L. *Proceedings* **2020**, *57*, 34.

18. Karaaslan Ayhan, N.; Karaaslan Tunc, M.G.; Noma, S.A.A.; Kurucay, A.; Ates, B. Characterization of the antioxidant activity, total phenolic content, enzyme inhibition, and anticancer properties of *Achillea millefolium* L.(yarrow). *Instrum. Sci. Technol.* **2022**, *50*, 654–667. [CrossRef]

19. Nemeth, E. Biological activities of yarrow species (*Achillea* spp.). *Curr. Pharm. Des.* **2008**, *14*, 3151–3167. [CrossRef]

20. Tadić, V.; Arsić, I.; Zvezdanović, J.; Zugić, A.; Cvetković, D.; Pavkov, S. The estimation of the traditionally used yarrow (*Achillea millefolium* L. Asteraceae) oil extracts with anti-inflamatory potential in topical application. *J. Ethnopharmacol.* **2017**, *199*, 138–148. [CrossRef]

21. Mozafari, N.; Hassanshahi, J.; Ostadebrahimi, H.; Shamsizadeh, A.; Ayoobi, F.; Hakimizadeh, E.; Pak Hashemi, M.; Kaeidi, A. Neuroprotective effect of *Achillea millefolium* aqueous extract against oxidative stress and apoptosis induced by chronic morphine in rat hippocampal CA1 neurons. *Acta Neurobiol. Exp.* **2022**, *82*, 179–186. [CrossRef] [PubMed]

22. Bashir, S.; Noor, A.; Zargar, M.I.; Siddiqui, N.A. Ethnopharmacology, Phytochemistry, and Biological Activities of *Achillea millefolium*: A Comprehensive Review. In *Edible Plants in Health and Diseases*; Masoodi, M.H., Rehman, M.U., Eds.; Springer: Singapore, 2022; pp. 457–481.

23. Zakeri, S. *Achillea millefolium* L. as a recommendation for the management of hysteria. *Tradit. Integr. Med.* **2020**, *5*, 19–25. [CrossRef]

24. Tilwani, K.; Patel, A.; Parikh, H.; Thakker, D.J.; Dave, G. Investigation on anti-Corona viral potential of Yarrow tea. *J. Biomol. Struct. Dyn.* **2022**, 1–13. [CrossRef] [PubMed]

25. Almadiy, A.A.; Nenaah, G.E.; Al Assiuty, B.A.; Moussa, E.A.; Mira, N.M. Chemical composition and antibacterial activity of essential oils and major fractions of four *Achillea* species and their nanoemulsions against foodborne bacteria. *LWT-Food Sci. Technol.* **2016**, *69*, 529–537. [CrossRef]

26. Stojanovic, G.; Radulovic, N.; Hashimoto, T.; Palic, R. In vitro antimicrobial activity of extracts of four *Achillea* species: The composition of *Achillea clavennae* L. (Asteraceae) extract. *J. Ethnopharmacol.* **2005**, *101*, 185–190. [CrossRef]

27. El-Kalamouni, C.; Venskutonis, P.R.; Zebib, B.; Merah, O.; Raynaud, C.; Talou, T. Antioxidant and antimicrobial activities of the essential oil of *Achillea millefolium* L. grown in France. *Medicines* **2017**, *4*, 30. [CrossRef]

28. Hughes, R.J.; Croley, T.R.; Metcalfe, C.D.; March, R.E. A tandem mass spectrometric study of selected characteristic flavonoids. *Int. J. Mass Spectrom.* **2001**, *210*, 371–385. [CrossRef]

29. Fabre, N.; Rustan, I.; de Hoffmann, E.; Quetin-Leclercq, J. Determination of flavone, flavonol, and flavanone aglycones by negative ion liquid chromatography electrospray ion trap mass spectrometry. *J. Am. Soc. Mass Spectrom.* **2001**, *12*, 707–715. [CrossRef]

30. Beszterda, M.; Frański, R. Electrospray ionisation mass spectrometric behaviour of flavonoid 5-O-glucosides and their positional isomers detected in the extracts from the bark of *Prunus cerasus* L. and *Prunus avium* L. *Phytochem. Anal.* **2021**, *32*, 433–439. [CrossRef]

31. Yang, W.Z.; Ye, M.; Qiao, X.; Wang, Q.; Bo, T.; Guo, D.A. Collision-induced dissociation of 40 flavonoid aglycones and differentiation of the common flavonoid subtypes using electrospray ionization ion-trap tandem mass spectrometry and quadrupole time-of-flight mass spectrometry. *Eur. J. Mass Spectrom.* **2012**, *18*, 493–503. [CrossRef]

32. Frański, R.; Gierczyk, B.; Kozik, T.; Popenda, Ł.; Beszterda, M. Signals of diagnostic ions in the product ion spectra of $[M-H]^-$ ions of methoxylated flavonoids. *Rapid Commun. Mass Spectrom.* **2019**, *33*, 125–132. [CrossRef] [PubMed]

33. Justesen, U. Collision-induced fragmentation of deprotonated methoxylated flavonoids, obtained by electrospray ionization mass spectrometry. *J. Mass Spectrom.* **2001**, *36*, 169–178. [CrossRef] [PubMed]

34. Lech, K.; Witkoś, K.; Jarosz, M. HPLC-UV-ESI MS/MS identification of the color constituents of sawwort (*Serratula tinctoria* L.). *Anal. Bioanal. Chem.* **2014**, *406*, 3703–3708. [CrossRef] [PubMed]

35. Dias, M.I.; Barros, L.; Dueñas, M.; Pereira, E.; Carvalho, A.M.; Alves, R.C.; Oliveira, B.P.P.; Santos-Buelga, C.; Ferreira, I.C.F.R. Chemical composition of wild and commercial *Achillea millefolium* L. and bioactivity of the methanolic extract, infusion and decoction. *Food Chem.* **2013**, *141*, 4152–4160. [CrossRef]

36. Benedek, B.; Rothwangl-Wiltschnigg, K.; Rozema, E.; Gjoncaj, N.; Reznicek, G.; Jurenitsch, J.; Kopp, B.; Glasl, S. Yarrow (*Achillea millefolium* L. sl): Pharmaceutical quality of commercial samples. *Pharmazie* **2008**, *63*, 23–26. [PubMed]

37. Yao, H.; Chen, B.; Zhang, Y.; Ou, H.; Li, Y.; Li, S.; Shi, P.; Lin, X. Analysis of the total biflavonoids extract from *Selaginella doederleinii* by HPLC-QTOF-MS and its in vitro and in vivo anticancer effects. *Molecules* **2017**, *22*, 325. [CrossRef]

38. Michler, H.; Laakmann, G.; Wagner, H. Development of an LC-MS method for simultaneous quantitation of amentoflavone and biapigenin, the minor and major biflavones from *Hypericum perforatum* L., in human plasma and its application to real blood. *Phytochem. Anal.* **2011**, *22*, 42–50. [CrossRef]

39. Beszterda, M.; Frański, R. Detection of flavone C-glycosides in the extracts from the bark of *Prunus avium* L. and *Prunus cerasus* L. *Eur. J. Mass Spectrom.* **2020**, *26*, 369–375. [CrossRef]

40. Keskes, H.; Belhadj, S.; Jlail, L.; El Feki, A.; Sayadi, S.; Allouche, N. LC-MS-MS and GC-MS analyses of biologically active extracts of Tunisian Fenugreek (*Trigonella foenum-graecum* L.) Seeds. *J. Food Meas. Charact.* **2018**, *12*, 209–220. [CrossRef]

41. Cao, J.; Yin, C.; Qin, Y.; Cheng, Z.; Chen, D. Approach to the study of flavone di-C-glycosides by high performance liquid chromatography-tandem ion trap mass spectrometry and its application to characterization of flavonoid composition in *Viola yedoensis*. *J. Mass Spectrom.* **2014**, *49*, 1010–1024. [CrossRef]

42. Simirgiotis, M.J.; Schmeda-Hirschmann, G.; Bórquez, J.; Kennelly, E.J. The *Passiflora tripartita* (Banana Passion) fruit: A source of bioactive flavonoid C-glycosides isolated by HSCCC and characterized by HPLC-DAD-ESI/MS/MS. *Molecules* **2013**, *18*, 1672–1692. [CrossRef] [PubMed]

43. Salles, B.C.C.; da Silva, M.A.; Taniguthi, L.; Ferreira, J.N.; da Rocha, C.Q.; Vilegas, W.; Dias, P.H.; Pennacchi, P.C.; da Silveira Duarte, S.M.; Rodrigues, M.R.; et al. *Passiflora edulis* leaf extract: Evidence of antidiabetic and antiplatelet effects in rats. *Biol. Pharm. Bull.* **2020**, *43*, 169–174. [CrossRef]

44. Cuyckens, F.; Claeys, M. Mass spectrometry in the structural analysis of flavonoids. *J. Mass Spectrum.* **2004**, *39*, 1–15. [CrossRef] [PubMed]

45. Śliwka-Kaszyńska, M.; Anusiewicz, I.; Skurski, P. The mechanism of a Retro-Diels-Alder fragmentation of luteolin: Theoretical studies supported by electrospray ionization tandem mass spectrometry results. *Molecules* **2022**, *27*, 1032. [CrossRef]

46. Beszterda, M.; Frański, R. Seasonal qualitative variations of phenolic content in the stem bark of *Prunus persica* var. nucipersica-implication for the use of the bark as a source of bioactive compounds. *ChemistrySelect* **2022**, *7*, e202200418.

47. Ramabulana, A.T.; Steenkamp, P.; Madala, N.; Dubery, I.A. Profiling of chlorogenic acids from *Bidens pilosa* and differentiation of closely related positional isomers with the aid of UHPLC-QTOF-MS/MS-based in-source collision-induced dissociation. *Metabolites* **2020**, *10*, 178. [CrossRef]

48. Clifford, M.N.; Johnston, K.L.; Knight, S.; Kuhnert, N. Hierarchical scheme for LC-MS^n identification of chlorogenic acids. *J. Agric. Food Chem.* **2003**, *51*, 2900–2911. [CrossRef]

49. Xie, C.; Yu, K.; Zhong, D.; Yuan, T.; Ye, F.; Jarrell, J.A.; Millar, A.; Chen, X. Investigation of isomeric transformations of chlorogenic acid in buffers and biological matrixes by ultraperformance liquid chromatography coupled with hybrid quadrupole/ion mobility/orthogonal acceleration time-of-flight mass spectrometry. *J. Agric. Food Chem.* **2011**, *59*, 11078–11087. [CrossRef]

50. Han, B.; Xin, Z.; Ma, S.; Liu, W.; Zhang, B.; Ran, L.; Yi, L.; Ren, D. Comprehensive characterization and identification of antioxidants in *Folium Artemisiae Argyi* using high-resolution tandem mass spectrometry. *J. Chromatogr. B* **2017**, *1063*, 84–92. [CrossRef]

51. Wianowska, D.; Gil, M. Recent advances in extraction and analysis procedures of natural chlorogenic acids. *Phytochem. Rev.* **2019**, *18*, 273–302. [CrossRef]

 *antioxidants*

Reply

Reply to Frański, R.; Beszterda-Buszczak, M. Comment on "Villalva et al. Antioxidant, Anti-Inflammatory, and Antibacterial Properties of an *Achillea millefolium* L. Extract and Its Fractions Obtained by Supercritical Anti-Solvent Fractionation against *Helicobacter pylori*. Antioxidants 2022, 11, 1849"

Marisol Villalva [1], Jose Manuel Silvan [1], Teresa Alarcón-Cavero [2,3], David Villanueva-Bermejo [4], Laura Jaime [4], Susana Santoyo [4] and Adolfo J. Martinez-Rodriguez [1,*]

[1]   Microbiology and Food Biocatalysis Group (MICROBIO), Department of Biotechnology and Food Microbiology, Institute of Food Science Research (CIAL, CSIC-UAM), C/ Nicolás Cabrera, 9, Cantoblanco Campus, Universidad Autónoma de Madrid, 28049 Madrid, Spain
[2]   Microbiology Department, Hospital Universitario de La Princesa, 28006 Madrid, Spain
[3]   Department of Preventive Medicine, Public Health and Microbiology, School of Medicine, Autonomous University of Madrid, 28029 Madrid, Spain
[4]   Department of Production and Characterization of Novel Foods, Institute of Food Science Research (CIAL, CSIC-UAM), C/ Nicolas Cabrera 9, Cantoblanco Campus, Universidad Autónoma de Madrid, 28049 Madrid, Spain
*   Correspondence: adolfo.martinez@csic.es; Tel.: +34-91-001-79-64

Citation: Villalva, M.; Silvan, J.M.; Alarcón-Cavero, T.; Villanueva-Bermejo, D.; Jaime, L.; Santoyo, S.; Martinez-Rodriguez, A.J. Reply to Frański, R.; Beszterda-Buszczak, M. Comment on "Villalva et al. Antioxidant, Anti-Inflammatory, and Antibacterial Properties of an *Achillea millefolium* L. Extract and Its Fractions Obtained by Supercritical Anti-Solvent Fractionation against *Helicobacter pylori*. Antioxidants 2022, 11, 1849". *Antioxidants* 2023, 12, 1384. https://doi.org/10.3390/antiox12071384

Academic Editors: Antonella D'Anneo and Marianna Lauricella

Received: 3 February 2023
Revised: 29 April 2023
Accepted: 1 June 2023
Published: 4 July 2023

Franski and Beszterda-Buszczak [1] report some errors made in the identification of compounds in *Achillea millefolium* extract by MS/MS analysis included in the supplementary material of our published article [2]. We thank them for the observations, and we are pleased to be able to clarify the doubts from these authors. The following response offers an analysis of the comments made, compound by compound.

Regarding apigenin identification, as shown in Figure 1, spectrum 269 remains as the main product after MS/MS analysis along with other product ions such as 151 or 117, which have been reported as characteristics of apigenin fragmentation elsewhere [3–6]. In addition, *m/z* at 112.9858 was found for apigenin after MS/MS analysis. According to this finding, it was the only fragment included in Table S1, although 151 and 117 could also be included. 113 ion is in accordance with Taamalli et al. [3], who found it to be one of the product ions for apigenin-*O*-glucuronide. Therefore, a typographical error would be attributable, in this case, by reflecting 112 instead of 113. Moreover, based on its accurate mass, $C_{15}H_{10}O_5$ (corresponding to the molecular formula of apigenin) was proposed for this product with an error of 3.2 ppm. In addition, the UV-Vis spectrum (data not included in this manuscript) and retention time matched those corresponding to the authentic apigenin standard.

As can be seen in Figure 2, diosmetin yielded the ions 299 (100), 284 (55), and 256 (12) as the main ion products, corresponding to a characteristic fragmentation ion from diosmetin [7]. No other ions were found in this analysis (e.g., 227, 151, or 107). However, based on the accurate mass $[M-H]^-$ at 299.0554, the molecular formula $C_{16}H_{12}O_6$ (error 1.2 ppm) was obtained, which corresponded to diosmetin. Moreover, further identification was performed according to the UV-Vis spectrum and the retention time compared to diosmetin's authentic standard. Therefore, 112 ion, currently registered in Table S1, is a typographical error, and the omitted product ions (284 and 256) should be included in Table S1.

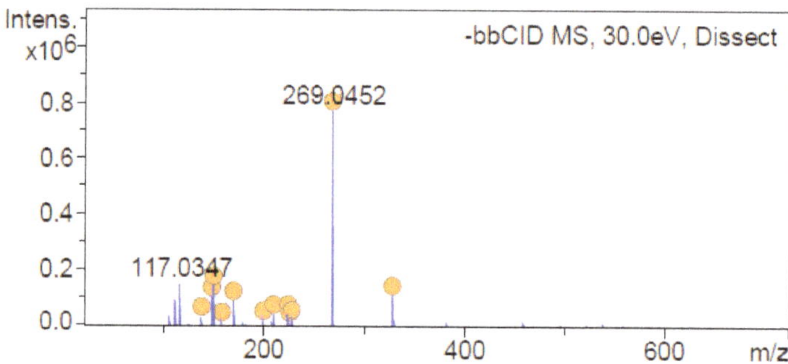

Figure 1. MS/MS spectrum peak identified as apigenin.

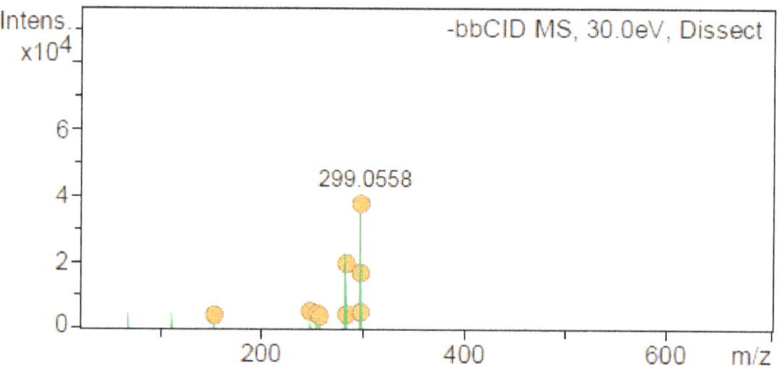

Figure 2. MS/MS spectrum peak identified as diosmetin.

Franski and Beszterda-Buszczak noted an incorrect assignment of $[M-H]^-$ at 315 as methoxyquercetin isomer. In the present study, the molecular formula $C_{16}H_{12}O_7$ resulted with the *m/z* of $[M-H]^-$ at 315.0508 with an error of -3.3 ppm. The fragmentation pattern resulted in 300 ion (100) as the main product. Accordingly, a loss of CH_3 (15 Da) could suggest the loss of $[M-H]^{2-}$ instead of 14 Da, and therefore the loss of $[M-H]$, corresponding to 301 as the fragmentation ion. Franksi and Beszterda-Buszczak also suggest the possibility of examining other product ions, such as isorhamnetin glycoside, which was reported by Dias et al. [8] in *Achillea millefolium* L. However, the mass $[M-H]^-$ at 477 was not detected in our case. Certainly, other product ions were considered for preliminary analysis. In this regard, we considered the authentic standard of isorhamnetin, which corresponds to 3'-*O*-methylquercetin (also known as 3'-methoxyquercetin in the literature), but its retention time did not correspond with any of the identified compounds for *Achillea millefolium* L. Therefore, we have used methoxyquercetin, as its generic name, instead of *O*-methylquercetin.

Regarding amentoflavone identification, there was an error about the fragmentation ions of this compound shown in the HPLC-MS/MS spectra in Table S1. As can be seen from the fragmentation pattern shown in Figure 3, the characteristic amentoflavone product ions were detected but not reported properly according to the *m/z* at 375 (100), 443 (10), and 417 (20). This is in accordance with the literature for amentoflavone product ion mass spectra [9,10]. In addition, amentoflavone was designated by comparing its UV-Vis spectra and retention time using an authentic standard. Hence, the fragmentation pattern for amentoflavone in Table S1 should be modified.

Franski and Beszterda-Buszczak also mentioned the product ions of three isomers of flavone *C*-glycosides: apigenin-*C*-hexoside-*C*-pentoside, schaftoside, and schaftoside isomer. They claimed that the product ions—and their abundance—are similar for the three reported compounds. Certainly, no other product ions were detected for these isomers, although the accuracy of the *m/z* product ions varied slightly (with an accuracy within four decimal places), along with their relative abundance. Hence, the relative abundance displayed in Table S1 for apigenin-*C*-hexoside-*C*-pentoside, schaftoside, and schaftoside isomer should be updated.

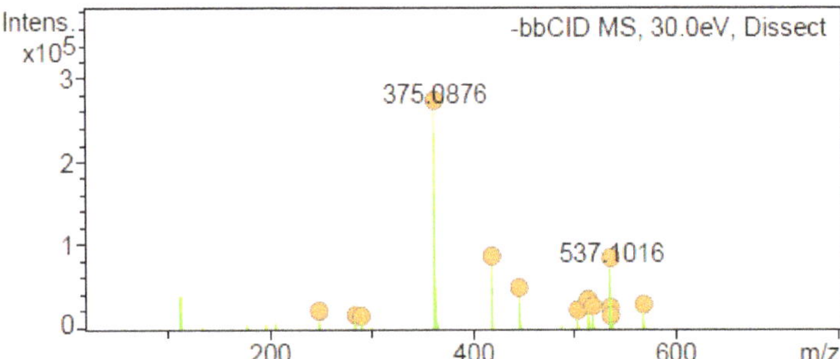

Figure 3. MS/MS spectrum peak identified as amentoflavone.

For luteolin-6,8-di-*C*-glucoside with an $[M-H]^-$ ion at *m/z* 609, the molecular formula $C_{27}H_{30}O_{16}$ was found for the most likely compound with an error of 1.6 ppm. The most reported MS/MS fragmentation pattern included a fragmentation ion of $[M-H-120]^-$ from *m/z* 609, corresponding to a neutral loss of sugar residue, which yields the main ion fragment at *m/z* 489 [3]. The second fragment ion yielded a mass (*m/z*) of 325, which Franski and Beszterda-Buszczak did not recognize as a characteristic product ion. Surely, common fragmentation patterns in a negative ion mode include an ion at *m/z* 327 for this compound [11]. We decided to include the questioned ion fragment (*m/z* at 325) since luteolin possesses four -OH radicals, and it is possible that further ionization can occur, resulting in three hydrolyzed -OH radicals by means of a loss of $[M-3H]^{3-}$.

For the product ions of cryptochlorogenic and chlorogenic acid, an inaccuracy exists in the reported fragmentation pattern in Table S1. The correct *m/z* of the main product ions correspond to 191 (100) for chlorogenic acid and 179 (67) and 173 (100) for cryptochlorogenic acid. Therefore, Table S1 should be updated. It is worth mentioning that these three chlorogenic acid isomers were also identified via a comparison with their authentic standards, as was already indicated in Table S1.

The identification of Vicenin 2 product ions was omitted by error but has now been updated and shown in Table S1. As can be observed in Table S1, the $[M-H]^-$ at *m/z* 593.1513, its corresponding molecular formula $C_{27}H_{30}O_{15}$ (with an error of −0.2 ppm), and its main ion product *m/z* at 473 (100) are suggested to be the correct identification of Vicenin 2. In addition, an authentic standard was used to elucidate the proper identification. Vitexin was misclassified in Table S1 since it appeared in the flavonols section. Now, vitexin can be found in the flavones section since it is a flavone glycoside derivative of apigenin. The accurate mass *m/z* of ferulic acid was reported with a typographical error; the correct mass $[M-H]^-$ corresponds to *m/z* 193.0504. This mass has been corrected in Table S1.

Thus, after the revisions and modifications reflected in this letter, we would like to state that we have clarified all doubts. In addition, the results already published have full rigor and quality according to the standards of the scientific community and the procedures of the journal itself.

Supplementary Materials: The following are available online at https://www.mdpi.com/article/10.3390/antiox12071384/s1, Table S1: Phenolic compounds identified in yarrow samples by using HPLC-ESI-QTOF-MS.

Author Contributions: Conceptualization, A.J.M.-R., L.J., S.S. and M.V.; methodology, A.J.M.-R. and J.M.S.; validation, A.J.M.-R., J.M.S. and M.V.; formal analysis, D.V.-B., J.M.S. and M.V.; investigation, J.M.S., D.V.-B. and M.V.; resources, A.J.M.-R., T.A.-C., L.J. and S.S.; data curation, A.J.M.-R., J.M.S. and M.V.; writing—original draft preparation, A.J.M.-R., J.M.S. and M.V.; writing—review and editing,

A.J.M.-R., J.M.S. and M.V.; visualization, A.J.M.-R. and J.M.S.; supervision, J.M.S. and M.V.; project administration, A.J.M.-R.; funding acquisition, A.J.M.-R. All authors have read and agreed to the published version of the manuscript.

Funding: This research was funded by Projects AGL2017-89566-R (HELIFOOD) (MCIN/AEI/10.13039/501100011033/ (SpanishMinistry of Science and Innovation) and Fondo Europeo de Desarrollo Regional (FEDER) "Una manera de hacer Europa") and ALIBIRD-CM2020 P2018/BAA-4343 (Comunidad de Madrid, Spain).

Data Availability Statement: The data presented in this study are available in the article and supplementary materials

Conflicts of Interest: The authors declare no conflict of interest.

References

1. Frański, R.; Beszterda-Buszczak, M. Comment on Villalva et al. Antioxidant, Anti-Inflammatory, and Antibacterial Properties of an *Achillea millefolium* L. Extract and Its Fractions Obtained by Supercritical Anti-Solvent Fractionation against *Helicobacter pylori*. *Antioxidants* 2022, 11, 1849. *Antioxidants* **2023**, *12*, 1226. [CrossRef]
2. Villalva, M.; Silvan, J.M.; Alarcón-Cavero, T.; Villanueva-Bermejo, D.; Jaime, L.; Santoyo, S.; Martinez-Rodriguez, A.J. Antioxidant, anti-inflammatory, and antibacterial properties of an *Achillea millefolium* L. extract and its fractions obtained by supercritical anti-solvent fractionation against *Helicobacter pylori*. *Antioxidants* **2022**, *11*, 1849. [CrossRef] [PubMed]
3. Taamalli, A.; Arráez-Román, D.; Abaza, L.; Iswaldi, I.; Fernández-Gutiérrez, A.; Zarrouk, M.; Segura-Carretero, A. LC-MS based metabolite profiling of methanolic extracts from the medicinal and aromatic species *Mentha pulegium* and *Origanum mejorana*. *Phytochem. Anal.* **2015**, *26*, 320–330. [CrossRef] [PubMed]
4. Vallverdú-Queralt, A.; Regueiro, J.; Alvarenga, J.F.; Martinez-Huelamo, M.; Leal, L.N.; Lamuela-Reventos, R.M. Characterization of the phenolic and antioxidant profiles of selected culinary herbs and spices: Caraway, turmeric, dill, marjoram and nutmeg. *Food Sci. Technol.* **2015**, *35*, 189–195. [CrossRef]
5. Kaiser, A.; Carle, R.; Kammerer, D.R. Effects of blanching on polyphenol stability of innovative paste-like parsley (*Petroselinum crispum* (Mill.) Nym ex A. W. Hill) and marjoram (*Origanum majorana* L.) products. *Food Chem.* **2013**, *138*, 1648–1656. [CrossRef] [PubMed]
6. Kenny, O.; Smyth, T.J.; Walsh, D.; Kelleher, C.T.; Hewage, C.M.; Brunton, N.P. Investigating the potential of under-utilised plants from the Asteraceae family as a source of natural antimicrobial and antioxidants extracts. *Food Chem.* **2014**, *161*, 79–86. [CrossRef] [PubMed]
7. Mitreski, I.; Stanoeva, J.P.; Stefova, M.; Stefkov, G.; Kulevanova, S. Polyphenols in Representative Teucrium Species in the Flora of R. Macedonia: LC/DAD/ESI-MS^n Profile and Content. *Nat. Prod. Commun.* **2014**, *9*, 1934578X1400900211. [CrossRef]
8. Dias, M.I.; Barros, L.; Dueñas, M.; Pereira, E.; Carvalho, A.M.; Alves, R.C.; Oliveira, B.P.P.; Santos-Buelga, C.; Ferreira, I.C.F.R. Chemical composition of wild and commercial *Achillea millefolium* L. and bioactivity of the methanolic extract, infusion and decoction. *Food Chem.* **2013**, *141*, 4152–4160. [CrossRef] [PubMed]
9. Liao, S.; Ren, Q.; Yang, C.; Zhang, T.; Li, J.; Wang, X.; Qu, X.; Zhang, X.; Zhou, Z.; Zhang, Z.; et al. Liquid chromatography–tandem mass spectrometry determination and pharmacokinetic analysis of amentoflavone and its conjugated metabolites in rats. *J. Agric. Food Chem.* **2015**, *63*, 1957–1966. [CrossRef] [PubMed]
10. Yao, H.; Chen, B.; Zhang, Y.; Ou, H.; Li, Y.; Li, S.; Shi, P.; Lin, X. Analysis of the total biflavonoids extract from *Selaginella doederleinii* by HPLC-QTOF-MS and its in vitro and in vivo anticancer effects. *Molecules* **2017**, *22*, 325. [CrossRef] [PubMed]
11. Ozarowski, M.; Piasecka, A.; Paszel-Jaworska, A.; de Chaves, D.S.A.; Romaniuk, A.; Rybczynska, M.; Gryszczynska, A.; Sawikowska, A.; Kachlicki, P.; Mikolajczak, P.L.; et al. Comparison of bioactive compounds content in leaf extracts of Passiflora incarnata, P. caerulea and P. alata and in vitro cytotoxic potential on leukemia cell lines. *Rev. Bras. Farmacog.* **2018**, *28*, 179–191. [CrossRef]

antioxidants

Article

Polyphenols from Thinned Young Apples: HPLC-HRMS Profile and Evaluation of Their Anti-Oxidant and Anti-Inflammatory Activities by Proteomic Studies

Giulio Ferrario [1,†], Giovanna Baron [1,†], Francesca Gado [1], Larissa Della Vedova [1], Ezio Bombardelli [2], Marina Carini [1], Alfonsina D'Amato [1], Giancarlo Aldini [1] and Alessandra Altomare [1,*]

[1] Department of Pharmaceutical Sciences (DISFARM), Università degli Studi di Milano, Via Mangiagalli 25, 20133 Milan, Italy
[2] Plantex S.a.s., Galleria Unione 5, 20122 Milano, Italy
*   Correspondence: alessandra.altomare@unimi.it
†   These authors contributed equally to this work.

Abstract: The qualitative profile of thinned apple polyphenols (TAP) fraction ($\approx$24% of polyphenols) obtained by purification through absorbent resin was fully investigated by LC-HRMS in positive and negative ion mode and using ESI source. A total of 68 polyphenols were identified belonging to six different classes: flavanols, flavonols, dihydrochalchones, flavanones, flavones and organic and phenolic acids. The antioxidant and anti-inflammatory activities were then investigated in cell models with gene reporter for NRF2 and NF-κB and by quantitative proteomic (label-free and SILAC) approaches. TAP dose-dependently activated NRF2 and in the same concentration range (10–250 µg/mL) inhibited NF-κB nuclear translocation induced by TNF-α and IL-1α as pro-inflammatory promoters. Proteomic studies elucidated the molecular pathways evoked by TAP treatment: activation of the NRF2 signaling pathway, which in turn up-regulates protective oxidoreductases and their nucleophilic substrates such as GSH and NADPH, the latter resulting from the up-regulation of the pentose phosphate pathway. The increase in the enzymatic antioxidant cellular activity together with the up-regulation of the heme-oxygenase would explain the anti-inflammatory effect of TAP. The results suggest that thinned apples can be considered as a valuable source of apple polyphenols to be used in health care products to prevent/treat oxidative and inflammatory chronic conditions.

Keywords: thinned apples; polyphenols; anti-oxidant; anti-inflammatory; NRF2; NF-κB; proteomics

Citation: Ferrario, G.; Baron, G.; Gado, F.; Della Vedova, L.; Bombardelli, E.; Carini, M.; D'Amato, A.; Aldini, G.; Altomare, A. Polyphenols from Thinned Young Apples: HPLC-HRMS Profile and Evaluation of Their Anti-Oxidant and Anti-Inflammatory Activities by Proteomic Studies. *Antioxidants* **2022**, *11*, 1577. https://doi.org/10.3390/antiox11081577

Academic Editors: Antonella D'Anneo and Marianna Lauricella

Received: 5 July 2022
Accepted: 10 August 2022
Published: 15 August 2022

Publisher's Note: MDPI stays neutral with regard to jurisdictional claims in published maps and institutional affiliations.

1. Introduction

Epidemiological studies indicate that consumption of apples and derivatives, such as apple juice, are beneficial in the treatment of some human diseases including CVD and related events, cancer and diabetes [1]. Beneficial effects have also been confirmed by intervention studies. Vallée Marcotte et al. [2] recently reviewed 20 intervention studies using apple juice, concluding that its consumption could exert some benefits on a variety of markers associated with the risk of developing chronic diseases including cardiovascular, cancer, and neurodegenerative diseases.

Recently, a regular consumption of 2–3 apples per day was associated to beneficial effects. In a randomized, controlled, crossover intervention study, in healthy subjects with mildly raised serum cholesterol concentrations, consumption of two Renetta Canada apples for eight weeks improved CVD risk factors, by reducing total and LDL cholesterol and ICAM-1 and increasing microvascular vasodilation [3]. Liddle et al. recently reported that consumption of three whole Gala apples per day for 6 weeks may be an effective strategy to mitigate inflammation in overweight and obese subjects, by reducing circulating biomarkers of inflammation and endotoxin exposure, including CRP, IL-6 and LBP and increasing the plasma antioxidant capacity [4].

Most of the documented beneficial effects of apples are attributed to the fraction of polyphenols represented by five main groups, namely, flavanols (catechins, epicatechin and procyanidins), flavonols (quercetin glycosides), phenolic acids (chlorogenic, gallic and coumaric acids), dihydrochalcones (phloretin glycosides) and anthocyanins (cyanidin) [5–7]. Apple polyphenols exert antioxidant, anti-inflammatory and lipid lowering effects as also confirmed by pre-clinical studies. In particular, isolated apple polyphenols (AP) have been proven to be effective in preventing/treating hypercholesterolemia [8,9], atherosclerosis in ApoE-deficient mice [10], colorectal cancer [11], non-alcoholic hepatitis [12], ulcerative colitis [13] and indomethacin-induced gastric damage [14].

At a molecular level, the antioxidant and anti-inflammatory activities of apple polyphenols can be partially explained by considering the role of polyphenols as activators of the NRF2 pathway [15]. NRF2 is a transcriptional factor associated with antioxidant enzymes playing a master role in redox homeostasis in cells. NRF2 and its principal negative regulator, KEAP1, play a central role in the maintenance of intracellular redox homeostasis and regulation of inflammation. NRF2 is proved to contribute to the regulation of the heme oxygenase-1 (HMOX1) axis, which is a potent anti-inflammatory target [16], and to inhibit oxidative stress by up-regulating antioxidant enzymes and co-factors [17]. Recently, there is an increase in the research literature regarding the regulation of NRF2 signaling pathways in different aspects of inflammation such as cytokines, chemokine releasing factors, MMPs and other inflammatory mediators affecting the NF-κB and MAPK networks to control inflammation, as recently reviewed by Saha et al. [18]. Activation of NRF2 by polyphenol compounds is mainly mediated by those compounds bearing an ortho-diphenol moiety which is oxidized to the corresponding quinone, which, being an electrophilic compound, reacts with the thiols of the KEAP1, thus releasing NRF2, which then translocates into the nucleus [19].

Hence, the phenolic enriched fraction from apples represents a valuable source of natural compounds with a beneficial effect against inflammation and oxidative stress, and they may be applied as a food supplement and/or functional ingredient for the treatment of chronic inflammatory diseases.

There is nowadays great interest in the bioactive compounds obtained from the waste products deriving from agriculture and the food industry (circular economy) [20]. Thinning young apples (around one month after blossom), which constitute a massive waste product of the apple production chain, is carried out in order to guarantee the output and to increase the quality of the harvested apples and are usually discarded in the orchard soil [21]. However, this has the negative effect of increasing the soil acidity and thus disturbing the microbial community, which in turn affects the growth of fruit trees [22]. Thinned young apples are particularly rich in polyphenols, more than 10-fold with respect to harvested apples. The total polyphenols and total antioxidant activity show a rapid decrease after 45 days from blossoming, stabilizing after 85 days, as found in the Fuji apples [23]. Recently, some papers have reported methods for purifying polyphenols from thinned apples by successive use of polyethylene and polyamide resins [24]. Moreover, studies have reported the beneficial effects of polyphenols from thinned apples which have a significant antibacterial activity and were found to be effective in inhibiting halitosis-related bacteria through damage to the cell membrane and hence should be a valuable waste material as a source of bioactive compounds [25]. Polyphenols from unripe apples were also found to exert anti-obesity activity in rats through the modulation of the fatty acid metabolism in the liver and the inhibition of the absorption of carbohydrates and fat [26].

Considering, on the one hand, the growing scientific interest in the apple and particularly in the apple polyphenol fraction as a source of bioactive compounds effective against inflammation and oxidative stress, and, on the other hand, taking into account that thinned apples are a waste product particularly rich in polyphenols, this work is aimed at fully characterizing the qualitative profile of polyphenols purified from thinned apples by a dual LC-HRMS approach (targeted and non-targeted), and at investigating their anti-inflammatory and antioxidant activities. Cell lines with gene reporters for NRF2 and NF-κB nuclear translocation were first used to assess the dose-dependent antioxidant and anti-inflammatory activities, and then to outline the molecular pathways involved by means of two integrated proteomic approaches, based on label-free and SILAC methodologies.

2. Material and Methods

2.1. Chemicals

Thinned Golden, Fuji, Bella del Bosco and Rosa Mantovana apples were sourced from farms located in Trentino-Alto Adige, Italy. Ultrapure water was prepared by a Milli-Q purification system (Millipore, Bedford, MA, USA). Protocatechuic acid, caffeic acid, *p*-coumaric acid, (+)-catechin, (−)-epicatechin, prunin (naringenin glucoside), phlorizin, phloretin, luteolin, cysteine (Cys), iodoacetamide (IAA), tris(2-carboxyethyl)phosphine (TCEP), tetraethylammonium bromide (TEAB), 3-(4,5-Dimethyl-2-thiazolyl)-2,5-diphenyl-2H-tetrazolium bromide (MTT), IL-1α, TNF-α, 6-hydroxy-2,5,7,8-tetramethyl-3,4-dihydrochromene-2-carboxylic acid (Trolox), ascorbic acid, catechin, sodium carbonate, Folin–Ciocalteu reagent, ferulic acid, naringenin-7-*O*-glucoside, quercetin, ethanol, methanol, phlorizin dihydrate, sodium acetate, acetic acid, 2,2-diphenyl-1-picrylhydrazyl (DPPH), CDDO-Me, rosiglitazone, formic acid (FA), trifluoroacetic acid (TFA), acetonitrile (ACN) and all ultra-pure-grade (99.5%) solvents used in LC-MS analysis were obtained from Merck KGaA, Darmstadt, Germany. Kaempferol, quercetin, quercetin-3-*O*-rhamnoside and quercetin-3-*O*-galactoside were purchased from Extrasynthese (Genay Cedex, France). S-TRAP™ columns were provided by Protifi (Huntington, NY, USA).

2.2. Isolation of Thinned Apple Polyphenols (TAP) Fraction

A total of 500 kg of thinned Golden, Fuji, Bella del Bosco and Rosa Mantovana apples were harvested 1 month after blossoming and stored at 2 °C for 1 month. The apples were thoroughly washed in a mixer with an aqueous solution containing 0.05% citric acid and then coarsely ground in a hammer mill. A 0.1% solution of pectinase was added to the resulting mush, which was then heated in a linear tunnel at 25 °C for 20 min.

The mass was then continuously forced through a filter with pressure of 200 bar and the squeezed juice was enzymatically treated with pectinase and centrifuged until a colorless solution was obtained. The liquid was then eluted through 20 L of XAD7 absorbent resin and washed with demineralized water until elimination of all the substances not retained by the resin. At the end of the elution with water, the retained substances (polyphenols) were eluted with 95% ethanol until all the bound components were recovered. Elution of polyphenols was checked by TLC. The hydro-alcoholic solution was concentrated under vacuum and then micronized.

2.3. Qualitative Analysis by LC-HRMS

The phytochemical profile of TAP fraction was performed by LC-HRMS as described by Baron et al. [27]. The extract was dissolved in methanol and diluted with mobile phase A to a final concentration of 2 mg/mL, added with Trolox (50 μM, final concentration) as internal standard (IS). The mixture was separated on a RP Agilent Zorbax SB-C18 column (150 × 2.1 mm i.d., 3.5 μm, CPS analitica, Milan, Italy) by an Ultimate 3000 system (Dionex, Sunnyvale, CA, USA) with a multistep program (80 min) of mobile phase A (H_2O/HCOOH, 100/0.1, %*v/v*) and B (CH_3CN/HCOOH, 100/0.1, %*v/v*). An LTQ Orbitrap XL mass spectrometer (Thermo Fisher Scientific, San Jose, CA, USA) was set to perform the analysis in data-dependent scan mode, and acquired the spectra in positive and negative ion mode. Full MS spectra were acquired by the FT analyzer (resolution 30,000 FWHM at *m/z* 400) in profile mode and in the range of *m/z* 120–1800. The MS/MS spectra of the 3 most intense ions exceeding 1×10^4 counts of the full MS scan were acquired by the linear ion trap (LTQ) by using a normalized collision energy (CID) of 40 eV. A database containing the known components of apples and derivatives was built for the targeted data analysis [28–44]: the putative identification was obtained by comparing the accurate mass (5 ppm mass tolerance), the isotopic pattern and the fragmentation pattern with the compounds in the database. The identity of some molecules was confirmed by means of pure standards available in our laboratory. An untargeted data analysis was performed of the most intense ions not identified in the targeted analysis, as described by Baron et al. [27].

2.4. Quantitative Analysis of Total Polyphenol Content
2.4.1. Colorimetric Analysis

The total polyphenol content was determined by the Folin–Ciocalteu colorimetric method, as reported by Baron et al. [45]; the calibration curve was built using catechin as a standard in a 1–1000 µg/mL range.

2.4.2. HPLC Analysis

High-performance liquid chromatography (HPLC) coupled with a PDA detector was performed to evaluate the total polyphenol content. For the analysis, a methanolic solution of thinned apple polyphenols (TAP) fraction was diluted 1:4 in H_2O/HCOOH, 100/0.1, %v/v (mobile phase A) to obtain a final concentration of 1 mg/mL.

The sample (injection volume, 10 µL) was analyzed in triplicate with a HPLC system (Surveyor, ThermoFinnigan Italy, Milan, Italy), equipped with a PDA detector (Surveyor, ThermoFinnigan Italy, Milan, Italy) and an RP Agilent Zorbax SB-C18 column (150 × 2.1 mm i.d., 3.5 µm, CPS analitica, Milan, Italy). The same gradient program described in Section 2.3 was used for the separation and quantification of TAP polyphenols, setting the PDA detector in a 200–600 nm range.

For the quantification, five calibration curves were built using a standard for each class of polyphenols characterizing the TAP extract: catechin for flavanols (10–100 µg/mL; λ_{max} 278 nm), ferulic acid for phenolic acids (1–10 µg/mL; λ_{max} 323 nm), naringenin-7-glucoside for flavanones (10–100 µg/mL; λ_{max} 283 nm), phlorizin for dihydrochalcones (10–100 µg/mL; λ_{max} 284 nm) and quercetin for flavonols (10–100 µg/mL; λ_{max} 371 nm).

Each chromatographic peak of the sample was assigned to the corresponding polyphenolic class on the basis of the UV spectrum. For the quantification, the AUC was interpolated using the calibration curve of the relative standard and the sum of the concentrations of all the compounds belonging to the same polyphenol class was calculated. The total polyphenol content was then expressed as a percentage (%), that is, mg of polyphenols present in 100 mg of extract.

2.5. Direct Radical Scavenging Activity

The antioxidant activity was evaluated with the DPPH assay. A solution of thinned apple polyphenols (TAP) fraction was prepared with H_2O:EtOH 50:50 (%v/v) to obtain final concentrations in the range 1–20 µg/mL. An aliquot of TAP extract solution (500 µL) was mixed with 1 mL of acetate buffer (pH 5.5, 100 mM) and 1 mL of EtOH. Then, 500 µL of DPPH (500 µM, ethanolic solution) was added and samples were kept in the dark for 90 min. A Shimadzu UV 1900 spectrophotometer (Shimadzu, Milan, Italy) was used for reading the absorbance at 517 nm. Trolox and ascorbic acid were used as reference antioxidant compounds. The percentage of inhibition (I%) was calculated as expressed by Equation (1) and the results are reported as mean ± SD.

$$I\% = \frac{Abs\ (blank\ sample) - Abs\ (sample)}{Abs\ (blank\ sample)} \times 100 \tag{1}$$

2.6. Cell-Based Assays
2.6.1. MTT Assay

The effect of thinned apple polyphenols (TAP) fraction on the cell viability for all the concentrations tested in the anti-inflammatory and antioxidant assays was verified by MTT assay on R3/1-NF-κB cells and HEK293 cells in transparent 96-well plates seeded with 4000 cells/well and 10,000 cells/well, respectively. Cells were treated with different concentration of the extract (1 μg/mL–250 μg/mL) for 18 h in complete medium (DMEM 10% FBS, 1% penicillin/streptomycin). Subsequently, media were removed and only for R3/1-NF-κB cell line, one wash with 100 μL PBS occurred. Then, 100 μL of DMEM, not supplemented with FBS and penicillin/streptomycin, was added to each well and the 4 h incubation started after the addition of 11 μL, 5 mg/mL MTT reagent. After medium removal, cells were lysed using 100 μL of a solution composed of DMSO, 8 mM HCl and 5% TWEEN20. The 96-well plate was shaken for 15 min in a plate shaker in the dark and the absorbance at 575 nm and 630 was measured using a plate reader (BioTek's PowerWave HT, Winooski, VT, USA). Cells incubated with DMSO (0.1%) were used as a control for 100% cell proliferation, while cells incubated with DMSO (3%) were used as a negative control.

2.6.2. NRF2 Gene Reporter Cell Model

Thinned apple polyphenols (TAP) extract was evaluated for its ability to modulate the antioxidant response mediated by NRF2 activation using NRF2/ARE Responsive Luciferase Reporter HEK293 stable cell line (Signosis, Santa Clara, CA, USA) as previously described [46]. Briefly, HEK293 cells (10,000 cells/well) were treated with the extract (concentrations between 1 and 250 μg/mL). CDDO-Me 75 (bardoxolone methyl) 75 nM was used as a positive control [47]. After adding ONE-Glo™ Luciferase Assay Substrate (purchased from Promega Corporation, Madison, WI, USA) (100 μL/well), luciferase measurement was performed with a luminometer (Wallac Victor2 1420, Perkin-Elmer™ Life Science, Monza, Italy). Experiments were carried out with biological and technical replicates; values are reported as mean ± SD compared to untreated control cells. One-way ANOVA with Bonferroni's multiple comparisons test ($p < 0.05$ was considered significant) was used for the statistical analysis.

2.6.3. NF-κB Gene Reporter Cell Model

The in vitro anti-inflammatory activity of the TAP extract was evaluated by using a cell model previously described [27]. Briefly, R3/1 NF-κB cells (5000 cells/well) were pre-treated for 18 h with different concentrations of the extract (1–250 μg/mL) in complete medium (DMEM 10% FBS, 1% L-glutamine, 1% Penicillin/Streptomycin). Rosiglitazone 10 μM was used as a positive control [27]. Then, cells were stimulated for 6 h with 10 ng/mL IL-1α, and for 6 and 24 h with 10 ng/mL TNF-α. Luciferase measurements was performed with a luminometer (Wallac Victor2 1420, Perkin-Elmer™ Life Science, Monza, Italy) after adding 100 μL of ONE-Glo™ Luciferase Assay Substrate (purchased from Promega Corporation, Madison, WI, USA). Experiments were carried out with biological and technical replicates; values are reported as mean ± SD compared to untreated control cells. One-way ANOVA with Bonferroni's multiple comparisons test ($p < 0.05$ was considered significant) was used for the statistical analysis.

2.7. Quantitative Proteomic Studies

2.7.1. SILAC Culture

The R3/1 NF-κB reporter cell line used for SILAC experiments was cultured in DMEM for SILAC supplemented with 10% FBS, 1% pen/strep and 1% sodium pyruvate; the medium was completed by adding 0.5 mL of heavy or light L-lysine and L-arginine 1000X stock solution diluted in PBS. The final concentration of the light amino acids was 84 mg/mL for arginine and 146 mg/mL for lysine. For the heavy condition, both amino acids ($^{13}C_6$ $^{15}N_2$ lysine and $^{13}C_6$ $^{15}N_4$ arginine) were diluted in the medium up to the working concentration of 88 mg/mL and 151.3 mg/mL for the heavy arginine and lysine, respectively; this difference in concentration is due to the different molecular weights of the amino acids [48,49]. The cell line was cultivated for at least 9 passages and the incorporation rate was checked: more than 95% of all peptides are required to be labeled.

SILAC Experimental Design

Each condition tested was cultivated in biological triplicate, and once 70% confluence was reached in a T75 flask, the cells were treated according to the planned experimental design, as shown in Figure 1. Overall, the conditions chosen were as follows: (i) *Control (CTR)*, i.e., untreated cells; (ii) *Thinned apple polyphenols (TAP) extract treatment (CTR-TAP)*, i.e., cells that underwent a double 24 h treatment with the extract at a 200 µg/mL concentration; (iii) *Inflammation* (CTR-TNF), condition achieved treating cells for 24 h with TNF-α 0.01 µg/mL; (iv) *Thinned apple polyphenols (TAP) extract treatment of inflamed cells (CTR-TNF-TAP)*, consisting of the 24 h pre-treatment with the extract at 200 µg/mL and then with TNF-α at 0.01 µg/mL for another 24 h.

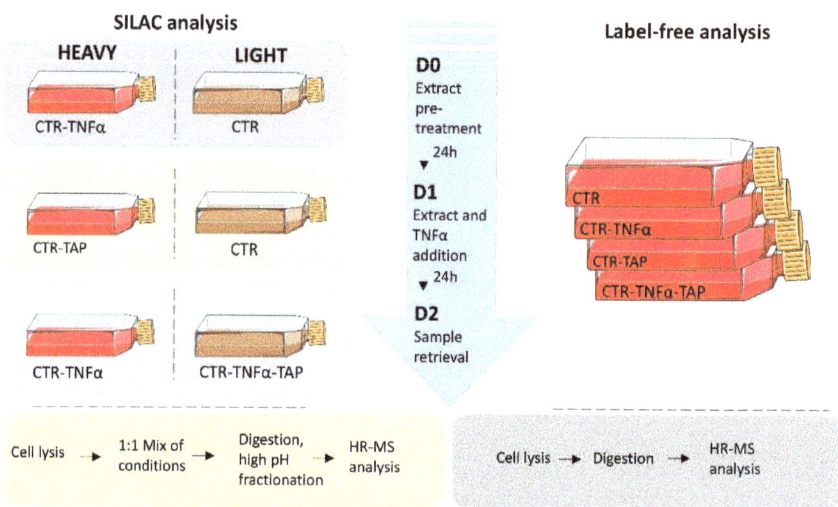

Figure 1. Experimental design of the quantitative proteomic analysis based on the two orthogonal approaches applied, namely, SILAC (**left side**) and LFQ (**right side**), interspersed by a timeline describing the three basic treatment steps. Below is a schematic of the sample processing protocol.

The metabolic labeling strategy was planned so that all conditions are compared with each other in order to have a comprehensive and meaningful pathway modulation overview (Figure 1). All the experiments were planned to finish in the same day.

After light or heavy amino acid incorporation, the flasks were washed twice with warmed PBS and the cell was detached using 2 mL of warmed trypsin incubated at 37 °C for 5 min. Cells were then recovered. After diluting the trypsin with 8 mL of complete medium, the suspension was centrifuged for 5 min at $300\times g$ at room temperature and the pellet was then washed 5 times with 5 mL of 4 °C PBS. The cell pellets were then lysed using a suitable S-trap (Protifi) buffer composed of SDS 5%, TEAB 50 mM, MgCl 2 mM, one complete™ EDTA-free protease inhibitor cocktail tablet (Sigma-Aldrich, Milan, Italy) and 100 Unit of benzonase to reduce nucleic acids content that could occlude the S-trap column. The lysates completely solubilized after 30 min of mechanic resuspension in a tube rotator at 4 °C and were centrifuged at $10,000\times g$ for 15 min at 4 °C in a refrigerated centrifuge. Once the lysates were ready, the protein concentration of differentially labeled samples was assessed using the BCA kit and the conditions were mixed in a 1:1 ratio to obtain 100 µg of proteins before any further processing.

Each mixture was prepared by mixing one specific biological replicate with another at random.

S-Trap Digestion

Samples collected from all the prepared incubation mixtures were then processed according to the bottom-up proteomics procedure. Sample preparation begins with the solubilization of samples in 5% SDS followed by further denaturation by acidification and subsequent exposure to a high concentration of methanol. The collected incubation mixtures were then solubilized 1:1 in lysis buffer (10% SDS, 100 mM TEAB). The reduction of disulfide bridges was performed by adding 5 μL of the reducing solution of tris(2-carboxyethyl)phosphine (c.f. 5 mM TCEP in 50 mM AMBIC) and incubating the mixtures under slow shaking in a Thermomixer for 10 min at 95 °C. Next, a volume of 5 μL of iodoacetamide solution (c.f. 20 mM IAA in 50 mM AMBIC) was added, with the aim of alkylating the free thiol residues; incubation in this case was carried out for 45 min at room temperature in the dark. Proteins were further denatured by acidification to pH < 1 by adding a 12% phosphoric acid solution in water (1:10 relative to sample volume). The next step consisted of the sample loading: 165 μL of the binding buffer (90% methanol, 10% TEAB 1 M) and 25 μL of sample were simultaneously added onto the spin columns, then centrifuged at a speed of $4000 \times g$ for 1 min at 15 °C; this step was repeated until the protein sample was fully loaded onto the columns.

After that, three washing steps by adding 150 μL of binding buffer onto the S-TRAP columns, followed by centrifugation (1 min, $4000 \times g$, 15 °C), were performed to remove all the excess of unbound sample. At this point, the proteolytic digestion step started by adding a volume of 25 μL containing 1 μg of trypsin (sequencing-grade trypsin, Roche) diluted in 50 mM AMBIC. Upon addition of the protease, the physical confinement within the submicron pores of the trap forces substrate and protease interaction to yield rapid digestion; therefore, protein digestion requires much shorter incubation times, i.e., 1.5 h at 47 °C under slow stirring (400 rpm). The peptide mixture was recovered by loading two different solutions—40 μL of elution buffer 1 (10% H_2O, 90% ACN, 0.2% FA) and 35 μL of elution buffer 1 (60% H_2O, 40% ACN, 0.2% FA)—onto the columns (elution and 1′ centrifugation, $4000 \times g$, 15 °C). The collected peptide mixtures were dried in the SpeedVac (Martin Christ., Osterode am Harz, Germany) at 37 °C and stored at −80 °C until analysis.

High pH Fractionation

The digested peptide mixtures obtained by means of the S-Trap™ micro-spin column digestion strategy were fractionated using the Pierce™ High pH Reversed-Phase Peptide Fractionation Kit according to the manufacturer's instructions, with one additional fraction at 80% ACN.

nLC-HRMS Orbitrap Elite™ Mass Spectrometer Analysis

Tryptic peptides, resuspended in an appropriate volume (18 μL, sufficient for two technical replicates) of 0.1% TFA mobile phase, were analyzed using a Dionex Ultimate 3000 nano-LC system (Sunnyvale, CA, USA) connected to the Orbitrap Elite™ Mass Spectrometer (Thermo Scientific, Brema, Germania) equipped with an ionization source, the nanospray (Thermo Scientific Inc., Milano, Italia).

For each sample, 5 μL of solubilized peptides was injected in triplicate onto the Acclaim PepMap™ C18 column (75 μm × 25 cm, 100 Å pores, Thermo Scientific, Waltham, MA, USA), "protected" by a pre-column, the Acclaim PepMap™ (100 μm × 2 cm, 100 Å pores, Thermo Scientific, Waltham, MA, USA), thermostatically controlled at 40 °C. The chromatographic method used a binary pump system (LC/NC pumps) and started from sample loading onto the pre-column (3 min) using the loading pump with a flow rate of 5 μL/min of mobile phase consisting of 99% buffer A_LC, 0.1% TFA/1% buffer B_LC and 0.1% FA in ACN. After the loading valve switching, peptide separation was performed by the Nano Column Pump (NC_pump) with a 117 min linear gradient (0.3 μL/min) of buffer B_NC_pump (0.1% FA in ACN) from 1% to 40%, and a further 8 min of linear gradient from 40% to 95% (Buffer B_NC_pump). Then, 5 min at 95% of buffer B_NC_pump to rinse the column followed the separative gradient, and the last 7 min served to re-equilibrate the column to initial conditions. The total run time was 144 min. A washout injection with pure acetonitrile (5 µL) was performed between sample injections.

The nanospray ionization source was set as follows: positive ion mode, spray voltage at 1.7 kV; capillary temperature at 220 °C, capillary voltage at 35 V; tube lens offset at 120 V. The orbitrap mass spectrometer operating in data-dependent acquisition (DDA) mode was set to acquire full MS spectra in "profile" mode over a scan range of 250–1500 *m/z*, with the AGC target at 5×10^5, and resolution power at 120,000 (FWHM at 400 *m/z*). Tandem mass spectra were instead acquired by the linear ion trap (LTQ), set to automatically fragment in CID mode the ten most intense ions for each full MS spectra (over 1×10^4 counts) under the following conditions: centroid mode, normal mode, isolation width of the precursor ion of 2.5 *m/z*, AGC target 1×10^4 and normalized collision energy of 35 eV. Dynamic exclusion was enabled (exclusion dynamics for 45 s for those ions observed 2 times in 10 s). Charge state screening and monoisotopic precursor selection were enabled, and singly charged and unassigned charged ions were not fragmented. Xcalibur software (version 3.0.63, Thermo Scientific Inc., Milan, Italy) was used to control the mass spectrometer.

2.7.2. LFQ Analysis

For the LFQ (Label-Free Quantitative Proteomics) experiment, the same cell line (R3/1 NF-κB reporter cell line) was cultivated in biological triplicate in T-25 flasks and the same experimental conditions were tested for SILAC ((i) *Control, CTR*; (ii) *Thinned apple extract treatment (CTR-TAP)*; (iii) *Inflammation (CTR-TNF)*; (iv) *Thinned apple extract treatment of inflamed cells (CTR-TNF-TAP)*, as shown in Figure 1). Except for the labeling strategy, the lysate preparation procedure is the same, as is the proteolytic digestion performed exploiting the potential of the S-Trap™ Micro Spin Column Digestion Protocol.

nLC-HRMS Orbitrap Fusion™ Tribrid™ Mass Spectrometer Analysis

Tryptic peptides were analyzed using a Dionex Ultimate 3000 nano-LC system (Sunnyvale, CA, USA) connected to an Orbitrap Fusion Tribrid Mass Spectrometer (Thermo Scientific, Bremen, Germany) equipped with a nano-electrospray ion source according to the procedure previously described [50].

2.7.3. Data Analysis

For both the SILAC and label-free analysis, the instrumental raw files were processed by MaxQuant software v.1.6.6.0 set on *Rattus_Norvegicus* database (Uniprot taxonomy ID: 10116) against the Andromeda search engine. Protein quantification using SILAC (3 biological $\times$ 2 technical replicates for each condition) was based on the ratio of peptides' peak intensities in the mass spectrum reflecting the relative protein abundance, while for the label-free approach (3 biological $\times$ 3 technical replicates for each condition), the quantification was based on LFQ intensity. In both analyses, trypsin was specified as proteolytic enzyme, cleaving after lysine and arginine except when followed by proline, and up to two missed cleavages were allowed along with match between run option. The precursor ion tolerance was set to 5 ppm while the fragment tolerance was set to 0.5 Da. Carbamidomethylation of cysteine was defined as fixed modification, while oxidation of methionine and acetylation at the protein N-terminus were specified as variable modifications. For the SILAC experiment, only the multiplicity of the labels was set to 2, and Arg10 and Lys8 were selected as heavy aminoacidic residues. Interpretation of the results was performed using Perseus (v.1.6.1.43, Max Plank Institute of Biochemistry, Martinsried, Germany). For the SILAC analysis, the normalized ratio count was selected, transformed in log2, filtered for minimum of 3 valid values and then a one sample *t*-test was performed with Benjamini–Hochberg FDR for truncation with a threshold of 0.05. LFQ analysis was validated by applying a two-sample t-test of the log2 LFQ intensities. The network protein analyses related to significantly altered proteins were carried out by means of Cytoscape v.3.9.1 and the ingenuity pathways analysis (IPA) (QIAGEN Aarhus Prismet, Aarhus, Denmark, September 2021) licensed software based on the Gene Ontology database. All statistical analyses of the inflammatory assays were conducted using GraphPad prism 8. The Venn diagrams with proportional areas were obtained using the bioVenn package for R [51].

Cytoscape Analysis

The protein–protein interaction network of the up-regulated proteins (log2 *Fold-Change* > 0.5) and down-regulated proteins (log2 *Fold-Change* < −0.5) was obtained by importing a list of Protein IDs into Cytoscape v.3.9.1 (http://cytoscape.org, accessed on 15 February 2022). We used the embedded STRING interaction database (http://apps.cytoscape.org/apps/stringApp, accessed on 15 February 2022) with a default confidence cut-off score of 0.4. *Rattus Norvegicus* was selected as a reference database with GO, KEGG, PFAM, WikiPathways, Reactome and InterPro cluster terms selected as functional annotations for the enrichment analyses.

Ingenuity Pathways Analysis (IPA)

The core analyses performed by IPA, using the differentially expressed proteins in the uploaded dataset, assess signaling pathways, molecular interaction network and biological functions that can likely be perturbed. The overall activation/inhibition states of canonical pathways are predicted through a z-score algorithm. This z-score is used to statistically compare the uploaded dataset with the pathway patterns. The pathways are colored to indicate their activation z-scores: orange predicts a gain of function, and blue a loss of function. The pathway is activated when molecules' causal relationships with each other (i.e., activation edge and the inhibition edge between the molecules based on literature findings) generate an activity pattern for the molecules and the end-point functions in the pathway.

3. Results and Discussion
3.1. Qualitative Profile of Polyphenols Determined by Targeted LC-HRMS Analysis

The qualitative profile of polyphenol components of TAP was firstly evaluated by a targeted and untargeted metabolomic approach by HPLC-HRMS in negative and positive ion mode. Figure 2 shows the chromatograms of the extract as total ion current (TIC) recorded in negative (A) and positive (B) ion mode where the identified peaks are numbered progressively (co-eluting peaks share the same number), according to the elution order. The peak of the internal standard (Trolox) is indicated by "IS". A total of 68 compounds were identified, 68 of which are polyphenols, 52 by the targeted approach and 16 by the untargeted approach.

The 52 compounds identified or putatively identified with the targeted analysis are listed in Table S1 of the Supplementary Materials; for each identified compound, the relative peak number (Figure 2), the retention time (RT), the experimental mass (as $[M-H]^-$, $[M+H]^+$, $[M+2H]^{2+}$ or $[M+Na]^+$), the mass accuracy (Δppm), MS/MS fragments and identification method are reported. Thirty-nine compounds were detected in both polarities, seven in negative ion mode and six in positive ion mode. Among the 52 components, 20 are phenolic and organic acids, 11 are flavanols, 10 are flavonols, 5 are flavanones, 4 are dihydrochalcones, 1 is a flavone (luteolin) and 1 is a triterpenoid (euscaphic acid). Thirteen compounds were confirmed by means of in-house available pure standards: three phenolic acids, protocatechuic acid, caffeic acid and *p*-coumaric acid; two flavanols, catechin and epicatechin; two flavanones, naringenin glucoside and naringin; four flavonols, quercetin-3-O-galactoside, quercetin-3-O-rhamnoside, quercetin and kaempferol; two dihydrochalcones, phlorizin and phloretin; one flavone, luteolin. For the compounds for which the standard was not available (39 molecules), a putative identification was carried out by matching the accurate mass, the isotopic pattern and the fragmentation pattern with data reported in the literature. Among phenolic acids, the glucoside forms of protocatechuic, ferulic, caffeic and coumaric acids, and also quinic acid esters of caffeic and coumaric acids were identified. Flavanols are represented in the monomeric forms (catechin and epicatechin) up to the nonamer oligomer, detected as $[M+2H]^{2+}$. Naringenin and eriodictyol derivatives (and aglycones) were the flavanones identified in the extract. The untargeted analysis allowed the putative identification of 16 compounds (Table S2 of Supplementary Materials): nine flavonols, three phenolic acids, one flavanone, one dihydrochalchone and two lipids. As an example, 3-(benzoyloxy)-2-hydroxypropyl glucopyranosiduronic acid (compound **11**) was tentatively annotated, firstly by calculating the possible molecular formula with the QualBrowser tool of Xcalibur (mass tolerance of 5 ppm and using C, H, O, N, S as possible atoms). The resulting formulae were searched in databases and in the literature and we found a match with 3-(benzoyloxy)-2-hydroxypropyl glucopyranosiduronic acid ($C_{16}H_{20}O_{10}$, 0.037 ppm, in negative ion mode) [52]. The MS/MS spectrum was also compared: the main fragment was at *m/z* 249 in negative ion mode deriving from the loss of benzoic acid (−122 Da). Glycosides of quercetin, methoxyquercetin (patuletin, detected with the untargeted approach) and kaempferol are the most representative flavonols identified, but they were also present as aglycones, and a coumaroylglucoside derivative of quercetin was annotated with the untargeted approach. Methoxyquercetin (patuletin, compound **36**) was putatively identified through HMDB database: the molecular formula $C_{16}H_{12}O_8$ (0.107 ppm in negative ion mode; 1.862 ppm in positive ion mode) and MS/MS fragments at *m/z* 285, 209 and 181 (in negative ion mode) matched with those found in the database, and in addition, we detected an additional fragment at *m/z* 316, corresponding to the loss of a methyl group (−15 Da). We detected also five patuletin derivatives: three hexoside isomers (compounds **20**, **22** and **23**) at 20.8, 22.5 and 23.8 min, tentatively annotated with PubChem; one patuletin pentoside derivative (compound **27**), tentatively identified through the matching of the molecular formula ($C_{21}H_{20}O_{12}$, −0.179 ppm and −1.591 ppm in negative and positive ion mode, respectively); and characteristic fragments of patuletin aglycone at *m/z* 331 and 316 due to a neutral loss of −132 Da, corresponding to a pentoside moiety and a subsequent loss of a methyl moiety. Similarly, patuletin rhamnoside (compound **29**) was annotated through the molecular formula ($C_{22}H_{22}O_{12}$, 0.161 ppm and −2.066 ppm in negative and positive ion mode, respectively) and the MS/MS fragments (at *m/z* 331 and 316) deriving from the neutral loss of a rhamnoside moiety (−146 Da) and a subsequent loss of a methyl (−15 Da). Patuletin is a methoxy derivative of quercetin which, among the bioactivities reported in the literature, was found effective in the reduction of serum TNF-α in a rodent model of adjuvant-induced arthritis [53]. CFM-ID online software gave for compound **38** eriodictyol 7-(6-trans-*p*-coumaroylglucoside) as the best match, using the compound identification tool which compares the molecular ion (mass tolerance of 5 ppm) and the MS/MS fragments to data present in online databases. Coumaroyl glucosides seems to show a higher antioxidant effect with respect to the relative glucoside derivative as reported by Li X. et al. [54]: in TAP extract, we found eriodictyol 7-(6-trans-*p*-coumaroylglucoside) and quercetin 3-(3-*p*-coumaroylglucoside) (putatively identified with HMDB). The phloretin derivatives typically identified in apples were also found, with a new putative annotation (through HMDB) of hydroxyphloretin glucoside (peak **25**) detected with the untargeted approach. The untargeted approach also revealed two lipids (peaks **41** and **42**), annotated as (10E,15Z)-9,12,13-trihydroxy-10,15-octadecadienoic acid and trihydroxy-octadecenoic acid.

3.2. Quantitative Analyses (Total Polyphenols and HPLC)

The total polyphenol content was determined both by spectrophotometry and HPLC analysis as reported in the Section 2. Overall, the results derived by these two methods are superimposable, being 24.14 ± 1.58 and 27.97 ± 0.68 mg/100 mg, as determined by the Folin–Ciocalteu colorimetric test and HPLC analysis, respectively.

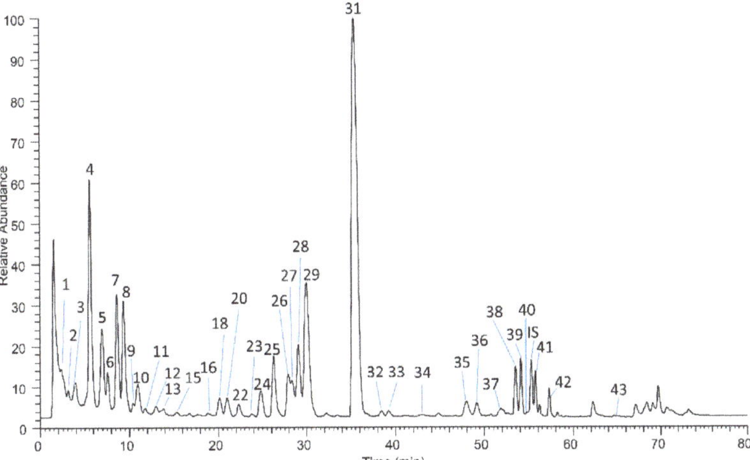

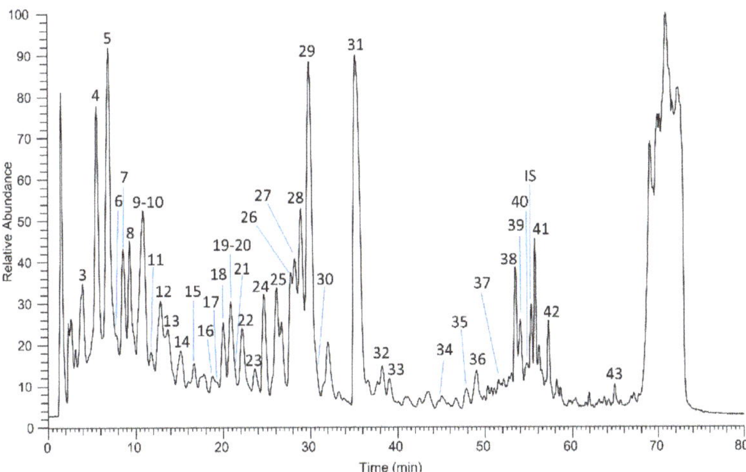

Figure 2. Total ion currents (TICs) of PAT fraction acquired in (**A**) negative ion mode and (**B**) positive ion mode. Identified peaks are numbered progressively, according to the elution order, and their identity or putative identity is shown in Table S1 of the Supplementary Materials with the experimental mass, MS/MS fragments and identification method. IS: internal standard.

By considering the values reported by Sun et al., who found that the total extractable phenolic content from fresh apples is between 110 and 357 mg/100 g [55], we can say that the preparation of the TAP extract resulted in a significant increase in the recovery yield of the polyphenolic fraction.

3.3. Direct Radical Scavenging Activity (DPPH)

The direct radical scavenging activity was evaluated by the DPPH assay and the results are reported in Table 1 and expressed as IC_{50}. The lower the value obtained, the higher the direct radical scavenging activity. The TAP extract was found to exert a significant radical scavenging activity, which was found to be significantly higher than the reference compounds when expressed on the basis of the polyphenol content as determined by HPLC.

Table 1. Direct radical scavenging activity of TAP extract, TAP expressed on the basis of polyphenol content, Trolox, and ascorbic acid. Results are reported as mean ± standard deviation.

Sample	Radical Scavenging Activity IC$_{50}$ µg/mL (Mean ± SD)
TAP	11.4 ± 1.1
TAP (expressed on polyphenol content)	3.2 ± 0.3
Trolox	5.0 ± 0.3
Ascorbic acid	3.9 ± 0.05

3.4. NRF2 Activation and Anti-Inflammatory Activity

We first tested the effect of TAP on cell viability using the MTT assay up to a concentration of 250 µg/mL. Cell viability was found to be higher than 95% at all the tested doses (data not shown). Figure 3 reports the dose-dependent effect of TAP on NRF2 activation after 6 and 18 h of incubation in a concentration range between 1 and 250 µg/mL. After 6 h, the effect started to be significant at a concentration of 50 µg/mL and induced a 2.3-fold increase at a 250 µg/mL concentration. The fold increase was higher after an incubation time of 18 h to reach more than a 5-fold increase at the highest tested dose.

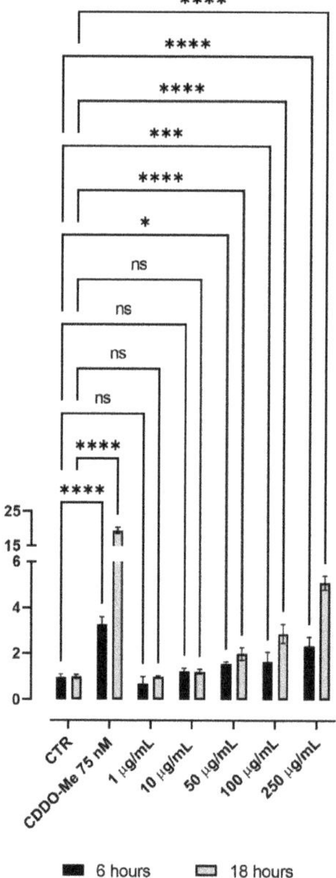

Figure 3. Dose-dependent effect of TAP on NRF2 nuclear translocation. NRF2 activation was tested using NRF2/ARE Responsive Luciferase Reporter HEK293 stable cell line incubated for 6 and 18 h with TAP in a concentration range between 1 and 250 µg/mL. CDDO-Me 75 nM was used as a positive control. Statistical analysis was calculated by two-way ANOVA with Dunnett's multiple comparison test, with individual variances computed for each comparison (* $p < 0.05$, *** $p < 0.001$, **** $p < 0.0001$, ns: not significant).

The anti-inflammatory activity of TAP extract was then tested in the same concentration range (1–250 μg/mL) using R3/1 NF-κB cells and two different inflammatory inducers: TNF-α and IL-1α. The results are summarized in Figure 4. (A) shows the NF-κB-dependent luciferase activity in cells incubated in the absence (black columns) and presence (gray columns) of IL-1α and treated with TAP. Results are reported as luciferase fold increase with respect to untreated cells. IL-1α induced more than a 5-fold increase in NF-κB-induced luciferase, which was dose-dependently reduced by TAP incubation. Luciferase activity as not affected by TAP in the absence of IL-1α. TAP was then tested using TNF-α as a stimulus (Figure 4B), and in this condition, it was found to dose-dependently reduce the NF-κB-dependent luciferase activity after both 6 and 24 h of incubation. The TAP anti-inflammatory activity was greater when IL-1α was used as an inflammatory stimulus with respect to TNF-α.

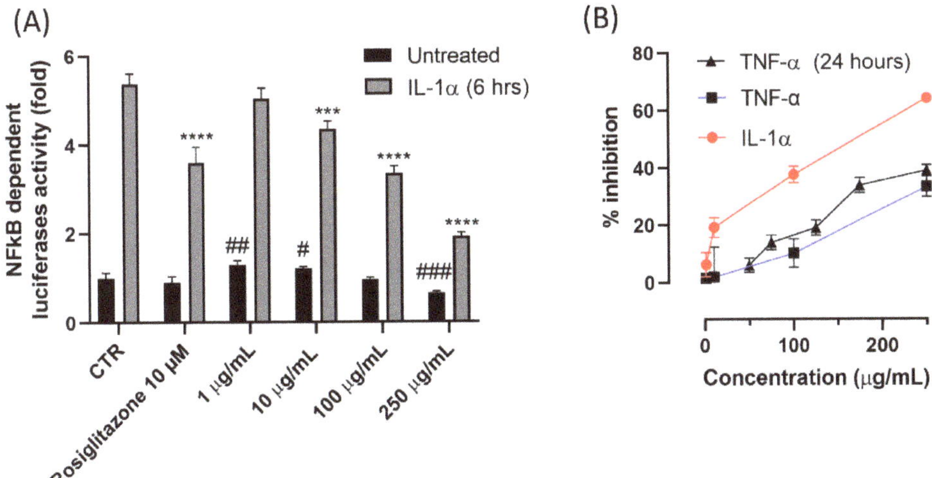

Figure 4. Dose-dependent anti-inflammatory activity of TAP. Activity was tested in R3/1 with the luciferase gene reported for NF-κB nuclear translocation. (**A**) The NF-κB-dependent luciferase activity in cells incubated in the absence (black columns) and presence (gray columns) of IL-1α and treated with TAP in a 1–250 μg/mL concentration range. Rosiglitazone 10 μM was used as a positive control. (**B**) The dose-dependent effect of TAP on the NF-κB-dependent luciferase activity of cells challenged for 6 h with IL1-α, and for 6 and 24 h with TNF-α, (*** $p < 0.001$, **** $p < 0.0001$—# $p < 0.05$, ## $p < 0.005$, ### $p < 0.001$).

3.5. Quantitative Proteomic Studies

The effect of TAP on the proteome of control cells and cells stimulated with TNF-α was then studied by using two different quantitative proteomic approaches, SILAC and label-free, and the following four different experimental groups: control cells, cells stimulated with TNF-α, and cells incubated in the absence and presence of TAP.

3.5.1. SILAC Proteomic Studies

Figure 5 shows the volcano plots obtained from the three comparative analyses: (i) cells incubated in the absence (CTR) and presence of TAP (CTR-TAP), (ii) cells incubated in the absence (CTR) and presence of TNF-α (CTR-TNF); (iii) cells stimulated with TNF-α and incubated in the absence (CTR-TNF) and presence of TAP (CTR-TNF-TAP). The analyses were conducted through two orthogonal proteomic approaches, SILAC and the label-free method.

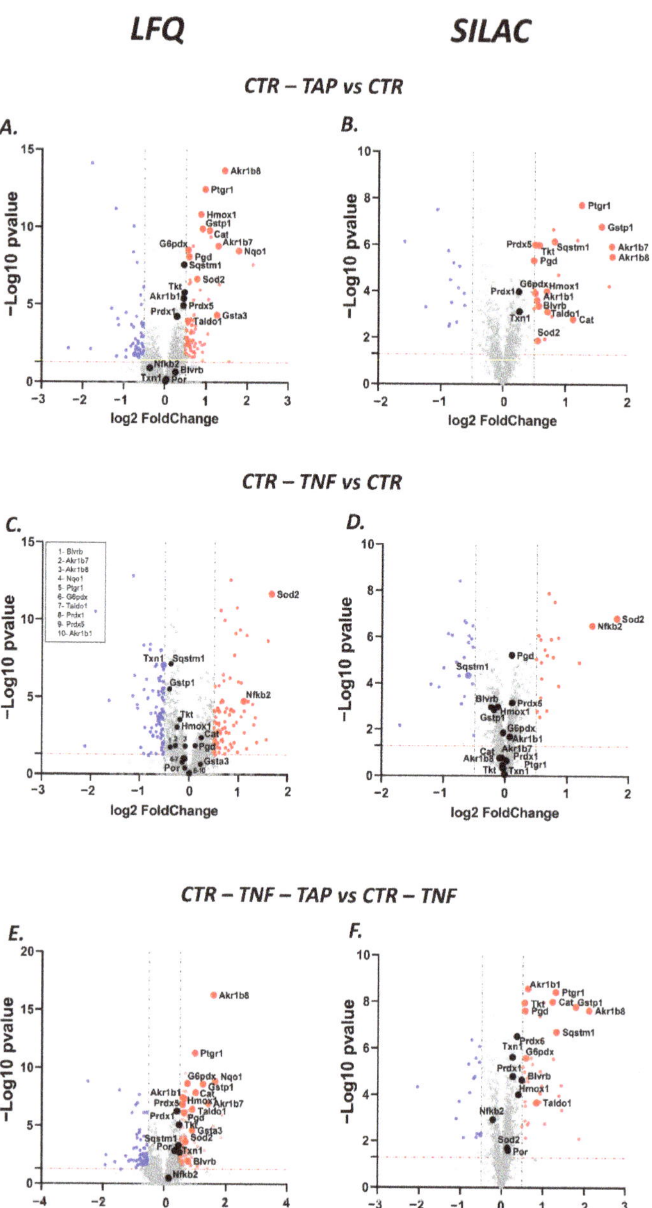

Figure 5. Volcano plot of the LFQ and SILAC analyses. The two columns represent the LFQ (left) or SILAC (right) analyses. In the first row, the two volcano plots (**A**,**B**) illustrate the analyses performed to understand the impact of TAP extract on the cell proteome. On the second row, (**C**,**D**) show the volcano plot for the comparative condition achieved by treating cells with TNFα. Lastly, (**E**,**F**) are the volcano plots showing the differential proteome expression for the inflamed cells treated with TAP extract. In each volcano plot, the most representative proteins identified in the various analyses are labeled in red if up-regulated (*p* value < 0.05, log2 *Fold-change* > 0.5), in blue if down-regulated (*p* value < 0.05, log2 *Fold-change* < −0.5) and in black when they do not fall into the two aforementioned conditions. A more comprehensive analysis of the log2 *Fold-change* values is available in Table 2.

Table 2. Key proteins identified by quantitative proteomics analysis; fold-change values calculated by each of the experimental approaches used are reported by gene product. The fold-change values are classified according to the comparative analysis of the conditions tested. ND indicates a not-detected protein identification with the corresponding proteomic approach. The value 0 indicates a fold-change not significantly different from zero. Each fold-change set, reported per gene product and per proteomic approach, is colored on a green (minimum value) to light orange (maximum value) color scale according to the conditional formatting function of excel. The gene products involved in the nuclear factor erythroid-derived 2 signaling pathway are highlighted in light violet and those involved in the pentose phosphate pathway in light blue.

	CTR-TNF vs. *CTR*		*CTR-TAP* vs. *CTR*		*CTR-TNF-TAP* vs. *CTR-TNF*	
Gene Name	log2 *Fold-change* LFQ	log2 *Fold-change* SILAC	log2 *Fold-change* LFQ	log2 *Fold-change* SILAC	log2 *Fold-change* LFQ	log2 *Fold-change* SILAC
Akr1b1	0	0	0.4508	0.5497	0.5960	0.6343
Akr1b3	0	ND	0.4508	ND	0.5960	ND
Akr1b7	0	0	1.2905	1.7506	1.4273	ND
Akr1b8	0	0	1.4440	1.7608	1.5898	2.1034
Blvrb	−0.3809	−0.2301	0	0.5764	0.7862	0.4835
Cat	0	0	1.0800	1.1262	1.0148	1.2222
G6pdx	0	0	0.5651	0.5211	0.7523	0.5877
Gsta3	0	ND	1.2664	ND	0.9136	ND
Gstp1	−0.3949	−0.1888	0.9105	1.5843	1.2616	1.7841
Hmox1	−0.2424	−0.1233	0.8703	0.7048	0.6607	0.4045
Nfkb2	1.1143	1.3995	0	ND	0	−0.2200
Nqo1	0	ND	1.7907	ND	1.6534	ND
Pgd	0	0.1012	0.5913	0.4975	0.6452	0.5666
Por	−0.0862	ND	0.0429	ND	0.3405	0.1561
Prdx1	0	0	0.2915	0.2540	0.4062	0.2665
Prdx5	0	0.1132	0.4444	0.5167	0.6037	0.3645
Ptgr1	0	0	0.9705	1.2600	0.9978	1.2999
Sqstm1	−0.3753	−0.6061	0.4559	0.8246	0.4579	1.3186
Taldo1	0	ND	0.5603	0.7171	0.9184	0.8525
** Tkt	−0.1891	0	0.4719	0.5824	0.4815	0.5555
Txn1	−0.5145	0	0	0.2655	0.4939	0.2589
Sod2	1.6735	1.8043	0.7816	0.5633	0.6954	0.1391

** $p < 0.005$.

The first experimental comparison analysis permits the evaluation of the general effect of TAP on the cell proteome in homeostatic conditions. From the SILAC experiment, 24 and 14 proteins were found to be up- and down-regulated, respectively, while 1235 were found to be unchanged. Ingenuity pathway analysis identified two main up-regulated pathways: the nuclear factor, erythroid-derived 2 signaling pathway with a z-Score of +2.9 (Figure 6A), and the pentose phosphate pathway with a z-Score of +2.0 (Figure 6B).

In particular, the up-regulated enzymes involved in the first pathway include HMOX1, CAT, GSTP1, AKR(1–7), NQO1 and SQSTM1, while for the second one, TKT, PGD, G6PDX and TALDO1 are identified. In addition to these proteins, PTGR1 and BLVRB were also found to be up-regulated, as reported in the volcano plots (Figure 5). PTGR1 is an NADPH-dependent oxidoreductase which is found to be induced by NRF2 activators [56] and directly regulated by NRF2 [57]. BLVRB is found to be an NRF2 target gene and together with HMOX is critical in the heme metabolism [58] (Figure 11A).

The volcano plot relative to the second experimental comparison analysis (CTR-TNF vs. CTR, Figure 5C,D) mainly identifies two overexpressed proteins involved in inflammation, NF-KB2 and SOD2. As expected, the ingenuity pathway analysis identifies, with a z-Score of 1.8, the TNF-α as up-stream regulator (data not shown).

The volcano plot displaying the third comparison analysis (CTR-TNF-TAP vs. CTR-TNF, Figure 5E,F) reports that TAP treatment down-regulates the two proteins over-expressed in the inflammatory conditions (NF-KB2 and SOD2), while the two main pathways up-regulated by the TPA treatment were confirmed to be the NRF2 and the pentose phosphate pathway.

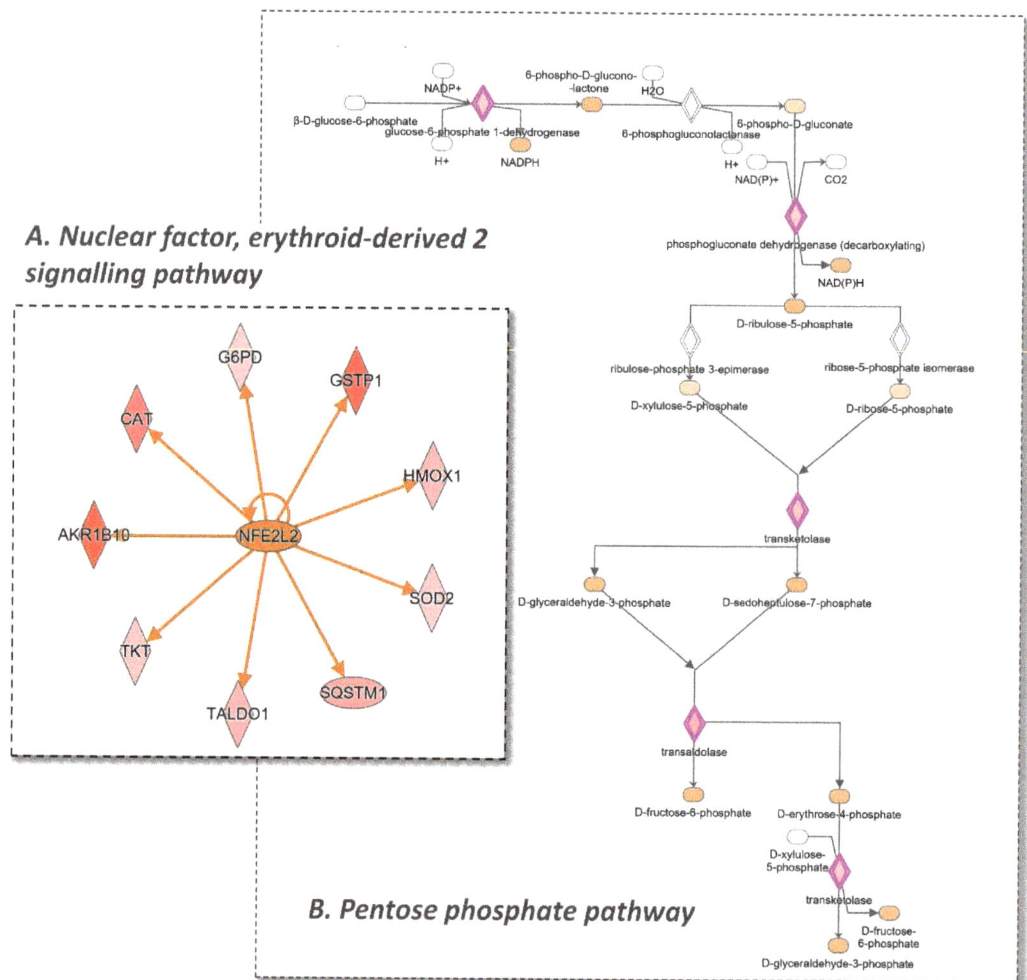

Figure 6. IPA analysis of the effect of the extract on the cell proteome. (**A**) illustrates that the differentially regulated proteins in the analysis are consistent with an NRF2 activation. (**B**) illustrates the Pentose Phosphate Pathway; in red are the proteins that were found to be up-regulated by TAP, resulting in a increment in the NADPH pool in the cell. Color legend: red represents the increased genes, green the decreased (not present in the figure). The intensity of the color is related to the intensity of up- or down-regulation. The orange line leads to activation and a blue line (not present in the figure) leads to inactivation. The yellow line indicates that findings that are not consistent with the proteomics results obtained.

Finally, protein–protein interaction network analysis using the Cytoscape interface in STRING confirmed the modulation of a protein set involved in the biological processes/pathways already highlighted in IPA, as a result of TAP treatment (Figure 7A,B). Assigned GO annotations lead firstly to confirmation of the activation of the NRF2 and pentose phosphate pathways and the involvement of several oxidoreductive enzymes as well as of proteins involved in the metabolism of glutathione (GSH). It should be taken into account that STRING–Cytoscape network analysis does not distinguish between up- and down-regulated proteins but only considers significant expression modulation.

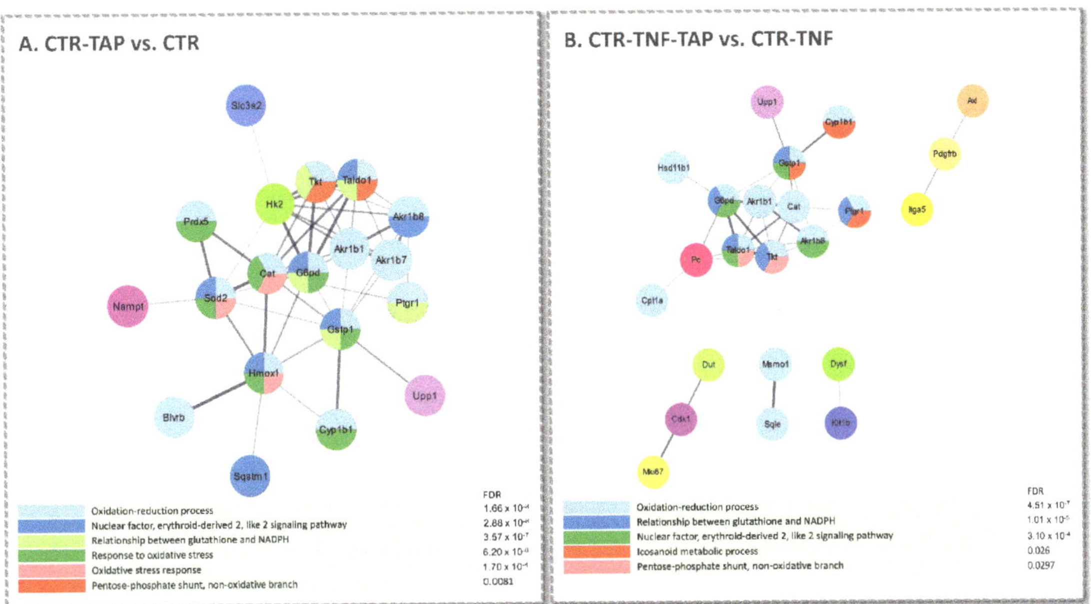

Figure 7. Cytoscape and STRING analysis of the up-regulated gene products determined by SILAC. The panels illustrate the protein–protein interaction networks obtained by STRING in the CTR-TAP vs. CTR (**A**) and CTR-TNF-TAP vs. CTR (**B**) conditions. The color/s of each node (gene product) reflect/s the functional enrichment analysis performed; the color code describing the enriched biological processes with the corresponding FDR value is shown below each network. In both analyses, the singlets were left out.

3.5.2. Label-Free Quantitative Proteomic Studies

The label-free quantitative proteomic study, through the use of a Fusion MS analyzer, identified a larger number of proteins with respect to those identified by the SILAC approach, which was based on an LTQ Orbitrap system. In particular, LFQ identified 2340 more proteins than SILAC in the CTR-TNF vs. CTR mixture, 2412 more in the CTR-TAP vs. CTR mixture and 2259 more in the CTR-TNF-TAP vs. CTR-TNF mixture. As shown by the volcano plots relative to the three experimental conditions' comparison, depicted in Figure 5 (lower panels), besides a larger number of identified proteins, the LFQ approach also permitted the identification of more up- and down-regulated proteins in all three conditions. Details about the comparison of the number of up- and down-regulated proteins identified by the two approaches in the three experimental groups are shown by Venn diagrams in Figure 8.

Pathway analysis (IPA) of the LFQ results, besides confirming the results achieved by the SILAC approach, identified further regulated pathways. As shown in Figure 9, a better description of the NF-κB activation was observed because two upstream regulators were found to be compatible with the over-expressed proteins, AKT (protein kinase B, Figure 9B) and IKBKB (inhibitor of nuclear factor kappa-B kinase subunit beta, Figure 9C). AKT regulates the transcriptional activity of NF-κB by inducing phosphorylation and the subsequent degradation of the inhibitor of κB (IκB). IKBKB phosphorylates the inhibitor in the inhibitor/NF-κB complex, causing dissociation of the inhibitor and activation of NF-κB. As a result, the number of proteins identified in the TNF-a pathway significantly increases. Besides AKT and IKBKB, another obvious upstream regulator is reported in Figure 9A, the TNF pathway, which simply confirms the effectiveness of the method used to induce inflammation.

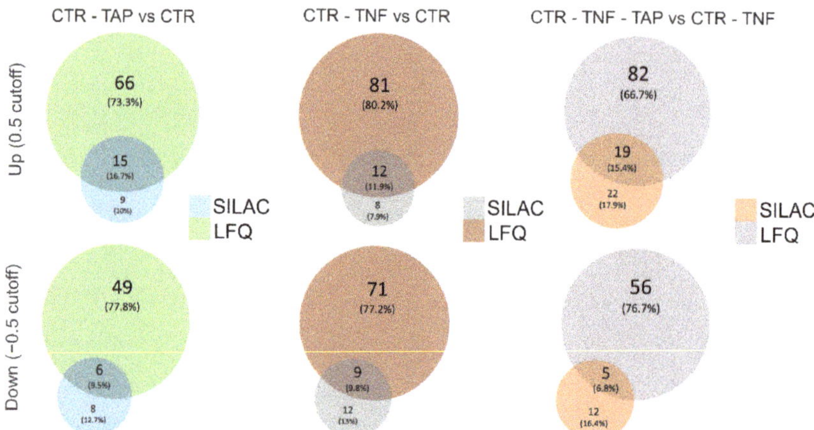

Figure 8. Venn diagrams showing technical differences in terms of the number of significantly (log ratio cutoff ± 0.5) up- (first line diagrams) and down-regulated (second line diagrams) proteins identified by the two approaches in the three comparison analyses (CTR-TAP vs. CTR; CTR-TNF vs. CTR; CTR-TNF-TAP vs. CTR-TNF). Additionally, the percentage distribution of identifications in the two approaches and the percentage of common assignments are shown for each diagram.

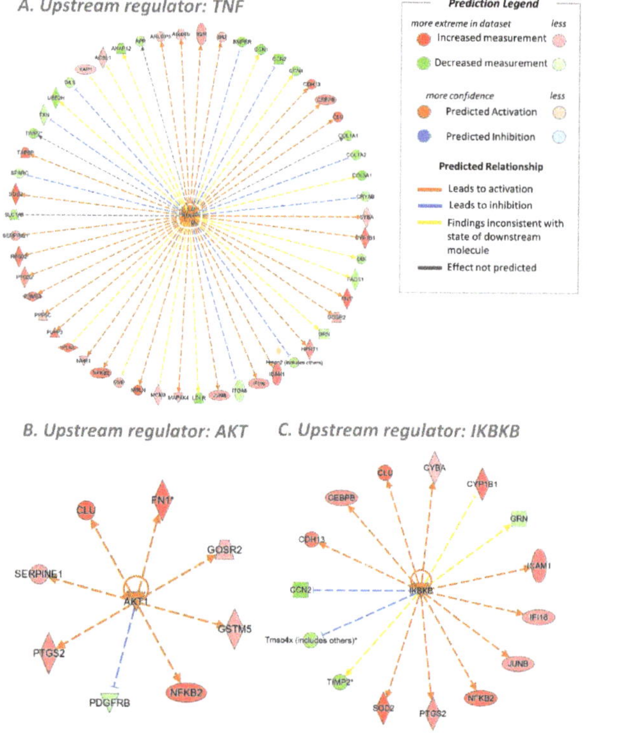

Figure 9. Upstream regulators determined in the CTR-TNF vs. CTR condition by IPA analysis (LFQ). (**A**) The wheel of proteins involved in the TNFα stimuli. (**B**,**C**) The wheels of proteins associated with AKT and IKBKB as upstream regulators. Above is the prediction legend (IPA) essential for understanding networks. * multiple isoforms.

Regarding the CTR-TAP vs. CTR comparison, the activation of the NRF2 pathway by TAP, as already evidenced by the SILAC approach, was confirmed and the number of proteins involved in the signaling was greatly extended. Furthermore, TAP treatment was associated with a reduction in ferroptosis, apoptosis and oxidative stress (data not shown).

Regarding the third group, CTR-TNF-TAP vs. CTR-TNF, the LFQ approach permitted a better description of the upstream regulators relative to a reduction in oxidative stress and NRF2 pathway due to the increased number of the identified proteins involved in the pathways (Figure 10).

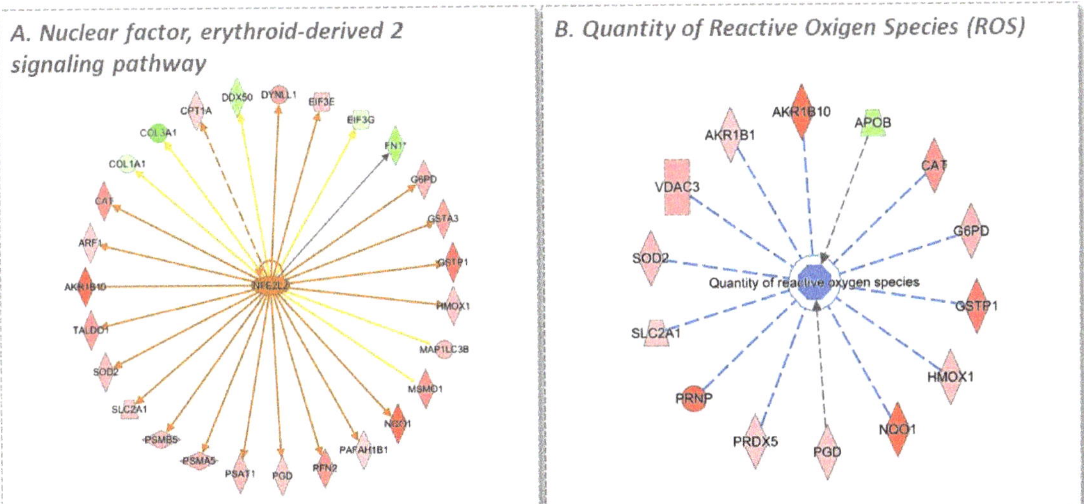

Figure 10. IPA analysis in the CTR-TNF-TAP vs. CTR-TNF condition. (**A**) illustrates the different regulated proteins present in the analysis which correlate with NRF2 activation. (**B**) illustrates the effect that the differentially regulated proteins have on the quantity of reactive oxygen species, confirming a trend of reduction in oxidative stress. The prediction legend (IPA) is reported in Figure 9.

4. Discussion

Epidemiological and intervention studies indicate that consumption of apples and derivatives, such as apple juice, exert beneficial effects relating to some human diseases including CVD and related events, e.g., cancer and diabetes. Some of the health effects, and in particular, the antioxidant and anti-inflammatory activities, have been attributed to the polyphenol fraction, as also confirmed by pre-clinical studies. Hence, there is a growing interest in the use of apple polyphenols as health care products, as well as on their production with green methods. Apple polyphenols are contained in large quantities in thinned apples which represent a large waste material of the apple chain. Based on these premises, the aim of the present study was to evaluate thinned apples as a source for the isolation of bioactive apple polyphenols, effective as anti-inflammatory and antioxidant compounds to be used in health care products. To reach this goal, we first set up a scalable procedure for polyphenol isolation from thinned apple. The polyphenol content of the purified fraction obtained by absorption resin accounted for 24%, as determined by colorimetric analysis, and almost 28% by HPLC, indicating the suitability of the isolation process. The isolated polyphenols were then fully identified by a targeted and untargeted LC-HRMS approach which is, to our knowledge, the first qualitative profile of polyphenols from thinned apples.

The targeted and untargeted metabolomics approaches allowed us to identify a total of 68 polyphenols belonging to the six polyphenolic classes. Moreover, 16 compounds were found in the TAP fraction, though not yet reported in apple or derivatives. All the characteristic polyphenols identified in harvested apples as reported in the most recent literature were identified.

The biological evaluation of the extract followed the qualitative characterization; TAP was first evaluated in two cell models in order to evaluate the antioxidant and anti-inflammatory properties of the extract, the two main activities ascribed to apple polyphenols as reported by the most recent literature. The antioxidant activity was firstly studied by the DPPH assay, which demonstrated a greater radical scavenging activity of the polyphenol fraction with respect to reference compounds, i.e., Trolox and ascorbic acid. For several years, the in vivo antioxidant activity of plant polyphenols has been explained by considering their direct radical scavenging activity towards ROS and carbon centered radicals. However, as pointed out by Forman and Ursini [59], kinetic constraints indicate that in vivo scavenging of radicals is ineffective in antioxidant defense (except for vitamin E towards peroxyl radicals). Instead, enzymatic removal of non-radical electrophiles, such as hydroperoxides, in two-electron redox reactions, is the major antioxidant mechanism. Based on this overview, the major mechanism of action for nutritional antioxidants, and in particular, for plant polyphenols such as those contained in TAP, is the paradoxical oxidative activation of the NRF2 (NF-E2-related factor 2) signaling pathway, which maintains protective oxidoreductases and their nucleophilic substrates. As stated by Ursini, the maintenance of the "Nucleophilic Tone" by a mechanism called "Para-Hormesis" provides a means for regulating physiological non-toxic concentrations of the non-radical oxidant electrophiles that boost antioxidant enzymes, damage removal and repair systems (for proteins, lipids and DNA) at the optimal levels consistent with good health [59].

Hence, based on this new vision of the antioxidant activity of natural antioxidants, as a next step, we tested the ability of TAP to act as an indirect antioxidant by activating the NRF2 pathway.

Results well indicate that TAP dose-dependently activates the NRF2 pathway, as determined by measuring the activity of luciferase, which is the product of the gene reporter activated by the nuclear binding of NRF2. Activation of NRF2 by TAP can be firstly ascribed to the polyphenol compounds bearing an ortho-diphenol moiety which is oxidized to the corresponding quinone, which, being an electrophilic compound, reacts with the thiols of KEAP1, thus releasing NRF2, which then translocates into the nucleus. TAP contains a set of polyphenols containing the catechol moiety, including derivatives of caffeic acid, and quercetin. TAP also contains phenols or methoxy derivatives, such as derivatives of coumaric acid, phloretin and naringenin, which can also be NRF2 activators, but in this case, metabolic activation is required to form a di-phenol moiety. In particular, the metabolic activation of phenols requires the insertion of a hydroxyl group in ortho or para position mediated by the cytochrome enzymes, as found for resveratrol [60], while a CYP mediated O-demethylation activates methoxy derivatives, as reported for silybin [61]. Such metabolic reactions usually occur in the liver tissue and hence are unlikely to occur in the cell model used in the in vitro assay. Hence, we presume that the potency of NRF2 activation is underestimated and that presumably it is potentiated in vivo by metabolic activation occurring in the organism, as well as by the microbiota in the gastrointestinal tract.

NRF2 activation paralleled the dose-dependent anti-inflammatory activity, as demonstrated in the tested cell model (luciferase gene reporter for NF-κB nuclear translocation), and using two different pro-inflammatory agents, IL1-α and TNF-α. A strict interdependence between oxidative stress and inflammation and between the two transcriptional factors involved, NF-κB and NRF2, is well established [62,63]. When oxidative stress appears as a primary disorder, inflammation develops as a secondary disorder and further enhances oxidative stress. On the other hand, inflammation as a primary disorder can induce oxidative stress as a secondary disorder, which can further enhance inflammation [64]. Based on this evidence, a well-established strategy to block the inflammatory chronic condition consists of inhibiting oxidative stress, and in turn, the best approach to inhibit oxidative stress is to activate the nucleophilic tone by inducing NRF2 nuclear translocation.

Hence, the anti-inflammatory activity of TAP can be first of all ascribed to the NRF2-2-dependent antioxidant effect. To gain a deeper insight into this mechanism, and in particular, to confirm the activation of antioxidant enzymes and to search for other enzymes involved in the anti-inflammatory activity, a proteomic investigation was then carried out.

Quantitative proteomic studies were carried out using the cell line with the gene reporter for NF-κB, used for the anti-inflammatory activity evaluation, and TNF-α as an inflammatory inducer. Two orthogonal proteomic approaches were applied, namely, a label-free method, using a Fusion MS analyzer, and a stable isotope in vivo labeling method, i.e., SILAC, coupled to a LTQ Orbitrap system. In general, the label-free approach identified many more proteins with respect to SILAC and was revealed to be more efficient in the identification of up- and down-regulated pathways. These differences could be easily explained by considering the intrinsic potential in terms of sensitivity and resolution characteristics of the mass analyzers used for each of the experiments. Besides the different performances, the two approaches identified the same modulated pathways, although with a different number of proteins, making the results more robust, as confirmed by two orthogonal approaches. This very interesting technical aspect is particularly evident in the following table (Table 2), which shows the experimentally calculated fold-change values for each key protein identified, whose expression is significantly modulated in the comparative analysis between the tested conditions.

The first relevant observation emerging from the comparison of the calculated fold-change values in the different conditions tested is the apparent shift in the expression of some of the proteins involved in the NRF2 and pentose phosphate pathways. A fitting example could be the HMOX1 expression modulation, having a negative fold-change (LFQ: -0.2424, SILAC: -0.1233) in the first comparison, meaning that inflammation causes its down-regulation, and showing a significant up-regulation when the cells are treated with TAP extract, in both conditions.

Targeting a more comprehensive biological significance, rather than investigating individual key proteins, the global effect of TAP treatment in physiological conditions (CTR), to identify the set of genes whose expression is variably modulated by treatment, was evaluated. As expected, many up-regulated genes involved in the NRF2 pathway activation were detected, particularly antioxidant enzymes, or those directly involved in the detoxification of peroxides and in two-electron redox reactions, e.g., SOD, CAT, GXP and redox proteins such as thioredoxins. Furthermore, a set of enzymes required for the synthesis of GSH and NADPH, co-factors for the enzymes involved in the antioxidant and redox regulation, were also identified. GCLC, GSTA3 and GSTP were the identified up-regulated enzymes involved in the GSH synthesis and metabolism while the enzymes belonging to the oxidative pentose-phosphate pathway are those up-regulated for NADPH synthesis, such as malic enzyme-1 (ME-1), isocitrate dehydrogenase-1 (IDH1), glucose-6-phosphate dehydrogenase (G6PDX) and 6-phosphogluoconate dehydrogenase (PGD).

The crucial finding is that TAP treatment over-expresses the inducible isoform of heme oxygenase (HMOX1), a well-established immunomodulator [65–67]; it has been proven that the induction of HMOX1 protects against the cytotoxicity caused by oxidative stress and apoptotic cell death, making HMOX1 an appealing target for the treatment of several chronic inflammatory diseases, including osteoporosis [68], cancer [69], acute kidney injury [70], retinal pigment epithelium degeneration [71] and Parkinson's disease [56,67].

In physiological conditions, HMOX1 is involved in rate-limiting heme degradation using NADPH–cytochrome P450 reductase (POR) and oxygen to generate linear tetrapyrrole biliverdin (further converted into linear tetrapyrrole bilirubin by the enzyme biliverdin reductase, BLVRB), ferrous iron (Fe^{2+}) and carbon monoxide (CO) (Figure 11A). The enzymatic heme-degradation by-products are known to be the main cause of the beneficial protective effects promoted by HMOX1; bilirubin, in particular, is one of the most potent antioxidant and is particularly effective in protecting against lipid peroxidation.

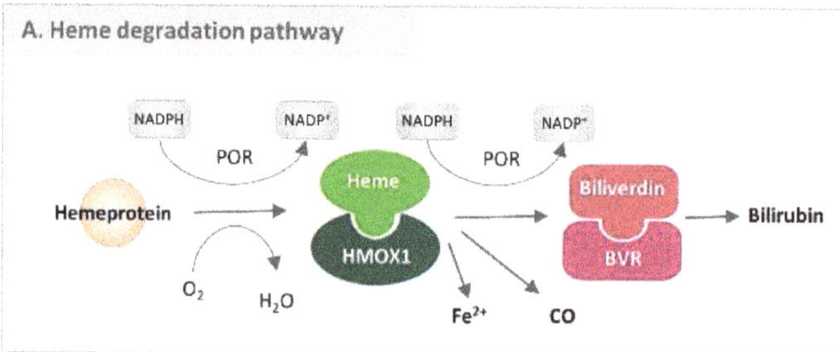

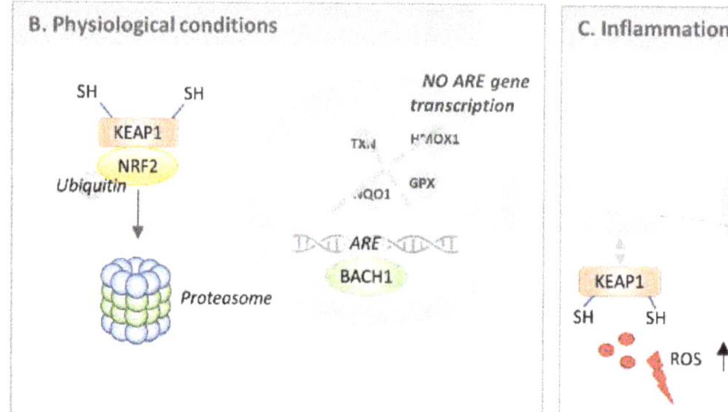

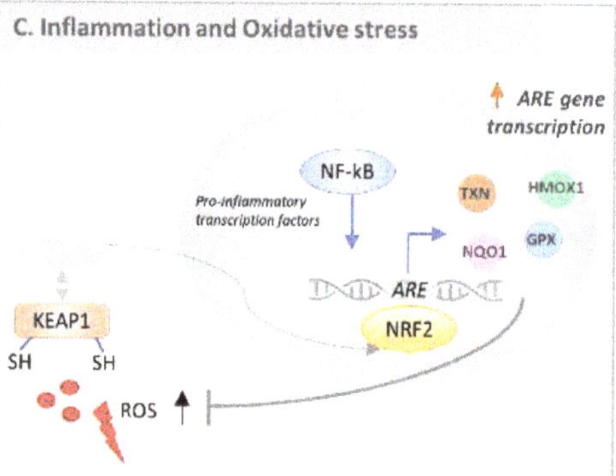

Figure 11. Signaling pathways in which HMOX1 is involved. (**A**) Heme degradation pathway promoted by HMOX1 occurring in the cytosol of the cell. Heme degradation occurs in the cytosol by the enzyme HMOX1; the reaction involves cytochrome P450 reductase, NADPH as cofactor and O_2, leading to the production of $NADP^+$, and an equimolar amount of biliverdin, carbon monoxide and iron (Fe^{2+}). Specifically, biliverdin can be further converted to bilirubin by the enzyme biliverdin reductase (BVR). (**B**) Regulation of HMOX1 expression under homeostatic conditions. NRF2 is bound to KEAP1, ubiquitinated and targeted for proteasomal degradation. BACH1 binds the antioxidant response element (ARE) at the promoter of HMOX1, inhibiting transcription of the gene. (**C**) Regulation of HMOX1 expression during inflammation and oxidative stress. Increased intracellular heme levels lead BACH1 to dissociate from ARE, and reactive oxygen species (ROS), interacting with cysteine residues on KEAP1, cause dissociation from NRF2, which in turn migrates into the nucleus, binds ARE and promotes transcription of HMOX1 (adapted from [67]).

HMOX1 can be up-regulated in response to numerous different stress stimuli, including various pro-oxidant and pro-inflammatory mediators; most of them do not interact directly with the transcription factors, but activate them via intermediate signaling pathways. HMOX1 expression is largely under the control of NRF2, a well-known binder of the ARE (antioxidant response element), aimed at promoting many antioxidant genes, including HMOX1. Under homeostatic conditions, NRF2 is bound to Kelch-like ECH-associated protein 1 (KEAP1), ubiquitylated and targeted by the proteosome for degradation (Figure 11B); meanwhile, BACH1 binds to ARE and inhibits gene transcription. An increased level of reactive oxygen species (ROS) during cell stress implies a concomitant KEAP1-NRF2 dissociation due to their reactivity against thiols in KEAP1; in turn, NRF2 translocates to the nucleus and binds the ARE to promote transcription of HMOX1 (Figure 11C). Basically, NRF2, KEAP1 and BACH1 constitute an intricate feedback system enabling cells to respond to oxidative stress ending at the up-regulation of HMOX1.

The pathway activation in this study is supported not only by the significant modulation of HMOX1 but also by the up-regulation of BLVRB and POR (NAPH-Cytochrome P450 reductase), which, although not significantly up-regulated (fold-change: 0.340528912, *p*-value n.s.), exactly matches the trend of the other associated gene products. In addition, activation of the pentose phosphate pathway leads to a plausible increase in circulating NADPH, a cofactor of both POR and BLVRB (Table 2).

The phenotypic effect described here, in terms of protein expression induced by TAP treatment, is completely concordant when comparing CTR-TNF-TAP to CTR-TNF conditions. The protective effect manifested, such as the triggering of the two pathways of interest, is perhaps even more evident looking at the volcano plots shown in Figure 5A,B: AKR1B1, AKR1B8, BLVRB, CAT, G6PDX, GSTP1, HMOX1, PGD, PTGR1, SQSTM1, TALDO1 and TKT (listed alphabetically) switch to a fold-change value indicating significant up-regulation, explaining the anti-inflammatory activity overall.

Besides the up-regulated set of NRF2-dependent pathways, another key point is the regulation of TAP towards the increased content of NF-κB in cells exposed to TNF-α with respect to control (Figure 5C,D), and the overexpression of a set of proteins which are known to be overexpressed by the TNF-signaling. Hence, NF-κB, in the cell model used, besides being activated and translocated at the nuclear level, as evidenced by the gene reporter, is also overexpressed, thus sustaining the inflammatory process. As stated above, TAP treatment of the inflamed cells not only inhibits NF-κB activation but also its overexpression: NF-κB2, together with SOD2, returns to homeostatic conditions, even if the fold-change was not significantly different from zero (Figure 5E,F). An interesting aspect to stress is that NF-κB, together with AP-1 (Figure 11C), can also bind the HMOX1 promoter to enhance its expression; indeed, some instances of HMOX1 up-regulation in vitro have been shown to be dependent on or correlated with NF-κB or AP-1 expression [72].

5. Conclusions

In conclusion, thinned apples represent a valuable source of apple polyphenols whose qualitative profile was fully elucidated by targeted and untargeted metabolomic approaches. The anti-inflammatory and antioxidant activities of TPA were first evaluated by cell-based assays using engineered cells with gene reporters for NRF2 and NF-κB nuclear translocation. Quantitative proteomic studies were then used to deeply investigate the molecular mechanisms involved in the proven biological activity. Two different proteomic approaches, i.e., label-free and SILAC, both confirmed TAP ability in improving the antioxidant cellular defense, by activating the NRF2 signaling pathway which up-regulates protective oxidoreductases and their nucleophilic substrates. A clear increase in enzymatic antioxidants together with the up-regulation of the heme-oxygenase explain the anti-inflammatory effect of TAP. Taken together, these results suggest that thinned apples can be effectively considered a valuable source of apple polyphenols to be used in health care products to prevent/treat oxidative and inflammatory chronic conditions.

Supplementary Materials: The following supporting information can be downloaded at: https://www.mdpi.com/article/10.3390/antiox11081577/s1, Table S1: Compounds identified or putatively identified by LC-HRMS in negative and positive ion modes with the targeted method ordered on the basis of the retention time; Table S2: Compounds putatively identified by LC-HRMS in negative and positive ion modes with the untargeted method ordered on the basis of the retention time.

Author Contributions: Conceptualization, A.A., G.B. and G.A.; Formal analysis, A.A., G.F., G.B., F.G. and L.D.V.; Funding acquisition, M.C., E.B. and G.A.; Investigation, A.A., G.F., G.A. and A.D.; Methodology, A.A., G.F. and G.B.; Project administration, A.A., E.B., A.D. and G.A.; Resources, M.C. and G.A.; Supervision, A.A., G.B. and G.A.; Validation, A.A., G.F. and G.B.; Writing—original draft, A.A., G.F., G.B. and G.A.; Writing—review and editing, A.A., G.F., G.B. and G.A. All authors have read and agreed to the published version of the manuscript.

Funding: LDV and GF are supported as Ph.D student and temporary researcher (RTDA) by Ministero dell'Università e della Ricerca PON "Ricerca e Innovazione" 2014–2020, Azione IV.4—"Dottorati e contratti di ricerca su tematiche dell'innovazione" and Azione IV.6—"Contratti di ricerca su tematiche Green". This research is part of the project "MIND FoodS HUB (Milano Innovation District Food System Hub): Innovative concept for the eco-intensification of agricultural production and for the promotion of dietary patterns for human health and longevity through the creation in MIND of a digital Food System Hub", cofunded by POR FESR 2014–2020_BANDO Call HUB Ricerca e Innovazione, Regione Lombardia. This work was supported by the University of Milan through the APC initiative.

Institutional Review Board Statement: Not applicable.

Informed Consent Statement: Not applicable.

Data Availability Statement: The data presented in this study are available on request from the corresponding author. The data are not publicly available.

Acknowledgments: The authors are grateful to AKHYNEX S.r.l. POLISTENA (RC) for the partial financial support.

Conflicts of Interest: The authors declare no conflict of interest.

Abbreviations

LC-HRMS	Liquid chromatography–high-resolution mass spectrometry
ESI	Electrospray ionization
NRF2	Nuclear factor erythroid 2-related factor 2
NF-κB	Nuclear factor kappa-light-chain-enhancer of activated B cells
SILAC	Stable isotope labeling by amino acids in cell culture
TAP	Thinned apple polyphenols
TNF-α	Tumor necrosis factor
IL1-α	Interleukin 1 alpha
GSH	Glutathione
NADPH	Reduced nicotinamide adenine dinucleotide phosphate
CVD	Cardiovascular diseases
LDL	Low-density lipoprotein
ICAM-1	Intercellular adhesion molecule 1
CRP	C-reactive protein
IL-6	Interleukin 6
LBP	Lipopolysaccharide-binding protein
KEAP1	Kelch-like ECH-associated protein 1
HMOX1	Heme oxygenase 1
MMPs	Matrix metalloproteinases
MAPK	Mitogen-activated protein kinases
TLC	Thin-layer chromatography
CAT	Catalase
AKR	Aldo-keto reductases
NQO1	NAD(P)H dehydrogenase quinone 1
SQSTM1	Sequestosome 1
TKT	Transketolase
PGD	Phosphogluconate dehydrogenase
G6PDX	Glucose-6-phosphate 1-dehydrogenase X
TALDO1	Transaldolase 1
PTGR1	Prostaglandin reductase 1
BLVRB	Biliverdin reductase B
SOD2	Superoxide dismutase 2
AKT	Serine/threonine kinase 1
IKKB	Inhibitor of nuclear factor kappa-B kinase subunit beta
GSTA3	Glutathione S-transferase alpha 3
GSTP1	Glutathione S-transferase Pi 1
POR	Cytochrome P450 oxidoreductase
PRDXs	Peroxiredoxins
TXN1	Thioredoxin 1
GXP	Glutathione peroxidases
ME-1	Malic enzyme-1
IDH1	Isocitrate dehydrogenase-1

<table>
<tr><td>CO</td><td>Carbon monoxide</td></tr>
<tr><td>BACH1</td><td>BTB domain and CNC homolog 1</td></tr>
<tr><td>AP-1</td><td>Activator protein 1</td></tr>
</table>

References

1. Boyer, J.; Liu, R. Apple phytochemicals and their health benefits. *Nutr. J.* **2004**, *3*, 5. [CrossRef] [PubMed]
2. Vallée Marcotte, B.; Verheyde, M.; Pomerleau, S.; Doyen, A.; Couillard, C. Health benefits of apple juice consumption: A review of interventional trials on humans. *Nutrients* **2022**, *14*, 821. [CrossRef] [PubMed]
3. Koutsos, A.; Riccadonna, S.; Ulaszewska, M.; Franceschi, P.; Trošt, K.; Galvin, A.; Braune, T.; Fava, F.; Perenzoni, D.; Mattivi, F.; et al. Two apples a day lower serum cholesterol and improve cardiometabolic biomarkers in mildly hypercholesterolemic adults: A randomized, controlled, crossover trial. *Am. J. Clin. Nutr.* **2019**, *111*, 307–318. [CrossRef] [PubMed]
4. Liddle, D.; Lin, X.; Cox, L.; Ward, E.; Ansari, R.; Wright, A.; Robinson, L. Daily apple consumption reduces plasma and peripheral blood mononuclear cell–secreted inflammatory biomarkers in adults with overweight and obesity: A 6-week randomized, controlled, parallel-arm trial. *Am. J. Clin. Nutr.* **2021**, *114*, 752–763. [CrossRef]
5. Łata, B.; Trampczynska, A.; Paczesna, J. Cultivar variation in apple peel and whole fruit phenolic composition. *Sci. Hortic.* **2009**, *121*, 176–181. [CrossRef]
6. Patocka, J.; Bhardwaj, K.; Klimova, B.; Nepovimova, E.; Wu, Q.; Landi, M.; Kuca, K.; Valis, M.; Wu, W. *Malus domestica*: A Review on nutritional features, chemical composition, traditional and medicinal value. *Plants* **2020**, *9*, 1408. [CrossRef] [PubMed]
7. Wojdyło, A.; Oszmiański, J.; Laskowski, P. Polyphenolic compounds and antioxidant activity of new and old apple varieties. *J. Agric. Food Chem.* **2008**, *56*, 6520–6530. [CrossRef] [PubMed]
8. Aprikian, O.; Busserolles, J.; Manach, C.; Mazur, A.; Morand, C.; Davicco, M.-J.; Besson, C.; Rayssiguier, Y.; Rémésy, C.; Demigné, C. Lyophilized apple counteracts the development of hypercholesterolemia, oxidative stress, and renal dysfunction in obese zucker rats. *J. Nutr.* **2002**, *132*, 1969–1976. [CrossRef] [PubMed]
9. Sekhon-Loodu, S.; Catalli, A.; Kulka, M.; Wang, Y.; Shahidi, F.; Rupasinghe, H. Apple flavonols and N-3 polyunsaturated fatty acid–rich fish oil lowers blood c-reactive protein in rats with hypercholesterolemia and acute inflammation. *Nutr. Res.* **2014**, *34*, 535–543. [CrossRef]
10. Soleti, R.; Trenteseaux, C.; Fizanne, L.; Coué, M.; Hilairet, G.; Kasbi-Chadli, F.; Mallegol, P.; Chaigneau, J.; Boursier, J.; Krempf, M.; et al. Apple supplementation improves hemodynamic parameter and attenuates atherosclerosis in high-fat diet-fed apolipoprotein e-knockout mice. *Biomedicines* **2020**, *8*, 495. [CrossRef] [PubMed]
11. Marzo, F.; Milagro, F.; Barrenetxe, J.; Díaz, M.; Martínez, J. Azoxymethane-induced colorectal cancer mice treated with a polyphenol-rich apple extract show less neoplastic lesions and signs of cachexia. *Foods* **2021**, *10*, 863. [CrossRef]
12. Skinner, R.; Warren, D.; Naveed, M.; Agarwal, G.; Benedito, V.; Tou, J. Apple pomace improves liver and adipose inflammatory and antioxidant status in young female rats consuming a western diet. *J. Funct. Foods* **2019**, *61*, 103471. [CrossRef]
13. Yeganeh, P.; Leahy, J.; Spahis, S.; Patey, N.; Desjardins, Y.; Roy, D.; Delvin, E.; Garofalo, C.; Leduc-Gaudet, J.-P.; St-Pierre, D.; et al. Apple peel polyphenols reduce mitochondrial dysfunction in mice with dss-induced ulcerative colitis. *J. Nutr. Biochem.* **2018**, *57*, 56–66. [CrossRef]
14. Lee, Y.-C.; Cheng, C.-W.; Lee, H.-J.; Chu, H.-C. Apple polyphenol suppresses indomethacin-induced gastric damage in experimental animals by lowering oxidative stress status and modulating the MAPK signaling pathway. *J. Med. Food* **2017**, *20*, 1113–1120. [CrossRef] [PubMed]
15. Sharma, S.; Rana, S.; Patial, V.; Gupta, M.; Bhushan, S.; Padwad, Y.S. Antioxidant and hepatoprotective effect of polyphenols from apple pomace extract via apoptosis inhibition and Nrf2 activation in mice. *Hum. Exp. Toxicol.* **2016**, *35*, 1264–1275. [CrossRef] [PubMed]
16. Ryter, S. Heme Oxygenase-1: An anti-inflammatory effector in cardiovascular, lung, and related metabolic disorders. *Antioxidants* **2022**, *11*, 555. [CrossRef] [PubMed]
17. Tonelli, C.; Chio, I.; Tuveson, D. Transcriptional regulation by Nrf2. *Antioxid. Redox Signal.* **2018**, *29*, 1727–1745. [CrossRef]
18. Saha, S.; Buttari, B.; Panieri, E.; Profumo, E.; Saso, L. An overview of Nrf2 signaling pathway and its role in inflammation. *Molecules* **2020**, *25*, 5474. [CrossRef]
19. Dinkova-Kostova, A.; Wang, X. Induction of the Keap1/Nrf2/ARE pathway by oxidizable diphenols. *Chem. Biol. Interact.* **2011**, *192*, 101–106. [CrossRef] [PubMed]
20. Osorio, L.L.D.R.; Flórez-López, E.; Grande-Tovar, C.D. The potential of selected agri-food loss and waste to contribute to a circular economy: Applications in the food, cosmetic and pharmaceutical industries. *Molecules* **2021**, *26*, 515. [CrossRef] [PubMed]
21. Abdhul, K.; Ganesh, M.; Shanmughapriya, S.; Kanagavel, M.; Anbarasu, K.; Natarajaseenivasan, K. Antioxidant activity of exopolysaccharide from probiotic strain enterococcus faecium (BDU7) from Ngari. *Int. J. Biol. Macromol.* **2014**, *70*, 450–454. [CrossRef] [PubMed]
22. Mazzola, M. Elucidation of the microbial complex having a causal role in the development of apple replant disease in washington. *Phytopathology* **1998**, *88*, 930–938. [CrossRef] [PubMed]
23. Zheng, H.-Z.; Kim, Y.-I.; Chung, S.-K. A profile of physicochemical and antioxidant changes during fruit growth for the utilisation of unripe apples. *Food Chem.* **2012**, *131*, 106–110. [CrossRef]

24. Sun, L.; Liu, D.; Sun, J.; Yang, X.; Fu, M.; Guo, Y. Simultaneous separation and purification of chlorogenic acid, epicatechin, hyperoside and phlorizin from thinned young qinguan apples by successive use of polyethylene and polyamide resins. *Food Chem.* **2017**, *230*, 362–371. [CrossRef] [PubMed]

25. Liu, T.; Shen, H.; Wang, F.; Zhou, X.; Zhao, P.; Yang, Y.; Guo, Y. Thinned-young apple polyphenols inhibit halitosis-related bacteria through damage to the cell membrane. *Front. Microbiol.* **2022**, *12*, 745100. [CrossRef]

26. Azuma, T.; Osada, K.; Aikura, E.; Imasaka, H.; Handa, M. Anti-obesity effect of dietary polyphenols from unripe apple in rats. *Nippon Shokuhin Kagaku Kogaku Kaishi* **2013**, *60*, 184–192. [CrossRef]

27. Baron, G.; Altomare, A.; Mol, M.; Garcia, J.; Correa, C.; Raucci, A.; Mancinelli, L.; Mazzotta, S.; Fumagalli, L.; Trunfio, G.; et al. Analytical profile and antioxidant and anti-inflammatory activities of the enriched polyphenol fractions isolated from bergamot fruit and leave. *Antioxidants* **2021**, *10*, 141. [CrossRef]

28. Wojdyło, A.; Oszmiański, J. Antioxidant activity modulated by polyphenol contents in apple and leaves during fruit development and ripening. *Antioxidants* **2020**, *9*, 567. [CrossRef]

29. López, V.; Les, F.; Mevi, S.; Nkuimi Wandjou, J.G.; Cásedas, G.; Caprioli, G.; Maggi, F. Phytochemicals and enzyme inhibitory capacities of the methanolic extracts from the italian apple cultivar *mela rosa dei monti sibillini*. *Pharmaceuticals* **2020**, *13*, 127. [CrossRef]

30. Fernandes, P.; Ferreira, S.; Bastos, R.; Ferreira, I.; Cruz, M.; Pinto, A.; Coelho, E.; Passos, C.; Coimbra, M.; Cardoso, S.; et al. Apple pomace extract as a sustainable food ingredient. *Antioxidants* **2019**, *8*, 189. [CrossRef]

31. Sánchez-Rabaneda, F.; Jáuregui, O.; Lamuela-Raventós, R.; Viladomat, F.; Bastida, J.; Codina, C. Qualitative analysis of phenolic compounds in apple pomace using liquid chromatography coupled to mass spectrometry in tandem mode. *Rapid Commun. Mass Spectrom.* **2004**, *18*, 553–563. [CrossRef] [PubMed]

32. Mcdougall, G.; Foito, A.; Dobson, G.; Austin, C.; Sungurtas, J.; Su, S.; Wang, L.; Feng, C.; Li, S.; Wang, L.; et al. Glutathionyl-S-Chlorogenic acid is present in fruit of vaccinium species, potato tubers and apple juice. *Food Chem.* **2020**, *330*, 127227. [CrossRef] [PubMed]

33. da Silva, L.; Souza, M.; Sumere, B.; Silva, L.; da Cunha, D.; Barbero, G.; Bezerra, R.; Rostagno, M. Simultaneous extraction and separation of bioactive compounds from apple pomace using pressurized liquids coupled on-line with solid-phase extraction. *Food Chem.* **2020**, *318*, 126450. [CrossRef] [PubMed]

34. Bars-Cortina, D.; Macià, A.; Iglesias, I.; Garanto, X.; Badiella, L.; Motilva, M.-J. Seasonal variability of the phytochemical composition of new red-fleshed apple varieties compared with traditional and new white-fleshed varieies. *J. Agric. Food Chem.* **2018**, *66*, 10011–10025. [CrossRef]

35. Groth, S.; Budke, C.; Neugart, S.; Ackermann, S.; Kappenstein, F.-S.; Daum, D.; Rohn, S. Influence of a selenium biofortification on antioxidant properties and phenolic compounds of apples (*Malus Domestica*). *Antioxidants* **2020**, *9*, 187. [CrossRef] [PubMed]

36. Gorjanović, S.; Micić, D.; Pastor, F.; Tosti, T.; Kalušević, A.; Ristić, S.; Zlatanović, S. Evaluation of apple pomace flour obtained industrially by dehydration as a source of biomolecules with antioxidant, antidiabetic and antiobesity effects. *Antioxidants* **2020**, *9*, 413. [CrossRef] [PubMed]

37. Hollands, W.J.; Voorspoels, S.; Jacobs, G.; Aaby, K.; Meisland, A.; Garcia-Villalba, R.; Tomas-Barberan, F.; Piskula, M.K.; Mawson, D.; Vovk, I.; et al. Development, validation and evaluation of an analytical method for the determination of monomeric and oligomeric procyanidins in apple extracts. *J. Chromatogr. A* **2017**, *1495*, 46–56. [CrossRef]

38. Lee, J.; Chan, B.; Mitchell, A. Identification/Quantification of free and bound phenolic acids in peel and pulp of apples (*Malus Domestica*) using High Resolution Mass Spectrometry (HRMS). *Food Chem.* **2017**, *215*, 301–310. [CrossRef]

39. Wen, C.; Wang, D.; Li, X.; Huang, T.; Huang, C.; Hu, K. Targeted isolation and identification of bioactive compounds lowering cholesterol in the crude extracts of crabapples Using UPLC-DAD-MS-SPE/NMR based on pharmacology-guided PLS-DA. *J. Pharm. Biomed. Anal.* **2018**, *150*, 144–151. [CrossRef]

40. Bestwick, C.; Scobbie, L.; Milne, L.; Duncan, G.; Cantlay, L.; Russell, W. Fruit-based beverages contain a wide range of phytochemicals and intervention targets should account for the individual compounds present and their availability. *Foods* **2020**, *9*, 891. [CrossRef] [PubMed]

41. Sut, S.; Zengin, G.; Maggi, F.; Malagoli, M.; Dall'acqua, S. Triterpene acid and phenolics from ancient apples of friuli venezia giulia as nutraceutical ingredients: Lc-ms study and in vitro activities. *Molecules* **2019**, *24*, 1109. [CrossRef] [PubMed]

42. Yousefi-Manesh, H.; Dehpour, A.; Ansari-Nasab, S.; Hemmati, S.; Sadeghi, M.; Shahraki, R.; Shirooie, S.; Nabavi, S.; Nkuimi Wandjou, J.; Sut, S.; et al. Hepatoprotective effects of standardized extracts from an ancient Italian apple variety (*Mela Rosa Dei Monti Sibillini*) against carbon tetrachloride (CCl4)-Induced hepatotoxicity in Rats. *Molecules* **2020**, *25*, 1816. [CrossRef]

43. Liang, X.; Zhu, T.; Yang, S.; Li, X.; Song, B.; Wang, Y.; Lin, Q.; Cao, J. Analysis of phenolic components and related biological activities of 35 apple (*Malus Pumila Mill.*) cultivars. *Molecules* **2020**, *25*, 4153. [CrossRef] [PubMed]

44. Kim, I.; Ku, K.-H.; Jeong, M.-C.; Kwon, S.-I.; Lee, J. metabolite profiling and antioxidant activity of 10 new early- to mid-season apple cultivars and 14 traditional cultivars. *Antioxidants* **2020**, *9*, 443. [CrossRef] [PubMed]

45. Baron, G.; Ferrario, G.; Marinello, C.; Carini, M.; Morazzoni, P.; Aldini, G. Effect of extraction solvent and temperature on polyphenol profiles, antioxidant and anti-inflammatory effects of red grape skin by-product. *Molecules* **2021**, *26*, 5454. [CrossRef]

46. Della Vedova, L.; Ferrario, G.; Gado, F.; Altomare, A.; Carini, M.; Morazzoni, P.; Aldini, G.; Baron, G. Liquid chromatography–high-resolution mass spectrometry (LC-HRMS) profiling of commercial enocianina and evaluation of their antioxidant and anti-inflammatory activity. *Antioxidants* **2022**, *11*, 1187. [CrossRef]

47. Wen, X.; Thorne, G.; Hu, L.; Joy, M.S.; Aleksunes, L.M. Activation of NRF2 signaling in HEK293 Cells by a first-in-class direct KEAP1-NRF2 inhibitor. *J. Biochem. Mol. Toxicol.* **2015**, *29*, 261–266. [CrossRef] [PubMed]

48. Ong, S.-E.; Mann, M. A practical recipe for stable isotope labeling by amino acids in cell culture (SILAC). *Nat. Protoc.* **2006**, *1*, 2650–2660. [CrossRef]

49. Deng, J.; Erdjument-Bromage, H.; Neubert, T. Quantitative comparison of proteomes using SILAC. *Curr. Protoc. Protein Sci.* **2018**, *95*, e74. [CrossRef]

50. Aiello, G.; Rescigno, F.; Meloni, M.; Baron, G.; Aldini, G.; Carini, M.; D'amato, A. Oxidative stress modulation by carnosine in scaffold free human dermis spheroids model: A proteomic study. *Int. J. Mol. Sci.* **2022**, *23*, 1468. [CrossRef] [PubMed]

51. Hulsen, T. BioVenn—An r and python package for the comparison and visualization of biological lists using area-proportional venn diagrams. *Data Sci.* **2021**, *4*, 51–61. [CrossRef]

52. Wang, Y.; Vorsa, N.; Harrington, P.; Chen, P. Nontargeted metabolomic study on variation of phenolics in different cranberry cultivars using UPLC-IM-HRMS. *J. Agric. Food Chem.* **2018**, *66*, 12206–12216. [CrossRef] [PubMed]

53. Jabeen, A.; Mesaik, M.A.; Simjee, S.U.; Lubna; Bano, S.; Faizi, S. Anti-TNF-α and anti-arthritic effect of patuletin: A rare flavonoid from *Tagetes patula*. *Int. Immunopharmacol.* **2016**, *36*, 232–240. [CrossRef] [PubMed]

54. Li, X.; Tian, Y.; Wang, T.; Lin, Q.; Feng, X.; Jiang, Q.; Liu, Y.; Chen, D. Role of the p-coumaroyl moiety in the antioxidant and cytoprotective effects of flavonoid glycosides: Comparison of astragalin and tiliroside. *Molecules* **2017**, *22*, 1165. [CrossRef] [PubMed]

55. Sun, L.; Guo, Y.; Fu, C.; Li, J.; Li, Z. Simultaneous separation and purification of total polyphenols, chlorogenic acid and phlorizin from thinned young apples. *Food Chem.* **2013**, *136*, 1022–1029. [CrossRef] [PubMed]

56. Wang, Y.; Gao, L.; Chen, J.; Li, Q.; Huo, L.; Wang, Y.; Wang, H.; Du, J. Pharmacological modulation of Nrf2/HO-1 signaling pathway as a therapeutic target of parkinson's disease. *Front. Pharmacol.* **2021**, *12*, 757161. [CrossRef] [PubMed]

57. Sánchez-Rodríguez, R.; Torres-Mena, J.E.; Quintanar-Jurado, V.; Chagoya-Hazas, V.; Rojas del Castillo, E.; del Pozo Yauner, L.; Villa-Treviño, S.; Pérez-Carreón, J.I. Ptgr1 expression is regulated by NRF2 in rat hepatocarcinogenesis and promotes cell proliferation and resistance to oxidative stress. *Free Radic. Biol. Med.* **2017**, *102*, 87–99. [CrossRef] [PubMed]

58. Kerins, M.; Ooi, A. The roles of NRF2 in modulating cellular iron homeostasis. *Antioxid. Redox Signal.* **2018**, *29*, 1756–1773. [CrossRef] [PubMed]

59. Forman, H.; Davies, K.; Ursini, F. How do nutritional antioxidants really work: Nucleophilic tone and para-hormesis versus free radical scavenging in vivo. *Free Radic. Biol. Med.* **2014**, *66*, 24–35. [CrossRef] [PubMed]

60. Potter, G.; Patterson, L.; Wanogho, E.; Perry, P.; Butler, P.; Ijaz, T.; Ruparelia, K.; Lamb, J.; Farmer, P.; Stanley, L.; et al. The cancer preventative agent resveratrol is converted to the anticancer agent piceatannol by the cytochrome P450 enzyme CYP1B1. *Br. J. Cancer* **2002**, *86*, 774–778. [CrossRef] [PubMed]

61. Xie, Y.; Zhang, D.; Zhang, J.; Yuan, J. Metabolism, transport and drug–drug interactions of silymarin. *Molecules* **2019**, *24*, 3693. [CrossRef] [PubMed]

62. Wardyn, J.; Ponsford, A.; Sanderson, C. Dissecting molecular cross-talk between Nrf2 and NF-KB response pathways. *Biochem. Soc. Trans.* **2015**, *43*, 621–626. [CrossRef] [PubMed]

63. Ahmed, S.; Luo, L.; Namani, A.; Wang, X.; Tang, X. Nrf2 Signaling pathway: Pivotal roles in inflammation. *Biochim. Biophys. Acta. Mol. Basis. Dis.* **2017**, *1863*, 585–597. [CrossRef] [PubMed]

64. Biswas, S.K. Does the interdependence between oxidative stress and inflammation explain the antioxidant paradox? *Oxid. Med. Cell. Longev.* **2016**, *2016*, 7432797. [CrossRef] [PubMed]

65. Lee, T.-S.; Chau, L.-Y. Heme Oxygenase-1 mediates the anti-inflammatory effect of interleukin-10 in mice. *Nat. Med.* **2002**, *8*, 240–246. [CrossRef] [PubMed]

66. Vijayan, V.; Wagener, F.; Immenschuh, S. The macrophage heme-heme Oxygenase-1 system and its role in inflammation. *Biochem. Pharmacol.* **2018**, *153*, 159–167. [CrossRef] [PubMed]

67. Campbell, N.; Fitzgerald, H.; Dunne, A. Regulation of inflammation by the antioxidant haem oxygenase 1. *Nat. Rev. Immunol.* **2021**, *21*, 411–425. [CrossRef]

68. Che, J.; Yang, J.; Zhao, B.; Shang, P. HO-1: A new potential therapeutic target to combat osteoporosis. *Eur. J. Pharmacol.* **2021**, *906*, 174219. [CrossRef]

69. Chau, L.-Y. Heme Oxygenase-1: Emerging target of cancer therapy. *J. Biomed. Sci.* **2015**, *22*, 22. [CrossRef] [PubMed]

70. Bolisetty, S.; Zarjou, A.; Agarwal, A. Heme Oxygenase 1 as a therapeutic target in acute kidney injury. *Am. J. Kidney Dis.* **2017**, *69*, 531–545. [CrossRef]

71. Tang, Z.; Ju, Y.; Dai, X.; Ni, N.; Liu, Y.; Zhang, D.; Gao, H.; Sun, H.; Zhang, J.; Gu, P. HO-1-Mediated ferroptosis as a target for protection against retinal pigment epithelium degeneration. *Redox Biol.* **2021**, *43*, 101971. [CrossRef]

72. Alam, J.; Cook, J. How many transcription factors does it take to turn on the heme Oxygenase-1 gene? *Am. J. Respir. Cell Mol. Biol.* **2007**, *36*, 166–174. [CrossRef] [PubMed]

antioxidants

Review

Pressurized Liquid Extraction for the Recovery of Bioactive Compounds from Seaweeds for Food Industry Application: A Review

Ana Perez-Vazquez [1,†], Maria Carpena [1,†], Paula Barciela [1], Lucia Cassani [1,2,*], Jesus Simal-Gandara [1,*] and Miguel A. Prieto [1,2,*]

1 Nutrition and Bromatology Group, Department of Analytical Chemistry and Food Science, Faculty of Science, Universidade de Vigo, E32004 Ourense, Spain
2 Centro de Investigação de Montanha (CIMO), Instituto Politécnico de Bragança, Campus de Santa Apolonia, 5300-253 Bragança, Portugal
* Correspondence: luciavictoria.cassani@uvigo.es (L.C.); jsimal@uvigo.es (J.S.-G.); mprieto@uvigo.es (M.A.P.)
† These authors contributed equally to this work.

Abstract: Seaweeds are an underutilized food in the Western world, but they are widely consumed in Asia, with China being the world's larger producer. Seaweeds have gained attention in the food industry in recent years because of their composition, which includes polysaccharides, lipids, proteins, dietary fiber, and various bioactive compounds such as vitamins, essential minerals, phenolic compounds, and pigments. Extraction techniques, ranging from more traditional techniques such as maceration to novel technologies, are required to obtain these components. Pressurized liquid extraction (PLE) is a green technique that uses high temperatures and pressure applied in conjunction with a solvent to extract components from a solid matrix. To improve the efficiency of this technique, different parameters such as the solvent, temperature, pressure, extraction time and number of cycles should be carefully optimized. It is important to note that PLE conditions allow for the extraction of target analytes in a short-time period while using less solvent and maintaining a high yield. Moreover, the combination of PLE with other techniques has been already applied to extract compounds from different matrices, including seaweeds. In this way, the combination of PLE-SFE-CO$_2$ seems to be the best option considering both the higher yields obtained and the economic feasibility of a scaling-up approximation. In addition, the food industry is interested in incorporating the compounds extracted from edible seaweeds into food packaging (including edible coating, bioplastics and bio-nanocomposites incorporated into bioplastics), food products and animal feed to improve their nutritional profile and technological properties. This review attempts to compile and analyze the current data available regarding the application of PLE in seaweeds to determine the use of this extraction technique as a method to obtain active compounds of interest for food industry application.

Keywords: pressurized liquid extraction; seaweeds; green extraction technique; bioactive compounds; functional ingredients; food packaging; future trends

Citation: Perez-Vazquez, A.; Carpena, M.; Barciela, P.; Cassani, L.; Simal-Gandara, J.; Prieto, M.A. Pressurized Liquid Extraction for the Recovery of Bioactive Compounds from Seaweeds for Food Industry Application: A Review. *Antioxidants* **2023**, *12*, 612. https://doi.org/10.3390/antiox12030612

Academic Editors: Antonella D'Anneo and Marianna Lauricella

Received: 31 December 2022
Revised: 14 February 2023
Accepted: 16 February 2023
Published: 1 March 2023

1. Introduction

1.1. Seaweeds as a Bioactive Compound Matrix

Seaweeds, also known as macroalgae, are eukaryotic, photosynthetic, pluricellular organisms found in the marine environment. They are divided into three groups: green (Chlorophyta), brown (Phaeophyta) and red algae (Rhodophyta). They are widely consumed in Asia, with China being the world's largest producer [1]. In recent decades, Western countries have become interested in seaweeds due to their high nutritional value [2]. Seaweeds are distinguished by their high quality profile of lipids, proteins, essential minerals, phenolic compounds and pigments. Several species, for example, such as *Palmaria palmata*, *Vertebrata lanosa* and *Enteromorpha intestinalis*, have been reported to have high quality profiles of essential amino acids and lipids when compared to other food matrices such as rice, corn, or wheat. Seaweed are an interesting matrix for the industry due to the bioactive compounds and the hydrocolloids found in them [3]. It is important to note that the species, harvest season, and eco-habitat are all factors that influence the composition of seaweed [2]. Aside from their nutritional value, some compounds found in seaweeds have different technological and biochemical properties that can be used in the food industry, either to improve the food process or to increase the nutritional value of food products. The industrial functionality as well as the principal components and bioactive properties of seaweed compounds will be briefly described to demonstrate the potential benefits of their incorporation into food products.

Seaweed carbohydrates (CH) are classified in hydrocolloids (which include carrageenan, alginates, fucoidans and laminarin) and phycocolloids (agar being the most relevant example). These carbohydrates typically account for 4–76% of the dried weight (DW), with *Ulva lactuca* reporting an 65% of DW as one of the major CH contents reported in seaweeds [4]. The lipidic content of these organisms is typically less than 5% [3], but they have been reported to present a high-quality profile of fatty acids (FAs), which may vary between 1–5% [5] and 1–8% [6]. For instance, the FA content of *Laurencia filiformis*, *Cystoseira baccata* and *Fucus vesiculosus* is 6.2%, 6.7% and 6.6%, respectively, which proves that algae genera influence the FA content. Moreover, brown and red algae have higher lipid content than green algae [7]. Despite having a lower FA content than microalgae, seaweeds are interesting to the food industry for several reasons. For example, seaweed treatment is simpler than microalgae treatment, and seaweeds contain a significant amount of unsaturated fatty acids (USFA) [7]. Protein values range between 3 and 47% with brown algae having the lowest DW of this compound. The most important seaweed protein for the food industry is lectin. Because lectin is a glycoprotein with carbohydrate-binding properties, it can agglutinate yeasts, tumor cells and erythrocytes. Because lectin has antimicrobial, antitumor and antiviral activity, it could be used as a functional ingredient [4]. The main micronutrients in algae are inorganic minerals and vitamins. These organisms are high in potassium, sodium, magnesium and calcium, as well as vitamins A, B and E, and particularly vitamin B12.

Certain organisms, including algae, produce secondary metabolites as defensive and/or adaptive responses to environmental stresses. Among these compounds, pigments and phenolic compounds are the most investigated. Regarding pigments, chlorophylls, carotenoids and phycobilins are the three major classes of photosynthetic pigments in seaweeds. Carotenoids have higher industrial value due to their ability to provide color naturally as well as their bioactive function. Carotenoids (β-carotene, lutein and astaxanthin) have antioxidant activity, immune system effects and can help to prevent cardiovascular diseases and non-alcoholic fatty liver diseases [8]. Fucoxanthin, the pigment found primarily in brown algae has attracted industry attention in the recent years due to a wide range of biological properties that may be of interest to the food and nutraceutical industries [8]. Seaweed polyphenols have also been studied, with phlorotannins being the most extensively studied. Phlorotannins are polymers and oligomers composed of several phloroglucinol units linked in different ways [9,10]. These highly hydrophilic secondary metabolites are only produced by brown algae and range in molecular size from 162 Da to 650 kDa [11,12]. Phlorotannins have been linked to several biological functions including antioxidant, anticoagulant, antibacterial, anti-inflammatory and anti-diabetic activities [12]. Therefore, the incorporation of these compounds in food formulations may be noteworthy.

1.2. Extraction Techniques Applied in Seaweeds Matrices

The first step in extracting target compounds from any matrix is extraction. A solvent that can penetrate the solid matrix is required for this process so the target compound can be dissolved in and extracted, and the solute of interest can be separated from the raw material [13]. Different extraction techniques have been used to release the previously mentioned target compounds. The extraction methodologies can be classified in two groups: traditional extraction techniques and new extraction techniques. Traditional methods are maceration, percolation and reflux extraction. In these techniques, an organic solvent is usually used, and large volumes and a long time of extraction are needed. These drawbacks motivated the search for new alternatives to traditional extraction techniques, which is usually related to the green revolution concept. Food technologies for preservation, processing, extraction and analysis have evolved from those conventional procedures to more innovative and environmentally friendly processes by reducing fossil energy use and hazardous solvents while avoiding water loss and residues generation. Therefore, the design of green and sustainable processes, and particularly, green extraction processes, remains a hot topic in the food industry [14].

Maceration is a solid-extraction technique characterized by its low cost and simple equipment requirements. The solubility of the target compound is determined by agitation and temperature in this technique. Moreover, the protocols used in this process are easily adaptable to obtain a wide range of compounds of interest. This is possible because different solvents, temperatures and agitation conditions can be selected, increasing mass transfer selectivity and efficiency. Unfortunately, several cycles of filtration or centrifugation are required to separate the compounds from the matrix [15]. Percolation is an extraction technique that works continuously. As a result, the saturated solvent is continuously replaced by a fresh one, increasing efficiency when compared to maceration. The operating conditions are typically room temperature and atmospheric pressure, but heating can also be used. Finally, compared to maceration and percolation, reflux extraction has a higher efficiency, despite the fact that the time of extraction and the amount of solvent required are slightly lower. Moreover, the operational parameters are atmospheric pressure and heating [13].

New extraction techniques are characterized by shorter extraction times, lower operational temperatures, reduced solvent amount and process automation. Moreover, because of the previously mentioned benefits, these techniques are regarded as environmentally friendly. Different parameters should be optimized to obtain higher yields of the target analyte. Thus, in these new methodologies, common parameters to be optimized include solvent ratio, extraction solvent, extraction time, pH, temperature and particle size [16,17]. New methodologies for extracting seaweed compounds include ultrasound-assisted extraction (UAE), supercritical fluid extraction (SFE), microwave-assisted extraction (MAE), enzyme-assisted extraction (EAE) and pressurized liquid extraction (PLE).

UAE is based on the application of low-frequency (16–100 kHz) and high-power (8–20 W) waves to disrupt cells, releasing target analytes, accelerating the diffusion and increasing mass transfer [17]. This technique has been used successfully to extract pigments, phenolic compounds and carbohydrates from different seaweed species [18]. SFE is based on the use of solvents at pressures and temperatures above their critical points, so solvents are denser than gases but have comparable viscosity and intermediate diffusivities to liquids and gases. This method has been used to extract carotenoids, chlorophylls, PUFAs and polyphenols [17]. MAE is based on the heat produced by the direct interaction of electromagnetic waves (usually 2.45 GHz) with polar solvent molecules through dipole rotation and ionic conduction [17]. MAE has been used to extract several seaweed compounds, including carbohydrates and proteins [19]. EAE is a very selective and specific method since enzymes are used to degrade the cell wall of algae cells. For this process, the enzyme concentration and the optimal enzyme reaction conditions should be optimized to improve specificity and selectivity [17]. This extraction method has been used to extract phlorotannin, proteins or hydrocolloids from various seaweeds [20].

PLE is a green extraction technique that involves extracting analytes from a solid matrix using high temperature and pressure, typically between 50–20 °C and 3.5–20 MPa [21]. With these conditions, both solubility and mass transfer rates are increased, leading to a solvent diffusivity increment and, therefore, meliorating matrix kinetics [17]. In this way, an experimental design is needed so all the parameters can be selected to guarantee optimal operational conditions [22]. Moreover, PLE allows the use of several solvents, including green extraction solvents such as water, and a mixture of water with ionic liquids or eutectic solvents. When water is used as the solvent, this technique is also known as high-pressure solvent extraction

(HPSE), accelerated solvent extraction (ASE), enhanced solvent extraction, pressurized fluid extraction (PFE), pressurized hot solvent extraction (PHSE) or subcritical water extraction (SWE) [23]. When compared to the SFE technique, PLE operates without reaching the critical point of the liquid solvent and allows the use of broader range of solvents [21,22]. Moreover, PLE requires less extraction time than other traditional extraction techniques, such as Soxhlet extraction. In fact, the extraction time ranges from 5 to 20 min [24]. Furthermore, the use of PLE allows the achievement of higher yields, despite the fact that is not suitable for the extraction of thermolabile compounds and is not as selective as SFE [21,25]. However, PLE may be considered as a suitable green extraction technique to extract different bioactive compounds including polysaccharides, proteins and PUFAs since non-toxic solvents and high extraction yields are obtained. Considering the non-toxic solvents used, the extraction of these compounds with PLE may be an interesting pathway for the food industry.

The aim of this study was to compile, analyze and organize the available data on seaweeds components and the use of PLE as a potential extraction technique to obtain active compounds of interest to the food industry. Furthermore, the effects of combining PLE with other novel techniques for increasing extraction yield were revised. Finally, the use of different compounds extracted from edible seaweeds in the food industry was summarized to identify a potential new pathway in this production sector.

2. Matrix Components

This section compiles the main characteristics and interest of different compounds found in seaweeds. It is important to highlight that the information described is primarily aimed at the food industry, although pharmaceutical approximation was also considered. Moreover, the nutritional composition of different microalgae is compiled in Table 1 to provide comprehensive information.

Table 1. Bioactive compounds of microalgae, expressed as percentage dry weight, and their functional properties.

Compound	Algae	Dry Weight (%)	Functional Properties	Ref.
Polysaccharides	*Nannochloropsis oceanica* *Nannochloropsis oculata* *Nannochloropsis granulata* *Nannochloropsis limnetica* *Microchloropsis salina* *Microchloropsis gaditana*	8.33 6.4 14.9 10 17.8–36.2 21.7	- Prebiotic activity [1]* - Immuno-modulatory [1]* - Low blood sugar and lipid levels in vitro [1] *	[26–28]
Sulfated polysaccharides	*Arthrospira platensis* *Chlorella ellipsoidea* *Phaeodactylum* sp. *Schizochytrium* spp. *Thraustochytrium* spp.	5–14.6	-Antiviral - Anti-tumor - Anti-inflammatory - Immuno-modulatory	[29,30]
Lipids	*Nannochloropsis oceanica* *Nannochloropsis oculata* *Nannochloropsis granulate* *Nannochloropsis limnetica* *Microchloropsis salina* *Microchloropsis gaditana*	18.40–46.12 8.2 28.5 24 6.2–26 16.5	-	[27,31–36]

Table 1. *Cont.*

Compound	Algae	Dry Weight (%)	Functional Properties	Ref.
Essential fatty acids	*Nannochloropsis oceanica* *Nannochloropsis oculata* *Nannochloropsis limnetica* *Microchloropsis gaditana* *Phaeodactylum tricornutum* *Porosira glacialis* *Monodopsis subterranean* *Nannochloris* spp.	2.34 2.33 2.81 4.4–11 0.7–6.1 [3*]	- Reduce cardiovascular disease - Improve mental health - Anti-inflammatory - Anti-diabetes - Anti-thrombotic activity	[26,27,31–37]
(EPA)				
(ARA)	*Ceramium rubrum* *Parietochloris incisa* *Phormidium pseudoprsleyi* *Prphyridium purpureum*	24–77 [3*]	- Anti-inflammatory - Anti-cancer - Prevention of neurological disorders	[32,36]
(DHA)	*Amphidiunum carterae* *Aurantochytrium* spp. *Phaeodactylum tricurnutum* *Schizochytrium* spp. *Thraustochytrium* spp.	17.5–30.2 [3*]	- Decreasing preterm birth - Improving cognitive - Prevention of cardiovascular disease - Promotion of eye health - Slowing Alzheimer's disease	[36,38]
Proteins	*Nannochloropsis oceanica* *Nannochloropsis oculata* *Nannochloropsis granulata* *Nannochloropsis limnetica* *Microchloropsis salina* *Microchloropsis gaditana*	14.5 22.6 45.8 37 18.1–36.2 47	- Gelling properties [2*] - Foaming properties [2*]	[26,27,39]
Bioactive peptides	*Chlorella elipsoidea* *D. salina* *Nitzschia* sp. *Bellerochae* sp. *Tetraselmis suecica*	-	- Antihypertensive - Antibiotic - Antiviral	[40,41]
Phenolic compounds	*Chlorella vulgaris* *Nannochloropsis* sp. *Phaedactylum tricornutum* *Scenedesmus obliquus* *Tetraselmis* sp.	0.54–4.57	- Antioxidant - Anti-inflammatory - Antimicrobial - Anti-cancer - Prevention of cardiovascular and neurodegenerative diseases	[42,43]
Vitamins	*Arthrospira* *Chlorella* *Isochrysis galbana* *Pavlova* *Porphyridium cruentum* *Tetraselmis* sp.	-	- Blood coagulation - Modulating inflammation - Neuroprotection, promoting eye and bone health	[44,45]

Abbreviations: EPA (eicosapentaenoic acid); DHA (docosahexaenoic acid); ARA (arachidonic acid). [1*] is referred to *N. oculata* polysaccharides; [2*] is referred to *A. platensis* proteins; [3*] expressed as total fatty acids.

2.1. Proteins

Proteins are large molecules composed of smaller units known as amino acids that are linked together with aminoacidic bonds. Considering the amino acid profile of seaweeds, it is important to note that most of them contain all the essential amino acids, with aspartic and glutamic acid being particularly abundant. Figure 1 depicts the chemical structure of these amino acids found in edible seaweeds. *Ulva lactuca* (a green edible seaweed) has an amino acid profile that is like that recommended by the Food and Agricultural Organization (FAO) and the World Health Organization (WHO). Seaweed proteins are suitable for inclusion in food formulations due to their amino acid profile and agglutination properties [4]. Several authors have proposed producing antioxidant hydrolysates by hydrolyzing seaweed proteins. Researchers used commercial enzymes to hydrolyze *Ecklonia cava, Ishige okamurae, Sargassum fullvelum, Sargassum horneri, Sargassum corearum, Sargassum thunbergii* and *Pyropia columbina* to obtain bioactive peptides. *P. columbina* increased its antioxidant activity after a simulated gastrointestinal digestion [34]. Thus, since some studies have found drawbacks in the digestibility of proteins from some seaweeds, the use of cellulases, xylanases and β-glucanases has been studied to improve the digestibility of protein from

Palmaria palmata [15]. Furthermore, seaweed proteins such as phycobiliproteins have been linked with anti-inflammatory, hepatoprotective and antioxidant activities [46].

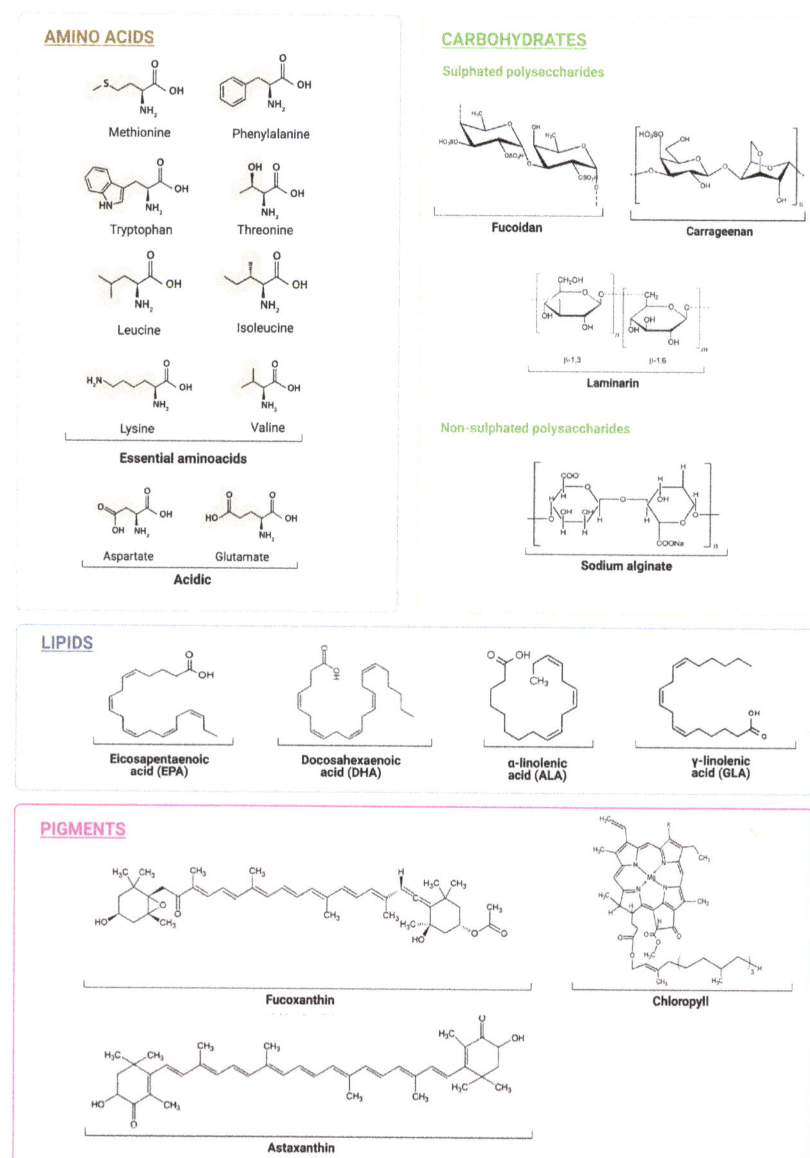

Figure 1. Chemical structure of the main amino acids, polysaccharides, lipids, and pigments extracted from edible seaweeds.

Moreover, a relationship has been discovered between seaweed bioactive peptides and metabolic syndrome. The metabolic syndrome is a collection of medical conditions that can lead to different cardiovascular diseases, and it can be avoided with functional foods. Indeed, NMEKGSSSVVSSRMKQ is the first antithrombotic peptide produced by hydrolyzing *Porphyra yezoensis* proteins with pepsin. This peptide binds to the coagulation pathway and inhibits it. Furthermore, seaweed-derived bioactive peptides inhibit some enzymes involved in the renin–angiotensin–aldosterone system (RAAS), which plays a key role in the hypertension treatment [4]. Microalgae proteins have been reported to be good gel and foam formers, with *Arthosphira platensis* being particularly notable. Moreover, microalgae are also an interesting source of proteins and bioactive peptides, as shown in Table 1. A study published in 2022 found that the protein composition of *Nannochloropsis granulata* and *Microchloropsis gaditana* was high, with 45.8 and 47% expressed as DW, respectively. Furthermore, studies revealed that microalgae peptides from different genera, including *Chlorella*, *Nitzschia* and *Bellerochae* are distinguished by their antihypertensive, antibiotic and antiviral activities [40,41]. Figure 2 depicts a graphical representation of different micro and macroalgae protein composition, except for *Nannochloropsis granulata*, *Nannochloropsis limnetica* and *Microchloropsis gaditana*, which had higher protein content. Based on the data presented, seaweeds are a suitable and complete source of proteins that can be incorporated into a variety food products.

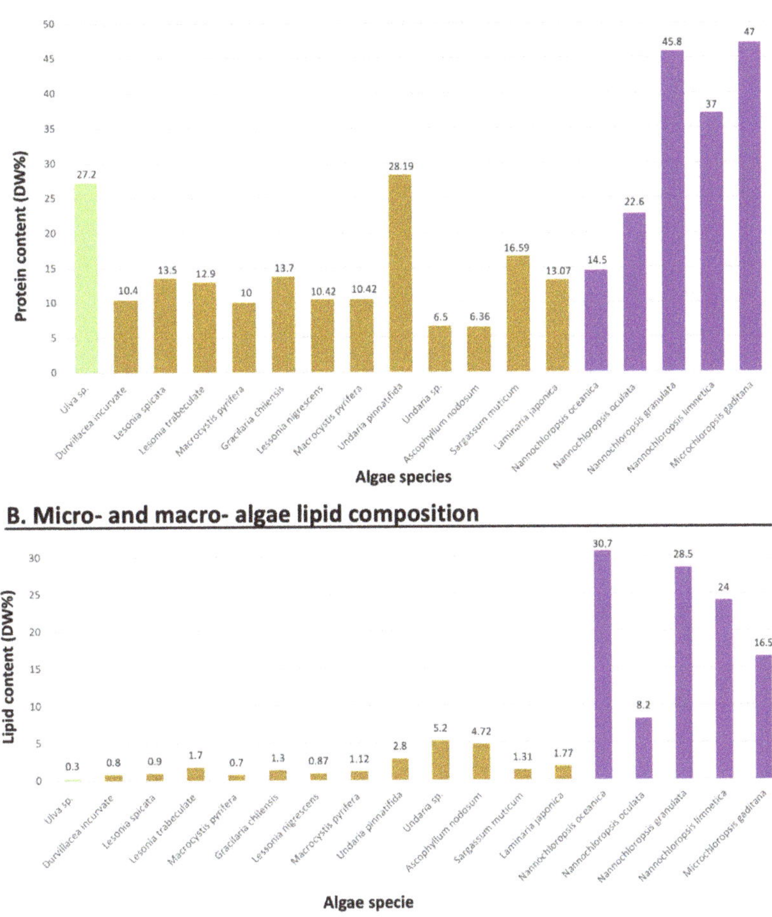

Figure 2. Graphical comparison between different microalgae and macroalgae protein (**A**) and lipid (**B**) composition, expressed as dried weight (%). Green color refers to green seaweeds, brown color refers to brown seaweeds and purple color refers to microalgae species.

2.2. Carbohydrates

Seaweed contains monosaccharides, disaccharides and polysaccharides. Polysaccharides found in seaweeds are classified into two groups: sulphated and non-sulphated. As is shown in Figure 1, sulphated polysaccharides include fucoidans, carrageenan and laminarin, while non-sulphated polysaccharides are mainly alginates [47]. Seaweed polysaccharides and fiber are not digestible by humans. Moreover, in vitro studies have shown that polysaccharides from *Undaria pinnatifida*, *Laminaria japonica* and *Hizkia fusiformis* inhibit pepsin activity by 21%, 55% and 41%, respectively [15]. Despite their inability to be digested, these compounds serve as prebiotics in the human body because they can be degraded by intestinal bacteria [3]. However, it is important to highlight that some studies have linked prebiotic consumption to human flatulence. In fact, 14 women were studied for 4 weeks after consuming inulin, and 12% of the volunteers experienced severe flatulence [48]. Moreover, because seaweeds are composed of soluble fiber with high capacity to retain water (CRW), they can be used as hydrocolloids in food formulations. Some of the functions of hydrocolloids include thickening, stabilizing and emulsifying [3].

Carrageenan is a sulphated linear galactant found in 71% and 88% of *Chondrus crispus* and *Kappaphycus* spp., respectively. Carrageenan gel form from *Chondrus crispus* has been shown to have antiviral and anticoagulant properties against HIV and herpes simplex virus (HSV). The anticoagulant capacity of this compound has been related to the sulphate molecules in the polysaccharide chains [4].

Alginates are non-sulphated linear unbranched polysaccharides found in the intercellular spaces of brown algae [4,49]. *Laminaria*, *Saccharina*, *Lessonia*, *Macrocystis*, *Durvillaea*, *Eckonia* and *Ascophyllum* are the main seaweeds used to obtain this polysaccharide. Alginates are used in the food industry due to their technological properties, as these compounds are characterized by their gelation ability and become insoluble as a result of the formation of a cross-linked structure. Because of this, alginate is an excellent material for the active edible coating systems on foods such as fruits and vegetables [50]. Furthermore, the Food and Drug Administration (FDA) designated alginic acid and its salts as generally regarded as safe (GRAS) ingredients for oral administration [4].

Laminarins (Figure 1) are the main storage polysaccharides found in the cytoplasm of brown algae [40]. This small glucan is isolated from brown seaweeds and has a molecular weight that ranges from 1 to 10 kDa [51]. Laminarin can be found in a variety of seaweed species, including *Laminaria* spp., *Ascophyllum*, *Fucus* and *Undaria* [4]. This compound is well known for its gelling and emulsifying properties and is commonly used in the food industry as an additive [51]. Moreover, several studies have shown that laminarin has anti-apoptotic, anti-inflammatory, immunoregulatory, antitumor, anticoagulant and antioxidant activities [52]. These biomedical properties may be related to the sulphated composition of seaweed polysaccharides, which are not found in the terrestrial plants [4]. When the average molecular weight of laminarin was reduced to six kDa, its antioxidant activity increased from 7.5 to 79.7%. This could be explained by the fact that as the average molecular weight decreases, the number of carbonyl groups increases, interacting with transition metal ions and enhancing lipid oxidation protection [51]. Moreover, laminarin has a positive effect on the biochemistry and microbiology of the human gut microflora, modulating the intestinal metabolism in a positive way [4].

Fucoidans are sulphated polysaccharides (SP) found in brown seaweeds such as *Saccharina japonica*, *Laminaria ochroleuca* and *Himanthalia elongata*. The anti-inflammatory, anticoagulant, antitumoral, anti-thrombotic, antioxidant and antiviral activities of fucoidans have been associated with their sulphation level [34,40]. An in vitro and in vivo study of *Fucus evanescens* fucoidan's anticoagulant activity revealed that it has similar anticoagulant properties to heparin [4]. The anticoagulant activity has been related to the sulphation level, carbohydrate content and the position of sulphated groups on sugar residues [53]. In addition, fucoidans may alter the cellular surface properties effectively preventing virus penetration, and as a result, antiviral activity [53]. A study using *Laminaria japonica* as a source of fucoidans showed its ability to scavenge superoxide radicals and hypochlorous acid. The low molecular fraction of *L. japonica* fucoidans also had a significant inhibitory effect on low-density lipid (LDL) oxidation induced by Cu^{2+} [53]. Thus, fucoidans may be used to prevent free radical-mediated diseases. Finally, the antitumor activity of fucoidans from brown seaweed was studied, and it was discovered that this sulphated polysaccharide could inhibit the proliferation of tumor cells by stimulating the apoptosis, blocking tumor cell metastasis, and enhancing immune response. These antitumor properties may lead to the use of fucoidans as functional ingredients or nutraceuticals [53]. Considering all the data presented and the increased interest of consumers in bioactive compounds in recent years, fucoidans may be considered as functional ingredients in the food industry [53].

The composition of microalgae polysaccharide (Table 1) has also been studied, showing immunomodulatory activity, low blood sugar and lipid levels in vitro and prebiotic activity [27,28]. Moreover, sulphated polysaccharides derived from *Arthosphira*, *Chlorella*, *Phaeodactylum*, *Schizochytrium* and *Thrautochytrium* showed antiviral, antitumor, and anti-inflammatory activity [29,30].

2.3. Lipids

The lipidic fraction in seaweeds varies between 1–8%, with the most common long-chain polyunsaturated fatty acids (PUFAs) found in seaweeds being γ-linolenic, α-linolenic, eicosapentaenoic and docosahexaenoic acid, whose chemical structure is shown in Figure 1. Because PUFAs have been linked to the prevention of cardiovascular, diabetes and hypertension diseases, their presence in seaweed has an interesting functional activity [3]. Moreover, the w-6/w-3 ratio is commonly used to define lipid quality [7]. According to the data, the w-6/w-3 ratio is 15/1 in Western diets, while the FAO recommends a ratio of less than 10. Because a high w-6/w-3 ratio is associated with the progression of various coronary diseases, the WHO recommends substituting saturated fatty acids for polyunsaturated fatty acids [6]. Furthermore, the atherogenic index (AI) and thrombotic index (TI) are parameters that indicate the lipid deposition in the artery wall as well as the thrombotic effect of saturated fatty acids (SFAs), respectively. Both indices are widely used to assess the quality of SFAs. In 2019, a study on the bioactive fatty acids extracted from *Laminaria ochroleuca* was conducted, with AI and TI results comparable to those obtained from some fish species [7]. It is important to note that the fatty acids in seaweeds are in a phospholipid and glycolipid form, which confer the cell wall membrane. This allows for very little degradation during digestion. Hence, a mechanical disruption of the cell wall is needed in order for the seaweed lipid content to be released and absorbed [3]. To demonstrate the differences between microalgae and macroalgae species, the lipid composition of each was compared. As is shown in Figure 2, the lipid composition of microalgae is richer than that of seaweed. In fact, the lipid composition of microalgae varies depending on the genus (Table 1), reaching 46.12% of DW in some species of *Nannochloropsis*. Moreover, the lipid composition of microalgae is rich in PUFAs such as EPA, ARA and DHA, which has been linked to different biological properties such as anti-inflammatory, anti-diabetes, antithrombotic and anticancer activities, as well as a high capacity to prevent cardiovascular diseases [31–33,38]. Although the lipid content of microalgae is higher than that of seaweed, based on the data presented, seaweed may be a valuable source of quality fatty acids for the food industry, capable of being used in vegetarian and vegan formulations.

2.4. Pigments

Chlorophyll (Figure 1) is one of the most abundant pigments on the planet, and it has mainly been studied in higher plants [54]. The significance of this phytochemical is due to its antimutagenic and antigenotoxic activity [54] as well as its potential use in the food industry as a natural pigment. A study conducted in 2017 identified the chlorophyll profile, distinguishing over 31 pigments. Although extraction protocols for these pigments in seaweeds are not well developed [55], the abundance of these phytochemicals in edible seaweeds makes this matrix a good source of them. Moreover, the stability of chlorophyll in fresh and cooked *Porphyra* seaweeds was studied during in vitro digestion. The bioaccessibility of cooked *Laminaria* chlorophyll seaweed improved, while processing decreased this parameter [54].

Fucoxanthin (FUCO) is a carotenoid found in the glycoglycerolipids of brown algae chloroplasts and is involved in the photosynthesis process [8,56]. FUCO accounts for 10% of total carotenoids in nature and has a market with a 2.47% annual growth rate. This pigment is present in several algae genera, with *Undaria pinnatifida* having the highest concentration. The main bioactive properties of FUCO are the antioxidant, anticancer and anti-inflammatory activity, as well as the cytoprotective and skin protective effects [8]. The antioxidant effect of this compound is explained by the presence of an allenic bond and an acetyl functional group in its structure (Figure 1), both of which can scavenge different free radicals. Thus, it has been demonstrated that FUCO reduces the production of intracellular ROS and DNA damage while increasing glutathione levels, which is a key molecule in oxidant defense and the maintenance of the redox cell homeostasis. All these actions contribute to the prevention of apoptotic processes [8]. The skin protective effect of FUCO has also been studied using oral administration. Results showed that this compound suppresses transcription of the melanogenesis factor by inhibiting mRNA expression. Therefore, FUCO could be used to prevent harmful effects of ultraviolet (UV) radiation, such as melanomas [8]. The ideal conditions for FUCO incorporation into food formulations have already been studied, with the conclusion that an encapsulation should be made with a solution with pH 5–7 and it should be stored at 4 °C [8].

2.5. Metals

As previously stated, seaweeds are a rich source of bioactive compounds that can be used in the food industry. However, seaweeds also contain significant amounts of metals. In fact, seaweed consumption has been considered as a high-risk route for heavy metals and metalloids due to their high capacity to bioaccumulate these compounds. Thus, the main metals found in these matrices are lead (Pb), cadmium (Cd), mercury (Hg), and arsenic (As) [57]. Red algae are high in selenium (Se), manganese (Mn), nickel (Ni) and silver (Ag), while brown seaweeds are high in copper (Cu), cobalt (Co), chromium (Cr), As and iron (Fe). Finally, zinc and Pb are commonly found in green seaweeds [58]. These metals are toxic and persistent, and their consumption may cause endocrine disruption and carcinogenic activity [59]. However, the potential risk of consuming seaweeds grown in Saint Martin's Island was studied, which is potentially a risk zone for heavy metal accumulation. A total of 21 heavy metals and metalloids were analyzed, and no health risk was found because bioaccumulation was below the established limits (Hazard Quantities < 1) [57]. Another study analyzed 11 species of seaweeds grown in South China Sea, showing a high degree of variability and complexity [58]. In this way, considering the importance of ensuring the consumers' health, heavy metal and metalloid analysis of the species used is required.

3. Pressurized Liquid Extraction (PLE)

3.1. General Aspects of PLE

PLE is an extraction technique that consists of the removal of analytes present in a solid matrix by applying high temperatures (T_{extr}) and pressure (P_{extr}), usually up to 200 °C and over 200 bar, respectively according to Nieto et al. [22], without reaching the critical point using liquid solvents [23]. These conditions increase solubility and mass transfer rates, resulting in increased solvent diffusivity and, as a result, improved matrix kinetics [17].

Temperature, pressure, time, number of cycles, sample weight and solvent all influence extraction yield. To improve the efficacy of PLE, these parameters should be optimized by using a proper experimental design [22].

Figure 3 depicts a schematic representation of the PLE extraction equipment's operation. A high-pressure pump feeds the solvent into an extraction cell and the P_{extr} in the system is kept constant [24,60]. Because operational T_{extr} and P_{extr} control is critical in this method, the extraction cell is kept in an oven with different valves and restrictors [60]. Moreover, an extract cooler system, a back pressure regulator and a vial to collect the extract are required [24]. Finally, it is important to keep in mind that the equipment must be resistant to corrosion and high pressure [24].

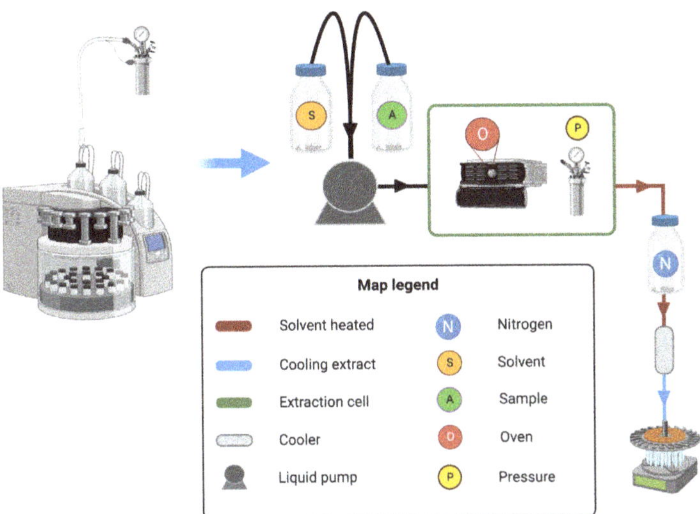

Figure 3. Pressurized liquid extraction equipment and schematic representation of its operation. First, the solvent (S) and the sample (A) are injected into the extraction cell. The extraction cell is composed of an oven (O) and a pressure valve (P) which together allow the achievement of the temperature and pressure selected to extract the compound present in the sample. Then, the extracted compound is cooled and collected in a carousel.

PLE can be used in three modes of action: static mode, dynamic mode and a combination of the two. The static mode is characterized by the use of constant temperature and pressure values, resulting in the sample being in contact with the solvent for a set period. On the contrary, in the dynamic mode, the solvent (usually water) flows continuously through the sample. As a result of the higher volume of the extract obtained, the analytes are diluted in the liquid extract. Analytes are typically pre-concentrated by liquid–liquid extraction or by solid-phase extraction to address this issue. Finally, a combination of both modes of action can be used, which may improve analyte extraction [23].

3.2. Sample Pre-Treatments

Before using PLE, samples must be pre-treated to increase the contact surface between the solvent and the matrix during the extraction [23]. Pre-treatment can be compiled into four steps, as explained below and in Figure 4:

— Drying: the objective of this step is to remove water from the matrix, increasing extraction efficiency [23]. Air-drying, oven heating or lyophilization can all be used, with the latter being the most advantageous because it takes less time and does not degrade the compounds. Indeed, when non-polar solvents are used, this step is critical and a desiccant is commonly included in the PLE cell [22].
— Homogenization: by grinding, the sample should be distributed in a homogeneous manner [22].
— Sieving: this step increases the surface area of the analyte as well as the diffusion of the analyte from the matrix to the solvent [23]. This step yielded a similar particle size in which 2 mm is commonly used for PLE [22]. After sieving you can carry out grinding of the separated portion with greater particle size.
— Dispersion with an inert material: this step is recommended for some samples to avoid aggregation of particles that may lead to alteration in the extraction efficiency [23].

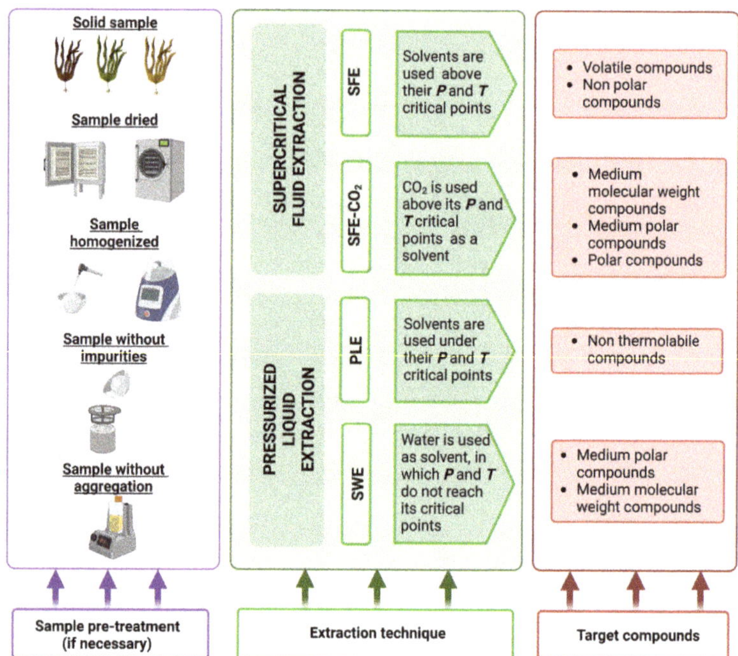

Figure 4. Pre-treatment steps, extraction techniques and target compounds from seaweeds using PLE and SWE. In purple, a schematic representation of the steps that should be followed to prepare the sample before the extraction technique is applied. In green, PLE and SPE operational conditions considering water and CO_2 as solvents, respectively. In red, a comparison between the compounds extracted using each extraction technique.

3.3. Relevant Parameters in PLE

3.3.1. Solvent

One of the most important parameters to optimize is the extraction solvent [22]. The solvent's function is to solubilize the target analytes while minimizing the extraction of other components [23]. Therefore, it is important to choose a solvent that has the same polar behavior of the target analytes [22]. Non-polar and water immiscible solvents or a combination of non-polar with medium-polarity solvents are used for non-polar or lipophilic compound extraction. Consequently, solvents with high polarity are used to extract polar and hydrophilic compounds. Finally, when extracting analytes with different polarities, a mixture of solvents with high and low polarity is commonly used. Indeed, some authors suggest following two PLE extractions when the target is for high and low polar analytes, so they can be removed in two steps [23]. As a result, because affinity and miscibility are the two parameters used in predictive approaches to determine the solubility of the target compound in green solvents at different temperatures, experimental trials may be limited [24].

Regarding the application of PLE on seaweeds, using water as a solvent is the most common green technique applied for the carbohydrates extraction since they are more soluble in water at 100–150 °C and the dielectric constant of water is reduced at this temperature. Moreover, subcritical water acts as an acid or an alkali, helping the polysaccharides extraction [47]. A study conducted in 2022 in which PLE was optimized by varying different parameters, showed that temperature was the most critical for the extraction of carbohydrates in microalgae, which can be extrapolated to seaweeds. Furthermore, bioactive polysaccharides extracted from seaweeds using PLE with water are not degraded because temperatures are kept below 200 °C, avoiding caramelization and other degradation reactions [47]. Although water is a good green solvent, it is important to consider that its use may result in unwanted reactions or interference coextraction, affecting the procedure's selectivity [17]. Therefore, other eco-friendly solvent alternatives, such as deep eutectic solvents, are being considered for carbohydrate extraction. Deep eutectic solvents (DES) are eutectic mixtures composed of hydrogen bonding acceptors (HBAs) and hydrogen bonding donors (HBDs) [61]. Due to their stability, cost-competitiveness, and ease of synthesis, DES have been proposed to dissolve different polysaccharides such as cellulose, starch, chitin and lignin for biomass processing. Moreover, DES was recently used as a solvent to extract fucoidans and alginates from brown algae. The results of this study showed that DES functioned as a catalyst, yielding twice as much as acidified water extraction. Table 2 summarizes the information explained above by showing different solvents used in PLE extraction, and their target analytes.

Table 2. Solvents used in PLE for the extraction of different compounds.

Solvent	Extracted Compound	Ref.
Water	Phenolic compounds//Di-, triterpenes//Proteins//Polysaccharides	[24,62]
Water + ionic liquids	Carrageenan//Alginates	[24,63]
Water + eutectic solvents	Carrageenan//Alginates	[24,63]
Ethanol	Polyphenols//Carotenoids//Alkaloids//Lipids	[24,64–66]
Aqueous ethanol	Polyphenols//Carotenoids//Alkaloids//Lipids	[24,64–66]
Ethyl acetate	Polyphenols//Carotenoids//Alkaloids//Lipids	[24,64–66]
Ethyl lactate	Fatty acids	[24,67,68]
(+)-limonene	Fatty acids	[24,67,68]

When selecting a solvent, it is also important to consider subsequent steps of the process, such as the clean-up step or concentration step if necessary. Selectivity is the parameter that determines whether or not purification and concentration are required, and it is critical when developing a green technique process [69]. Finally, the solvent used must be both physically and chemically stable. Water, ethanol, organic esters such as ethyl acetate and ethyl lactate, (+)-limonene and their mixtures are the most commonly used solvents in PLE [24].

The operational conditions used to extract bioactive compounds from seaweeds using PLE are shown in Table 3. In 2017, for example, a study on the accuracy of some green solvents with PLE for the fucoxanthin extraction was conducted. Limonene, ethyl lactate and ethyl acetate were selected as green solvents and their ability to extract fucoxanthin was compared to that of ethanol [69,70]. The highest yields were obtained for each solvent when the operating temperature was set to 100 °C. None of the green solvents reached ethanol's yield, with ethyl lactate had the highest percentage. Despite the yield results, limonene had the highest selectivity (expressed as the ratio of total carotenoids to total chlorophylls), proving that limonene is a good alternative green solvent for fucoxanthin extraction.

3.3.2. Temperature and Pressure

As previously stated, T_{extr} and P_{extr} are two important parameters to be optimized when using PLE. Elevated temperatures are used to reduce the viscosity of the liquid solvent used, allowing it to a better wet the matrix, and solubilize the analytes of interest. In addition, diffusion of analytes in the matrix surface is facilitated because high temperatures aid in the breakdown of the analyte–matrix bonds [23]. T_{extr} varies between 50 and 200 °C and is dependent on the target analyte. Thus, lower T_{extr} are selected for extraction of certain bioactive compounds due to their thermolability. Because high temperatures above the atmospheric boiling point are required in order to keep the solvent liquid, a high operational P_{extr} is required [23]. Furthermore, high pressure increases the extraction yield by forcing the solvent to enter the matrix pores [60]. In PLE methodology, P_{extr} usually varies from 5.0 to 15 MPa. These high-pressure and temperature conditions allow for the extraction of the target analytes in a short period of time while using less solvent and showing a recovering ability in terms of extraction yield similar to other techniques [22].

Table 3. Operational conditions for the PLE extraction of different compounds from seaweeds.

Seaweed	Compound	Solvent	T (°C)	P (bar)	t (min)	*Yield* (%)	Ref.
Saccharina japonica (B)	Alginate	NaOH 0.1%; H₂Od; H₂O	80; 110; 140	5; 25; 50	5; 12	3–27.21	[49]
	Fucoidan	NaOH 0.1%; H₂Od; CH₂O₂ 0.1%	80; 110; 125; 150	5; 25; 50	5; 25	2.5–15.7	[49,62]
	TPC	H₂O	200	50	15	39.52	
Eisinia bicyclis (B)	Fucoxanthin	EtOH 90%	110	103.42	5	0.39	[71]
Laminaria ochroleuca (B)	Fatty acids	Hexane; ethyl acetate; EtOH; EtOH 50%	120	100	10	7.42–47.16	[7]
Sargassum muticum (B)	Phlorotannin	EtOH- H₂O 95:5; 75:25; 25:75.	120; 160	103.42	20	0.77–2.93	[61,72]
Sargassum thumbergii (B)	Phlorotannin	H₂O	180	30	30	1.35	[73]
Sargassum cristaelefolium (B)	Fucoidan	H₂O	121	1.0133	20	12.60	[74]
Kappaphycus alvarezii (R)	Carrageenan	H₂O	150	50	5	71	[75]
Ascophyllum nodosum (B)	TPC	H₂O; EtOH: H₂O 60:40; EtOH: H₂O 80:20; MeOH: H₂O 60:40	90; 100; 120	68.95	90 s	34.5- 114	[76]
Fucus serratus (B)	TPC	H₂O; EtOH: H₂O 60:40; EtOH: H₂O; 80:20; MeOH: H₂O 60:40;	90; 100; 120	68.95	90 s	19.7–40.7	[76]
Fucus vesiculosus (B)	TPC	H2O; EtOH: H₂O; 60:40; EtOH: H₂O; 80:20; MeOH: H₂O; 60:40	90; 100; 120	68.95	90 s	114.0–110	[76]
	Fatty acids	EtOH: H₂O; 50:50	120; 160	100	10	34.85–57.19	[77]
Ulva lactuca (G)	Fatty acids	EtOH: H₂O; 50:50	80; 120; 160	100	10	34.85; 41.49; 57.19	[77]
Ulva intestinalis (G)	Fatty acids	EtOH: H₂O; 50:50	80; 120; 160	100	10	34.85; 41.49; 57.19	[77]
Himanthalia elongata (B)	Fatty acids	EtOH: H₂O; 50:50	80; 120; 160	100	10	34.85; 41.49; 57.19	[77]

Abbreviations: (G): green algae; (R): reed algae; (B): brown algae. TPC (total phenol compounds), EtOH (ethanol), MeOH (methanol), H₂O (water), NaOH (sodium hydroxide).

3.3.3. Time and Number of Cycles

The time of extraction is defined as the duration of direct solvent contact with the sample for a given T_{extr} and P_{extr} [24]. This value is determined by a variety of factors, including the mode of action. When using static mode d, extraction time is reduced ($t_{extr} = 5$–20 min) [24]. On the contrary, when the dynamic mode is established, the flow of the solvent must be determined to select t_{extr}. Furthermore, it is significant to notice that low flows cause PLE system blockages while high flows result in diluted extracts. Finally, it is known that several cycles with low volume lead to higher yields of the target analyte, while a single extraction with a large amount of solvent does not correspond with higher extraction yields [24].

3.4. Post-Extraction Treatment (Clean-Up)

During PLE, some compounds of the matrix could be co-extracted causing interferences, so a clean-up step could be necessary to decrease the limit of detection (LOD) value [22]. Extraction and clean-up steps can be carried out simultaneously, which leads to a reduction in time and quantity of solvents used, between 15% and 52% [78]. Different clean-up techniques can be used:

- On-cell clean-up technique: the solvent is passed through the cell to elute interferences prior to the extraction step. This reduces the extraction time required and enables the process to be automated. On the other hand, the analyte must have a different polarity than the compounds that cause the interference. Moreover, finding a solvent capable of eliminating the interferences without causing damage to the analytes can be difficult [78].
- Liquid chromatography techniques: they are used to remove interferences from complex matrices. The most used techniques are normal phase liquid extraction (NPLE) and gel permeation chromatography (GPC). On the one hand, NPLE is used as a preparative chromatography in a glass column filled in-house with the stationary phase [78], whereas GPC is a purifying technique based in the separation depending on the molecular size of the compounds. The main advantages of GPC are its ability to be automated and the clean-up capacity maintenance for months. Divinylbenzene-linked polystyrene gel is the main material used for GPC [22].

4. Combinatorial Approaches of PLE with New Extraction Methodologies

As shown in Table 3, there have been few studies on the use of PLE in seaweeds due to its novelty. However, following the green technology trends, PLE could be combined with other methodologies to improve bioactive compound extraction efficiency and reduce solvent and time consumption. Moreover, because combinatorial approaches of PLE applied in the extraction of bioactive compounds of seaweeds have not been thoroughly studied, results of trials where PLE and other techniques are applied in other matrices are shown in this section and compiled in Figure 5 as case studies for future applications of these techniques combined.

4.1. PLE Combined with SPE

The combination of PLE and SPE has been used for the separation of specific phenolic compounds [79]. The mode of action of this combination is based on the ability of PLE to extract bioactive compounds from the matrix and the ability of SPE to purify the extracted compounds [80]. Thus, SPE is mainly used as a post-extraction technique since PLE is a non-selective methodology.

There are no data on the use of PLE and SPE for the extraction of phenolic compounds from seaweeds, but it was applied in other matrices such as apple pomace, mate leaves and lemon peel. Higher yields of total flavonoids were obtained in all the three matrices, when compared to the extraction using PLE alone [79–81]. On the contrary, when lemon peel was used as the matrix, the yields of the polar compounds were lower, while total phenolic acids and flavonoids showed no statistical difference between PLE and PLE combined with SPE in mate leaves [80,81].

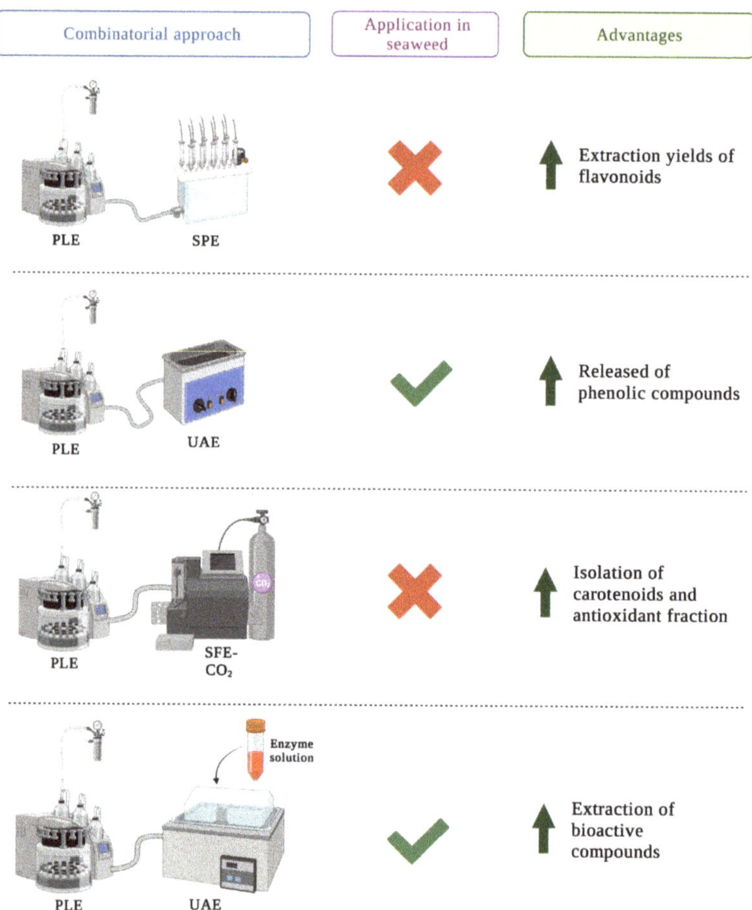

Figure 5. Improvements achieved in the recovery of compounds from different matrices when PLE is combined with other extraction techniques.

4.2. PLE Combined with UAE

Some studies have combined the use of PLE with UAE (UAPLE) in different matrices, including seaweeds. A recent study combined PLE and UAE to extract phenolic compounds from three brown and one red algae. The operational conditions (solvent 80% MeOH:H_2O (v/v); 10 mL; 130 °C; 130 bar; two static cycles of 10 min) were able to increase the release of phenolic compounds from the matrix due to the high and stable pressure [82]. Nevertheless, further research is needed to know if scale-up is available from the economic point of view [83], since many companies have difficulties because of the high expense with facilities, extraction time and ultrasound power [84].

4.3. PLE Combined with SFE-CO₂

The combination of PLE with SFE-CO_2 is a sequential process based on the ability of SFE-CO_2 for the extraction of the lipophilic fraction of the matrix, and the ability of PLE for the extraction of the antioxidant or high polarity compounds [85]. Because there is no information available about this sequential process used in seaweeds, results from other studies were compared to determine if this methodology could be useful for extracting seaweed compounds. For the recovery of bioactive compounds from rowanberry pomace using SPE-CO_2 and PLE consecutively, results showed that this is an effective method for the isolation of carotenoid-rich and antioxidant-rich fractions [86]. Same conclusions were achieved when this sequential process was applied in *Viburnum opulus* pomace and berries [85]. Moreover, an economic evaluation of this process applied in passion fruit by-products was carried out in Brazil. This study showed that the combination of these techniques is economically applicable in large-scale production since it increases process productivity and decreases the cost of manufacturing [87].

4.4. PLE Combined with EAE

The application of EAE as an extraction technique leads to some disadvantages, such as the high cost of enzymes, the limitation of cell disruption because of the specificity of the enzyme and the inactivation of enzymes with parameters such as temperature and pH change [88]. To solve these limitations, EAE studies combined with other new methodologies have been performed. For example, when EAE was combined with alkaline hydrolysis and PLE for the extraction of bioactive compounds from *Sargassum muticum*, the extraction yields were higher than when PLE was used alone [89]. This could be explained by the formation of a protein–polyphenol complex which results in decreased polyphenol recovery. This complex may be formed when the enzyme disrupts the cell of the seaweed, releasing proteins and other compounds. Thus, these compounds may form complexes with polyphenols, resulting in aggregation and precipitation and ultimately, lower yields [89]. Because no additional information was discovered, more research is required to determine whether combining EAE and SPE could result in higher yields.

Given the current data on PLE combined with various new methods, a sequential process using PLE and SFE-CO$_2$ should be considered because extraction yields are increasing, and an industrial scale appears to be feasible. However, because there has been no research on the application of a sequential process involving PLE in seaweeds and different parameters affecting the percentage of recovery, including the matrix, more research is required. Furthermore, because UAPLE produces intriguing results, scale-up trials should be conducted to determine whether this sequential process is economically feasible.

5. Evaluation of Pressurized Liquid Extraction (PLE) Applications

As mentioned in previous sections, using PLE to recover seaweed compounds results in the extraction of various bioactive compounds. Because of their technological function, importance as functional ingredients, or application in innovative food packaging systems, these compounds can be used in the food industry.

5.1. Technological Function and Functional Ingredients

Different compounds extracted from seaweeds are already used in the food industry for the technological improvement of food products, such as carbohydrates from seaweeds which are mainly used for its functional properties. Thus, agar is applied in the confectionary industry for its hydration maintenance capacity. Moreover, the addition of agar in meat products allows the reduction of the fat content in the final product [90]. The use of seaweed extracts in meat emulsions is interesting from a technological point of view since a harder and chewier structure with better water and fat binding is achieved [91]. For example, due to the increase in vegan and vegetarian diets, meat analogues are increasingly in demand and carrageenan is used because of their stability properties [90]: it is already used in low-fat sausages, beef burgers and beef patties as a thickener and stabilizer agent [92]. In addition, since algae have essential micronutrients such as Mg, K and Fe, and low Na content, the addition of these extracts into meat products may be a good opportunity to increase the nutritional value of these products [93]. Carrageenan and alginates have been added in bakery products such as bread, being able to reduce the moisture loss during storage and the dehydration rate of the crumb. Additionally, alginate was able to retard the hardening of the crumb [94]. Fruit jellies, donuts and cakes are also examples of products where agar is added [90]. At last, the use of alginates has an antimicrobial growth activity in the vegetable industry and is a good choice for encapsulation and delivery systems of probiotics, according to the bibliography [90].

On the other hand, proteins, peptides and amino acids are mainly used as stabilizers, thickener agents, protein replacements and gelling agents [95]. Peptides extracted from different seaweeds were incorporated in pasta products, showing superior pasta quality and antioxidant properties over the control [95]. Furthermore, the addition of *Palmaria palmata* hydrolysate in bread improved texture and sensory acceptability. In addition, several studies of peptides extracted from seaweeds have been carried out, and Wakame peptide jelly and Nori peptide S are two bioactive peptides included in some Japan foods due to their antihypertension activity [4]. Nutritional supplements are also considered in the scope of the food application of seaweed extracts. As seaweeds are a good source of proteins, not only because of the quantity but for the quality, the use of this extract as supplementation would be a good option for those athletes following a vegan diet [96]. Thus, there are already products on the market, such as Solaray, that contain extracts of *Rhodymenia palmata*, which helps in the maintenance of the immune system health [77].

The incorporation of phlorotannin into food formulations may be limited because of their astringency and bitter taste. Thus, these compounds were encapsulated into nanofibers made of sodium alginate and polyethylene oxide and successfully incorporated into chicken breasts. The chicken was stored, and thanks to the phlorotannin's encapsulation, *Salmonella* growth was prevented, while the sensorial characteristics of the product were unaffected [97]. Moreover, the preservation ability against polyphenol oxidase activity and melanosis formation was proved during white shrimps' ice storage when phlorotannin extracted from *S. tenerimum* were added. Furthermore, when shrimps were immersed in 5% phlorotannin solution, shelf life was extended by 4 days and higher scores on overall sensory acceptability when compared to control [97]. Finally, different studies proved that the addition of seaweeds to the food formulation of different products leads to a reduction in cooking loss and in an improvement in the texture, as it is shown in Table 4 [98].

Finally, due to the current trend of changing the soy- and animal-derived protein sources for animal feed, seaweed extracts have been also incorporated into these products [95]. According to one study, incorporating algae extracts into dairy cattle feed resulted in higher I and Se content [99]. In fact, it has been demonstrated that including seaweed extracts in animal feed is a good way to achieve the I daily intake for those people with I deficiency, since milk excretion meets the needs of this mineral [95]. The incorporation of red seaweed extracts to poultry feed was also studied. The results show that incorporating *Sarchodiotheca guadichaudii* and *Chondrus crispus* extracts into layer feed improves the growth of beneficial bacteria and reduces *Clostridium perifringens* proliferation in the gut, thus improving the safety of the products obtained. Moreover, the egg yolk and weight were increased by adding 1% of *Sarchodiotheca guadichaudii* into the feed without altering the color of the yolk and the shell thickness [100]. Extracts from seaweeds are also being studied for feeding fish, especially protein extracts, since they are the most expensive dietary requirement for fish and shellfish aquaculture. Moreover, seaweed is also a good source of highly unsaturated fatty acids. Considering the requirements for fish nutrition, studies suggest that seaweed extracts may be a good option for fish feed, allowing the substitution of animal meal by plant meal in these products [95,101].

Considering the advantages of using PLE alone or in combination with other green techniques to obtain previously exposed seaweed bioactive compounds, and how food products may improve with the addition of these biomolecules, the incorporation of seaweed extracts may be a good strategy to enhance their nutritional profile and the technological properties.

Table 4. Application of seaweed extracts in food products.

Product	Seaweed	Form	Technological/Nutritional Function	Ref.
Pasta	*Undaria pinnatifida*	Powder	Fucoxanthin was not altered Lower cooking loss	[102]
Pasta	*Sargassum marginatum*	Powder	Better gluten network when 2.5% of seaweed was added Lower cooking loss	[103]
Beef patty	*Undaria pinnatifida*	Dried and ground	Better texture Higher quantities of mineral and fiber when 3% of seaweed was added Red and yellow color were increased	[104]
Chicken breast meat	*Undaria pinnatifida*	Carotenoid pigment Fucoxanthin	Inhibition of lipid peroxidation after cooking	[98]
Restructured poultry steak	*Hymanthalia elongata*	Powder	Lower cooking loss	[105]
Pork emulsion meat	*Hymanthalia elongata, Undaria pinnatifida, Porphyra umbilicalis*	Dried and ground	Higher antioxidant ability Omega-3 was increased Higher mineral profile Higher amino acid profile (*P. umbilicalis*)	[93]
Low-fat Frankfurters	*Hymanthalia elongata*	Dried and ground	Higher levels of w-3 Higher quantity of fiber	[106]
Cod	*Fucus vesiculosus*	Extract and subfractions of phlorotannin	Inhibition of lipid oxidation in fish model systems	[107]
Fish oil-enriched granola bar	*Fucus vesiculosus*	Phlorotannin extract	Inhibition of lipid oxidation	[108]
Canola oil stored under favorable oxidation conditions	*Fucus vesiculosus, Ascophyllum nodosum, Bifurcaria bifurcata*	Phenolic compounds	Reduction of lipid oxidation	[109]
Pork homogenates	*Laminaria digitata*	Fucoidan extract	Reduction of lipid oxidation Oxidation of myoglobin	[110]
Fish oil-enriched mayonnaise	*Fucus vesiculosus*	Phenolic compounds	Prevention of lipid oxidation	[111]
Fish oil-enriched milk	*Fucus vesiculosus*	Phlorotannin	Prevention of lipid oxidation	[112]
Cookies	*Fucus vesiculosus, Ascophyllum nodosum, Bifurcaria bifurcata*	Phenolic compounds	Antioxidant effect	[113]
Bread	*Sargassum fulvellum*	Powder	The shelf life is increased Less hardness and gumminess	[114]
Bread	*Kappaphycus alvarezii*	Powder	High dietary fiber content	[115]

5.2. Application as Innovative Food Packaging

Some compounds extracted from seaweeds such as laminarin, phlorotannin, flavonoids, terpenes, lactones and proteins are active against bacteria and fungi cells and bacteria biofilm formation (which is more difficult to remove) [116]. The correct preservation of organoleptic characteristics, while avoiding microbial growth during the storage as well as the need to extend the shelf life of the products, are critical for the food industry because these are factors deeply involved with the increase in food waste. Therefore, the use of antimicrobial compounds extracted from seaweeds may be a good option for increasing the shelf life of food products.

Sensory analyses of different products with some of these antimicrobial compounds were carried out to identify their impact on different parameters such as flavor, taste, color and smell. The results showed that edible film made of chitosan and seaweed extracts from *H. longata* and *P. palmata* inhibit the growth of mesophilic and psychrophilic microorganisms by maintaining the pH and water activity without affecting the sensorial characteristics of fish burgers. On the contrary, the sensory evaluation of pork patties with fucoidan and laminarian extract in a ratio of 0.5 w/w proved an adverse impact on the product. To avoid a possible negative impact on the organoleptic characteristics of the products, adding the antimicrobial compounds onto the packaging instead of adding them directly in the product could be an option. In fact, considering that most of the spoilage and contamination of food occurs on its surface, adding these compounds in the packaging may extend the shelf life of the product without affecting its organoleptic characteristics [116].

In terms of packaging, seaweed derivatives were studied to determine their suitability for bioplastic production. Bioplastics are synthetic plastics derived from biodegradable sources and their main disadvantage is their hygroscopicity [117], which affects the mechanical and storage properties required for food packaging. Furthermore, edible coatings are thin membranes composed of GRAS such as polysaccharides, lipids and proteins. Moreover, edible coatings can act as carriers of different bioactive compounds useful in food preservation. Thus, these special coatings maintain firmness, inhibit microbial growth and prevent food weight loss during long-term storage [50]. Carrageenan is a polysaccharide that can be used as an additive combined with other compounds for bioplastic production since it has low water vapor permeability (WVP) [117]. A biodegradable film made of olive extract, glycerol and 1% of carrageenan (w/v) showed good mechanical and antimicrobial properties. Moreover, a bionanocomposite made of 10% of gelatin (w/v), 0.5% (w/v) of k-carrageenan and 1, 3 and 5% of nano-SiO_2 showed a drop in the WVP from 100% to 68%. Finally, the bioplastic production using starch, glycerol and 5% of carrageenan showed that the addition of carrageenan enhanced the moisture resistance, brittleness and the tensile properties of the polymer. Thus, carrageenan can be used to produce edible food packaging [117].

Alginates are another type of polysaccharides that could be used in the formulation of biodegradable films because their main properties are tensile strength, elongation and WVP that are suitable for biodegradable packaging [117]. To determine how mechanical properties were affected, alginate biofilms were compared using hydrophilic and hydrophobic plasticizers. Tributyl citrate (TC) showed better results because TC and alginate secondary interactions improved mechanical resistance. Furthermore, it was proved that elongation at break can be increased by using hydrophilic plasticizers such as glycerol [117]. A study using alginate with aloe vera (AV) and garlic oil (GO) in different proportions to produce a brand-new edible coating was performed to determine their UV shielding, thermal and antimicrobial properties. The results showed that after 16 days of storage, tomatoes with the edible coating made of 33.3% alginate, 66.7% AV and 5% GO showed an 8% mass loss while tomatoes without edible coating showed a 47% of mass loss. In addition, tomatoes with 33.3% alginate, 66.7% AV and 5% GO as edible coating suffered less damaged in the UV light measurement and showed better elongation break properties. For the inhibition growth of *Staphylococcus aureus*, *Escherichia coli* and *Syncephalastrum racemosum*, better results were obtained when tomatoes were coated with 33.3% alginate, 66.7% AV and 5% GO. On the contrary, better tensile strength results were obtained in tomatoes with 33.3% alginate and 67.3% AV [50].

Based on the data presented, the incorporation of alginates, carrageenan, laminarin, phlorotannin, flavonoids, terpenes and proteins onto innovative food packaging is a viable option for this innovative pathway.

6. Advantages and Drawbacks of the Application of PLE as an Extraction Technique of the Bioactive Compounds from Seaweeds

Although PLE is a better option for extracting compounds than traditional extraction techniques, the main disadvantage of this methodology is the high cost of the equipment. This is mainly due to the high requirements in terms of temperature and pressure, as well as the fact that the equipment must be made from materials that can withstand these conditions while avoiding corrosion [24,118].

Despite the high cost of the equipment, PLE has many advantages. On the one hand, one of the primary benefits of using PLE is that it has a lower environmental impact than other conventional extraction methodologies such as maceration and Soxhlet. This improvement is primarily due to a reduction in extraction time and a lower amount of solvent required. As previously stated, high temperatures allow for a decrease in solvent viscosity, resulting in faster solubilization and diffusion of the target compound [17]. Generally, extraction takes about 15 min [118]. This reduces the amount of energy required in the extraction process, making this methodology greener than other conventional methods. Solvents, particularly organic solvents, have traditionally been a problem in conventional extraction methods from an environmental standpoint. This is due to their classification as

Volatile Organic Compounds (VOCs). VOCs are organic pollutants that contribute to the photochemical smog formation in the troposphere and ozone depletion in the stratosphere as a source of radical sources [119]. Furthermore, given that the food industry accounts for 2% of total global solvent consumption [119], primarily for extraction processes, it is necessary to develop techniques that reduce the amount of solvent used. Moreover, biobased solvents are commonly used in this technique. These are defined as solvents derived from biomass and characterized for their biodegradability, lower VOC content and near-zero carbon balance. The use of these solvents is even more critical because it is known that a portion of the solvent used remains in food, food additive excipients and packaging [119]. In PLE, the mainly biobased solvents used are alcohol, ethanol, ethyl acetate, methyl lactate, ethyl lactate and D-limonene. Then, PLE is considered as one of the novel techniques in which solvent consumption is not only low, but it is also better in terms of biodegradability and toxicity.

7. Conclusions and Future Perspectives

Nowadays, the scientific community is increasingly interested in obtaining bioactive compounds from novel matrices using less aggressive environmental methodologies. As explained in this review, PLE is a green extraction technique that allows the separation of target active compounds such as polysaccharides, lipids, proteins and bioactive compounds using short time cycles and low quantities of solvents due to the high-pressure and temperature operating conditions. In this way, since edible seaweeds are becoming more important in the Western world due to both their nutritional profile and their technological properties, using PLE as an extraction technique appears to be a viable option, according to the bibliography. Moreover, data from the combination of PLE with other extraction techniques were evaluated to determine if they were appropriate, and satisfactory results were obtained. There are little data available on the combination of PLE with other extraction techniques using edible seaweeds, but comparable results are expected based on the results obtained with other matrices.

However, even though the data available today are primarily focused on the pharmaceutical industry and PLE has not been applied to edible seaweeds, this review attempts to provide an approximation of the PLE technique applied to seaweeds to generate knowledge that could potentially be applied in the food industry. Thus, compounds derived from edible seaweeds using PLE appear to be suitable for the bioplastic production and edible coating required in packaging; the synthesis of bio-nanocomposites that can be incorporated into food packaging to improve the mechanical properties of bioplastics; the incorporation into nutritional supplements; and the improvement of the nutritional profiles of different food products and animal feed. More experimental approaches of the PLE use for the extraction of seaweed compounds used in food products are required to determine whether this technique is appropriate for this matrix and the final products.

Author Contributions: Conceptualization, A.P.-V., M.C. and M.A.P.; methodology, A.P.-V. and P.B.; software, A.P.-V. and P.B., M.C. and L.C.; formal analysis, M.C. and L.C.; investigation, A.P.-V., M.C. and L.C.; resources, M.A.P. and J.S.-G.; writing—original draft preparation, A.P.-V., M.C., P.B. and L.C.; writing—review and editing, M.A.P. and J.S.-G.; visualization, A.P.-V., M.C. and M.A.P.; supervision, L.C., M.A.P. and J.S.-G.; project administration, M.C. and M.A.P.; funding acquisition, M.A.P. and J.S.-G. All authors have read and agreed to the published version of the manuscript.

Funding: The authors are grateful to the Ibero–American Program on Science and Technology (CYTED—AQUA-CIBUS, P317RT0003) and the Bio Based Industries Joint Undertaking (JU) under grant agreement No. 888003 UP4HEALTH Project (H2020-BBI-JTI-2019). The JU receives support from the European Union's Horizon 2020 research and innovation program and the Bio Based Industries Consortium. The project SYSTEMIC Knowledge Hub on Nutrition and Food Security has received funding from national research funding parties in Belgium (FWO), France (INRA), Germany (BLE), Italy (MIPAAF), Latvia (IZM), Norway (RCN), Portugal (FCT) and Spain (AEI) in a joint action of JPI HDHL, JPI-OCEANS and FACCE-JPI launched in 2019 under the ERA-NET ERA-HDHL (n° 696295).

Acknowledgments: The research leading to these results was supported by MICINN supporting the Ramón y Cajal grant for M.A. Prieto (RYC-2017-22891); by Xunta de Galicia for supporting the program EXCELENCIA-ED431F 2020/12, the post-doctoral grant of L. Cassani (ED481B-2021/152), and the pre-doctoral grant of M. Carpena (ED481A 2021/313).

Conflicts of Interest: The authors declare no conflict of interest.

References

1. Gutiérrez Cuesta, A.; García, G.; Rivera, H.; Suárez, A.; Delange, M. Algas marinas, fuente potencial de macronutrientes. Item Type Journal Contribution. *Rev. Investig. Mar.* **2017**, *37*, 14. Available online: http://hdl.handle.net/1834/12438 (accessed on 17 July 2022).

2. Blikra, M.J.; Altintzoglou, T.; Løvdal, T.; Rognså, G.; Skipnes, D.; Skåra, T.; Sivertsvik, M.; Noriega Fernández, E. Seaweed products for the future: Using current tools to develop a sustainable food industry. *Trends Food Sci. Technol.* **2021**, *118*, 765–776. [CrossRef]

3. Demarco, M.; Oliveira de Moraes, J.; Matos, Â.P.; Derner, R.B.; de Farias Neves, F.; Tribuzi, G. Digestibility, bioaccessibility and bioactivity of compounds from algae. *Trends Food Sci. Technol.* **2022**, *121*, 114–128. [CrossRef]

4. Lafarga, T.; Acién-Fernández, F.G.; Garcia-Vaquero, M. Bioactive peptides and carbohydrates from seaweed for food applications: Natural occurrence, isolation, purification, and identification. *Algal Res.* **2020**, *48*, 101909. [CrossRef]

5. Meng, W.; Mu, T.; Sun, H.; Garcia-Vaquero, M. Evaluation of the chemical composition and nutritional potential of brown macroalgae commercialised in China. *Algal Res.* **2022**, *64*, 102683. [CrossRef]

6. Melo, T.; Alves, E.; Azevedo, V.; Martins, A.S.; Neves, B.; Domingues, P.; Calado, R.; Abreu, M.H.; Domingues, M.R. Lipidomics as a new approach for the bioprospecting of marine macroalgae—Unraveling the polar lipid and fatty acid composition of chondrus crispus. *Algal Res.* **2015**, *8*, 181–191. [CrossRef]

7. Otero, P.; López-Martínez, M.I.; García-Risco, M.R. Application of pressurized liquid extraction (PLE) to obtain bioactive fatty acids and phenols from Laminaria ochroleuca collected in Galicia (NW Spain). *J. Pharm. Biomed. Anal.* **2019**, *164*, 86–92. [CrossRef]

8. Lourenço-Lopes, C.; Fraga-Corral, M.; Jimenez-Lopez, C.; Carpena, M.; Pereira, A.G.; Garcia-Oliveira, P.; Prieto, M.A.; Simal-Gandara, J. Biological action mechanisms of fucoxanthin extracted from algae for application in food and cosmetic industries. *Trends Food Sci. Technol.* **2021**, *117*, 163–181. [CrossRef]

9. Kim, S.M.; Kang, S.W.; Jeon, J.S.; Jung, Y.J.; Kim, W.R.; Kim, C.Y.; Um, B.H. Determination of major phlorotannins in Eisenia bicyclis using hydrophilic interaction chromatography: Seasonal variation and extraction characteristics. *Food Chem.* **2013**, *138*, 2399–2406. [CrossRef]

10. Shin, T.; Ahn, M.; Hyun, J.W.; Kim, S.H.; Moon, C. Antioxidant marine algae phlorotannins and radioprotection: A review of experimental evidence. *Acta Histochem.* **2014**, *116*, 669–674. [CrossRef]

11. Lopes, G.; Barbosa, M.; Vallejo, F.; Gil-Izquierdo, Á.; Andrade, P.B.; Valentão, P.; Pereira, D.M.; Ferreres, F. Profiling phlorotannins from *Fucus spp.* of the Northern Portuguese coastline: Chemical approach by HPLC-DAD-ESI/MSn and UPLC-ESI-QTOF/MS. *Algal Res.* **2018**, *29*, 113–120. [CrossRef]

12. Lee, S.H.; Jeon, Y.J. Anti-diabetic effects of brown algae derived phlorotannins, marine polyphenols through diverse mechanisms. *Fitoterapia* **2013**, *86*, 129–136. [CrossRef]

13. Zhang, Q.-W.; Lin, L.-G.; Ye, W.-C. Techniques for extraction and isolation of natural products: A comprehensive review. *Chin. Med.* **2018**, *13*, 20. [CrossRef]

14. Chemat, F.; Rombaut, N.; Meullemiestre, A.; Turk, M.; Périno, S.; Fabiano-Tixier, A.-S.; Abert-Vian, M. Review of Green Food Processing techniques. Preservation, transformation, and extraction. *Innov. Food Sci. Emerg. Technol.* **2017**, *41*, 357–377. [CrossRef]

15. Garcia-Vaquero, M.; Rajauria, G.; Tiwari, B. Conventional extraction techniques: Solvent extraction. In *Sustainable Seaweed Technologies*; Elsevier: Amsterdam, The Netherlands, 2020; pp. 171–189. [CrossRef]

16. Pereira, T.; Barroso, S.; Mendes, S.; Amaral, R.A.; Dias, J.R.; Baptista, T.; Saraiva, J.A.; Alves, N.M.; Gil, M.M. Optimization of phycobiliprotein pigments extraction from red algae Gracilaria gracilis for substitution of synthetic food colorants. *Food Chem.* **2020**, *321*, 126688. [CrossRef]

17. Mena-García, A.; Ruiz-Matute, A.I.; Soria, A.C.; Sanz, M.L. Green techniques for extraction of bioactive carbohydrates. *TrAC Trends Anal. Chem.* **2019**, *119*, 115612. [CrossRef]

18. Carreira-Casais, A.; Otero, P.; Garcia-Perez, P.; Garcia-Oliveira, P.; Pereira, A.G.; Carpena, M.; Soria-Lopez, A.; Simal-Gandara, J.; Prieto, M.A. Benefits and Drawbacks of Ultrasound-Assisted Extraction for the Recovery of Bioactive Compounds from Marine Algae. *Int. J. Environ. Res. Public Health* **2021**, *18*, 9153. [CrossRef]

19. Carreira-Casais, A.; Lourenço-Lopes, C.; Otero, P.; Carpena, M.; Gonzalez Pereira, A.; Echave, J.; Soria-Lopez, A.; Chamorro, F.; Prieto, M.A.; Simal-Gandara, J. Application of Green Extraction Techniques for Natural Additives Production. In *Natural Food Additives*; IntechOpen: London, UK, 2022. [CrossRef]

20. Sabeena, S.F.; Alagarsamy, S.; Sattari, Z.; Al-Haddad, S.; Fakhraldeen, S.; Al-Ghunaim, A.; Al-Yamani, F. Enzyme-assisted extraction of bioactive compounds from brown seaweeds and characterization. *J. Appl. Phycol.* **2020**, *32*, 615–629. [CrossRef]

21. Kadam, S.U.; Álvarez, C.; Tiwari, B.K.; O'Donnell, C.P. Extraction of biomolecules from seaweeds. In *Seaweed Sustainability: Food and Non-Food Applications. Manchester*; Academic Press: Cambridge, MA, USA, 2015; pp. 243–269. [CrossRef]

22. Nieto, A.; Borrull, F.; Pocurull, E.; Marcé, R.M. Pressurized liquid extraction: A useful technique to extract pharmaceuticals and personal-care products from sewage sludge. *TrAC Trends Anal. Chem.* **2010**, *29*, 752–764. [CrossRef]

23. Carabias-Martínez, R.; Rodríguez-Gonzalo, E.; Revilla-Ruiz, P.; Hernández-Méndez, J. Pressurized liquid extraction in the analysis of food and biological samples. *J. Chromatogr. A* **2005**, *1089*, 1–17. [CrossRef]

24. Ballesteros-Vivas, D.; Ortega-Barbosa, J.P.; del Pilar Sanchez-Camargo, A.; Rodríguez-Varela, L.I.; Parada-Alfonso, F. Pressurized Liquid Extraction of Bioactives. In *Comprehensive Foodomics*; Elsevier: Amsterdam, The Netherlands, 2020; pp. 754–770. ISBN 9780128163955.

25. Pavkovich, A.M.; Bell, D.S. Extraction. In *Encyclopedia of Analytical Science*; Elsevier: Amsterdam, The Netherlands, 2019. [CrossRef]

26. Ashour, M.; Kamel, A.E.-W. Enhance Growth and Biochemical Composition of Nannochloropsis oceanica, Cultured under Nutrient Limitation, Using Commercial Agricultural Fertilizers. *J. Mar. Sci. Res. Dev.* **2017**, *07*, 233. [CrossRef]

27. Zanella, L.; Vianello, F. Microalgae of the genus Nannochloropsis: Chemical composition and functional implications for human nutrition. *J. Funct. Foods* **2020**, *68*, 103919. [CrossRef]

28. Pandeirada, C.O.; Maricato, É.; Ferreira, S.S.; Correia, V.G.; Pinheiro, B.A.; Evtuguin, D.V.; Palma, A.S.; Correia, A.; Vilanova, M.; Coimbra, M.A.; et al. Structural analysis and potential immunostimulatory activity of Nannochloropsis oculata polysaccharides. *Carbohydr. Polym.* **2019**, *222*, 114962. [CrossRef] [PubMed]

29. Qi, J.; Kim, S.M. Characterization and immunomodulatory activities of polysaccharides extracted from green alga Chlorella ellipsoidea. *Int. J. Biol. Macromol.* **2017**, *95*, 106–114. [CrossRef] [PubMed]

30. Priscila Barros de Medeiros, V.; Karoline Almeida da Costa, W.; Tavares da Silva, R.; Colombo Pimentel, T.; Magnani, M. Microalgae as source of functional ingredients in new-generation foods: Challenges, technological effects, biological activity, and regulatory issues. *Crit. Rev. Food Sci. Nutr.* **2021**, *62*, 4929–4950. [CrossRef]

31. Patil, V.; Källqvist, T.; Olsen, E.; Vogt, G.; Gislerød, H.R. Fatty acid composition of 12 microalgae for possible use in aquaculture feed. *Aquac. Int.* **2007**, *15*, 1–9. [CrossRef]

32. Li, J.; Pora, B.L.R.; Dong, K.; Hasjim, J. Health benefits of docosahexaenoic acid and its bioavailability: A review. *Food Sci. Nutr.* **2021**, *9*, 5229–5243. [CrossRef]

33. Molino, A.; Martino, M.; Larocca, V.; Di Sanzo, G.; Spagnoletta, A.; Marino, T.; Karatza, D.; Iovine, A.; Mehariya, S.; Musmarra, D. Eicosapentaenoic acid extraction from nannochloropsis gaditana using carbon dioxide at supercritical conditions. *Mar. Drugs* **2019**, *17*, 132. [CrossRef]

34. Freire, I.; Cortina-Burgueño, A.; Grille, P.; Arizcun Arizcun, M.; Abellán, E.; Segura, M.; Witt Sousa, F.; Otero, A. Nannochloropsis limnetica: A freshwater microalga for marine aquaculture. *Aquaculture* **2016**, *459*, 124–130. [CrossRef]

35. Mitra, M.; Patidar, S.K.; George, B.; Shah, F.; Mishra, S. A euryhaline nannochloropsis gaditana with potential for nutraceutical (EPA) and biodiesel production. *Algal Res.* **2015**, *8*, 161–167. [CrossRef]

36. Li-Beisson, Y.; Thelen, J.J.; Fedosejevs, E.; Harwood, J.L. The lipid biochemistry of eukaryotic algae. *Prog. Lipid Res.* **2019**, *74*, 31–68. [CrossRef]

37. Krienitz, L.; Wirth, M. The high content of polyunsaturated fatty acids in Nannochloropsis limnetica (Eustigmatophyceae) and its implication for food web interactions, freshwater aquaculture and biotechnology. *Limnologica* **2006**, *36*, 204–210. [CrossRef]

38. Díaz, M.; Mesa-Herrera, F.; Marín, R. Dha and its elaborated modulation of antioxidant defenses of the brain: Implications in aging and ad neurodegeneration. *Antioxidants* **2021**, *10*, 907. [CrossRef] [PubMed]

39. Kumar, R.; Hegde, A.S.; Sharma, K.; Parmar, P.; Srivatsan, V. Microalgae as a sustainable source of edible proteins and bioactive peptides—Current trends and future prospects. *Food Res. Int.* **2022**, *157*, 111338. [CrossRef] [PubMed]

40. Skjånes, K.; Aesoy, R.; Herfindal, L.; Skomedal, H. Bioactive peptides from microalgae: Focus on anti-cancer and immunomodulating activity. *Physiol. Plant.* **2021**, *173*, 612–623. [CrossRef] [PubMed]

41. Barkia, I.; Al-Haj, L.; Abdul Hamid, A.; Zakaria, M.; Saari, N.; Zadjali, F. Indigenous marine diatoms as novel sources of bioactive peptides with antihypertensive and antioxidant properties. *Int. J. Food Sci. Technol.* **2019**, *54*, 1514–1522. [CrossRef]

42. Chen, L.; Cao, H.; Xiao, J. Polyphenols: Absorption, bioavailability, and metabolomics. In *Polyphenols: Properties, Recovery, and Applications*; Elsevier: Amsterdam, The Netherlands, 2018; pp. 45–67. ISBN 9780128135723.

43. Goiris, K.; Muylaert, K.; Fraeye, I.; Foubert, I.; De Brabanter, J.; De Cooman, L. Antioxidant potential of microalgae in relation to their phenolic and carotenoid content. *J. Appl. Phycol.* **2012**, *24*, 1477–1486. [CrossRef]

44. Del Mondo, A.; Smerilli, A.; Sané, E.; Sansone, C.; Brunet, C. Challenging microalgal vitamins for human health. *Microb. Cell Factories* **2020**, *19*, 201. [CrossRef]

45. Tarento, T.D.C.; McClure, D.D.; Vasiljevski, E.; Schindeler, A.; Dehghani, F.; Kavanagh, J.M. Microalgae as a source of vitamin K1. *Algal Res.* **2018**, *36*, 77–87. [CrossRef]

46. Raja, K.; Kadirvel, V.; Subramaniyan, T. Seaweeds, an aquatic plant-based protein for sustainable nutrition—A review. *Futur. Foods* **2022**, *5*, 100142. [CrossRef]

47. Sarkar, S.; Sarkar, S.; Manna, M.S.; Gayen, K.; Bhowmick, T.K. Extraction of carbohydrates and proteins from algal resources using supercritical and subcritical fluids for high-quality products. In *Innovative and Emerging Technologies in the Bio-Marine Food Sector*; Academic Press: Cambridge, MA, USA, 2022; pp. 249–275. [CrossRef]

48. Cummings, J.H.; Macfarlane, G.T.; Englyst, H.N. Prebiotic digestion and fermentation. *Am. J. Clin. Nutr.* **2001**, *73* (Suppl. 2), 415–420. [CrossRef] [PubMed]

49. Saravana, P.S.; Cho, Y.J.; Park, Y.B.; Woo, H.C.; Chun, B.S. Structural, antioxidant, and emulsifying activities of fucoidan from *Saccharina japonica* using pressurized liquid extraction. *Carbohydr. Polym.* **2016**, *153*, 518–525. [CrossRef] [PubMed]
50. Abdel Aziz, M.S.; Salama, H.E. Developing multifunctional edible coatings based on alginate for active food packaging. *Int. J. Biol. Macromol.* **2021**, *190*, 837–844. [CrossRef] [PubMed]
51. Rajauria, G.; Ravindran, R.; Garcia-Vaquero, M.; Rai, D.K.; Sweeney, T.; O'Doherty, J. Molecular characteristics and antioxidant activity of laminarin extracted from the seaweed species Laminaria hyperborea, using hydrothermal-assisted extraction and a multi-step purification procedure. *Food Hydrocoll.* **2021**, *112*, 106332. [CrossRef]
52. Sellimi, S.; Maalej, H.; Rekik, D.M.; Benslima, A.; Ksouda, G.; Hamdi, M.; Sahnoun, Z.; Li, S.; Nasri, M.; Hajji, M. Antioxidant, antibacterial and in vivo wound healing properties of laminaran purified from Cystoseira barbata seaweed. *Int. J. Biol. Macromol.* **2018**, *119*, 633–644. [CrossRef] [PubMed]
53. Vo, T.S.; Kim, S.K. Fucoidans as a natural bioactive ingredient for functional foods. *J. Funct. Foods* **2013**, *5*, 16–27. [CrossRef]
54. Chen, K.; Roca, M. Cooking effects on bioaccessibility of chlorophyll pigments of the main edible seaweeds. *Food Chem.* **2019**, *295*, 101–109. [CrossRef]
55. Chen, K.; Ríos, J.J.; Pérez-Gálvez, A.; Roca, M. Comprehensive chlorophyll composition in the main edible seaweeds. *Food Chem.* **2017**, *228*, 625–633. [CrossRef]
56. Sharma, P.P.; Baskaran, V. Polysaccharide (laminaran and fucoidan), fucoxanthin and lipids as functional components from brown algae (Padina tetrastromatica) modulates adipogenesis and thermogenesis in diet-induced obesity in C57BL6 mice. *Algal Res.* **2021**, *54*, 102187. [CrossRef]
57. Siddique, M.A.M.; Hossain, M.S.; Islam, M.M.; Rahman, M.; Kibria, G. Heavy metals and metalloids in edible seaweeds of Saint Martin's Island, Bay of Bengal, and their potential health risks. *Mar. Pollut. Bull.* **2022**, *181*, 113866. [CrossRef]
58. Peng, Z.; Guo, Z.; Wang, Z.; Zhang, R.; Wu, Q.; Gao, H.; Wang, Y.; Shen, Z.; Lek, S.; Xiao, J. Species-specific bioaccumulation and health risk assessment of heavy metal in seaweeds in tropic coasts of South China Sea. *Sci. Total Environ.* **2022**, *832*, 155031. [CrossRef] [PubMed]
59. Kibria, G.; Haroon, A.K.; Rose, G.; Hossain, M.M.; Nugegoda, D. *Pollution Risks, Impacts and Management: Social, Economic, and Enviromental Perspectives*; Scientific Publishers: Jodhpur, India, 2021; 833p, ISBN 9789389184969.
60. Herrero, M.; de Paula Sánchez-Camargo, A.; Cifuentes, A.; Ibáñez, E. Plants, seaweeds, microalgae and food by-products as natural sources of functional ingredients obtained using pressurized liquid extraction and supercritical fluid extraction. *TrAC Trends Anal. Chem.* **2015**, *71*, 26–38. [CrossRef]
61. MS, J.; Soylak, M. Deep eutectic solvents-based adsorbents in enviromental analysis. *TrAC Trends Anal. Chem.* **2022**, *157*, 116762. [CrossRef]
62. Vo Dinh, T.; Saravana, P.S.; Woo, H.C.; Chun, B.S. Ionic liquid-assisted subcritical water enhances the extraction of phenolics from brown seaweed and its antioxidant activity. *Sep. Purif. Technol.* **2018**, *196*, 287–299. [CrossRef]
63. Plaza, M.; Marina, M.L. Pressurized hot water extraction of bioactives. *TrAC Trends Anal. Chem.* **2019**, *116*, 236–247. [CrossRef]
64. Okiyama, D.C.G.; Soares, I.D.; Cuevas, M.S.; Crevelin, E.J.; Moraes, L.A.B.; Melo, M.P.; Oliveira, A.L.; Rodrigues, C.E.C. Pressurized liquid extraction of flavanols and alkaloids from cocoa bean shell using ethanol as solvent. *Food Res. Int.* **2018**, *114*, 20–29. [CrossRef]
65. Mustafa, A.; Trevino, L.M.; Turner, C. Pressurized hot ethanol extraction of carotenoids from carrot by-products. *Molecules* **2012**, *17*, 1809–1818. [CrossRef]
66. Ballesteros-Vivas, D.; Alvarez-Rivera, G.; Ibánez, E.; Parada-Alfonso, F.; Cifuentes, A. Integrated strategy for the extraction and profiling of bioactive metabolites from Passiflora mollissima seeds combining pressurized-liquid extraction and gas/liquid chromatography–high resolution mass spectrometry. *J. Chromatogr. A* **2019**, *1595*, 144–157. [CrossRef]
67. Golmakani, M.T.; Mendiola, J.A.; Rezaei, K.; Ibáñez, E. Pressurized limonene as an alternative bio-solvent for the extraction of lipids from marine microorganisms. *J. Supercrit. Fluids* **2014**, *92*, 1–7. [CrossRef]
68. Pereira, C.S.M.; Silva, V.M.T.M.; Rodrigues, A.E. Ethyl lactate as a solvent: Properties, applications and production processes—A review. *Green Chem.* **2011**, *13*, 2658–2671. [CrossRef]
69. del Pilar Sánchez-Camargo, A.; Pleite, N.; Herrero, M.; Cifuentes, A.; Ibáñez, E.; Gilbert-López, B. New approaches for the selective extraction of bioactive compounds employing bio-based solvents and pressurized green processes. *J. Supercrit. Fluids* **2017**, *128*, 112–120. [CrossRef]
70. Gilbert-López, B.; Barranco, A.; Herrero, M.; Cifuentes, A.; Ibáñez, E. Development of new green processes for the recovery of bioactives from *Phaeodactylum tricornutum*. *Food Res. Int.* **2017**, *99*, 1056–1065. [CrossRef] [PubMed]
71. Shang, Y.F.; Kim, S.M.; Lee, W.J.; Um, B.H. Pressurized liquid method for fucoxanthin extraction from *Eisenia bicyclis* (Kjellman) Setchell. *J. Biosci. Bioeng.* **2011**, *111*, 237–241. [CrossRef] [PubMed]
72. Anaëlle, T.; Serrano Leon, E.; Laurent, V.; Elena, I.; Mendiola, J.A.; Stéphane, C.; Nelly, K.; Stéphane, L.B.; Luc, M.; Valérie, S.P. Green improved processes to extract bioactive phenolic compounds from brown macroalgae using Sargassum muticum as model. *Talanta* **2013**, *104*, 44–52. [CrossRef]
73. Park, J.S.; Han, J.M.; Surendhiran, D.; Chun, B.S. Physicochemical and biofunctional properties of Sargassum thunbergii extracts obtained from subcritical water extraction and conventional solvent extraction. *J. Supercrit. Fluids* **2022**, *182*, 105535. [CrossRef]

74. Lin, E.T.; Lee, Y.C.; Wang, H.M.D.; Chiu, C.Y.; Chang, Y.K.; Huang, C.Y.; Chang, C.C.; Tsai, P.C.; Chang, J.S. Efficient fucoidan extraction and purification from Sargassum cristaefolium and preclinical dermal biological activity assessments of the purified fucoidans. *J. Taiwan Inst. Chem. Eng.* **2022**, *137*, 104294. [CrossRef]

75. Gereniu, C.R.N.; Saravana, P.S.; Chun, B.S. Recovery of carrageenan from Solomon Islands red seaweed using ionic liquid-assisted subcritical water extraction. *Sep. Purif. Technol.* **2018**, *196*, 309–317. [CrossRef]

76. O'Sullivan, A.M.; O'Callaghan, Y.C.; O'Grady, M.N.; Hayes, M.; Kerry, J.P.; O'Brien, N.M. The effect of solvents on the antioxidant activity in Caco-2 cells of Irish brown seaweed extracts prepared using accelerated solvent extraction (ASE®). *J. Funct. Foods* **2013**, *5*, 940–948. [CrossRef]

77. Otero, P.; Quintana, S.E.; Reglero, G.; Fornari, T.; García-Risco, M.R. Pressurized Liquid Extraction (PLE) as an innovative green technology for the effective enrichment of galician algae extracts with high quality fatty acids and antimicrobial and antioxidant properties. *Mar. Drugs* **2018**, *16*, 156. [CrossRef]

78. Fontanals, N.; Pocurull, E.; Borrull, F.; Marcé, R.M. Clean-up techniques in the pressurized liquid extraction of abiotic environmental solid samples. *Trends Environ. Anal. Chem.* **2021**, *29*, e00111. [CrossRef]

79. da Silva, L.C.; Souza, M.C.; Sumere, B.R.; Silva, L.G.S.; da Cunha, D.T.; Barbero, G.F.; Bezerra, R.M.N.; Rostagno, M.A. Simultaneous extraction and separation of bioactive compounds from apple pomace using pressurized liquids coupled on-line with solid-phase extraction. *Food Chem.* **2020**, *318*, 126450. [CrossRef]

80. Souza, M.C.; Silva, L.C.; Chaves, J.O.; Salvador, M.P.; Sanches, V.L.; da Cunha, D.T.; Foster Carneiro, T.; Rostagno, M.A. Simultaneous extraction and separation of compounds from mate (Ilex paraguariensis) leaves by pressurized liquid extraction coupled with solid-phase extraction and in-line UV detection. *Food Chem. Mol. Sci.* **2021**, *2*, 100008. [CrossRef] [PubMed]

81. Chaves, J.O.; Sanches, V.L.; Viganó, J.; de Souza Mesquita, L.M.; de Souza, M.C.; da Silva, L.C.; Acunha, T.; Faccioli, L.H.; Rostagno, M.A. Integration of pressurized liquid extraction and in-line solid-phase extraction to simultaneously extract and concentrate phenolic compounds from lemon peel (*Citrus limon* L.). *Food Res. Int.* **2022**, *157*, 111252. [CrossRef] [PubMed]

82. Klejdus, B.; Plaza, M.; Šnóblová, M.; Lojková, L. Development of new efficient method for isolation of phenolics from sea algae prior to their rapid resolution liquid chromatographic–tandem mass spectrometric determination. *J. Pharm. Biomed. Anal.* **2017**, *135*, 87–96. [CrossRef] [PubMed]

83. Viganó, J.; de Paula Assis, B.F.; Náthia-Neves, G.; dos Santos, P.; Meireles, M.A.A.; Veggi, P.C.; Martínez, J. Extraction of bioactive compounds from defatted passion fruit bagasse (*Passiflora edulis* sp.) applying pressurized liquids assisted by ultrasound. *Ultrason. Sonochem.* **2020**, *64*, 104999. [CrossRef] [PubMed]

84. Dias, A.L.B.; de Aguiar, A.C.; Rostagno, M.A. Extraction of natural products using supercritical fluids and pressurized liquids assisted by ultrasound: Current status and trends. *Ultrason. Sonochem.* **2021**, *74*, 105584. [CrossRef]

85. Kraujalis, P.; Kraujalienė, V.; Kazernavičiūtė, R.; Venskutonis, P.R. Supercritical carbon dioxide and pressurized liquid extraction of valuable ingredients from Viburnum opulus pomace and berries and evaluation of product characteristics. *J. Supercrit. Fluids* **2017**, *122*, 99–108. [CrossRef]

86. Bobinaitė, R.; Kraujalis, P.; Tamkutė, L.; Urbonavičienė, D.; Viškelis, P.; Venskutonis, P.R. Recovery of bioactive substances from rowanberry pomace by consecutive extraction with supercritical carbon dioxide and pressurized solvents. *J. Ind. Eng. Chem.* **2020**, *85*, 152–160. [CrossRef]

87. Viganó, J.; Zabot, G.L.; Martínez, J. Supercritical fluid and pressurized liquid extractions of phytonutrients from passion fruit by-products: Economic evaluation of sequential multi-stage and single-stage processes. *J. Supercrit. Fluids* **2017**, *122*, 88–98. [CrossRef]

88. Das, S.; Nadar, S.S.; Rathod, V.K. Integrated strategies for enzyme assisted extraction of bioactive molecules: A review. *Int. J. Biol. Macromol.* **2021**, *191*, 899–917. [CrossRef]

89. Sánchez-Camargo, A.D.P.; Montero, L.; Stiger-Pouvreau, V.; Tanniou, A.; Cifuentes, A.; Herrero, M.; Ibáñez, E. Considerations on the use of enzyme-assisted extraction in combination with pressurized liquids to recover bioactive compounds from algae. *Food Chem.* **2016**, *192*, 67–74. [CrossRef] [PubMed]

90. Otero, P.; Carpena, M.; Garcia-Oliveira, P.; Echave, J.; Soria-Lopez, A.; Garcia-Perez, P.; Fraga-Corral, M.; Cao, H.; Nie, S.; Xiao, J.; et al. Seaweed polysaccharides: Emerg-ing extraction technologies, chemical modifications and bioactive properties. In *Critical Reviews in Food Science and Nutrition*; Taylor and Francis Ltd.: London, UK, 2021. [CrossRef]

91. Cofrades, S.; López-López, I.; Solas, M.T.; Bravo, L.; Jiménez-Colmenero, F. Influence of different types and proportions of added edible seaweeds on characteristics of low-salt gel/emulsion meat systems. *Meat Sci.* **2008**, *79*, 767–776. [CrossRef] [PubMed]

92. Rajauria, G.; Cornish, L.; Ometto, F.; Msuya, F.E.; Villa, R. Identification and selection of algae for food, feed, and fuel applications. In *Seaweed Sustainability*; Academic Press: Cambridge, MA, USA, 2015; pp. 315–345. [CrossRef]

93. López-López, I.; Bastida, S.; Ruiz-Capillas, C.; Bravo, L.; Larrea, M.T.; Sánchez-Muniz, F.; Cofrades, S.; Jiménez-Colmenero, F. Composition and antioxidant capacity of low-salt meat emulsion model systems containing edible seaweeds. *Meat Sci.* **2009**, *83*, 492–498. [CrossRef] [PubMed]

94. Kadam, S.U.; Prabhasankar, P. Marine foods as functional ingredients in bakery and pasta products. *Food Res. Int.* **2010**, *43*, 1975–1980. [CrossRef]

95. Garcia-Vaquero, M.; Hayes, M. Red and green macroalgae for fish and animal feed and human functional food development. *Food Rev. Int.* **2016**, *32*, 15–45. [CrossRef]

96. Bleakley, S.; Hayes, M. Algal Proteins: Extraction, Application, and Challenges Concerning Production. *Foods* **2017**, *6*, 33. [CrossRef] [PubMed]

97. Cassani, L.; Gomez-Zavaglia, A.; Jimenez-Lopez, C.; Lourenço-Lopes, C.; Prieto, M.A.; Simal-Gandara, J. Seaweed-based natural ingredients: Stability of phlorotannins during extraction, storage, passage through the gastrointestinal tract and potential incorporation into functional foods. *Food Res. Int.* **2020**, *137*, 109676. [CrossRef]

98. Mahadevan, K. Seaweeds: A sustainable food source. In *Seaweed Sustainability: Food and Non-Food Applications*; Elsevier Inc.: Manchester, UK, 2015; ISBN 9780124199583.

99. Rey-Crespo, F.; López-Alonso, M.; Miranda, M. The use of seaweed from the Galician coast as a mineral supplement in organic dairy cattle. *Animal* **2014**, *8*, 580–586. [CrossRef]

100. Kulshreshtha, G.; Rathgeber, B.; Stratton, G.; Thomas, N.; Evans, F.; Critchley, A.; Hafting, J.; Prithiviraj, B. Feed supplementation with red seaweeds, Chondrus crispus and Sarcodiotheca gaudichaudii, affects performance, egg quality, and gut microbiota of layer hens. *Poult. Sci.* **2014**, *93*, 2991–3001. [CrossRef]

101. Fleurence, J. Seaweed proteins: Biochemical, nutritional aspects and potential uses. *Trends Food Sci. Technol.* **1999**, *10*, 25–28. [CrossRef]

102. Prabhasankar, P.; Ganesan, P.; Bhaskar, N.; Hirose, A.; Stephen, N.; Gowda, L.R.; Hosokawa, M.; Miyashita, K. Edible Japanese seaweed, wakame (Undaria pinnatifida) as an ingredient in pasta: Chemical, functional and structural evaluation. *Food Chem.* **2009**, *115*, 501–508. [CrossRef]

103. Prabhasankar, P.; Ganesan, P.; Bhaskar, N. Influence of Indian brown seaweed (Sargassum marginatum) as an ingredient on quality, biofunctional, and microstructure characteristics of pasta. *Food Sci. Technol. Int.* **2009**, *15*, 471–479. [CrossRef]

104. López-López, I.; Cofrades, S.; Yakan, A.; Solas, M.T.; Jiménez-Colmenero, F. Frozen storage characteristics of low-salt and low-fat beef patties as affected by Wakame addition and replacing pork backfat with olive oil-in-water emulsion. *Food Res. Int.* **2010**, *43*, 1244–1254. [CrossRef]

105. Cofrades, S.; López-López, I.; Ruiz-Capillas, C.; Triki, M.; Jiménez-Colmenero, F. Quality characteristics of low-salt restructured poultry with microbial transglutaminase and seaweed. *Meat Sci.* **2011**, *87*, 373–380. [CrossRef] [PubMed]

106. López-López, I.; Cofrades, S.; Ruiz-Capillas, C.; Jiménez-Colmenero, F. Design and nutritional properties of potential functional frankfurters based on lipid formulation, added seaweed and low salt content. *Meat Sci.* **2009**, *83*, 255–262. [CrossRef] [PubMed]

107. Wang, T.; Jónsdóttir, R.; Kristinsson, H.G.; Thorkelsson, G.; Jacobsen, C.; Hamaguchi, P.Y.; Ólafsdóttir, G. Inhibition of haemoglobin-mediated lipid oxidation in washed cod muscle and cod protein isolates by Fucus vesiculosus extract and fractions. *Food Chem.* **2010**, *123*, 321–330. [CrossRef]

108. Hermund, D.B.; Andersen, U.; Jónsdóttir, R.; Kristinsson, H.G.; Alasalvar, C.; Jacobsen, C. Oxidative Stability of Granola Bars Enriched with Multilayered Fish Oil Emulsion in the Presence of Novel Brown Seaweed Based Antioxidants. *J. Agric. Food Chem.* **2016**, *64*, 8359–8368. [CrossRef]

109. Agregán, R.; Munekata, P.E.; Domínguez, R.; Carballo, J.; Franco, D.; Lorenzo, J.M. Proximate composition, phenolic content and in vitro antioxidant activity of aqueous extracts of the seaweeds Ascophyllum nodosum, Bifurcaria bifurcata and Fucus vesiculosus. Effect of addition of the extracts on the oxidative stability of canola oil under accelerated storage conditions. *Food Res. Int.* **2017**, *99*, 986–994. [CrossRef]

110. Moroney, N.C.; O'Grady, M.N.; Lordan, S.; Stanton, C.; Kerry, J.P. Seaweed polysaccharides (laminarin and fucoidan) as functional ingredients in pork meat: An evaluation of anti-oxidative potential, thermal stability and bioaccessibility. *Mar. Drugs* **2015**, *13*, 2447–2464. [CrossRef]

111. Honold, P.J.; Jacobsen, C.; Jónsdóttir, R.; Kristinsson, H.G.; Hermund, D.B. Potential seaweed-based food ingredients to inhibit lipid oxidation in fish-oil-enriched mayonnaise. *Eur. Food Res. Technol.* **2016**, *242*, 571–584. [CrossRef]

112. Baun, D. *General Rights Extraction, Characterization and Application of Antioxidants from the Nordic Brown Alga Fucus vesiculosus*; National Food Institute (DTU Food), Technical University of Denmark (DTU): Lyngby, Denmark, 2016.

113. Arufe, S.; Sineiro, J.; Moreira, R. Determination of thermal transitions of gluten-free chestnut flour doughs enriched with brown seaweed powders and antioxidant properties of baked cookies. *Heliyon* **2019**, *5*, e01805. [CrossRef] [PubMed]

114. Quitral, V.; Sepúlveda, M.; Gamero-Vega, G.; Jiménez, P. Seaweeds in bakery and farinaceous foods: A mini-review. *Int. J. Gastron. Food Sci.* **2022**, *28*, 100403. [CrossRef]

115. Komatsuzaki, N.; Arai, S.; Fujihara, S.; Shima, J.; Wijesekara, R.S.; Dileepa, M.; De Croos, S.T. Development of Novel Bread by Combining Seaweed Kappaphycus alvarezii from Sri Lanka and Saccharomyces cerevisiae Isolated from Nectarine. *J. Agric. Sci. Technol. B* **2019**, *9*, 339–346. [CrossRef]

116. Surendhiran, D.; Li, C.; Cui, H.; Lin, L. Marine algae as efficacious bioresources housing antimicrobial compounds for preserving foods—A review. *Int. J. Food Microbiol.* **2021**, *358*, 109416. [CrossRef]

117. Dang, B.T.; Bui, X.T.; Tran, D.P.H.; Hao Ngo, H.; Nghiem, L.D.; Hoang, T.K.D.; Nguyen, P.T.; Nguyen, H.H.; Vo, T.K.Q.; Lin, C.; et al. Current application of algae derivatives for bioplastic production: A review. *Bioresour. Technol.* **2022**, *347*, 126698. [CrossRef]

118. Pavkovich, A.M.; Bell, D.S. Extraction ǀ Pressurized liquid extraction. In *Encyclopedia of Analytical Science*; Elsevier: Amsterdam, The Netherlands, 2019; pp. 78–83. ISBN 9780081019832.
119. Calvo-Flores, F.G. Green Processes in Foodomics. Green Solvents for Sustainable Processes. In *Comprehensive Foodomics*; Elsevier: Amsterdam, The Netherlands, 2020; pp. 690–709. ISBN 9780128163955.

 *antioxidants*

Article

Shedding Light on the Hidden Benefit of *Porphyridium cruentum* Culture

Davide Liberti [1], Paola Imbimbo [1,*], Enrica Giustino [1], Luigi D'Elia [1], Mélanie Silva [2], Luísa Barreira [2] and Daria Maria Monti [1,*]

[1] Department of Chemical Sciences, University of Naples Federico II, Via Cinthia 4, 80126 Naples, Italy
[2] Centre of Marine Sciences, University of Algarve, 8005-139 Faro, Portugal
* Correspondence: paola.imbimbo@unina.it (P.I.); mdmonti@unina.it (D.M.M.)

Abstract: Microalgae can represent a reliable source of natural compounds with different activities. Here, we evaluated the antioxidant and anti-inflammatory activity of sulfated exopolysaccharides (s-EPSs) and phycoerythrin (PE), two molecules naturally produced by the red marine microalga *Porphyridium cruentum* (CCALA415). *In vitro* and cell-based assays were performed to assess the biological activities of these compounds. The s-EPSs, owing to the presence of sulfate groups, showed biocompatibility on immortalized eukaryotic cell lines and a high antioxidant activity on cell-based systems. PE showed powerful antioxidant activity both *in vitro* and on cell-based systems, but purification is mandatory for its safe use. Finally, both molecules showed anti-inflammatory activity comparable to that of ibuprofen and helped tissue regeneration. Thus, the isolated molecules from microalgae represent an excellent source of antioxidants to be used in different fields.

Keywords: microalgae; exopolysaccharides; phycoerythrin; antioxidant activity; anti-inflammatory activity; biocompatibility; wound healing

Citation: Liberti, D.; Imbimbo, P.; Giustino, E.; D'Elia, L.; Silva, M.; Barreira, L.; Monti, D.M. Shedding Light on the Hidden Benefit of *Porphyridium cruentum* Culture. *Antioxidants* **2023**, *12*, 337. https://doi.org/10.3390/antiox12020337

Academic Editors: Antonella D'Anneo and Marianna Lauricella

Received: 9 January 2023
Revised: 25 January 2023
Accepted: 26 January 2023
Published: 31 January 2023

1. Introduction

Microalgae are ubiquitous eukaryotic photosynthetic microorganisms that are able to live in different environments, in single colonies, chains, or groups; depending on the species, their size can vary from a few to hundreds of micrometers [1–3]. The biodiversity of microalgae is mainly due to their unique ability to adapt and grow even under unfavorable growth conditions (e.g., extreme temperatures, variable salinity, and low or high light intensity) and to produce a wide range of interesting chemical compounds with novel structures and biological activities [4,5]. Among the microalgae, the red marine microalga *Porphyridium cruentum* could be pointed to as a commercial source of various high-value bioproducts [1], to be recovered from the same culture, in order to make the whole process economically feasible [6–10]. In particular, *P. cruentum* produces sulfated exopolysaccharides (s-EPSs) that are accumulated in a layer surrounding the cytoplasmic membrane. These exopolysaccharides act as a mucilage, because *P. cruentum* is without a well-defined cell wall [11]. They are composed of glucuronic acid and several major neutral monosaccharides, such as D- and L-Gal, D-Glc, D-Xyl, D-GlcA, and sulfate groups. S-EPSs from *P. cruentum* have antioxidant [12], immunomodulatory, anti-inflammatory, hypocholesterolemic, antimicrobial, and antiviral activity [13,14]. S-EPSs from *P. cruentum* also exhibit specific rheological properties that can be exploited in food applications [12,15]. In addition to exopolysaccharides, *P. cruentum* produces a broad range of colored pigments, including chlorophylls, carotenoids, and phycobilins, which are commercially utilized in the food, pharmaceutical, and cosmetic industries [16]. Amongst them, phycoerythrin (PE) is a light-harvesting protein with a structure of $(\alpha\beta)_6\gamma$ complex and a MW ranging from 240 to 260 kDa. Due to its unique biological properties, PE has gained much attention from the food and pharmaceutical industries and in the molecular biology field [17–22]. Here, starting from our recent results [10], a comprehensive study on the biological activities of s-EPSs and purified phycoerythrin was carried out in order to verify if the extraction techniques could affect their biological activities.

2. Materials and Methods

2.1. Reagents

All solvents, reagents, and chemicals were purchased from Sigma-Aldrich (St Louis, MO, USA).

2.2. Biocompounds Isolation

S-EPSs and PE were isolated and purified from the culture of *Porphyridium cruentum* (CCALA415) as previously described [10]. Briefly, at the end of cell growth, the culture was centrifuged to recover s-EPSs in the supernatant. The s-EPSs were precipitated by adding pure ethanol (1:2 v/v) and centrifuging the sample (12,000× g, 30 min, and 4 °C). The supernatant was discarded, and the precipitate was freeze-dried. S-EPSs yield was 300 ± 67 g/L, which corresponds to 0.53 g/$g_{d.w. biomass}$. In the case of PE, a crude aqueous extract was obtained *via* sonication (40% amplitude; 20 min, 30 s on and 30 s off) from the harvested biomass. PE was then isolated *via* a one-step purification procedure as reported by Liberti, up to a purity grade of 4 [10].

2.3. Eukaryotic Cell Culture and Biocompatibility Assay

Immortalized human keratinocytes (HaCaT, Innoprot, Derio, Spain) and immortalized murine fibroblasts Balb/c-3T3 (ATCC, Virginia, USA) were cultured in 10% foetal bovine serum in Dulbecco's modified Eagle's medium, in the presence of 1% penicillin/streptomycin and 2 mM L-glutamine, in a 5% CO_2 humidified atmosphere at 37 °C. To verify the biocompatibility of the crude extract of s-EPSs and of purified PE, cells were seeded in 96-well plates at a density of 2×10^3/well and, 24 h after seeding, were incubated with increasing concentrations of the extract/compounds (5 to 75 μg/mL for EPS, 5 to 500 μg/mL of total proteins for crude extracts, and 5 nM to 100 nM for purified PE) for 72 h. At the end of the incubation period, cell viability was assessed with the MTT assay. Cell survival is expressed as the percentage of viable cells in the presence of compounds compared with control cells (represented by the average obtained between untreated cells and cells supplemented with the highest concentration of buffer).

2.4. In Vitro Antioxidant Assays

The antioxidant activity of the extract/compounds was tested by measuring their ability to scavenge the free radicals 1,1-diphenyl-2-picrylhydrazyl radical and 2,2′-azinobis-[3-ethylbenzthiazoline-6-sulfonic acid] (DPPH and ABTS, respectively) and to reduce or chelate redox active iron and copper (ferric-reducing antioxidant power (FRAP); iron-chelating activity (ICA), and copper-chelating activity (CCA), respectively). DPPH and FRAP assays were performed following the procedure reported by Rodrigues et al. [23], and ascorbic acid and butylhydroxytoluene (BHT), respectively, were used as positive controls at the same concentrations of the sample under test. The ability of the extract/compounds to scavenge the ABTS radical was assessed as previously reported [24]. The results were compared to a calibration curve obtained using Trolox (6-hydroxy-2,5,7,8-tetramethylchromane-2-carboxylic acid) as the standard. ICA and CCA were determined by measuring the formation of the Fe^{2+}-ferrozine complex and by using pyrocatechol violet, respectively, according to the method reported by Megias [25]. EDTA was used as a standard at a final concentration of 100 μg/mL. S-EPS or purified PE was tested between 0.05 and 120 μg/mL and 0.2 and 270 nM, respectively. The results are expressed as IC_{50}, i.e., the concentration required to scavenge 50% of the free radical or as the highest percentage achieved.

2.5. Determination of Intracellular ROS Levels on Eukaryotic Cell Lines by DCFDA Assay

The protective effect of s-EPSs (from 5 to 75 µg/mL) or purified PE (10 nM) against oxidative stress was measured by determining the intracellular reactive oxygen species (ROS) levels, following the protocol used by Imbimbo [26].

2.6. Determination of Intracellular Glutathione Levels (DTNB Assay) and Lipid Peroxidation Levels (TBARS Assay) on Eukaryotic Cell Lines

Intracellular GSH levels and lipid peroxidation levels were measured by following the procedure described by Petruk [27] using 12 µg/mL of s-EPSs or 10 nM of purified PE.

2.7. Anti-Inflammatory Activity

The anti-inflammatory activity of the compounds was tested by their ability to inhibit cyclooxygenase-2 (COX-2). S-EPS or purified PE was tested at different concentrations (4 and 167 µg/mL for s-EPSs or 10 and 27 nM for purified PE) using a commercial inhibitory screening assay kit, Cayman test kit-560131 (Cayman Chemical Company, Ann Arbor, MI, USA). Ibuprofen was used as a positive control. Results are expressed as a percentage of inhibition of COX-2.

2.8. Wound Healing Assay

Wound healing was assessed with a scratch assay. HaCaT cells were seeded at a cell density of 3×10^5 cells/cm^2 for 24 h, to allow cells to reach about 95% of confluence. Then, cells were washed with PBS, scratched manually with a 200 µL pipet tip, and incubated with 12 µg/mL of s-EPSs or 10 nM of purified PE. The scratch size was monitored at 0 h and 24 h by acquiring images using optical microscopy (Zeiss LSM 710, Zeiss, Germany) at 10× magnification. The width of the wound was measured by using Zen Lite 2.3 software (Zeiss, Germany). Results are expressed as a reduction of the area (fold) compared with untreated cells.

2.9. Statistical Analyses

All the experiments were performed in triplicate. Results are presented as the mean of results obtained after three independent experiments (mean ±SD) and compared by one-way ANOVA according to Bonferroni's method (post hoc) using GraphPad Prism for Windows, version 6.01 (Dotmatics, California, USA).

3. Results

3.1. s-Exopolysaccharides Characterization

3.1.1. s-EPSs Biocompatibility on Cell-Based Model

s-EPSs were tested for their biocompatibility on two eukaryotic immortalized cell lines: HaCaT (human keratinocytes) and Balb/c-3T3 (murine fibroblasts). Twenty-four hours after seeding, cells were incubated with increasing amounts of s-EPSs (from 5 to 75 µg/mL). After 72 h of incubation, cell viability was assessed by the MTT assay; cell survival is expressed as the percentage of viable cells in the presence of s-EPSs compared with that of control samples (i.e., untreated cells). The results in Figure 1 show that, under all the experimental conditions, the s-EPSs were fully biocompatible on both the cell lines analyzed.

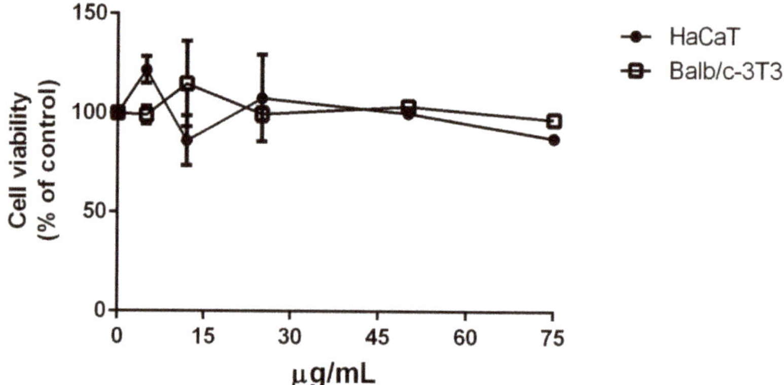

Figure 1. Effect of s-EPSs from *P. cruentum* on cell viability. Dose–response curves of HaCaT (black dots) and Balb/c-3T3 (empty squares) cells after 72 h of incubation with increasing concentrations of exopolysaccharides (5–75 µg/mL). Cell viability is reported as a function of s-EPS concentration.

3.1.2. s-EPSs *In Vitro* Antioxidant Activity

The antioxidant activity of s-EPS was evaluated with different *in vitro* analyses: ABTS, DPPH, FRAP, and iron and copper chelating assays. As shown in Table 1, s-EPSs were not able to scavenge the ABTS and DPPH radicals, whereas a slight but significant activity was observed for the chelation of iron and for ferric ion reduction assays. Both tests are based on the ability to act on iron: the former measures the ability of the compounds under test to bind Fe^{2+}, whereas the latter analyzes the ability to reduce Fe^{3+} to Fe^{2+}. As for the copper chelating assay, the highest activity reached, at the highest concentration tested, was $9 \pm 3\%$, a value much lower than the one obtained by testing the positive control molecule at the same concentration.

Table 1. *In vitro* antioxidant and chelating activity of s-EPSs. Results are expressed as percentage of inhibition. The concentration evaluated is referred to the final concentration of s-EPSs or positive control used in the well.

Test	Concentration (µg/mL)	s-EPSs Activity (%)	C+ Activity (%)
FRAP	120	34 ± 3	97 ± 1
ABTS	25	2 ± 2	98 ± 2
DPPH	50	1 ± 1	52 ± 1
ICA	55	66 ± 3	91 ± 1
CCA	45	9 ± 3	91 ± 2

3.1.3. s-EPSs Antioxidant Activity on a Cell-Based Model

The antioxidant activity of s-EPSs was also evaluated on HaCaT cells. For this purpose, cells were incubated with increasing concentrations of s-EPSs (from 5 to 50 µg/mL) for 2 h, and then oxidative stress was induced by UVA irradiation (100 J/cm^2). Immediately after irradiation, the intracellular ROS levels were measured by using H_2DCFDA as a probe. For each set of experiments, untreated cells were used as a control. As shown in Figure 2, UVA treatment significantly increased the DCF fluorescence (black bars, $p < 0.001$). In the absence of stress, s-EPSs induced a slight but significant increase in the intracellular ROS level (Figure 2, white, dashed grey, and dark grey bars on the left part of the graph). Interestingly, when cells were preincubated with s-EPSs prior to being stressed, only 5 and 12 µg/mL were able to protect the cells from ROS formation (Figure 2, light grey and white bars on the right part of the graph), whereas the higher concentrations had no protective effect. This result is in agreement with those of Giordano et al. [28], as antioxidants act at low concentrations, whereas, at high concentrations, they may work as pro-oxidants. Based on these results, s-EPSs were used at 12 µg/mL for further experiments.

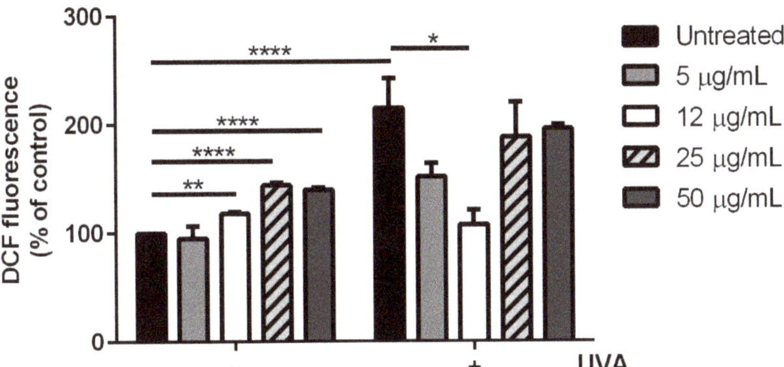

Figure 2. Antioxidant activity of s-EPSs on UVA-stressed HaCaT cells. Intracellular ROS levels were determined with DCFDA assay. Cells were preincubated in the presence of increasing amounts (from 5 to 50 µg/mL) of s-EPSs for 2 h prior to UVA irradiation (100 J/cm^2). Results are expressed as percentages compared with untreated cells. Black bars refer to untreated cells; light grey bars refer to cells incubated with 5 µg/mL of s-EPSs; white bars refer to cells incubated with 12 µg/mL; dashed bars refer to cells incubated with 25 µg/mL; dark grey bars refer to cells incubated with 50 µg/mL of s-EPSs in the absence (−) or presence (+) of UVA stress. Data shown are means ± S.D. of three independent experiments. * indicates $p < 0.05$, ** indicates $p < 0.01$, and **** indicates $p < 0.001$.

To deeply analyze the protective effect of s-EPSs, the intracellular glutathione levels and lipid peroxidation levels were determined with DTNB and TBARS assays, respectively. In the absence of any treatment, a significant decrease ($p < 0.01$) in GSH levels was observed after UVA exposure (Figure 3A), and s-EPSs (grey bars) were able to inhibit GSH oxidation, thus confirming a protective effect against oxidative stress. As for the TBARS assay, a significant increase ($p < 0.05$) in lipid peroxidation levels was observed after UVA treatment (black bars, Figure 3B), but, notably, this effect was inhibited upon pretreatment with s-EPSs (grey bars). Treatment of the cells with exopolysaccharides did not significantly alter either glutathione or lipid peroxidation levels in the absence of UVA treatment (−). Taken together, the results clearly indicate that s-EPSs are able to protect cells from oxidative damage.

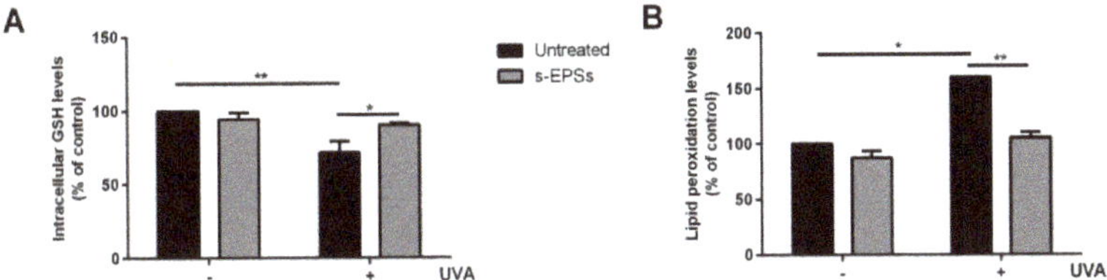

Figure 3. Protective effect of s-EPSs on HaCaT cells. Intracellular GSH levels were determined with a DTNB assay (**A**) and lipid peroxidation levels were determined with a TBARS assay (**B**). Cells were preincubated in the presence of 12 µg/mL of s-EPSs for 2 h prior to UVA irradiation (100 J/cm^2). GSH and lipid peroxidation levels were measured 90 min after UVA irradiation. Black bars refer to untreated cells, and grey bars refer to cells incubated with s-EPSs, in the absence (−) or in the presence (+) of UVA stress. Values are expressed as percentages compared with untreated cells. Data shown are means ± S.D. of three independent experiments. * Indicates $p < 0.05$; ** indicates $p < 0.01$.

3.1.4. *In Vitro* Anti-Inflammatory Activity of s-EPSs

As inflammation is a condition strictly linked to oxidative stress, the anti-inflammatory activity of s-EPSs was measured by evaluating their capacity to inhibit the enzyme COX-2. When inflammation occurs, COX-2 is able to enhance the prostanoid production [29]. As reported in Table 2, surprisingly, s-EPSs showed no significant differences compared with ibuprofen used as positive control when tested at the same concentration, thus suggesting a new role of s-EPSs in inflammation control.

Table 2. *In vitro* s-EPSs anti-inflammatory activity.

	Concentration (µg/mL)	Inhibition (%)
s-EPSs	167	77 ± 8
Ibuprofen	167	99 ± 1

3.2. Phycoerythrin Characterization

3.2.1. Phycoerythrin Biocompatibility on Immortalized Eukaryotic Cells

Following biomass lysis, phycoerythrin (PE) had a purity grade of 1.5 [10]. This value is considered as reagent-grade, thus indicating that the protein can be used as it is for food applications [30]. In order to verify the safety of the protein on eukaryotic cells, an MTT assay was performed by comparing the crude extract with the purified protein (purity grade of four). The results of the MTT assay, reported in Figure 4, clearly show that only pure PE was fully biocompatible with both cell lines (Figure 4B), while the crude extract exerted a dose-dependent toxicity (Figure 4A). These results clearly indicate that PE needs to be purified to a higher purity grade before being used on cell-based models, or, at least, that it cannot be applied when present in the extract at concentrations higher than a certain threshold (100 µg/mL).

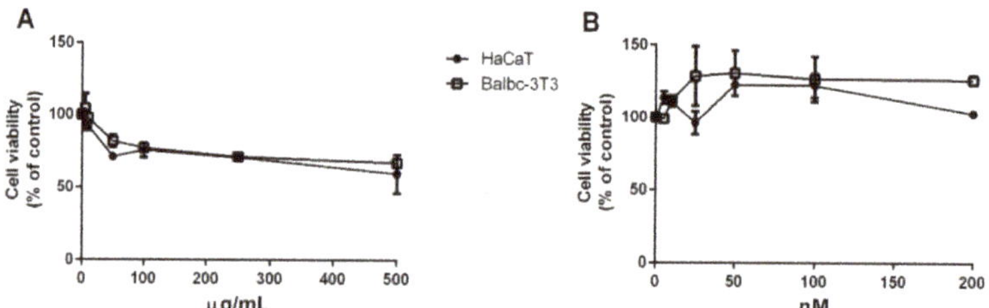

Figure 4. Biocompatibility of total extract (**A**) and purified PE (**B**) on eukaryotic cells. Dose–response curves of HaCaT (black dots) and Balb/c-3T3 cells (empty squares) after 72 h of incubation with increasing concentrations of total extract (**A**) and purified PE (**B**). Cell viability was assessed with an MTT assay and is reported as a function of extract/protein concentration.

3.2.2. *In Vitro* Antioxidant Activity

In vitro analysis of the antioxidant activity of purified PE was carried out with the abovementioned experimental procedures. As reported in Table 3, purified PE was not able to scavenge the DPPH radical or chelate copper ions. However, it demonstrated a high capacity to scavenge the ABTS radical ion and to reduce ferric iron or chelate iron with considerably low IC_{50} values (0.072 ± 0.004 and 0.084 ± 0.012 µM, 0.084 ± 0.004 µM, respectively). Noteworthy, the purified PE IC_{50} values were about 160, 1000, and 600 times lower than the IC_{50} values obtained with the positive control molecules (Trolox, 12 ± 1 µM in the ABTS; BHT, 90 ± 4 µM in the FRAP; and EDTA, 51 ± 3 µM in the ICA).

Table 3. *In vitro* antioxidant and chelating activity of purified PE. Results are expressed as IC$_{50}$ values, µM.

Test	Purified PE	Positive Control
	IC$_{50}$ (µM)	
ABTS	0.072 ± 0.004	12 ± 1
DPPH	>0.27	29 ± 2
FRAP	0.084 ± 0.012	90 ± 4
ICA	0.084 ± 0.004	51 ± 3
CCA	>0.1	63 ± 2

3.2.3. Cell-Based Antioxidant Activity of PE

Starting from the encouraging results obtained *in vitro*, purified PE was tested on the UVA-stressed HaCaT experimental system used for s-EPSs. Cells were treated with 2.5 µg/mL (10 nM) of purified PE for 2 h, and then oxidative stress was induced by UVA irradiation (100 J/cm^2). At the end of irradiation, the intracellular ROS levels were evaluated. As shown in Figure 5, UVA induced a significant increase in intracellular ROS levels (black bars, 200%) compared with untreated cells ($p < 0.001$). When cells were treated with purified PE (grey bars), no increase in intracellular ROS levels was observed. Interestingly, when cells were incubated with purified PE prior to UVA exposure, an inhibition of the intracellular ROS production was observed.

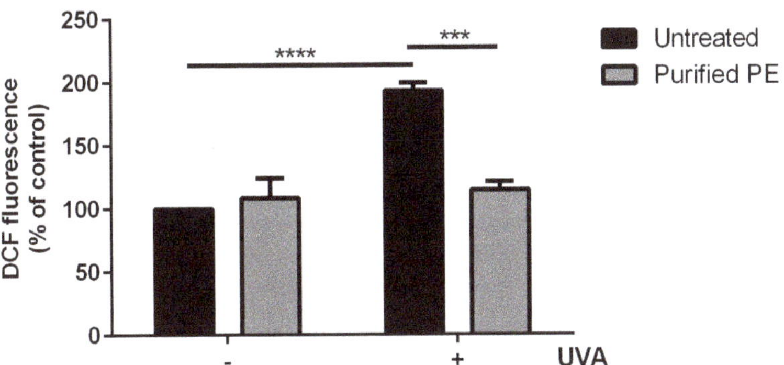

Figure 5. Protective effect of purified PE on UVA-stressed HaCaT cells. Intracellular ROS levels were determined with DCFDA assay. Cells were preincubated in the presence of 10 nM of purified PE (grey bars) for 2 h prior to UVA irradiation (100 J/cm^2). Black bars refer to untreated cells in the absence ($-$) or in the presence (+) of UVA stress. Values are expressed as percentages compared with untreated cells. Data shown are means $\pm$ S.D. of three independent experiments. *** indicates $p < 0.005$; **** indicates $p < 0.001$ with respect to UVA-treated cells.

The effect of purified PE on GSH and lipid peroxidation was also assessed. As shown in Figure 6, PE was able to fully protect cells from oxidative stress, as no alteration in either the GSH levels (Figure 6A) or in the lipid peroxidation levels (Figure 6B) was found when the cells were pretreated with purified PE prior to stress, thus confirming the protective effect of the protein against oxidative stress.

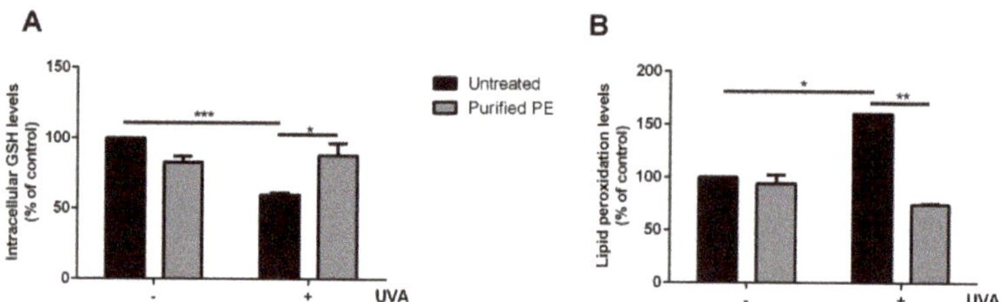

Figure 6. Analysis of intracellular GSH and lipid peroxidation levels on HaCaT cells. Cells were preincubated with 10 nM of purified PE for 2 h before UVA irradiation (100 J/cm^2). (**A**) determination of intracellular GSH levels; (**B**) analysis of lipid peroxidation levels. In both experiments, measurements were recorded 90 min after UVA-induced stress. Values are expressed as a percentage compared with control (i.e., untreated) cells. Data shown are means ± S.D. of three independent experiments. * indicates $p < 0.05$, ** indicates $p < 0.01$, and *** indicates $p < 0.005$.

3.2.4. *In Vitro* PE Anti-Inflammatory Activity

Purified PE was also able to inhibit COX-2 (Table 4) by about 75%, although the level of inhibition attained with ibuprofen 24 nM could not be achieved at any of the analyzed concentrations.

Table 4. COX-2 inhibition by purified PE.

Sample	Concentration (nM)	Inhibition (%)
Phycoerythrin	27	75 ± 8
	10	72 ± 8
Ibuprofen	24	96 ± 1

3.3. *Effect of s-EPSs and Purified PE on Wound Healing*

Finally, a scratch assay was carried out on HaCaT cells to test the ability of s-EPSs and purified PE to induce cell migration related to wound repairing. The results are reported in Figure 7 and in Table 5. In the absence of any treatment, the cells spontaneously migrated to induce the re-epithelialization. Interestingly, when the cells were treated with either s-EPSs or purified PE, a significant enhancement in the wound closure was observed after 24 h. Indeed, s-EPSs reduced the scratched area by 2.5 ± 0.17-fold and purified PE by 2.4 ± 0.1317-fold compared with untreated cells (1.80 ± 0.02-fold reduction).

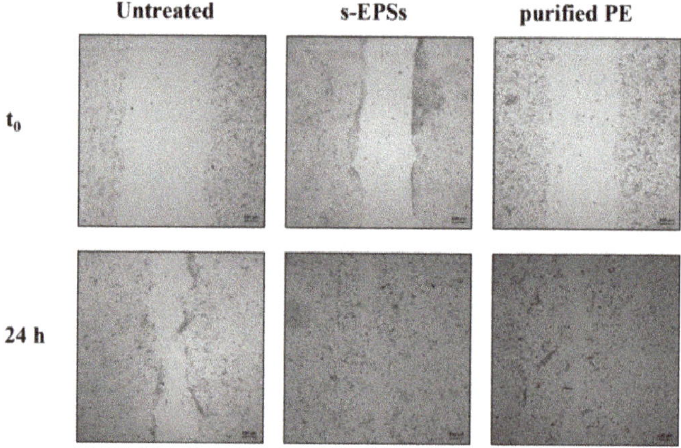

Figure 7. Effect of s-EPSs and purified PE on wound healing. Confluent HaCaT cells were scratched and treated with either 12 µg/mL s-EPSs or 10 nM purified PE for 24 h. Optical microscopy images were acquired at 10× magnification at the beginning (t$_0$) and end (24 h) of the incubation.

Table 5. Reduction of area (fold) of wound closure upon 24 h of incubation with either s-EPSs or purified PE. Data shown are means ± S.D. of three independent experiments. For each experiment, at least 10 images were acquired. * indicates $p < 0.05$.

Sample	Reduction in Area (Fold)
Untreated	1.80 ± 0.02
s-EPSs	2.50 ± 0.17 *
Purified PE	2.40 ± 0.13 *

* = $p < 0.05$.

4. Discussion

As the use of synthetic molecules is known to be harmful in the long run, the search for new natural compounds endowed with beneficial properties is urgent [31]. In this context, antioxidants from microalgae could represent an excellent alternative, but the costs of microalgae upstream and downstream processes are still too high [32].

We recently set up a cascade approach to recover four classes of molecules from *P. cruentum* culture: s-EPSs, PE, carotenoids, and saturated fatty acids. Among them, here, we evaluated the biological activity of s-EPSs and PE. S-EPSs were chosen as it is generally thought that polysaccharides with a high sulfated content have biological activities [33], such as antioxidant action [34]. It is known that antioxidant molecules can bind metal ions, forming metal–ion complexes. The presence of sulfate groups could increase the metal-binding capacity of the carbohydrates by donating an electron pair or by losing a proton, thus stabilizing the complex [35,36].

In agreement with the findings of Wang et al., we found that s-EPSs had no radical scavenging activity against DPPH, whereas they showed antioxidant activity in the ABTS assay, with IC_{50} values ranging from 6.59 to 8.92 mg/mL [37]. Interestingly, despite the low antioxidant activity observed *in vitro*, s-EPSs were active on a cell-based system at a concentration almost 600 times lower than that measured *in vitro*. This result is in agreement with literature, as it is well known that *in vitro* assays should not be compared with cell-based ones. Indeed, antioxidants provide their function by different mechanisms of action, so that bioavailability, stability, retention, or reactivity of the compound under test in a complex system, such as that of eukaryotic cells, cannot be either mimicked or evaluated *in vitro* [38]. Our results indicated that s-EPSs were able not only to inhibit the intracellular ROS production but also to prevent GSH depletion and lipid peroxidation.

Different is the case of PE, which was found to be a very powerful antioxidant agent *in vitro* and on a cell-based system. The ABTS assay was in line with that observed by Sonani on a PE from a different source (IC_{50} of 72 ± 4 nM vs. 101 nM, respectively) [39], whereas the PE prepared by this author had lower DPPH-scavenging (IC_{50} of 930 nM) and iron-chelating abilities (IC_{50} of 484 nM) than the purified PE prepared in this study. We hypothesize that the higher antioxidant activity measured in our experimental system may rely on the source or strain used. A different source may also affect the biocompatibility results: indeed, we found that only pure PE was biocompatible with eukaryotic cells, strongly suggesting the importance of purification of the protein for all the potential applications. Pure PE protected cells from UVA irradiation at a concentration in the low nanomolar range (10 nM). Generally, antioxidants prevent the generation of free radicals, which can significantly affect some physiological processes, including wound healing. In particular, ROS generation can damage tissues and slow down the regeneration process. The presence of antioxidants should counteract chronic inflammation and at the same time contribute to promoting tissue regeneration [40]. Considering that both s-EPSs and PE were able to inhibit one of the key enzymes in the inflammation process (COX-2) and to induce a significantly faster scratch closure compared with untreated cells, we can conclude that the bioproducts obtained by *P. cruentum* represent an excellent ingredient for new biomaterials, such as medical patches.

5. Conclusions

In this study, s-EPSs and PE, obtained from *P. cruentum* culture by a cascade approach described in a previous work [10], showed a remarkable antioxidant activity in a cell-based system, higher than that obtained by *in vitro* assays, thus suggesting that the reliability of *in vitro* assays has to be overhauled. Moreover, both molecules showed anti-inflammatory characteristics comparable with ibuprofen and a significant ability to promote cell proliferation.

Author Contributions: D.L., P.I. and D.M.M. designed the concept and supervised the experiments. D.L., P.I., E.G. and L.D. performed the experimental work with microalgae. D.L., P.I., E.G. and D.M.M. wrote the manuscript. L.B. supervised M.S. on the in vitro experiments. All authors have read and agreed to the published version of the manuscript.

Funding: This work was carried out with funding from the Portuguese Foundation for Science and Technology (FCT) through UIDB/04326/2020, UIDP/0436/2020, LA/P/0101/2020.

Institutional Review Board Statement: Not applicable.

Informed Consent Statement: Not applicable.

Data Availability Statement: The data presented in this study are available in the article.

Acknowledgments: P.I. would like to acknowledge the ALGAE4IBD project (From Natura to Bedside-Algae Based Bio Compound for Prevention) funded by the European Union's Horizon 2020 Research and Innovation program under grant agreement No. 101000501.

Conflicts of Interest: The authors declare no conflict of interest.

References

1. Mobin, S.; Alam, F. Some Promising Microalgal Species for Commercial Applications: A Review. *Energy Procedia* **2017**, *110*, 510–517. [CrossRef]
2. Alam, F.; Date, A.; Rasjidin, R.; Mobin, S.M.A.; Moria, H.; Baqui, A. Biofuel from Algae: Is It a Viable Alternative? In *Advances in Biofuel Production: Algae and Aquatic Plants*; Apple Academic Press: New York, NY, USA, 2014; pp. 32742–33487.
3. Suganya, T.; Varman, M.; Masjuki, H.H.; Renganathan, S. Macroalgae and Microalgae as a Potential Source for Commercial Applications along with Biofuels Production: A Biorefinery Approach. *Renew. Sustain. Energy Rev.* **2016**, *55*, 909–941. [CrossRef]
4. Herrero, M.; Mendiola, J.A.; Plaza, M.; Ibañez, E. Screening for Bioactive Compounds from Algae. In *Advanced Biofuels and Bioproducts*; Springer: New York, NY, USA, 2013; pp. 833–872.
5. Anbuchezhian, R.; Karuppiah, V.; Li, Z. Prospect of Marine Algae for Production of Industrially Important Chemicals. In *Algal Biorefinery: An Integrated Approach*; Springer: Cham, Switzerland, 2015; pp. 195–217.
6. Feller, R.; Matos, Â.P.; Mazzutti, S.; Moecke, E.H.S.; Tres, M.V.; Derner, R.B.; Oliveira, J.V.; Junior, A.F. Polyunsaturated ω-3 and ω-6 Fatty Acids, Total Carotenoids and Antioxidant Activity of Three Marine Microalgae Extracts Obtained by Supercritical CO_2 and Subcritical n-Butane. *J. Supercrit. Fluids* **2018**, *133*, 437–443. [CrossRef]
7. Di Lena, G.; Casini, I.; Lucarini, M.; Lombardi-Boccia, G. Carotenoid Profiling of Five Microalgae Species from Large-Scale Production. *Food Res. Int.* **2019**, *120*, 810–818. [CrossRef]
8. Rebolloso Fuentes, M.M.; Acién Fernández, G.G.; Sánchez Pérez, J.A.; Guil Guerrero, J.L. Biomass Nutrient Profiles of the Microalga *Porphyridium cruentum*. *Food Chem.* **2000**, *70*, 345–353. [CrossRef]
9. Gaignard, C.; Gargouch, N.; Dubessay, P.; Delattre, C.; Pierre, G.; Laroche, C.; Fendri, I.; Abdelkafi, S.; Michaud, P. New Horizons in Culture and Valorization of Red Microalgae. *Biotechnol. Adv.* **2019**, *37*, 193–222. [CrossRef]
10. Liberti, D.; Imbimbo, P.; Giustino, E.; D'elia, L.; Ferraro, G.; Casillo, A.; Illiano, A.; Pinto, G.; Di Meo, M.C.; Alvarez-Rivera, G.; et al. Inside out *Porphyridium Cruentum*: Beyond the Conventional Biorefinery Concept. *ACS Sustain. Chem. Eng.* **2023**, *11*, 381–389. [CrossRef]
11. Dvir, I.; Stark, A.H.; Chayoth, R.; Madar, Z.; Arad, S.M. Hypocholesterolemic Effects of Nutraceuticals Produced from the Red Microalga *Porphyridium* sp. in Rats. *Nutrients* **2009**, *1*, 156–167. [CrossRef]
12. Ben Hlima, H.; Smaoui, S.; Barkallah, M.; Elhadef, K.; Tounsi, L.; Michaud, P.; Fendri, I.; Abdelkafi, S. Sulfated Exopolysaccharides from *Porphyridium cruentum*: A Useful Strategy to Extend the Shelf Life of Minced Beef Meat. *Int. J. Biol. Macromol.* **2021**, *193*, 1215–1225. [CrossRef]
13. Mišurcová, L.; Škrovánková, S.; Samek, D.; Ambrožová, J.; Machů, L. Health Benefits of Algal Polysaccharides in Human Nutrition. *Adv. Food Nutr. Res.* **2012**, *66*, 75–145.

14. Gallego, R.; Martínez, M.; Cifuentes, A.; Ibáñez, E.; Herrero, M. Development of a Green Downstream Process for the Valorization of *Porphyridium cruentum* Biomass. *Molecules* **2019**, *24*, 1564. [CrossRef]

15. Medina-Cabrera, E.V.; Gansbiller, M.; Rühmann, B.; Schmid, J.; Sieber, V. Rheological Characterization of *Porphyridium sordidum* and *Porphyridium purpureum* Exopolysaccharides. *Carbohydr. Polym.* **2021**, *253*, 117237. [CrossRef]

16. Sudhakar, M.P.; Jagatheesan, A.; Perumal, K.; Arunkumar, K. Methods of Phycobiliprotein Extraction from *Gracilaria crassa* and Its Applications in Food Colourants. *Algal Res.* **2015**, *8*, 115–120. [CrossRef]

17. Kannaujiya, V.K.; Kumar, D.; Pathak, J.; Sinha, R.P. Phycobiliproteins and Their Commercial Significance. In *Cyanobacteria*; Elsevier: Amsterdam, The Netherlands, 2019; pp. 207–216.

18. Hsieh-Lo, M.; Castillo, G.; Ochoa-Becerra, M.A.; Mojica, L. Phycocyanin and Phycoerythrin: Strategies to Improve Production Yield and Chemical Stability. *Algal Res.* **2019**, *42*, 101600. [CrossRef]

19. Manirafasha, E.; Ndikubwimana, T.; Zeng, X.; Lu, Y.; Jing, K. Phycobiliprotein: Potential Microalgae Derived Pharmaceutical and Biological Reagent. *Biochem. Eng. J.* **2016**, *109*, 282–296. [CrossRef]

20. Simovic, A.; Combet, S.; Velickovic, T.C.; Nikolic, M.; Minic, S. Probing the Stability of the Food Colourant R-Phycoerythrin from Dried Nori Flakes. *Food Chem.* **2022**, *374*, 131780. [CrossRef]

21. Ganesan, A.R.; Subramani, K.; Shanmugam, M.; Seedevi, P.; Park, S.; Alfarhan, A.H.; Rajagopal, R.; Balasubramanian, B. A Comparison of Nutritional Value of Underexploited Edible Seaweeds with Recommended Dietary Allowances. *J. King Saud Univ.* **2020**, *32*, 1206–1211. [CrossRef]

22. García, A.B.; Longo, E.; Murillo, M.C.; Bermejo, R. Using a B-Phycoerythrin Extract as a Natural Colorant: Application in Milk-Based Products. *Molecules* **2021**, *26*, 297. [CrossRef]

23. Rodrigues, M.J.; Neves, V.; Martins, A.; Rauter, A.P.; Neng, N.R.; Nogueira, J.M.F.; Varela, J.; Barreira, L.; Custódio, L. In Vitro Antioxidant and Anti-Inflammatory Properties of *Limonium algarvense* Flowers' Infusions and Decoctions: A Comparison with Green Tea (*Camellia sinensis*). *Food Chem.* **2016**, *200*, 322–329. [CrossRef]

24. Re, R.; Pellegrini, N.; Proteggente, A.; Pannala, A.; Yang, M.; Rice-Evans, C. Antioxidant Activity Applying an Improved ABTS Radical Cation Decolorization Assay. *Free Radic. Biol. Med.* **1999**, *26*, 1231–1237. [CrossRef]

25. Megías, C.; Pastor-Cavada, E.; Torres-Fuentes, C.; Girón-Calle, J.; Alaiz, M.; Juan, R.; Pastor, J.; Vioque, J. Chelating, Antioxidant and Antiproliferative Activity of *Vicia sativa* Polyphenol Extracts. *Eur. Food Res. Technol.* **2009**, *230*, 353–359. [CrossRef]

26. Imbimbo, P.; Romanucci, V.; Pollio, A.; Fontanarosa, C.; Amoresano, A.; Zarrelli, A.; Olivieri, G.; Monti, D.M. A Cascade Extraction of Active Phycocyanin and Fatty Acids from *Galdieria phlegrea*. *Appl. Microbiol. Biotechnol.* **2019**, *103*, 9455–9464. [CrossRef] [PubMed]

27. Petruk, G.; Di Lorenzo, F.; Imbimbo, P.; Silipo, A.; Bonina, A.; Rizza, L.; Piccoli, R.; Monti, D.M.; Lanzetta, R. Protective Effect of *Opuntia ficus-indica* L. Cladodes against UVA-Induced Oxidative Stress in Normal Human Keratinocytes. *Bioorg. Med. Chem. Lett.* **2017**, *27*, 5485–5489. [CrossRef] [PubMed]

28. Giordano, M.E.; Caricato, R.; Lionetto, M.G. Concentration Dependence of the Antioxidant and Prooxidant Activity of Trolox in Hela Cells: Involvement in the Induction of Apoptotic Volume Decrease. *Antioxidants* **2020**, *9*, 1058. [CrossRef] [PubMed]

29. Landa, P.; Kokoska, L.; Pribylova, M.; Vanek, T.; Marsik, P. In Vitro Anti-Inflammatory Activity of Carvacrol: Inhibitory Effect on COX-2 Catalyzed Prostaglandin E2 Biosynthesisb. *Arch. Pharm. Res.* **2009**, *32*, 75–78. [CrossRef]

30. Fernández-Rojas, B.; Hernández-Juárez, J.; Pedraza-Chaverri, J. Nutraceutical Properties of Phycocyanin. *J. Funct. Foods* **2014**, *11*, 375–392. [CrossRef]

31. Montégut, L.; de Cabo, R.; Zitvogel, L.; Kroemer, G. Science-Driven Nutritional Interventions for the Prevention and Treatment of Cancer. *Cancer Discov.* **2022**, *12*, 2258–2279. [CrossRef]

32. Coulombier, N.; Jauffrais, T.; Lebouvier, N. Antioxidant Compounds from Microalgae: A Review. *Mar. Drugs* **2021**, *19*, 549. [CrossRef]

33. Wang, Y.; Xing, M.; Cao, Q.; Ji, A.; Liang, H.; Song, S. Biological Activities of Fucoidan and the Factors Mediating Its Therapeutic Effects: A Review of Recent Studies. *Mar. Drugs* **2019**, *17*, 183. [CrossRef]

34. de Souza, M.C.R.; Marques, C.T.; Dore, C.M.G.; da Silva, F.R.F.; Rocha, H.A.O.; Leite, E.L. Antioxidant Activities of Sulfated Polysaccharides from Brown and Red Seaweeds. *J. Appl. Phycol.* **2007**, *19*, 153–160. [CrossRef]

35. Bhunia, B.; Uday, U.S.P.; Oinam, G.; Mondal, A.; Bandyopadhyay, T.K.; Tiwari, O.N. Characterization, Genetic Regulation and Production of Cyanobacterial Exopolysaccharides and Its Applicability for Heavy Metal Removal. *Carbohydr. Polym.* **2018**, *179*, 228–243. [CrossRef]

36. Andrew, M.; Jayaraman, G. Structural Features of Microbial Exopolysaccharides in Relation to Their Antioxidant Activity. *Carbohydr. Res.* **2020**, *487*, 107881. [CrossRef]

37. Wang, W.-N.; Li, Y.; Zhang, Y.; Xiang, W.-Z.; Li, A.-F.; Li, T. Comparison on Characterization and Antioxidant Activity of Exopolysaccharides from Two Porphyridium Strains. *J. Appl. Phycol.* **2021**, *33*, 2983–2994. [CrossRef]

38. López-Alarcón, C.; Denicola, A. Evaluating the Antioxidant Capacity of Natural Products: A Review on Chemical and Cellular-Based Assays. *Anal. Chim. Acta* **2013**, *763*, 1–10. [CrossRef]

39. Sonani, R.R.; Singh, N.K.; Kumar, J.; Thakar, D.; Madamwar, D. Concurrent Purification and Antioxidant Activity of Phycobiliproteins from *Lyngbya* sp. A09DM: An Antioxidant and Anti-Aging Potential of Phycoerythrin in Caenorhabditis Elegans. *Process Biochem.* **2014**, *49*, 1757–1766. [CrossRef]
40. Alvarez, X.; Alves, A.; Ribeiro, M.P.; Lazzari, M.; Coutinho, P.; Otero, A. Biochemical Characterization of *Nostoc* sp. Exopolysaccharides and Evaluation of Potential Use in Wound Healing. *Carbohydr. Polym.* **2021**, *254*, 117303. [CrossRef]

 *antioxidants*

Article

A Deadly Liaison between Oxidative Injury and p53 Drives Methyl-Gallate-Induced Autophagy and Apoptosis in HCT116 Colon Cancer Cells

Antonietta Notaro [1,†], Marianna Lauricella [2,†], Diana Di Liberto [2], Sonia Emanuele [2], Michela Giuliano [1], Alessandro Attanzio [1], Luisa Tesoriere [1], Daniela Carlisi [2], Mario Allegra [1], Anna De Blasio [1], Giuseppe Calvaruso [1] and Antonella D'Anneo [1,*]

1   Laboratory of Biochemistry, Department of Biological, Chemical and Pharmaceutical Sciences and Technologies (STEBICEF), University of Palermo, 90127 Palermo, Italy; antonietta.notaro@unipa.it (A.N.); michela.giuliano@unipa.it (M.G.); alessandro.attanzio@unipa.it (A.A.); luisa.tesoriere@unipa.it (L.T.); mario.allegra@unipa.it (M.A.); anna.deblasio@unipa.it (A.D.B.); giuseppe.calvaruso@unipa.it (G.C.)
2   Section of Biochemistry, Department of Biomedicine, Neurosciences and Advanced Diagnostics (BIND), University of Palermo, 90127 Palermo, Italy; marianna.lauricella@unipa.it (M.L.); diana.diliberto@unipa.it (D.D.L.); sonia.emanuele@unipa.it (S.E.); daniela.carlisi@unipa.it (D.C.)
*   Correspondence: antonella.danneo@unipa.it; Tel.: +39-09123890650
†   These authors contributed equally to this work.

Abstract: Methyl gallate (MG), which is a gallotannin widely found in plants, is a polyphenol used in traditional Chinese phytotherapy to alleviate several cancer symptoms. Our studies provided evidence that MG is capable of reducing the viability of HCT116 colon cancer cells, while it was found to be ineffective on differentiated Caco-2 cells, which is a model of polarized colon cells. In the first phase of treatment, MG promoted both early ROS generation and endoplasmic reticulum (ER) stress, sustained by elevated PERK, Grp78 and CHOP expression levels, as well as an upregulation in intracellular calcium content. Such events were accompanied by an autophagic process (16–24 h), where prolonging the time (48 h) of MG exposure led to cellular homeostasis collapse and apoptotic cell death with DNA fragmentation and p53 and γH2Ax activation. Our data demonstrated that a crucial role in the MG-induced mechanism is played by p53. Its level, which increased precociously (4 h) in MG-treated cells, was tightly intertwined with oxidative injury. Indeed, the addition of N-acetylcysteine (NAC), which is a ROS scavenger, counteracted the p53 increase, as well as the MG effect on cell viability. Moreover, MG promoted p53 accumulation into the nucleus and its inhibition by pifithrin-α (PFT-α), which is a negative modulator of p53 transcriptional activity, enhanced autophagy, increased the LC3-II level and inhibited apoptotic cell death. These findings provide new clues to the potential action of MG as a possible anti-tumor phytomolecule for colon cancer treatment.

Keywords: oxidative stress; phytocompounds; methyl gallate; autophagy; apoptosis; p53

Citation: Notaro, A.; Lauricella, M.; Di Liberto, D.; Emanuele, S.; Giuliano, M.; Attanzio, A.; Tesoriere, L.; Carlisi, D.; Allegra, M.; De Blasio, A.; et al. A Deadly Liaison between Oxidative Injury and p53 Drives Methyl-Gallate-Induced Autophagy and Apoptosis in HCT116 Colon Cancer Cells. *Antioxidants* **2023**, *12*, 1292. https://doi.org/10.3390/antiox12061292

Academic Editor: Stanley Omaye

Received: 19 May 2023
Revised: 9 June 2023
Accepted: 13 June 2023
Published: 16 June 2023

1. Introduction

Nowadays, the identification of non-toxic drugs for normal cells capable of selectively targeting tumor systems represents one of the main challenges in the development of innovative and tailored therapies for cancer [1].

Cancer represents a global problem with a continued growing expansion, seriously affecting public health. For many decades, multimodal approaches based on surgery, radiation therapy and chemotherapy have been used. In recent years, a revolution in the development of tumor-targeting drugs has been based on the knowledge of neoplastic entities, which are hallmarks that have substantially improved the types of combinatorial strategies, thus opening the way for precision cancer medicine [2,3].

To achieve this goal, growing interest toward plant-derived chemicals has pushed cancer scientists to search for new unexplored molecules as preventative or anti-tumor compounds. Theoretically, each plant can represent a significant reservoir of bioactive compounds, such as secondary metabolites, phytonutrients, nutraceuticals and supplements [4], that can be obtained from the vegetal world, tested in a laboratory, and applied to ameliorate human health or fight chronic diseases. Plants have developed the ability to synthetize secondary metabolites through highly controlled biochemical pathways that act in response to environmental insults related to abiotic or biotic stresses. In general, these metabolites can be classified as phenolic, terpenoid and alkaloid compounds [5], and due to their multifaceted activities, have stimulated increasing interest from industries for their application as dietary supplements, biocides, pharmaceuticals or potential medicinal drugs.

Due to their intrinsic properties based on the action on multiple targets, limited side effects and excellent efficacies, plant-derived natural products have been applied as healing agents to treat a very wide range of human diseases [6]. A lot of related examples stem from natural drugs used for the treatment of different human malignancies. The main features of these compounds, such as terpenoids (artesunate, atractylodes, andrographolide, etc.), phenols (resveratrol, quercetin, curcumin, capsaicin, etc.) and alkaloids (piperine, berberine, matrine, etc.), rely on their ability to inhibit tumor proliferation and angiogenesis, trigger apoptosis and modulate immune responses [7,8]. A compound that is widely spread in chemotherapeutic regimens is paclitaxel, which is a diterpenoid molecule originally isolated from *Taxus brevifolia* [9,10]. It represents a classical natural phytochemical that has attracted the attention of researchers for its therapeutical ability as an anti-neoplastic agent in the treatment of lung, breast and ovary cancers [10–12]. An analog successful chemotherapeutic agent is campthothecin, which is a bioactive compound isolated from *Camptotheca acuminate* [13], that has shown significant efficacy against many different solid tumors when used alone or in combination treatment with cisplatin [14].

In light of these considerations, our research has recently focused on the chemical characterization of plant and fruit extracts or essential oils to discover anticancer phytomolecules to apply alone or as adjuvants to conventional therapies [15–22]. Their identification could also offer the chance to chemically modify plant-derived molecules through the introduction of active pharmacophores, creating novel and powerful lead compounds [23]. In particular, our interest was recently directed toward phytochemicals that are widely found in *Mangifera indica* L., where the peel fraction of the fruit is rich in many bioactive compounds, such as mangiferin; citric acid; quinic acid; digallic acids; gallic acid; and its esters, such as methyl gallate and pentagalloyl glucose [15]. We demonstrated that methyl gallate and pentagalloyl glucose were the most common phytocostituents and were found to be particularly effective in reducing the cell viability of three different colon cancer cell lines [16].

Methyl gallate (MG) is a natural methyl ester of gallic acid and is endowed with many different biological activities ranging from anti-inflammatory to antioxidant and anti-microbial properties [24–27]. MG was demonstrated to harbor a clear selective anti-neoplastic action in many tumor systems. It plays a crucial inhibitory role in the tumor infiltration of CD4+ CD25+ regulatory T cells and its administration was demonstrated to delay tumor progression and survival in an EL-4 lymphoma model [28]. Other studies provided substantial evidence that this phytomolecule exerts anti-tumor activity on glioma cells, inhibiting proliferation and migratory cell ability via the suppression of the ERK1/2, Akt and paxillin phosphorylation signaling pathways [29]. Huang et al. reported that MG can also exhibit antitumor potential in different HCC cells (Hep3B, Mahlavu and HepJ5) triggering ROS-mediated and caspase-dependent apoptotic cell death [30]. Additionally, MG was also shown to exert a tumor inhibitory effect in in vitro and in vivo mouse models of hepatocellular carcinoma. Furthermore, it did not exert any cytotoxic effect in human normal hepatocytes, while it significantly suppressed the migration, invasion and epithelial–mesenchymal transition in tumor systems of HCC via the AMPK/NF-κB pathway [31].

In light of these observations, and since the anti-neoplastic potential of MG has not been described in colon cancer systems, we explored its possible impact on this tumor. Our investigations provided evidence that this phytochemical starkly reduces colon cancer cell viability, sparing differentiated Caco-2 cells, which provide a model of polarized cells resembling enterocytes [32]. In addition, MG triggered both autophagy and apoptotic cell demise via intertwined crosstalk between oxidative stress and p53 activation as prime sources of the phytochemical action.

2. Materials and Methods

2.1. Cell Cultures and Chemicals

The Caco-2 and HCT116 colon cancer cells used in this study were obtained from Interlab Cell Line Collection (ICLC, Genoa, Italy) and cultured as monolayers in DMEM supplemented with 10% (v/v) heat-inactivated FCS and 2 mM glutamine and in the presence of a 1% penicillin/streptomycin solution. To obtain differentiated Caco-2 cells as enterocyte-like cells [33], Caco-2 cells were plated and cultured in a complete medium for 21 days as reported by Natoli et al. [34].

For the reported experiments, cells were seeded in a culture medium on 96-well microplates or 6-well plates as previously reported [16] and allowed to adhere overnight at 37 °C in a humidified atmosphere containing 5% CO_2 followed by treatment with MG or a vehicle only. MG stock solution was prepared in DMSO and stored at -20 °C according to vendor specifications. In each experiment, MG working solutions were prepared in DMEM, never exceeding 0.01% (v/v) DMSO. The vehicle condition reported in each experiment as control was represented by untreated cells incubated in the presence of the corresponding DMSO volume. All cell culture media and culture reagents were provided from Euroclone SpA (Pero, Italy). All other reagents and chemicals, except where differently indicated, were purchased from Millipore Sigma (Milan, Italy).

2.2. Cell Viability Assay

Cell viability was assessed using an MTT assay as previously reported [35]. Cells were plated in 96-well plates, and after 24 h, they were incubated with compounds for indicated periods. Since the measure of cell viability using an MTT assay is based on the reduction of MTT to formazan and many polyphenols may interfere with formazan production, we considered this aspect in our experimental conditions. In particular, after incubation with the compounds, the plate was centrifuged and the medium was withdrawn and replaced with a fresh one before proceeding with the assay. Afterward, 20 µL of 5 mg/mL MTT was added to each well and the plate was incubated at 37 °C for 2 h. Therefore, the media was removed from each well and replaced with 100 µL lysis buffer (20% SDS and 10% dimethylformamide) before reading at 450 nm. For the determination of IC_{50} values non-linear regression analysis with the equation of a sigmoidal dose response with a variable slope was performed using Graphpad Prism 7.0 software (San Diego, CA, USA).

The observation of cellular morphological changes was detected using a Leica DMR inverted microscope (Leica Microsystems, Wetzlar, Germany), while the pictures were taken using IM50 Leica software (Leica Microsystems, Wetzlar, Germany).

The cytotoxic action of MG was also evaluated using the LDH (lactate dehydrogenase) assay, which is a method based on the measure of the activity of a stable cytoplasmic enzyme commonly released upon cell damage. Cells were seeded in a 6-well plate at a density of 2×10^5 cells, and after incubation with the compound, they were collected and centrifuged at $120 \times g$ for 10 min. The supernatant medium of each sample was recovered and analyzed using ARCHITECT Lactate Dehydrogenase kit (Abbot Laboratories Diagnostics Division, IL, USA) according to vendor specifications. To detect the total LDH release, treated cells were compared with a positive control represented by cells incubated in the presence of 0.1% Triton 100×.

2.3. Colony Formation Assay

This assay measures cell proliferation in a cell-contact-independent way. Cells were plated in pre-tested appropriate densities yielding 500 cells per plate. The plates were cultured for 10 days in the presence or absence of different doses of MG. Then, the colony signals were measured after crystal violet staining as previously reported [15]. The clonogenic survival fraction was defined as the ratio of the signal intensity of the untreated group versus the MG-treated group. All assays were made in triplicate. The number of colonies for each experimental condition was determined using the "Colony Area" plugin for the open-source image analysis software ImageJ v 1.8.0 as reported by Guzman et al. [36].

2.4. ROS Measurement

To assess the intracellular generation of reactive oxygen species (ROS), cells were plated in 96-well plates and allowed to adhere overnight. Cells were treated with MG and incubation with 5-(and-6)-carboxy-2′,7′-dichlorodihydrofluorescein diacetate (H2DCFDA) fluorochrome (Molecular Probe; Thermo Fisher Scientific, Inc., Life Technologies Italia, Monza, Italy) was performed as previously reported [37]. For this purpose, stock solutions of H2DCFDA were dissolved in DMSO and aliquots were stored at $-20\ ^\circ$C until use. H2DCFDA working solution was prepared in a PBS solution containing 5 mM glucose and added to cells to the final concentration of 20 μM. Then, incubation was protracted for 30 min in the dark and ROS-positive cells were visualized using a fluorescein isothiocyanate (FITC) filter (excitation wavelength of 485 nm and emission wavelength of 530 nm) in a Leica inverted fluorescence microscope (Leica Microsystems S.r.l, Wetzlar, Germany) equipped with a DC300F camera. All pictures were captured using Leica Q Fluoro software (Leica Microsystems S.r.l, Wetzlar, Germany).

2.5. Measurement of Intracellular Calcium Levels

The intracellular calcium levels were assayed using the Ca^{2+}-sensitive fluorescent dye Fluo 3-AM following vendor instructions (Thermo Fisher Scientific, Ferentino, Italy). After incubation with the compounds, cells (2×10^5/well) were collected, washed in calcium-free PBS and incubated in the presence of Fluo 3-AM for 1 h at 37 $^\circ$C in the dark. Then, calcium generation was analyzed using flow cytometry on a FACSAria Cell Sorter (BD Biosciences Company, 283 Franklin Lakes, NJ, USA). At least 50,000 cells were analyzed for each experimental condition. The data analysis was then performed using FlowJo software workspace v10 (BD Biosciences).

2.6. Analysis of Autophagic Vacuoles

The generation of autophagic vacuoles was detected using monodansylcadaverine (MDC) staining according to Munafò et al.'s procedure [38]. Briefly, following to the treatment in the presence of MG, the medium was replaced and 50 mM MDC in PBS was added to cells. Then, cells were washed in PBS and the fluorescence was analyzed with a Leica fluorescence microscope (Leica Microsystems, Wetzlar, Germany) using a 4′,6-diamidino-2-phenylindole dihydrochloride (DAPI) filter (excitation wavelength of 372 nm and emission wavelength of 456 nm). The analysis of autophagic vacuoles was also performed using acridine orange (AO) staining that specifically detects acidic vesicular organelles (AVOs) producing a bright red fluorescence, whereas it generates a bright green fluorescence for cytoplasm and nucleus [39]. For these analyses, cells were incubated for 15 min with 1 μg/mL AO prepared in PBS. Then, cells were analyzed under a Leica fluorescence microscope equipped with an image system (Leica Microsystems, Wetzlar, Germany) using Rhodamine (excitation wavelength of 596 nm and emission wavelength of 620 nm) and FITC (excitation wavelength of 485 nm and emission wavelength of 530 nm) filters. Merged images were obtained by combining pictures of both channels using Leica Q Fluoro software (Leica Microsystems, Wetzlar, Germany).

2.7. Immunoblot Analyses

Protein analysis was performed via a Western blotting procedure as previously reported [40]. For these analyses, 30 μg protein/lane were resolved using SDS-PAGE and then electroblotted on a nitrocellulose membrane filter (Bio-Rad Laboratories Srl, Segrate, Italy). All primary antibodies were purchased from Santa Cruz Biotechnology Inc. (Santa

Cruz, CA, USA), except for Protein kinase R-like endoplasmic reticulum kinase (PERK), phospho-PERK, eukaryotic initiation Factor 2α (eiF2α) and anti-caspase-3, which were from Cell Signaling Technology (Cell Signaling Technology Inc., Beverly, MA, USA). Anti-rabbit IgG (H + L) HRP conjugate and anti-mouse IgG (H + L) HRP conjugate (dilution 1:10,000) secondary antibodies were from Promega (Milan, Italy). In all experiments performed, γ-tubulin (diluted 1:1000, Sigma-Aldrich, Milan, Italy) was used as the loading control.

For all analyses, protein band detection was performed with an ECL™ Prime Western Blotting System (Cytiva, Merck KGaA, Milan, Italy) using a ChemiDoc XRS System equipped with the Quantity One software 4.6.6 (Bio-Rad Laboratories, Inc., Hercules, CA, USA).

2.8. Analysis of Apoptotic Cell Death Using Hoechst and Annexin V/PI Staining

To detect apoptotic cell death, cells were pre-incubated with Hoechst 33342 (Invitrogen; Thermo Fisher Scientific, Inc.) for 30 min before treatment with compounds. Next, blue nuclei showing condensed or fragmented chromatin were analyzed using fluorescence microscopy (Leica Microsystems, Wetzlar, Germany) as reported [15]. The quantification of apoptotic cell death percentage was determined via flow cytometry analysis using an Allophycocyanin (APC) Annexin V conjugate and propidium iodide (Annexin V-APC/PI) staining. For these experiments, HCT116 cells (2×10^5/2 mL medium) were seeded into 6-well plates and then subjected to treatments with MG. At the end of the treatment, the cells were taken via trypsinization, centrifuged at $120 \times g$ for 10 min, resuspended in PBS and counted. Next, 10^5 cells were incubated with Annexin V-APC (BD Pharmingen™ APC Annexin V kit, BD Biosciences, Milan, Italy) and PI (Sigma-Aldrich) in the dark according to the manufacturers' instructions. At the end of the incubation, the samples were analyzed using a FACSAria Cell Sorter flow cytometer (BD Biosciences Company, 283 Franklin Lakes, NJ, USA), acquiring at least 50,000 cells for each sample analyzed. The data obtained were then examined with FlowJo software (BD Biosciences).

2.9. Preparation of Cytosolic and Nuclear Extracts

For the isolation of nuclear and cytosolic fractions, 2×10^6 HCT116 cells were plated in 100 mm cell culture dishes and after incubation with MG, lysates were prepared as previously reported [41]. Briefly, cells were washed in PBS and scraped in a lysis solution containing 250 mM sucrose, 20 mM HEPES, 10 mM KCl, 1.5 mM $MgCl_2$, 1 mM EDTA, 1 mM EGTA, 1 mM DTT and protease inhibitor cocktail, pH 7.4. Then, the homogenates were prepared by passing cells through a needle of 25 g on ice for 20 min (10 times) and centrifuging samples at $1000 \times g$ for 10 min at 4 °C. The pellets were recovered and resuspended in lysis solution, and homogenization was repeated by passing cells through a needle of 25 g for an additional 10 times. Therefore, samples were recentrifuged ($1000 \times g$ for 10 min at 4 °C) and pellets representing the nuclear fraction were resuspended in RIPA buffer (1% NP-40, 0.5% sodium deoxycholate, 0.1% SDS, inhibitors of proteases: 25 μg/mL aprotinin, 1 mM PMSF, 25 μg/mL leupeptin and 0.2 mM sodium pyrophosphate) before proceeding with sonication. The supernatants obtained at the first centrifugation were recentrifuged ($10,000 \times g$ for 30 min at 4 °C) and the supernatants obtained were used as the cytosolic fraction. Proteins from both fractions (nuclear and cytosolic) were quantified using a Bradford assay (Bio-Rad Laboratories, Inc.) and were resolved in a polyacrylamide gel to analyze the p53 cellular localization. To determine the purity of each cellular fraction obtained, GADPH and Lamin B were used as cytoplasmic and nuclear markers, respectively.

2.10. Statistical Analyses

The statistical analysis of the data was performed by using GraphPad Prism™ 7.0 software (Graph PadPrism™ Software Inc., San Diego, CA, USA) and data were reported as the mean $\pm$ S.E. The significant differences between the control (untreated) vs. treated samples were analyzed by applying Student's *t*-test, while the analysis of multiple groups of samples was conducted using the ANOVA test. The statistical significance threshold was considered to be $p < 0.05$.

3. Results

3.1. MG Affected Colon Cancer Cell Viability in a Dose-Dependent Manner

Studies on the cytotoxic effects of MG demonstrated that this phytochemical elicits remarkable cell viability inhibition in many tumor systems [29,31]. However, since no data are available for colon cancer, we undertook a study aimed at evaluating the MG antitumor potential on the viability of two colon cancer cell lines (HCT116 and Caco-2). Our results showed that the cytotoxic effects were visible after a lag phase of 24 h (not shown) and were clearly evident at 48 h. As can be observed from the response curves (Figure 1A) obtained using incremental doses of MG, the half-maximal inhibitory concentration of MG (IC_{50} value) was about 30 µg/mL in both cell lines after 48 h of incubation. A more consistent effect was observed in the presence of the 90 µg/mL dose, which caused a dramatic decrease in cell viability (about −80%). Differently, in comparison to the corresponding colon cancer Caco-2 cells, no cytotoxic effects were found when MG was administered to differentiated Caco-2 cells, which is a well-established model of polarized intestinal cells reproducing typical morphological and biochemical features of enterocytes [42].

The ability of MG in reducing colon cancer cell viability was also confirmed using both the lactate dehydrogenase (LDH) test and clonogenic assay, which is an in vitro survival test that estimates cell ability to maintain a reproductive potential over a prolonged period. As reported in panel B of Figure 1, MG enhanced the LDH release in HCT116 cells relative to the control. In addition, we also observed that the phytochemical reduced the colony-forming ability of colon cancer cells with doses spacing from 0.46 to 7.5 µg/mL range, while no colonies were found with higher doses. Based on cell viability tests, all further experiments were performed with HCT116 cells using those MG doses that caused cell reductions of about 50% (30 µg/mL) and 80% (90 µg/mL), respectively.

3.2. MG Cytotoxicity Was Mediated by Oxidative Injury, ER Stress and Upregulation of Intracellular Calcium

To clarify the underlying mechanism of MG cytotoxicity, we explored whether the observed cytotoxic effect could be ascribed to the induction of oxidative stress. For this purpose, we used NAC, which is a potent radical scavenger.

As reported in Figure 2A, when pre-incubating the cells in the presence of NAC for 2 h, the toxic effect of 90 µg/mL MG was consistently counteracted. Moreover, all morphological changes induced by MG, consisting of cell shrinkage and a reduction in cell number (Figure 2B), were also counteracted by NAC sustaining the induction of oxidative damage.

Such observations were confirmed by an evaluation of the ROS production assayed using H2DCFDA staining. Following MG exposure at different times, a dose-dependent ROS generation (green fluorescent cells) was observed. The effect, that was already visible after 2 h, reached a peak at 4 h to maintain lower levels for longer times of incubation in the presence of MG (Figure 2C).

Furthermore, we verified whether these effects were accompanied with the upregulation of stress-associated proteins at 24 h and 48 h. As it can be observed in Figure 3, the manganese superoxide dismutase (MnSOD) and catalase levels, which are two radical scavenger enzymes, increased at 24 h with the two doses of MG. Such an effect was counteracted by the pre-incubation of cells in the presence of NAC (Figure 3). No changes in the levels of these proteins were found at 48 h of treatment.

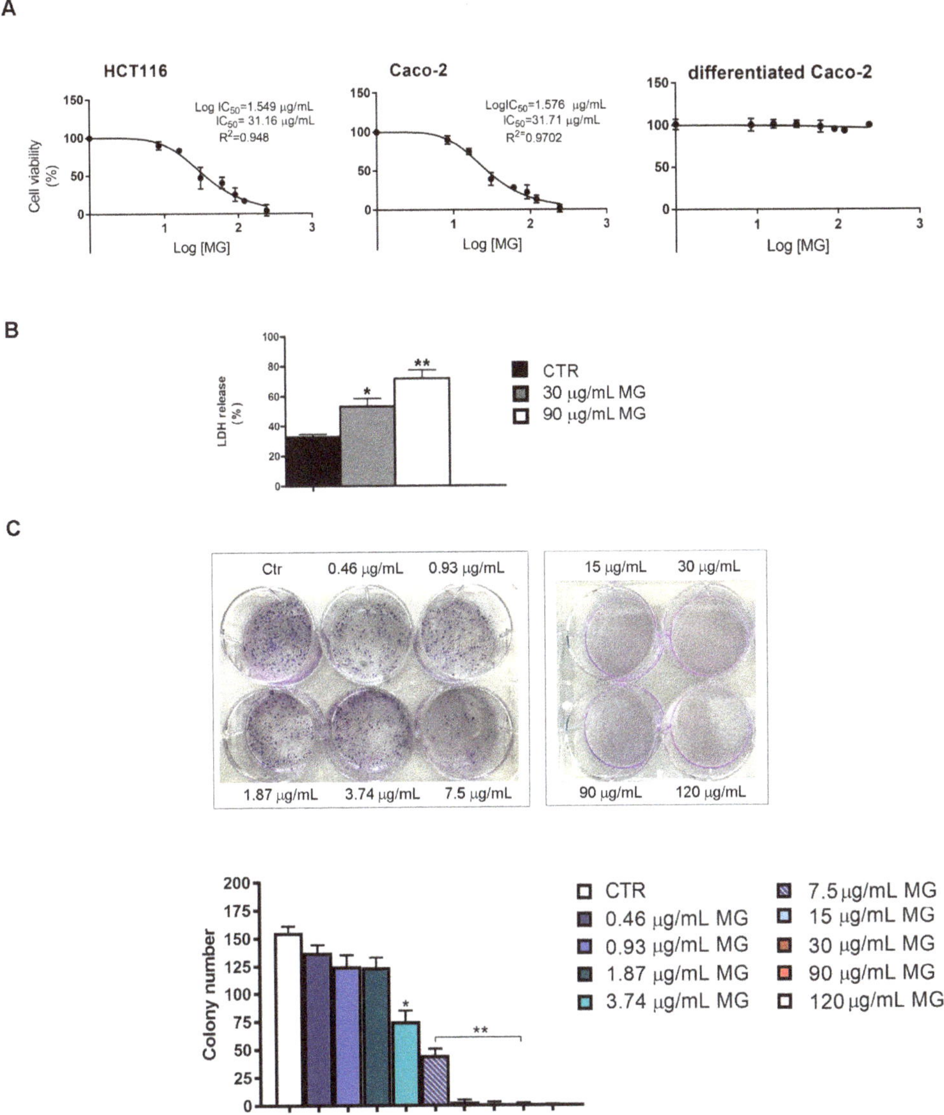

Figure 1. IC$_{50}$ determination and colony formation assay of MG-treated colon cancer cells. (**A**) MTT assay of colon cancer cells (HCT116 and Caco-2) and differentiated Caco-2 cells incubated with incremental doses of MG. Cell viability was determined at 48 h as reported in the Materials and Methods section. IC$_{50}$ values were assessed using GraphPad Prism 7 software. (**B**) LDH cytotoxicity test in MG-treated HCT116 cells. After incubation in the presence or absence of the phytocompound, cells were centrifuged and supernatants were used to assess the LDH content using a commercial kit. Data are reported as a percentage of the total LDH released from cells using as a positive control of cells incubated with 0.1% Triton 100×. (**C**) Clonogenic assay in MG-treated HCT116 cells. The colony formation inhibition was assessed by crystal violet staining after 10 days of exposure to MG. Representative images of colony formation are reported in the upper panel. The quantitative analysis of colonies (lower panel) was performed as reported in the Materials and Methods section. All experiments were performed in triplicate. (*) $p < 0.05$ and (**) $p < 0.01$ compared with the untreated sample.

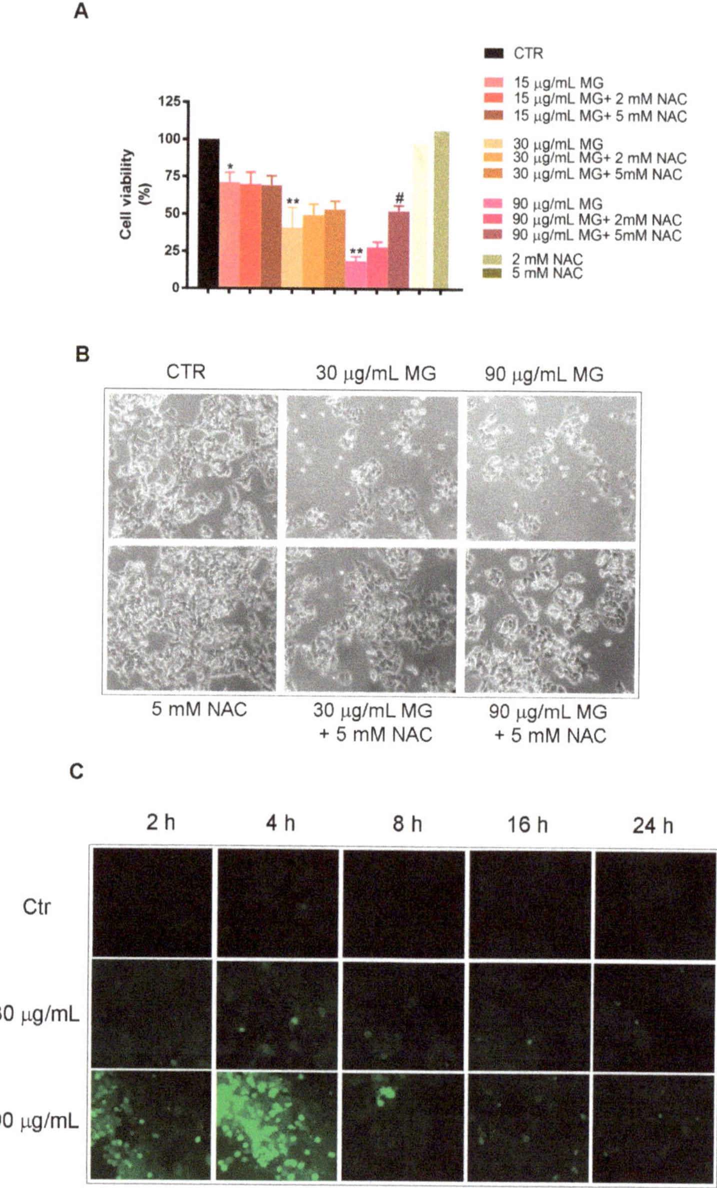

Figure 2. Oxidative stress is required for the cytotoxic efficacy of MG in colon cancer cells. (**A**) Effect of NAC pre-incubation on HCT116 cell viability of MG-treated cells for 48 h. Each value reported in the histogram represents the mean of three independent experiments $\pm$ SD. (*) $p < 0.05$ and (**) $p < 0.01$ compared with the untreated sample. (#) $p < 0.05$ compared with the MG-treated sample. (**B**) Phase-contrast micrographs of morphological changes of HCT116 cells treated for 48 h with MG and the protective effect of NAC pre-incubation (original magnification 200×). (**C**) ROS generation induced by MG treatment in HCT116 cells. The ROS level was measured using H2-DCFDA, which is a redox-sensitive fluorescent probe, as reported in the Materials and Methods section. Original pictures were taken using a Leica fluorescence microscope equipped with a CCD camera and FITC filter (original magnification 200×).

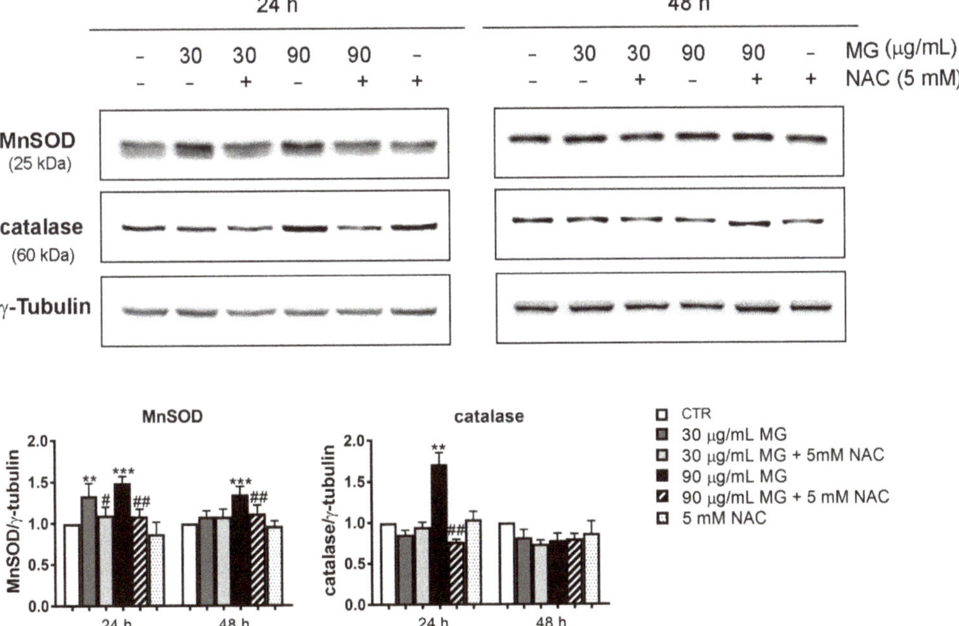

Figure 3. Upregulation of the antioxidant enzymatic systems in HCT116 cells treated with MG. After treatment with the indicated doses of the phytocompound in the presence or absence of NAC, the cell lysates were prepared and the level of stress-associated proteins (MnSOD and catalase) was detected using Western blotting. Representative blots from three independent experiments were considered and a densitometry analysis histogram was normalized to γ-tubulin, which was used as a loading control. (**) $p < 0.01$ and (***) $p < 0.001$ compared with the untreated sample. (#) $p < 0.05$, and (##) $p < 0.01$ compared with the MG-treated condition.

In addition, we also investigated whether the oxidative injury triggered by MG exposure could be also associated with ER stress. With this in mind, we conducted Western blotting analyses to explore the status of key factors involved in ER stress. As reported in Figure 4, the higher dose of MG provoked a modest increase in the level of PERK, phospho-PERK and eiF2α. Such an effect was also accompanied at 24 and 48 h with an upregulation of Glucose-Regulated Protein 78 (Grp78), which is an ER chaperone acting as a key regulator of the unfolded protein response (UPR) [43], as well as that of C/EBP Homologous Protein (CHOP).

As it is well known, another event that can contribute to oxidative injury and lead to cell death is a calcium surge released from different cellular compartments, such the ER and mitochondria [44–46]. On the other hand, the calcium homeostasis that lies at the heart of many cell signaling processes is under redox control.

In accordance with these observations, flow cytometry analyses using a Fluo 3-AM probe provided evidence that the MG provoked a remarkable increase in the intracellular calcium content (Figure 5). Interestingly, this event, which had already occurred at 16–24 h of exposure to MG when cells were found to be still alive, also remained high in treated conditions up to 48 h, a time at which cell death took place, in correlation with the high levels of both Grp78 and CHOP.

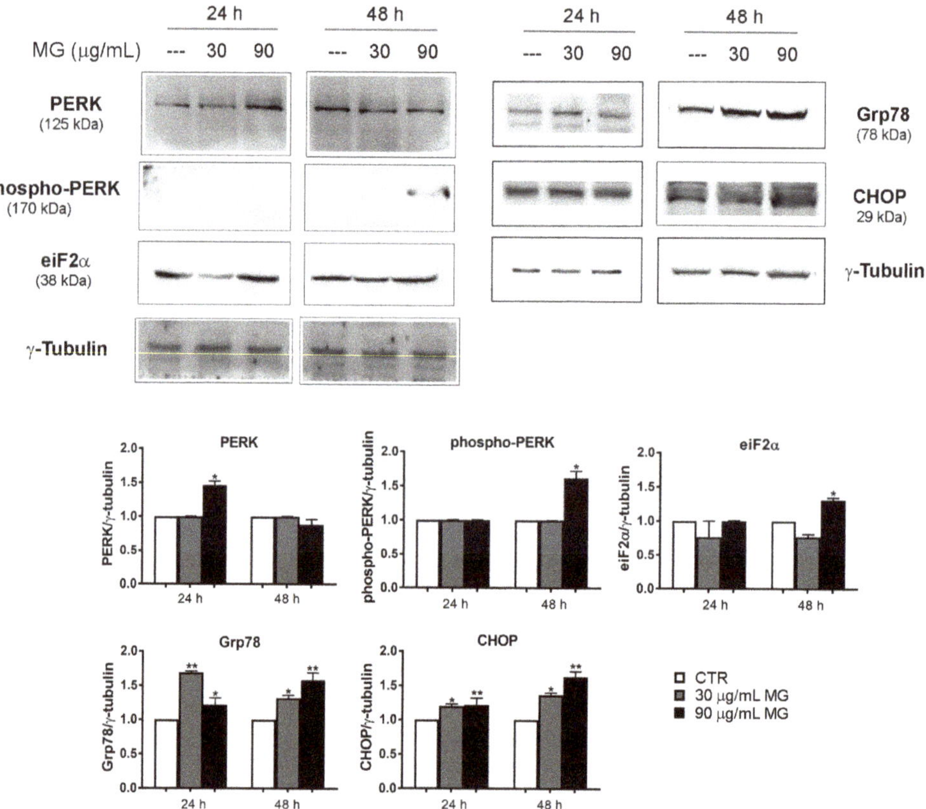

Figure 4. MG exposure activated the ER-stress-associated protein levels. HCT116 cells were treated in the presence of 30 and 90 µg/mL MG for 24 h and 48 h. Western blot analysis was performed to evaluate the protein expression of the ER stress markers PERK, phospho-PERK, eiF2α, Grp78 and CHOP. The amount of analyzed proteins was assessed using γ-tubulin as the loading control protein and for band density normalization. The data are presented as the mean ± SD; (*) $p < 0.05$ and (**) $p < 0.01$ compared with the untreated sample.

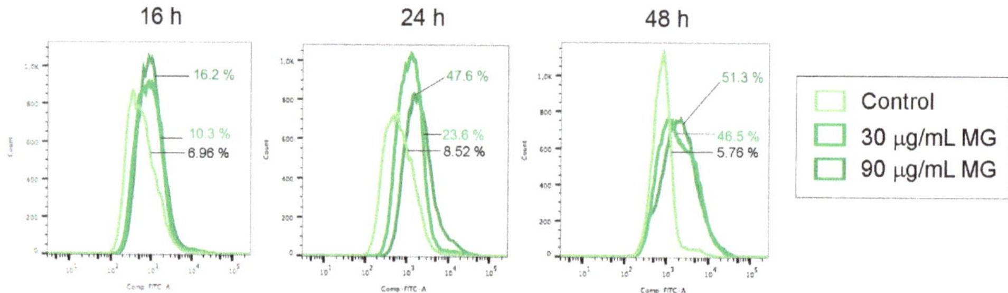

Figure 5. MG exposure increased the intracellular calcium level. Changes in the content of intracellular calcium analyzed at the indicated times via flow cytometry using Fluo 3-AM fluorochrome as reported in the Materials and Methods section.

3.3. Autophagy Was Upregulated in MG-Treated Cells

A growing number of studies indicated that many natural compounds, such as alkaloids, flavonoids, naphthoquinones, sequiterpene lactones and ginsenosides possess an anti-cancer potential acting as autophagy modulators [47]. On the basis of these observations, to further dissect the underlying mechanism of MG, we determined whether this phytochemical might act through the induction of autophagy [48]. Moreover, calcium release from ER storages can also contribute to the generation of autophagosomes [49], which are typical spherical structures endowed with double-layer membranes participating in the autophagic process. Using the monodansylcadaverine (MDC)-based staining, which is a selective fluorescent probe that accumulates in autophagosomes, we observed that MG promoted the generation of autophagic vacuoles, which appeared as dot-like structures.

The event, which was already visible at 16 h of incubation, further increased at 24 h, when MDC-positive cells amounted to almost 90% with a 90 µg/mL dose (Figure 6A). Differently, no MDC-stained structures were highlighted at 48 h of treatment.

The induction of autophagy was also sustained by the conversion of microtubule-associated protein 1A/1B-light chain 3 known as LC3-I to LC-3II, which is the phosphatidylethanolamine conjugated form that is recruited to autophagosomal membranes [50,51] and represents a crucial marker of the autophagic flux. An increasing trend of some autophagy-associated factors, such as p62, Beclin 1 and Atg7, was also observed in MG-treated conditions, while no changes were noticed for Atg1/Ulk1 (Figure 6B).

3.4. MG Treatment Induced DNA Damage and p53-Mediated Apoptotic cell Death

To evaluate whether the cytotoxic effect observed in the presence of MG could be ascribed to the induction of apoptotic cell death, we tested possible chromatin condensation and fragmentation using vital Hoechst staining. While MG administration triggered autophagy in the first phase of treatment, data reported in Figure 7A showed that when prolonging the exposure up to 48 h, remarkable nuclear modifications associated with cell death occurred in the presence of the phytocompound.

When exploring possible changes in DNA damage markers, we provided evidence that the phytochemical produced a consistent increase in γH2AX, as well as p53 (Figure 7B). Meanwhile, the pre-incubation with NAC counteracted MG effects on both chromatin condensation and γH2AX and p53 activation.

To elucidate the type of cell death induced by MG, we performed flow cytometry analyses using Annexin V/PI double staining (Figure 8A). At 48 h, about 19% of early (Q3 quadrant) and late (Q2 quadrant) apoptotic cells were found using 30 µg/mL of the compound. Such a value amounted to 60% apoptotic cells when the higher dose of MG was employed. We then checked the status of specific apoptotic markers, such as caspase-3 and its target PARP1. Specifically, we found that MG promoted a decrease in the pro-enzymatic form of caspase-3 and the appearance of the fragmented and activated forms at 19 and 17 kDa, respectively. Such an effect was accompanied with the fragmentation of PARP1, which is a caspase-3 target (Figure 8B).

3.5. MG Treatment Induced Early Upregulation of p53 Related to the Molecular Switch between Autophagy and Apoptotic Cell Death

In light of these results, we wondered about the possible role of p53 and γH2AX in the mechanism analyzed. Time course studies (4–24 h) provided evidence that both doses of MG promoted an early upregulation of these factors, which were already visible after 4 h of incubation with the compound (Figure 9A). As it is known, p53 protein is activated in key responses to genotoxic stresses and DNA damage. In these scenarios, p53 translocates to the nucleus, boosting a tumor suppressor program via the cell cycle arrest or apoptosis via the direct transcriptional activation of specific pro-apoptotic targets, such as Apaf-1, Puma, Bax and Noxa [52,53]. By performing subcellular fractionation experiments, we found p53 accumulation in the nuclear fraction already at 16 h of incubation with MG (Figure 9B).

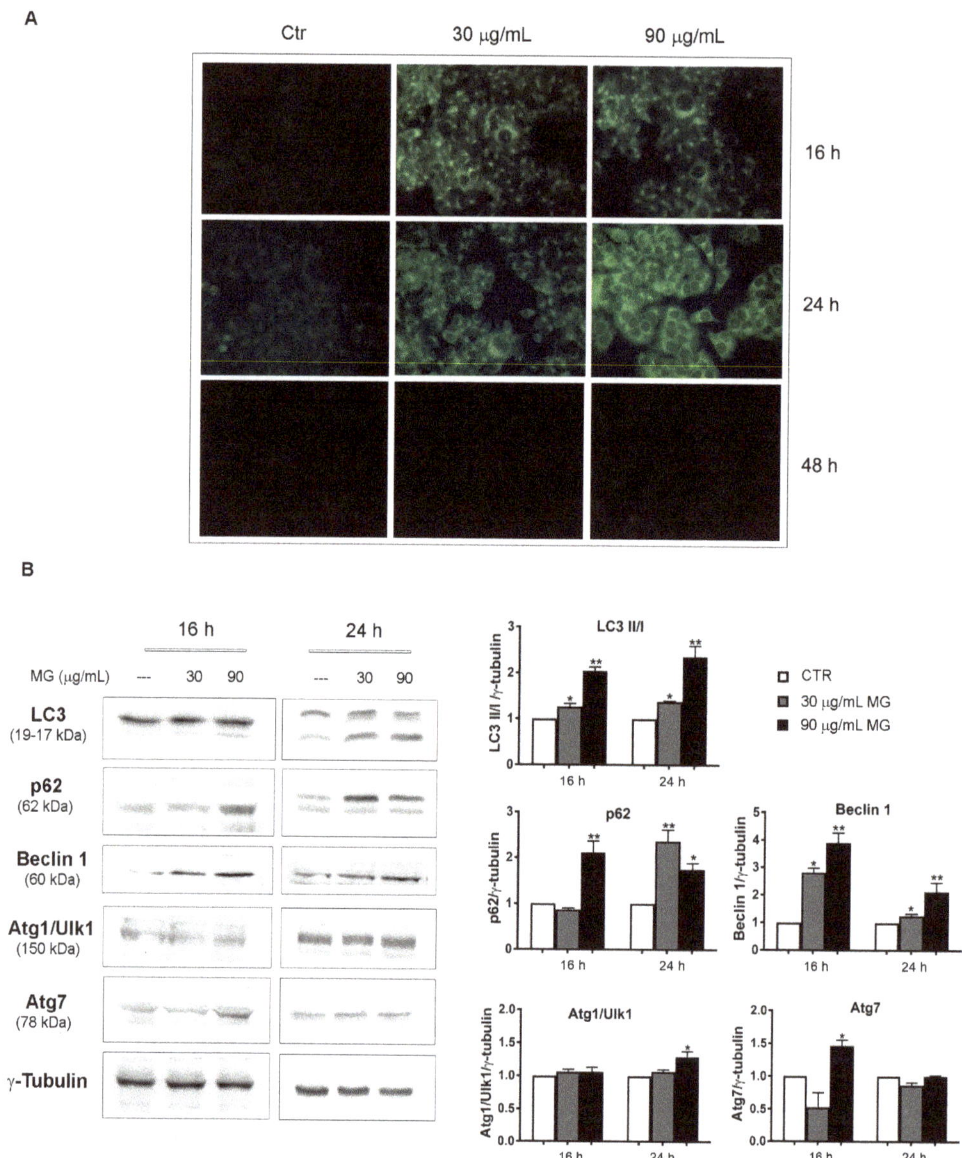

Figure 6. MG exposure provoked an autophagic process in the early phases of treatment. (**A**) Monodansylcadaverine (MDC) staining that enabled the visualization of autophagic vacuoles as dot-like structures was performed in MG-treated cells. HCT116 cells were incubated with MG for the indicated time periods and autophagic vacuoles were highlighted via fluorescence microscopy using a Leica microscope equipped with a DAPI filter. Representative fluorescence microscopy images were taken at a magnification of 400×, as reported in Section 2. (**B**) Immunoblots of autophagic markers performed in MG-treated HCT116 cells. Proteins were detected using different antibodies directed against the LC3-I and LC3-II forms, p62, Beclin 1, Atg1/Ulk1 and Atg7. γ-tubulin was used as the loading control. All graphs show the mean ± SD of three independent experiments. (*) $p < 0.05$ and (**) $p < 0.01$ compared with the untreated cells.

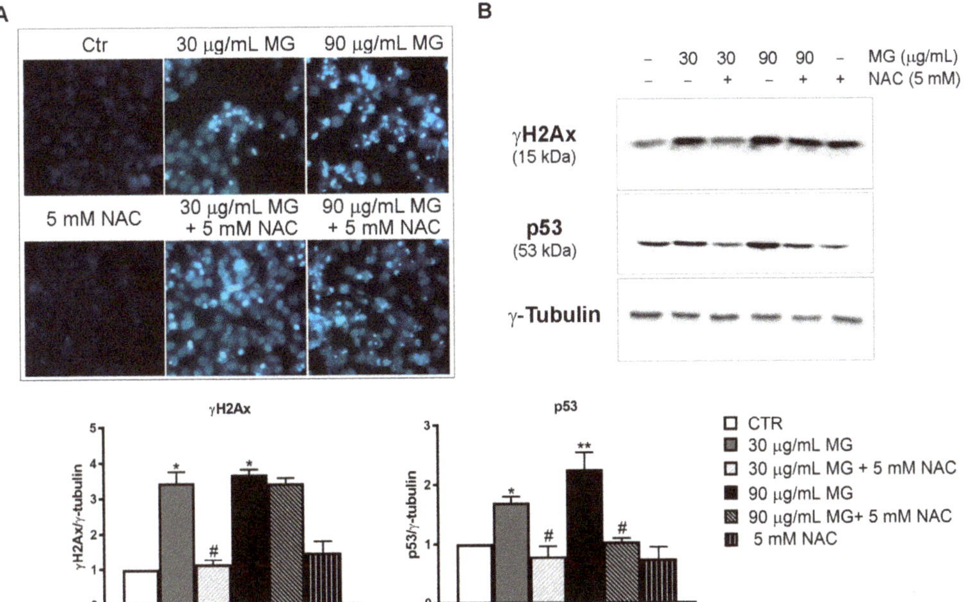

Figure 7. MG-induced DNA damage. (**A**) Morphological analysis of HCT116 cells after vital staining with Hoechst 33342. Following Hoechst staining, cells were treated for 48 h with different doses of MG in the presence or absence of NAC. Chromatin fragmentation and condensation were observed under a fluorescence microscope. The images (original magnification at 200×) were acquired with a DAPI filter using an inverted fluorescence microscope and processed with Leica Q Fluoro Software. (**B**) Analysis of DNA damage markers: γH2AX and p53. After treatment with MG in the presence or absence of NAC for 48 h, cells were lysed and proteins were analyzed using Western blotting. The γ-tubulin blot was reported as a loading control. The blots and histograms of densitometric analyses reported are representative of three independent experiments. (*) $p < 0.05$ and (**) $p < 0.01$ compared with untreated cells. (#) $p < 0.05$ compared with the MG-treated condition.

To further dissect the underlying role of p53 since its increase was already observed in the early phases of treatment (4–16 h) when oxidative injury and the autophagy process occurred, we explored a possible interplay between p53 and MG-induced autophagy. Compelling evidence showed that p53 can positively or negatively impact autophagy in a context-dependent manner or via its subcellular localization [54–57].

By using acridine orange (AO) staining, which corroborated the accumulation of acidic vesicular organelles (AVO, orange fluorescence) in the cytoplasm of MG-treated conditions, we observed that the addition of antioxidant NAC counteracted the autophagy, while Bafylomicin A1, which is an inhibitor of autophagosome-lysosome fusion, markedly inhibited the autophagic process (Figure 10A). On the other hand, pre-incubating the cells with pifithrin-α (PFT-α), which is a specific inhibitor of p53 transcriptional activity, the AVO accumulation induced by MG further increased (Figure 10A). Such an effect was also accompanied by an upregulation of the LC3-II form (Figure 10B), thus suggesting that p53 can negatively affect autophagy in response to MG.

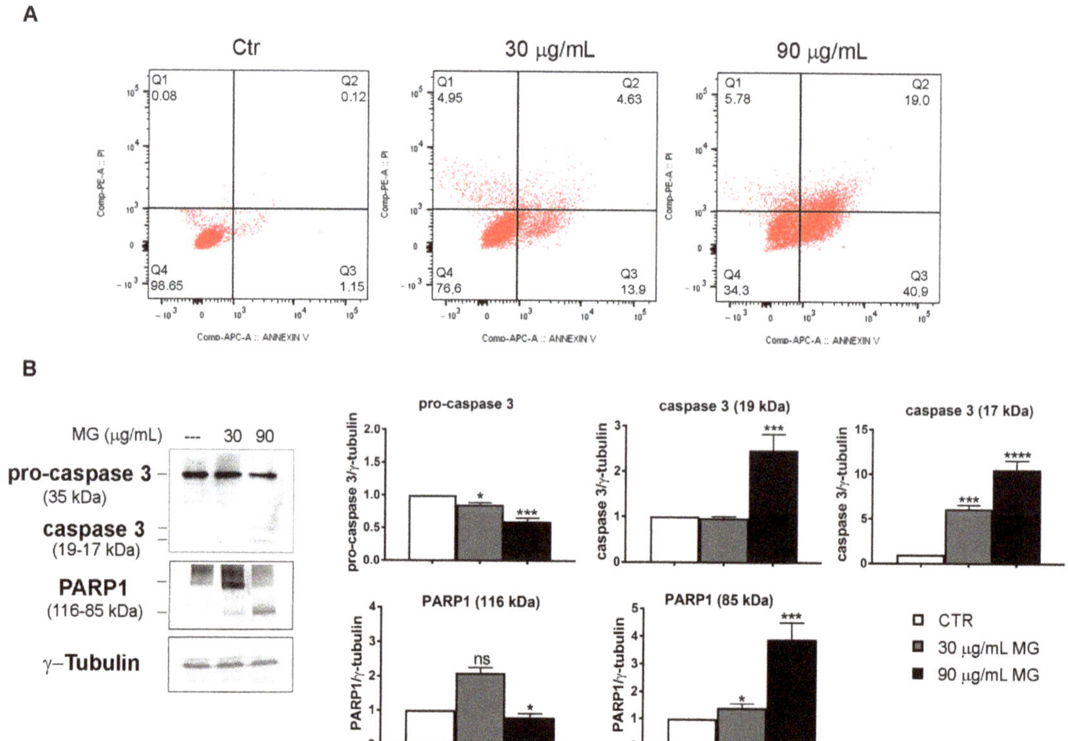

Figure 8. MG-induced apoptosis in colon cancer cells. (**A**) Annexin V/PI staining to evaluate apoptosis. HCT116 cells were treated with MG and compared with untreated cells. The rate of apoptosis was assessed via flow cytometry using the Annexin V/PI double staining assay. The data represent one of three independent experiments. (**B**) MG treatment evoked an increase in apoptotic markers. Caspase-3 activation and PARP1 fragmentation were analyzed using Western blotting. The relative quantification was assessed after densitometric analysis of bands and normalization to γ-tubulin used as a loading control. Histograms of densitometric analyses report the average values of three independent experiments. (*) $p < 0.05$, (***) $p < 0.001$ and (****) $p < 0.0001$ compared with the untreated sample. ns, not significant.

Accordingly, when we pre-incubated HCT116 cells for 2 h in the presence of PFT-α, followed by a co-treatment with MG for 48 h, PTF-α played a protective role against the cytotoxicity showed by the phytocompound alone. As shown in Figure 11A, the residual viability that amounted to only 22% with 90 µg/mL MG rose to about 60% when PFT-α was added.

Such prevention of MG cytotoxic effects was also found when PFT-α was co-administered and morphological changes were evaluated using light microscopy (Figure 11B).

To further ascertain whether PFT-α can inhibit MG-induced apoptosis, we explored its effect on both p53 and caspase-3. The results showed that compared with the control group, PFT-α counteracted p53 upregulation and, in the same experimental conditions, also suppressed caspase-3 activation, inhibiting the production of its active fragments (19–17 kDa) induced by the MG exposure (Figure 11C).

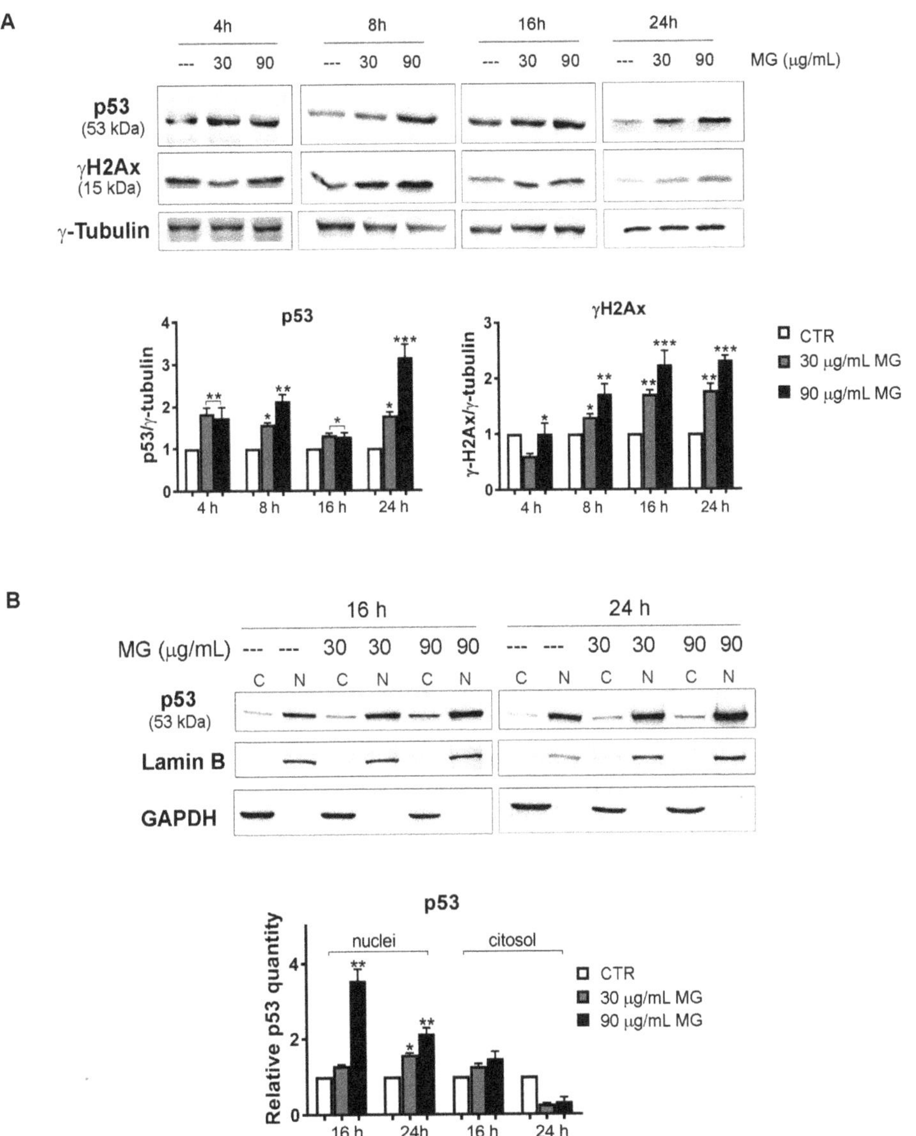

Figure 9. Time course analysis of the MG action on DNA damage markers and p53 nuclear accumulation. (**A**) MG treatment provoked an early upregulation of DNA damage markers p53 and γH2Ax in HCT116 cells. After treatment with MG for various periods, p53 and γH2Ax were detected using Western blotting analyses. Data were normalized to γ-tubulin, which was used as the loading control. The blots and histograms of densitometric analyses reported are representative of three independent experiments. (*) $p < 0.05$, (**) $p < 0.01$ and (***) $p < 0.001$ compared with the untreated sample. (**B**) Subcellular fractionation for cytosolic and nuclear protein extract displayed nuclear enrichment in the p53 content after the MG incubation. The relative quantification was assessed after densitometric analysis of the bands and normalization to the correspondent loading control. Lamin B or GADPH was used to assess the possible changes in the loaded protein amount for the cytosolic and nuclear fractions, respectively. (*) $p < 0.05$ and (**) $p < 0.01$ compared with the untreated sample.

A

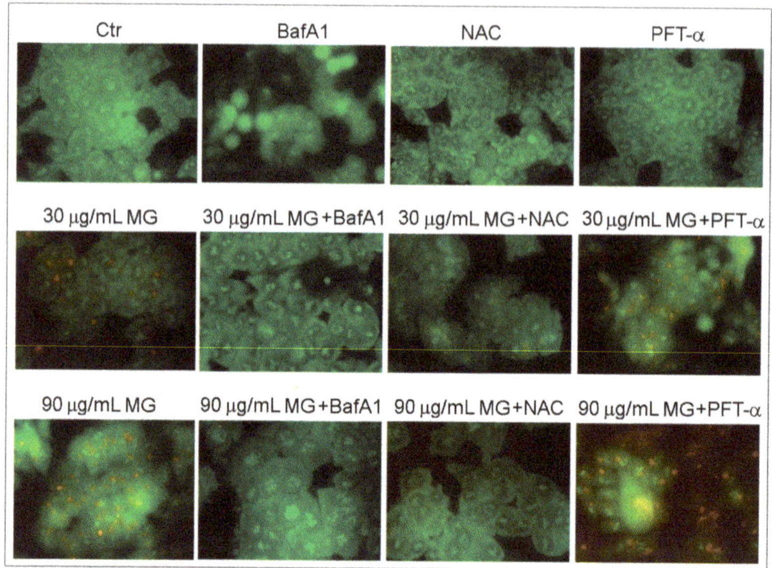

B

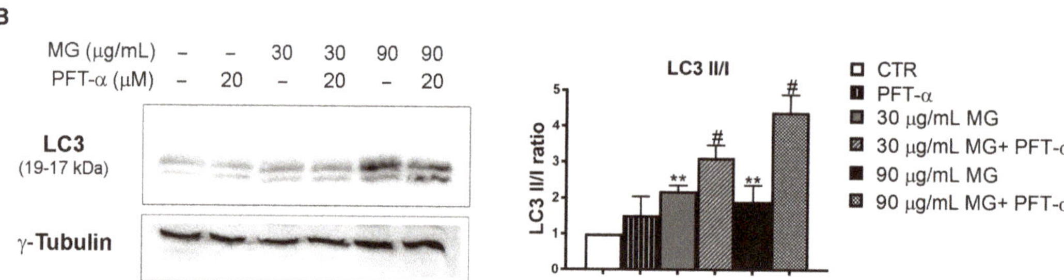

Figure 10. Effects of p53 inhibition on MG-induced autophagy. (**A**) The inhibition of p53 transcriptional activity by PFT-α enhanced the MG-induced autophagy in HCT116 cells. Cells were pre-incubated with NAC (5 mM), BafA1 (100 nM) or PFT-α (20 µM) in the presence or absence of MG for 24 h; then, the production of AVOs showing bright red fluorescence was evaluated via AO staining using Leica Q Fluoro software. (**B**) PFT-α/MG co-treatment enhanced the levels of the autophagic protein LC3. For the analysis of the LC3 forms, cells were treated with MG in the presence or absence of PFT-α (20 µM), followed by a Western blot analysis. The relative quantification was assessed after a densitometric analysis of bands and normalization to γ-tubulin. Data reported in the histograms were the average of three independent experiments. (**) $p < 0.01$ compared with the untreated sample. (#) $p < 0.05$ compared with the MG-treated sample.

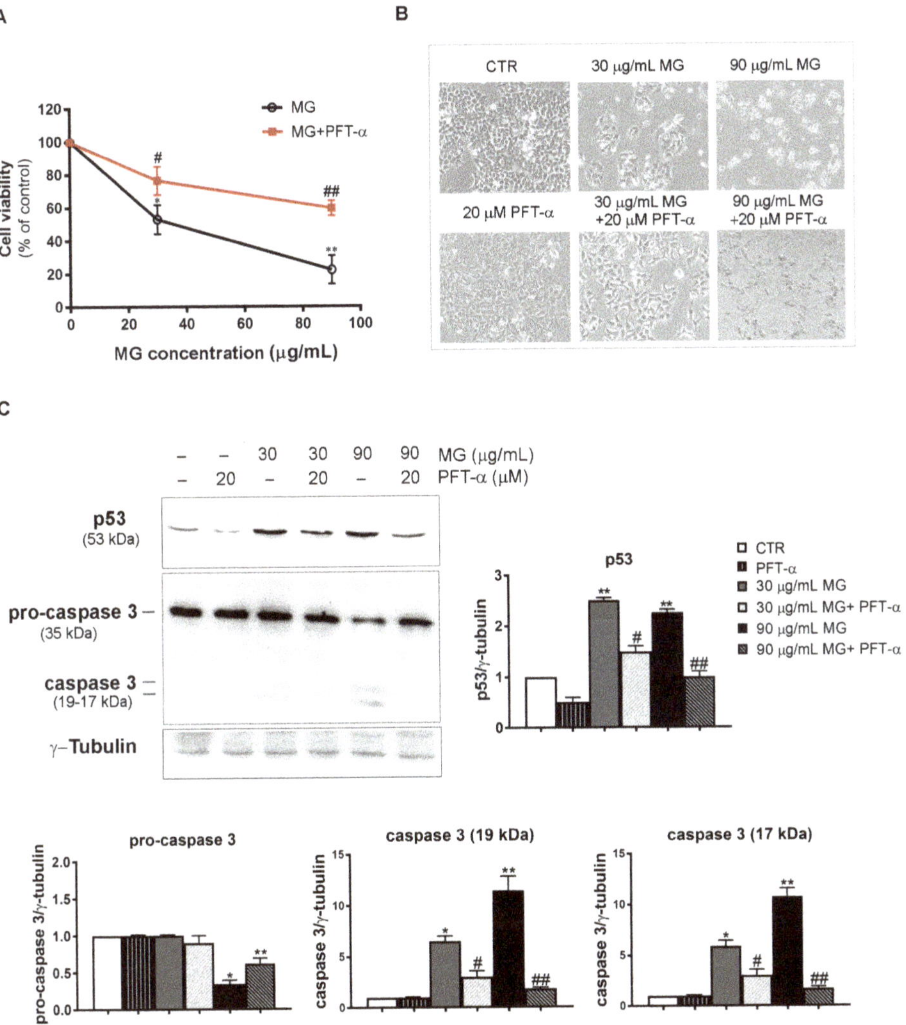

Figure 11. PFT-α, which is a p53 inhibitor, negatively affected the MG-induced apoptosis in HCT116 cells. (**A**) PFT-α inhibited the MG-induced cytotoxic effect. The pre-incubation of cells with PFT-α was performed as reported in the Results section; then, different MG concentrations were added and incubation was protracted for 48 h. Cell viability was analyzed using an MTT assay. Values reported in the line chart represent the mean of three independent experiments $\pm$ SD; (*) $p < 0.05$ and (**) $p < 0.01$ compared with the untreated conditions. (#) $p < 0.05$ and (##) $p < 0.01$ compared with the MG-treated sample. (**B**) Micrographs showing the PFT-α effect on morphological changes induced by the MG treatment. Pictures were taken using a Leica inverted microscope as reported in the Materials and Methods section. (**C**) Effect of PFT-α on the p53 and caspase-3 levels. Protein lysates were prepared as reported in the Materials and Methods section and resolved using SDS-PAGE. Blots were detected using specific antibodies directed against the proteins of interest and their level was normalized to γ-tubulin, which was used as the loading control. The blots and histograms of densitometric analyses reported are representative of three independent experiments. (*) $p < 0.05$ and (**) $p < 0.01$ compared with the untreated sample; (#) $p < 0.05$ and (##) $p < 0.01$ compared with the MG-treated sample.

4. Discussion

Moderate levels of ROS are crucial regulators of signaling pathways relevant for life in normal cells, while their increased generation can seriously impair cellular redox balance, contributing to many pathologic conditions, such as cancer. Many different studies highlighted that tumor cells, compared with their normal counterpart, exhibit an increased oxidative burst that favors cancer transformation, metabolic remodeling and increased generation of ROS [58]. Indeed, an escalated production of ROS positively impacts cancer through pro-tumorigenic signaling that enhances cell survival, DNA damage, genetic instability, hypoxia adaptation and resistance to the most common chemotherapeutics [59]. In this intricate scenario, malignant cells thrive in and counterbalance the ROS overload by boosting a plethora of enzyme-based scavenger systems to detoxify themselves with an oxidative burst and maintain their pro-tumorigenic profile.

However, despite these aspects, the role of ROS in cancer and anti-cancer strategies is still widely debated. A growing body of evidence supports the view that ROS conceal an oncojanus nature, either activating pro-tumorigenic or anti-tumorigenic signaling. Such pathways can be differently orchestrated in cancer treatment to preferentially kill tumor cells [58,60]. The fine-tuning of some anti-neoplastic therapies that promote an escalated ROS generation that overwhelms the scavenging tumor ability seems to be the other side of the coin that could be exploited as an Achilles' heel and drive tumors to different death pathways, such as apoptosis, autophagy or necroptosis [61].

Based on these rationales, some chemotherapeutics as natural or synthetic compounds (i.e., platinum-based compounds, anthracyclines, taxanes and sesquiterpenoids) have been extensively used with this purpose to treat tumors [62,63].

In particular, the investigations developed in this study focused on methyl gallate (MG), which is a well-known phytochemical that harbors strong anti-neoplastic properties in many different tumor systems, but no data are available on colorectal cancer. Here, we provide evidence for the first time that MG inhibits the growth of colon cancer cells, sparing differentiated Caco-2 cells, which is a model of polarized enterocytes. In particular, the cytotoxic action of MG could be ascribed to a deadly liaison occurring between ROS and p53 that, unavoidably, dictates the cell fate toward apoptosis. In the first phase of treatment, MG stimulated a stress-associated program characterized by the precocious increase in ROS content, along with an upsurge in both ER stress markers (PERK, phospho-PERK, Grp78 and CHOP) and the intracellular calcium level. As a consequence of this oxidative burst, probably to serve a stress defense response, MG-treated cells also upregulated the level of the ROS scavenger enzymes MnSOD and catalase.

In this complex scenario, we do not know the origin of intracellular calcium increase at this time. As it is well known, changes in intracellular calcium content that increase beyond the normal threshold can be ascribed to its release by the ER or dysfunctional mitochondria. Since some of our preliminary studies also suggested an involvement of mitochondria in MG-induced mechanism, we cannot exclude this possibility and aim to better clarify such an aspect in our future directions.

Overall, the early ROS generation drove cells in the first phases of treatment (16–24 h) along an autophagy process, as testified by the appearance of autophagic vacuoles to MDC and AO staining and significant changes in autophagic markers, such as LC3, p62, Beclin 1 and Atg7. However, when analyzing the timeline of MG exposure up to 48 h, such an autophagic flux was interrupted, leading cells to an apoptotic demise characterized by DNA fragmentation and caspase activation. These findings were in accordance with Huang's data [30], demonstrating that the extensive oxidative injury induced by MG treatment is a causative event in the apoptosis triggered by MG in HCC cells. Indeed, in our experimental condition, the addition of NAC, which is an antioxidant sulfidryl compound, prevented the toxic effect of MG in colon cancer cells, inhibited the autophagic process and counteracted the DNA fragmentation occurring during the apoptotic cell death. On the other hand, the observation that oxidative stress could play a role in the analyzed events was shown by data that reported a complex interconnection between ROS production, ER stress and autophagy [64,65].

Interestingly, our results also indicated that a pivotal role in the mode of action of MG was played by the tumor suppressor protein p53. p53 represents a key player that is capable of monitoring a plethora of cellular pathways related to the control of cell cycle progression, genome stability and apoptosis [66]. For all these multifunction properties, it has been named the "guardian of the genome". Accumulating evidence demonstrated that under normal conditions, p53 is maintained at very low levels by its negative regulator MDM2 that targets p53 degradation to 26S proteasome [67]. Differently, when cells are under stress conditions (nutrient deprivation, hypoxia, DNA damage), the p53 level increases and the protein translocates to the nucleus, where, as a transcription factor, it regulates the expression of a subset of genes functioning in cell cycle progression, cell metabolism, autophagy, tumor microenvironment and apoptosis. ROS and p53 were shown to establish a versatile partnership [68]. Indeed, ROS generated as by-products of cellular metabolism can act either upstream of p53, promoting its expression, or downstream, triggering apoptotic cell death pathways. In our experimental conditions, MG stimulated p53 upregulation, and such an effect seemed to be strictly intertwined with ROS generation. Its increased level, which was already visible in the first hours of exposure to MG (4–8 h), was maintained at a high level up to 48 h, when DNA damage occurred and cells collapsed via apoptosis. On the other hand, the addition of NAC counteracted the increase in both p53 and γH2Ax DNA damage markers, demonstrating that their upregulation can be ascribed to the impairment of redox balance induced by the phytocompound. Clearly, we believe that p53 was crucial in the context of the mechanism studied when monitoring both autophagy and apoptotic cell demise. Such an observation was supported by the experiments performed using PFT-α, which is a specific inhibitor of p53 transcriptional activity. Indeed, when PFT-α was co-administered with MG, we observed an enhancement of autophagy flux, as well as in LC3-II form, showing that p53 could inhibit autophagy in our condition. The relationship between p53 and autophagy has been widely discussed and still appears to be controversial since p53 can act as a rheostat system that adjusts the authophagy rate (through a positive or negative regulation) in a context-dependent fashion [56].

On the other hand, p53 also represents an active player in MG-induced apoptosis in colon cancer cells. Indeed, when exposing cells to a PFT-α/MG combo treatment, we observed that the p53 level dropped and the cytotoxic action of MG, as well as caspase-3 activation, were prevented by PFT-α.

A schematic representation describing the intricate mechanism of MG anticancer activity on colon cancer cells is reported in Figure 12.

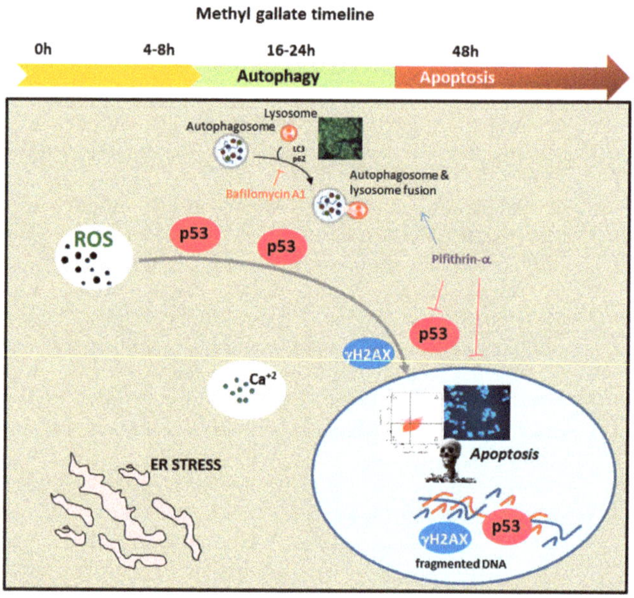

Figure 12. Timeline of the antitumor signaling pathway activated by MG in colon cancer cells. MG exposure triggered autophagy and apoptotic cell demise. The represented processes are orchestrated by an intertwined liaison between oxidative injury and p53. The early generation of ROS, accompanied with intracellular calcium increase and ER stress, stimulated p53 to switch the autophagy toward apoptotic cell death.

5. Conclusions

As a whole, the findings reported in this paper fit well with the current scientific literature on the anti-tumor potential of methyl gallate. In particular, our results shed light on the potential action of MG in preferentially targeting colon cancer cells with respect to enterocyte-like cell models. The biochemical characterization of MG signaling in these cells reveals a new clue to its underlying mechanism, thus highlighting an intertwined relationship between oxidative stress and p53 as a causative event in apoptotic demise. The data obtained here represent a reason to use MG in future investigations and explore whether this phytochemical can be used alone or in combinatorial studies as a possible adjuvant in traditional chemotherapy.

Author Contributions: Conceptualization, A.D. and M.L.; methodology, A.N., D.D.L. and A.A.; investigation, D.C. and S.E.; data curation, A.D.B. and S.E.; software, D.C., A.N. and D.D.L.; writing—original draft preparation, A.D. and M.L.; writing—review and editing, M.L., L.T., M.G., M.A. and A.D.; supervision, all authors; funding acquisition, G.C. All authors have read and agreed to the published version of the manuscript.

Funding: This work was partially sustained by Finalized Research Funding (FFR 2022), FFR-D15 D'Anneo and FFR-D03-Lauricella, Università degli Studi di Palermo, Palermo, Italy.

Institutional Review Board Statement: Not applicable.

Informed Consent Statement: Not applicable.

Data Availability Statement: Data reported in this paper are available on request from the corresponding author.

Acknowledgments: We thank Giovanni Perconti for providing the PERK, phospho-PERK and eiF2α antibodies and Roberto Chiarelli for the LC3 and p62 antibodies.

Conflicts of Interest: The authors declare no conflict of interest.

References

1. Debela, D.T.; Muzazu, S.G.; Heraro, K.D.; Ndalama, M.T.; Mesele, B.W.; Haile, D.C.; Kitui, S.K.; Manyazewal, T. New Approaches and Procedures for Cancer Treatment: Current Perspectives. *SAGE Open Med.* **2021**, *9*, 205031212110343. [CrossRef]
2. Abbas, Z.; Rehman, S. An Overview of Cancer Treatment Modalities. In *Neoplasm*; Shahzad, H.N., Ed.; InTech: London, UK, 2018; ISBN 978-1-78923-777-1.
3. Tsimberidou, A.M.; Fountzilas, E.; Nikanjam, M.; Kurzrock, R. Review of Precision Cancer Medicine: Evolution of the Treatment Paradigm. *Cancer Treat. Rev.* **2020**, *86*, 102019. [CrossRef] [PubMed]
4. Cisneros-Zevallos, L. The Power of Plants: How Fruit and Vegetables Work as Source of Nutraceuticals and Supplements. *Int. J. Food Sci. Nutr.* **2021**, *72*, 660–664. [CrossRef] [PubMed]
5. Cisneros-Zevallos, L. The Use of Controlled Postharvest Abiotic Stresses as a Tool for Enhancing the Nutraceutical Content and Adding-Value of Fresh Fruits and Vegetables. *J. Food Sci.* **2003**, *68*, 1560–1565. [CrossRef]
6. Bernardini, S.; Tiezzi, A.; Laghezza Masci, V.; Ovidi, E. Natural Products for Human Health: An Historical Overview of the Drug Discovery Approaches. *Nat. Prod. Res.* **2018**, *32*, 1926–1950. [CrossRef] [PubMed]
7. Nan, Y.; Su, H.; Zhou, B.; Liu, S. The Function of Natural Compounds in Important Anticancer Mechanisms. *Front. Oncol.* **2023**, *12*, 1049888. [CrossRef]
8. Pistollato, F.; Calderón Iglesias, R.; Ruiz, R.; Aparicio, S.; Crespo, J.; Dzul Lopez, L.; Giampieri, F.; Battino, M. The Use of Natural Compounds for the Targeting and Chemoprevention of Ovarian Cancer. *Cancer Lett.* **2017**, *411*, 191–200. [CrossRef]
9. Von Eiff, D.; Bozorgmehr, F.; Chung, I.; Bernhardt, D.; Rieken, S.; Liersch, S.; Muley, T.; Kobinger, S.; Thomas, M.; Christopoulos, P.; et al. Paclitaxel for Treatment of Advanced Small Cell Lung Cancer (SCLC): A Retrospective Study of 185 Patients. *J. Thorac. Dis.* **2020**, *12*, 782–793. [CrossRef]
10. Marupudi, N.I.; Han, J.E.; Li, K.W.; Renard, V.M.; Tyler, B.M.; Brem, H. Paclitaxel: A Review of Adverse Toxicities and Novel Delivery Strategies. *Expert Opin. Drug Saf.* **2007**, *6*, 609–621. [CrossRef]
11. Abu Samaan, T.M.; Samec, M.; Liskova, A.; Kubatka, P.; Büsselberg, D. Paclitaxel's Mechanistic and Clinical Effects on Breast Cancer. *Biomolecules* **2019**, *9*, 789. [CrossRef]
12. Kampan, N.C.; Madondo, M.T.; McNally, O.M.; Quinn, M.; Plebanski, M. Paclitaxel and Its Evolving Role in the Management of Ovarian Cancer. *BioMed Res. Int.* **2015**, *2015*, 413076. [CrossRef]
13. Garcia-Carbonero, R.; Supko, J.G. Current Perspectives on the Clinical Experience, Pharmacology, and Continued Development of the Camptothecins. *Clin. Cancer Res.* **2002**, *8*, 641–661. [PubMed]
14. Wall, M.E.; Wani, M.C. Camptothecin and Taxol: From Discovery to Clinic. *J. Ethnopharmacol.* **1996**, *51*, 239–254. [CrossRef] [PubMed]
15. Lauricella, M.; Lo Galbo, V.; Cernigliaro, C.; Maggio, A.; Palumbo Piccionello, A.; Calvaruso, G.; Carlisi, D.; Emanuele, S.; Giuliano, M.; D'Anneo, A. The Anti-Cancer Effect of *Mangifera indica* L. Peel Extract Is Associated to ΓH2AX-Mediated Apoptosis in Colon Cancer Cells. *Antioxidants* **2019**, *8*, 422. [CrossRef] [PubMed]
16. Lo Galbo, V.; Lauricella, M.; Giuliano, M.; Emanuele, S.; Carlisi, D.; Calvaruso, G.; De Blasio, A.; Di Liberto, D.; D'Anneo, A. Redox Imbalance and Mitochondrial Release of Apoptogenic Factors at the Forefront of the Antitumor Action of Mango Peel Extract. *Molecules* **2021**, *26*, 4328. [CrossRef] [PubMed]
17. Emanuele, S.; Lauricella, M.; Calvaruso, G.; D'Anneo, A.; Giuliano, M. Litchi Chinensis as a Functional Food and a Source of Antitumor Compounds: An Overview and a Description of Biochemical Pathways. *Nutrients* **2017**, *9*, 992. [CrossRef]
18. Emanuele, S.; Notaro, A.; Palumbo Piccionello, A.; Maggio, A.; Lauricella, M.; D'Anneo, A.; Cernigliaro, C.; Calvaruso, G.; Giuliano, M. Sicilian Litchi Fruit Extracts Induce Autophagy versus Apoptosis Switch in Human Colon Cancer Cells. *Nutrients* **2018**, *10*, 1490. [CrossRef]
19. D'Anneo, A.; Carlisi, D.; Lauricella, M.; Puleio, R.; Martinez, R.; Di Bella, S.; Di Marco, P.; Emanuele, S.; Di Fiore, R.; Guercio, A.; et al. Parthenolide Generates Reactive Oxygen Species and Autophagy in MDA-MB231 Cells. A Soluble Parthenolide Analogue Inhibits Tumour Growth and Metastasis in a Xenograft Model of Breast Cancer. *Cell Death Dis.* **2013**, *4*, e891. [CrossRef] [PubMed]
20. D'Anneo, A.; Carlisi, D.; Lauricella, M.; Emanuele, S.; Di Fiore, R.; Vento, R.; Tesoriere, G. Parthenolide Induces Caspase-Independent and AIF-Mediated Cell Death in Human Osteosarcoma and Melanoma Cells. *J. Cell. Physiol.* **2013**, *228*, 952–967. [CrossRef]
21. Carlisi, D.; D'Anneo, A.; Martinez, R.; Emanuele, S.; Buttitta, G.; Di Fiore, R.; Vento, R.; Tesoriere, G.; Lauricella, M. The Oxygen Radicals Involved in the Toxicity Induced by Parthenolide in MDA-MB-231 Cells. *Oncol. Rep.* **2014**, *32*, 167–172. [CrossRef]
22. Lauricella, M.; Maggio, A.; Badalamenti, N.; Bruno, M.; D'Angelo, G.; D'Anneo, A. Essential Oil of *Foeniculum vulgare* subsp. *piperitum* Fruits Exerts an Anti-tumor Effect in Triple-negative Breast Cancer Cells. *Mol. Med. Rep.* **2022**, *26*, 243. [CrossRef] [PubMed]
23. Seidel, T.; Wieder, O.; Garon, A.; Langer, T. Applications of the Pharmacophore Concept in Natural Product Inspired Drug Design. *Mol. Inf.* **2020**, *39*, 2000059. [CrossRef] [PubMed]
24. Correa, L.B.; Seito, L.N.; Manchope, M.F.; Verri, W.A.; Cunha, T.M.; Henriques, M.G.; Rosas, E.C. Methyl Gallate Attenuates Inflammation Induced by Toll-like Receptor Ligands by Inhibiting MAPK and NF-Kb Signaling Pathways. *Inflamm. Res.* **2020**, *69*, 1257–1270. [CrossRef]

25. Correa, L.B.; Pádua, T.A.; Alabarse, P.V.G.; Saraiva, E.M.; Garcia, E.B.; Amendoeira, F.C.; Ferraris, F.K.; Fukada, S.Y.; Rosas, E.C.; Henriques, M.G. Protective Effect of Methyl Gallate on Murine Antigen-Induced Arthritis by Inhibiting Inflammatory Process and Bone Erosion. *Inflammopharmacology* **2022**, *30*, 251–266. [CrossRef] [PubMed]

26. Asnaashari, M.; Farhoosh, R.; Sharif, A. Antioxidant Activity of Gallic Acid and Methyl Gallate in Triacylglycerols of Kilka Fish Oil and Its Oil-in-Water Emulsion. *Food Chem.* **2014**, *159*, 439–444. [CrossRef] [PubMed]

27. Choi, J.-G.; Kang, O.-H.; Lee, Y.-S.; Oh, Y.-C.; Chae, H.-S.; Jang, H.-J.; Shin, D.-W.; Kwon, D.-Y. Antibacterial Activity of Methyl Gallate Isolated from Galla Rhois or Carvacrol Combined with Nalidixic Acid Against Nalidixic Acid Resistant Bacteria. *Molecules* **2009**, *14*, 1773–1780. [CrossRef]

28. Lee, H.; Lee, H.; Kwon, Y.; Lee, J.-H.; Kim, J.; Shin, M.-K.; Kim, S.-H.; Bae, H. Methyl Gallate Exhibits Potent Antitumor Activities by Inhibiting Tumor Infiltration of CD4+ CD25+ Regulatory T Cells. *J. Immunol.* **2010**, *185*, 6698–6705. [CrossRef]

29. Lee, S.-H.; Kim, J.K.; Kim, D.W.; Hwang, H.S.; Eum, W.S.; Park, J.; Han, K.H.; Oh, J.S.; Choi, S.Y. Antitumor Activity of Methyl Gallate by Inhibition of Focal Adhesion Formation and Akt Phosphorylation in Glioma Cells. *Biochim. Biophys. Acta (BBA)—Gen. Subj.* **2013**, *1830*, 4017–4029. [CrossRef]

30. Huang, C.-Y.; Chang, Y.-J.; Wei, P.-L.; Hung, C.-S.; Wang, W. Methyl Gallate, Gallic Acid-Derived Compound, Inhibit Cell Proliferation through Increasing ROS Production and Apoptosis in Hepatocellular Carcinoma Cells. *PLoS ONE* **2021**, *16*, e0248521. [CrossRef]

31. Liang, H.; Chen, Z.; Yang, R.; Huang, Q.; Chen, H.; Chen, W.; Zou, L.; Wei, P.; Wei, S.; Yang, Y.; et al. Methyl Gallate Suppresses the Migration, Invasion, and Epithelial-Mesenchymal Transition of Hepatocellular Carcinoma Cells via the AMPK/NF-KB Signaling Pathway in Vitro and in Vivo. *Front. Pharmacol.* **2022**, *13*, 894285. [CrossRef]

32. Meunier, V.; Bourrié, M.; Berger, Y.; Fabre, G. The Human Intestinal Epithelial Cell Line Caco-2; Pharmacological and Pharmacokinetic Applications. *Cell Biol. Toxicol.* **1995**, *11*, 187–194. [CrossRef] [PubMed]

33. Marziano, M.; Tonello, S.; Cantù, E.; Abate, G.; Vezzoli, M.; Rungratanawanich, W.; Serpelloni, M.; Lopomo, N.F.; Memo, M.; Sardini, E.; et al. Monitoring Caco-2 to Enterocyte-like Cells Differentiation by Means of Electric Impedance Analysis on Printed Sensors. *Biochim. Biophys. Acta (BBA)—Gen. Subj.* **2019**, *1863*, 893–902. [CrossRef] [PubMed]

34. Natoli, M.; Leoni, B.D.; D'Agnano, I.; Zucco, F.; Felsani, A. Good Caco-2 Cell Culture Practices. *Toxicol. Vitr.* **2012**, *26*, 1243–1246. [CrossRef]

35. Lauricella, M.; Carlisi, D.; Giuliano, M.; Calvaruso, G.; Cernigliaro, C.; Vento, R.; D'Anneo, A. The Analysis of Estrogen Receptor-α Positive Breast Cancer Stem-like Cells Unveils a High Expression of the Serpin Proteinase Inhibitor PI-9: Possible Regulatory Mechanisms. *Int. J. Oncol.* **2016**, *49*, 352–360. [CrossRef]

36. Guzmán, C.; Bagga, M.; Kaur, A.; Westermarck, J.; Abankwa, D. ColonyArea: An ImageJ Plugin to Automatically Quantify Colony Formation in Clonogenic Assays. *PLoS ONE* **2014**, *9*, e92444. [CrossRef]

37. Pratelli, G.; Di Liberto, D.; Carlisi, D.; Emanuele, S.; Giuliano, M.; Notaro, A.; De Blasio, A.; Calvaruso, G.; D'Anneo, A.; Lauricella, M. Hypertrophy and ER Stress Induced by Palmitate Are Counteracted by Mango Peel and Seed Extracts in 3T3-L1 Adipocytes. *Int. J. Mol. Sci.* **2023**, *24*, 5419. [CrossRef] [PubMed]

38. Munafó, D.B.; Colombo, M.I. A Novel Assay to Study Autophagy: Regulation of Autophagosome Vacuole Size by Amino Acid Deprivation. *J. Cell Sci.* **2001**, *114*, 3619–3629. [CrossRef]

39. Yang, C.; Kaushal, V.; Shah, S.V.; Kaushal, G.P. Autophagy Is Associated with Apoptosis in Cisplatin Injury to Renal Tubular Epithelial Cells. *Am. J. Physiol. Ren. Physiol.* **2008**, *294*, F777–F787. [CrossRef] [PubMed]

40. Giuliano, M.; Bellavia, G.; Lauricella, M.; D'Anneo, A.; Vassallo, B.; Vento, R.; Tesoriere, G. Staurosporine-Induced Apoptosis in Chang Liver Cells Is Associated with down-Regulation of Bcl-2 and Bcl-XL. *Int. J. Mol. Med.* **2004**, *13*, 565–571. [CrossRef] [PubMed]

41. Cernigliaro, C. Ethanol-Mediated Stress Promotes Autophagic Survival and Aggressiveness of Colon Cancer Cells via Activation of Nrf2/HO-1 Pathway. *Cancers* **2019**, *11*, 505. [CrossRef]

42. Ding, X.; Hu, X.; Chen, Y.; Xie, J.; Ying, M.; Wang, Y.; Yu, Q. Differentiated Caco-2 Cell Models in Food-Intestine Interaction Study: Current Applications and Future Trends. *Trends Food Sci. Technol.* **2021**, *107*, 455–465. [CrossRef]

43. Ibrahim, I.M.; Abdelmalek, D.H.; Elfiky, A.A. GRP78: A Cell's Response to Stress. *Life Sci.* **2019**, *226*, 156–163. [CrossRef] [PubMed]

44. Prasad, A.; Bloom, M.S.; Carpenter, D.O. Role of Calcium and ROS in Cell Death Induced by Polyunsaturated Fatty Acids in Murine Thymocytes. *J. Cell. Physiol.* **2010**, *225*, 829–836. [CrossRef]

45. Hempel, N.; Trebak, M. Crosstalk between Calcium and Reactive Oxygen Species Signaling in Cancer. *Cell Calcium* **2017**, *63*, 70–96. [CrossRef] [PubMed]

46. Carreras-Sureda, A.; Pihán, P.; Hetz, C. Calcium Signaling at the Endoplasmic Reticulum: Fine-Tuning Stress Responses. *Cell Calcium* **2018**, *70*, 24–31. [CrossRef] [PubMed]

47. Law, B.Y.K.; Chan, W.K.; Xu, S.W.; Wang, J.R.; Bai, L.P.; Liu, L.; Wong, V.K.W. Natural Small-Molecule Enhancers of Autophagy Induce Autophagic Cell Death in Apoptosis-Defective Cells. *Sci. Rep.* **2014**, *4*, 5510. [CrossRef] [PubMed]

48. Kania, E.; Pająk, B.; Orzechowski, A. Calcium Homeostasis and ER Stress in Control of Autophagy in Cancer Cells. *BioMed. Res. Int.* **2015**, *2015*, 352794. [CrossRef]

49. Fazlul Kabir, M.; Kim, H.-R.; Chae, H.-J. Endoplasmic Reticulum Stress and Autophagy. In *Endoplasmic Reticulum*; Català, A., Ed.; IntechOpen: London, UK, 2019; ISBN 978-1-83880-087-1.

50. Tanida, I.; Ueno, T.; Kominami, E. LC3 and Autophagy. In *Autophagosome and Phagosome*; Deretic, V., Ed.; Methods in Molecular Biology™; Humana Press: Totowa, NJ, USA, 2008; Volume 445, pp. 77–88. ISBN 978-1-58829-853-9.

51. Emanuele, S.; Lauricella, M.; D'Anneo, A.; Carlisi, D.; De Blasio, A.; Di Liberto, D.; Giuliano, M. P62: Friend or Foe? Evidences for OncoJanus and NeuroJanus Roles. *Int. J. Mol. Sci.* **2020**, *21*, 5029. [CrossRef]

52. Appella, E.; Anderson, C.W. Post-Translational Modifications and Activation of P53 by Genotoxic Stresses: P53 Post-Translational Modifications. *Eur. J. Biochem.* **2001**, *268*, 2764–2772. [CrossRef]

53. Chen, J. The Cell-Cycle Arrest and Apoptotic Functions of P53 in Tumor Initiation and Progression. *Cold Spring Harb. Perspect. Med.* **2016**, *6*, a026104. [CrossRef]

54. White, E. Autophagy and P53. *Cold Spring Harb. Perspect. Med.* **2016**, *6*, a026120. [CrossRef] [PubMed]

55. Tasdemir, E.; Maiuri, M.C.; Morselli, E.; Criollo, A.; D'Amelio, M.; Djavaheri-Mergny, M.; Cecconi, F.; Tavernarakis, N.; Kroemer, G. A Dual Role of P53 in the Control of Autophagy. *Autophagy* **2008**, *4*, 810–814. [CrossRef] [PubMed]

56. Scherz-Shouval, R.; Weidberg, H.; Gonen, C.; Wilder, S.; Elazar, Z.; Oren, M. P53-Dependent Regulation of Autophagy Protein LC3 Supports Cancer Cell Survival under Prolonged Starvation. *Proc. Natl. Acad. Sci. USA* **2010**, *107*, 18511–18516. [CrossRef] [PubMed]

57. Tang, J.; Di, J.; Cao, H.; Bai, J.; Zheng, J. P53-Mediated Autophagic Regulation: A Prospective Strategy for Cancer Therapy. *Cancer Lett.* **2015**, *363*, 101–107. [CrossRef]

58. Pelicano, H.; Carney, D.; Huang, P. ROS Stress in Cancer Cells and Therapeutic Implications. *Drug Resist. Updates* **2004**, *7*, 97–110. [CrossRef] [PubMed]

59. Arfin, S.; Jha, N.K.; Jha, S.K.; Kesari, K.K.; Ruokolainen, J.; Roychoudhury, S.; Rathi, B.; Kumar, D. Oxidative Stress in Cancer Cell Metabolism. *Antioxidants* **2021**, *10*, 642. [CrossRef] [PubMed]

60. Perillo, B.; Di Donato, M.; Pezone, A.; Di Zazzo, E.; Giovannelli, P.; Galasso, G.; Castoria, G.; Migliaccio, A. ROS in Cancer Therapy: The Bright Side of the Moon. *Exp. Mol. Med.* **2020**, *52*, 192–203. [CrossRef]

61. Reczek, C.R.; Chandel, N.S. The Two Faces of Reactive Oxygen Species in Cancer. *Annu. Rev. Cancer Biol.* **2017**, *1*, 79–98. [CrossRef]

62. Sznarkowska, A.; Kostecka, A.; Meller, K.; Bielawski, K.P. Inhibition of Cancer Antioxidant Defense by Natural Compounds. *Oncotarget* **2017**, *8*, 15996–16016. [CrossRef]

63. Conklin, K.A. Chemotherapy-Associated Oxidative Stress: Impact on Chemotherapeutic Effectiveness. *Integr. Cancer Ther.* **2004**, *3*, 294–300. [CrossRef]

64. Li, L.; Tan, J.; Miao, Y.; Lei, P.; Zhang, Q. ROS and Autophagy: Interactions and Molecular Regulatory Mechanisms. *Cell Mol. Neurobiol.* **2015**, *35*, 615–621. [CrossRef]

65. Celesia, A.; Morana, O.; Fiore, T.; Pellerito, C.; D'Anneo, A.; Lauricella, M.; Carlisi, D.; De Blasio, A.; Calvaruso, G.; Giuliano, M.; et al. ROS-Dependent ER Stress and Autophagy Mediate the Anti-Tumor Effects of Tributyltin (IV) Ferulate in Colon Cancer Cells. *Int. J. Mol. Sci.* **2020**, *21*, 8135. [CrossRef] [PubMed]

66. Wang, H.; Guo, M.; Wei, H.; Chen, Y. Targeting P53 Pathways: Mechanisms, Structures, and Advances in Therapy. *Signal Transduct. Target. Ther.* **2023**, *8*, 92. [CrossRef] [PubMed]

67. Do Patrocinio, A.B.; Rodrigues, V.; Guidi Magalhães, L. P53: Stability from the Ubiquitin–Proteasome System and Specific 26S Proteasome Inhibitors. *ACS Omega* **2022**, *7*, 3836–3843. [CrossRef] [PubMed]

68. Liu, B.; Chen, Y.; St. Clair, D.K. ROS and P53: A Versatile Partnership. *Free Radic. Biol. Med.* **2008**, *44*, 1529–1535. [CrossRef] [PubMed]

Article

Evaluation of Proanthocyanidins from Kiwi Leaves (*Actinidia chinensis*) against Caco-2 Cells Oxidative Stress through Nrf2-ARE Signaling Pathway

Ji-Min Lv [1], Mostafa Gouda [1,2,*], Xing-Qian Ye [1], Zhi-Peng Shao [3] and Jian-Chu Chen [1,*]

[1] College of Biosystems Engineering and Food Science, National-Local Joint Engineering Laboratory of Intelligent Food Technology and Equipment, Zhejiang Key Laboratory for Agro-Food Processing, Zhejiang Engineering Laboratory of Food Technology and Equipment, Zhejiang University, Hangzhou 310058, China; lvjimin@zju.edu.cn (J.-M.L.); psu@zju.edu.cn (X.-Q.Y.)

[2] Department of Nutrition & Food Science, National Research Centre, Dokki, Giza 12622, Egypt

[3] School of Food Science and Biotechnology, Zhejiang Gongshang University, Hangzhou 310018, China; 15010000019@pop.zjgsu.edu.cn

* Correspondence: mostafa-gouda@zju.edu.cn (M.G.); jc@zju.edu.cn (J.-C.C.)

Abstract: Proanthocyanidins (PAs) are considered to be effective natural byproduct and bioactive antioxidants. However, few studies have focused on their mode of action pathways. In this study, reactive oxygen species (ROS), oxidative stress indices, real-time PCR, Western blotting, confocal microscopy, and molecular docking were used to investigate the protective effect of purified kiwi leaves PAs (PKLPs) on Caco-2 cells' oxidative stress mechanisms. The results confirmed that pre-treatment with PKLPs significantly reduced H_2O_2-induced oxidative damage, accompanied by declining ROS levels and malondialdehyde (MDA) accumulation in the Caco-2 cells. The PKLPs upregulated the expression of antioxidative enzymes (GSH-px, CAT, T-SOD) and the relative mRNA (Nrf, HO-1, SOD-1, CAT) of the nuclear factor erythroid 2-related factor (Nrf2) signaling pathway. The protein-expressing level of the Nrf2 and its relative protein (NQO-1, HO-1, SOD-1) were significantly increased ($p < 0.05$) in the PKLPs pre-treatment group compared to the model group. In conclusion, the novelty of this study is that it explains how PKLPs' efficacy on the Nrf2-ARE signaling pathway, in protecting vital cells from oxidative stress, could be used for cleaner production.

Keywords: antioxidant response element (ARE); Nrf2 signaling pathway; bioactive byproducts; proanthocyanidins; oxidative stress mechanisms

Citation: Lv, J.-M.; Gouda, M.; Ye, X.-Q.; Shao, Z.-P.; Chen, J.-C. Evaluation of Proanthocyanidins from Kiwi Leaves (*Actinidia chinensis*) against Caco-2 Cells Oxidative Stress through Nrf2-ARE Signaling Pathway. *Antioxidants* **2022**, *11*, 1367. https://doi.org/10.3390/antiox11071367

Academic Editors: Antonella D'Anneo and Marianna Lauricella

Received: 20 June 2022
Accepted: 12 July 2022
Published: 14 July 2022

Publisher's Note: MDPI stays neutral with regard to jurisdictional claims in published maps and institutional affiliations.

1. Introduction

With the increase in demand for natural safe alternatives to synthetic chemicals, scientists are searching for efficient techniques to evaluate the functionality of phytochemicals [1,2]. The correlation between phytochemicals and cellular antioxidant enzyme activity, based on their chemical and molecular gene expressions, can provide accurate information about their functionalities [3–5]. For instance, proanthocyanidins (PAs), which are abundant in kiwi (*Actinidia chinensis*) fruits and leaves, are a byproduct of kiwi fruit production, and are secondary plant metabolites that belong to the class of flavan-3-ols, with several biological activities and a wide range of health-related benefits [6,7].

For instance, as a natural safe extract, kiwi PAs have vital antioxidant, antidiabetic, and antimicrobial properties, and outstanding anticancer effects [8,9]. Our previous work found that purified kiwi leaves PAs (PKLPs) were relatively high, with a yield of 6.23% (dry weight), making kiwi leaves byproducts a commercially viable source of PAs. Moreover, the PAs isolated from kiwi leaves are mainly composed of (epi)afzelechin, (epi)catechin and (epi)gallocatechin [6]—in which regard, this structure is unique compared to other PAs' plants sources (like baobab seeds, grape seeds, and hazelnut skin).

The bioactive potential of PAs—such as their antioxidant, antimicrobial activity—make them applicable in functional food additives for the therapeutic intervention of human disorders [2]. Thus, this novel health-related natural molecule could be used in different medicinal foods applications: for instance, the effects of acid-hydrolyzed PKLPs on the viability of Caco-2 cancer cells, which indicate that the degree of polymerization (DP) of PAs has a significant effect on cancer cells [7]. However, the key antioxidative bioactive PAs fraction mode of action still requires further study.

Therefore, it is of great significance to explore the main bioactive ingredients of PKLPs as edible and potent viable antioxidant food ingredients. Based on the mean degree of polymerization (mDP) of PAs complexes, previous study has investigated the relationship between the antioxidant activity and PAs' structure compositions. The DP significantly affected the cellular absorption of PAs fractions. Ou and Gu [10] indicated that DP > 4 of PAs were not absorbable as a result of their large molecular dimensions. Additionally, a high DP influences the bioactivity of PAs. For instance, Li, et al. [11], reported a remarkable antiproliferative activity induction of PAs on Caco-2 cells with increasing PAs DP. Additionally, Li, Chen, Li, Liu, Liu, and Liu [11] reported that PAs' cellular antioxidant activities on Caco-2 cells decreased with the increase of their molecular weight. Under normal conditions, there is a balance between antioxidative defense and the generation of reactive oxygen species (ROS). Oxidative stress occurs when ROS production exceeds the extent of cellular antioxidative defense [12]. The increased ROS are able to disturb barrier integrity, damage cell membranes, and enhance incidences of endotoxemia and inflammation [13]. Thus, the over-production of ROS has a deleterious effect on human health, which needs to be controlled. As PAs have a significant impact on regulating central transcription factors, they have the ability to reduce ROS' generation of Caco-2 cells [14]. Koudoufio, et al. [15], reported that PAs have a protective impact on differentiated intestinal Caco-2/15 cells' oxidative stress (OxS) and inflammation. They reported that PAs significantly reduced malondialdehyde, as a lipid peroxidation biomarker, and raised the relative antioxidant enzymes via increasing the ratio of the Nrf2/Keap1. The uniqueness of using Caco-2 cells in studying vital antioxidant activity has been reported by Kellett, et al. [16], who noted that, due to differences in the active membrane transport among the cell types, Caco-2-based cellular antioxidant activity measurements appeared to be a more suitable method for phytochemicals bioactivities studies compared to other cell lines, like hepatocarcinoma (HepG2) cells.

The importance of studying the nuclear factor erythroid 2-related factor 2 (Nrf2) as a cellular antioxidant marker comes from its high ability to bind with Kelch-like ECH-associated protein (Keap1, the cysteine-based mammalian intracellular sensor for electrophiles and oxidants) in the cytoplasm, under normal circumstances [17]. Upon oxidative stress, the Nrf2 migrates to the nucleus, and combines with the antioxidant response element (ARE), which can upregulate the transcription of cell defense-related genes, including drug metabolizers, detoxifying enzymes, and antioxidant proteins [18]. Several studies have reported that the Nrf2 signaling pathway plays a crucial role in the activation of cytoprotective genes in response to xenobiotics, and protecting cells against oxidative stress. However, the underlying mechanisms of the antioxidative action were unclear. In particular, the relationship between the chemical structure of PAs and their antioxidant mechanisms has not been investigated before.

Arroyave-Ospina, et al. [19], noted that intracellular antioxidant activity is an effective assay for measuring ROS and oxidative stress indices. Moreover, real-time PCR, in combination with Western blot and florescent confocal microscopy, could reflect the molecular level of the genotype and phenotype relationship, through antioxidant enzymes gene expression [20]. Additionally, molecular docking is an effective tool for investigating the best intermolecular framework between bioactive phytochemicals, cellular proteins, and biological macromolecules, in order to clarify the potential mechanisms of their interactions [21,22].

Thus, the aim of this study was to characterize kiwi leaves PAs' viable antioxidant activity against Caco-2 cells, and to emphasize the potential pathways that cause that mode of action through their binding efficiency to the ARE from their impacts on NRF2–Keap1 complexes. This could enhance the functional application of these kinds of functional bioactive components that are produced as byproducts of leaves from kiwi fruit production. Thus, it could enhance the applicability of PAs in green and sustainable industrial applications.

2. Materials and Methods

2.1. Chemicals and Materials

RNase Free dH_2O, ethanol (analytical grade), and acetone (analytical grade) were purchased from Macklin Biochemical Technology Co., Ltd., (Shanghai, China). Hydrogen peroxide, Nile Red, Dichlorodihydrofluorescein diacetate (DCFH-DA), and Dihydroethidium (DHE) were purchased from Sigma–Aldrich (St. Louis, MO, USA). AB-8 Macroporous resin was obtained from Solarbio Science & Technology Company (Beijing, China), and Sephadex LH-20 was purchased from GE Healthcare Bio-Sciences (Uppsala, Sweden). Dulbecco's Modified Eagle Medium (DMEM), fetal bovine serum (FBS), penicillin/streptomycin (P/S), and phosphate-buffered solution (PBS), were supplied by Gibco company (Grand Island, NE, USA). SYBR Green PCR Master Mix and TRIzol^{TM} were obtained from Thermo Fisher Scientific, Inc., (Cleveland, OH, USA). The commercial PVDF membrane (0.45 μ, immobilon) was purchased from Solarbio Science & Technology Co., Ltd., Beijing, China.

Kiwi (*Actinidia chinensis*) leaves were collected from Zhuji farm, Shaoxing, Zhejiang Province, China, during October. Then, they were freeze-dried, and grounded into powder for extracting PAs.

2.2. Extraction and Purificantion of PAs from Kiwifruit Leaves

Fresh kiwifruit leaves PAs were extracted by using the optimized conditions of ultrasound-assisted extraction, following the method described by Lv, Gouda, Zhu, Ye, and Chen [6], with some modifications. Firstly, freeze-dried kiwi leaves powders (10 g) were sonicated by JY92-IIDN (Ningbo Scientz Biotechnology Co., Ningbo, China) under the following optimum conditions: 30 mL/g dry weight solvent to solid ration; 40% ultrasound-amplitude; and 70 °C sonication temperature for 15 min. Secondly, the crude PAs were extracted by 0.4 L aqueous acetone solvent (80%, *v/v*). Then, the acetone was removed through an evaporation process, under vacuum and 40 °C temperature, by using rotary evaporator (Dragon RE100-pro, Beijing, China). Afterwards, hexane was used to remove the non-polar components from the obtained aqueous phase. Thirdly, sugars, proteins, and pigments were removed from the extracted PAs by using an AB-8 Macroporous resin column (Solarbio Science & Technology, Beijing, China). Afterwards, to further purify the obtained PAs, according to the previous study (with some modifications) of Chai, et al. [23], the above-obtained PAs (1 g, freeze-dried) were loaded onto a Sephadex LH-20 column (GE Healthcare Bio-Sciences, Uppsala, Sweden); then, methanol (50%, *v/v*) was used to wash the column and remove the impurities. Subsequently, 90% methanol was used to elute the fraction A (FA) anthocyanins. After that, 50% acetone was used to collect as fraction B (FB). Both fractions (FA and FB) were freeze-dried as PKLPs.

2.3. Reversed-Phase HPLC-QTOF-MS/MS Analysis of PKLPs

The separation of the phenolic compounds was performed following our previous study [7]. The column used for analysis was the Luna HILIC column (Phenomenex, Torrance, CA, USA; 250 × 4.6 mm; 5 μm), with a flow rate of 0.35 mL/min at 30 °C. Injections of 10 μL of each purified extract were injected into a Waters 2489 HPLC with a UV-Vis detector (Waters Corp., Milford, MA, USA), using a mobile phase comprising a linear gradient of 99.5% acetonitrile and 0.5% acetic acid (solvent A),96.9% acetonitrile, 3% water, and 0.1% acetic acid (solvent B). The detection wavelength was set at 280 nm to monitor all phenolic compounds. For comparison, the elution conditions for solvent B were: 0–10 min, 15%; 10–20 min, 15–20%; 20–70 min, 20–60%; and 70–80 min, 60–100%. The separated compounds were fractionated and defined by mass spectra, using a Triple-TOF 5600+ ion trap mass spectrometer (AB scientific, Framingham, MA, USA). Three replicates of each sample were collected for data analysis.

2.4. Cell Culture and Treatment

The Caco-2 (human colonic carcinoma) cell line was provided by the Cell Resource Center, Shanghai Institutes for Biological Sciences, Chinese Academy of Sciences. The cells were cultured at 37 °C under 5% CO_2, in a DMEM medium (Gibco) with 1% P/S (Gibco) and 20% FBS (Gibco). The cells were sub-cultured 3–4 times a week, and replaced in a fresh medium to keep the cells in a good growth state.

2.5. Injured Cell Model Induced by H_2O_2

The antioxidant activity of the H_2O_2-induced cell death of the Caco-2 cells was determined according to a previous study [24]. Briefly, the Caco-2 cells were seeded in cell culture 96-well plates, with a density of 1×10^5 cells/mL, and cultivated for 24 h at 37 °C. Afterwards, the medium was removed, and the wells were washed 3 times with PBS, before a new, fresh medium (containing 200 µM/L H_2O_2) was added. The group treated by the same cell medium without H_2O_2 was taken as the control group. The H_2O_2 treatment lasted for 4 h, and the results were expressed as cell viability [7].

2.6. Intracellular Antioxidant Activity Assay

The effect of PKLPs' antioxidant activity on the Caco-2 cells was determined according to previous literature, with some modification [25]. Briefly, the Caco-2 cells were seeded in cell culture 96-well plates, with a density of 1×10^5 cells/mL, and cultivated for 24 h at 37 °C. The experimental groups were treated with various concentrations of samples (50 µg/mL of FA; 50 µg/mL of FB; 75 µg/mL of catechin; each for 100 µL) 24 h before H_2O_2 treatment. The control group was normally cultivated without H_2O_2 or PAs, and the model group was only treated with H_2O_2 to cause the oxidant damage. The result was expressed as the cell viability determined by the cell counting kit (CCK-8 assay), as described in the literature [7].

2.7. Determination of Reactive Oxygen Species (ROS)

The cellular ROS was determined according to previously described methods, with slight modifications [26]. Briefly, Caco-2 cells were seeded into 12-well plates, at a density of 1×10^5 cells/mL, for 24 h cultivation at 37 °C. The experimental groups were treated with various concentrations of samples (50 µg/mL of FA; 50 µg/mL of FB; 75 µg/mL of catechin; each for 100 µL) for 24 h cultivation, followed by treatment with (200 µM/L) H_2O_2 for another 4 h. The control group was normally cultivated at the same time, and the model group was only treated with (200 µM/L) H_2O_2 for 4 h. After incubation with 10 µM dichlorofluorescein diacetate (DCF-DA) at ambient temperature in the dark for 30 min, the cells were instantly washed 3 times by PBS. And then, the fluorescence value was measured by the fluorescence microscope (Nikon, Tokyo, Japan) at an excitation wavelength of 485 nm, and an emission wavelength of 525 nm. The fluorescence intensity was calculated by image analysis software ImageProPlus 6.0 (Media Cybernetics, Inc., Rockville, MD, USA), and expressed as mean DCF fluorescence intensity.

2.8. Determination of Oxidative Stress Indices

Malondialdehyde (MDA, #S0131M), catalase (CAT, # S0051), total superoxide dismutase (T-SOD, #S0101S), and glutathione peroxidase (GSH-Px, #S0056), were measured by the commercial kits purchased from Beyotime Biotechnology Company (Shanghai, China), according to the manufacturer's instructions.

2.9. RNA Extraction, Reverse Transcription, and Quantitative Real-Time PCR

Total RNA was extracted from the cells using TRIzol™ (Thermo Fisher Scientific, Inc.), and diluted to 1 μg/μL. qPCR was performed according to a previously described method. Firstly, cDNA was synthesized from 1.0 μg of total RNA using the PrimeScript RT reagent Kit (TaKaRa, Japan), in a final volume of 20 μL with 10 μL master mix (4 μL RNase Free dH$_2$O, 4 μL 5× PrimeScript Buffer, 1 μL RT Mix, and 1 μL PrimeScript RT Enzyme Mix). Then, 1 μL of cDNA template, 1 μL of upstream and downstream primers, 10 μL SYBR Green PCR MasterMix, and 7 μL of RNase Free dH$_2$O were mixed, before being carried out in a Stepone Plus qPCR instrument (Thermo Fisher Scientific, Inc.). The PCR conditions were set as: 95 °C maintained for 5 min, followed by 40 cycles of 95 °C for 15 s and 60 °C for 40 s. The primer sequences are listed in Table 1.

Table 1. Primer sets for quantitative real-time PCR.

Gene	Forward Sequence	Reverse Sequence	NCBI No.
Nrf2	TCACACGAGATGAGCTTAGGGCAA	TACAGTTCTGGGCGGCGACTTTAT	NM_010902.4
Keap1	CAGCAACTCTGTGACGTGACC	TCAATAAGCCTTTCCATGACCT	NM_016679.4
SOD-1	TGGTTTGCGTCGTAGTCTCC	CTTCGTCGCCATAACTCGCT	NM_000454.4
NQO-1	GGTGAGCTGAAGGACTCGAA	ACCACTGCAATGGGAACTGAA	NM_008706.5
HO-1	ATGGCCTCCCTGTACCACATC	TGTTGCGCTCAATCTCCTCCT	NM_002133.2
CAT	CCATTATAAGACTGACCAGGGC	AGTCCAGGAGGGGTACTTTCC	NM_001752.3
Bcl-2	ATGTGTGTGGAGCGTCAACC	CAGAGACAGCCAGGAGAAATC	NM_000633.3
Bax	GAGCTGCAGAGGATGATTGCT	TGATCAGCTCGGGCACTTTA	NM_007527.3
β-actin	CAAGAGAGGTATCCTGACCT	TGATCTGGGTCATCTTTTCAC	NM_007393.5

Relative mRNA expression was normalized to the control group. The $2^{-\Delta\Delta Ct}$ formula was used to quantify, using *β-actin* as a reference gene [27]. All the results were obtained from at least three independent experiments.

2.10. Immunofluorescence

Samples were blocked with bovine serum (5%, *w/v*) for 30 min at ambient temperature, and repaired with 10.2 mM sodium citrate buffer. Then, the following primary antibodies against the Nrf2 (1:200), and the HO-1 (1:200) were used for overnight incubation at 4 °C. Afterwards, the samples were washed with PBS before incubation with anti-rabbit secondary antibodies DAPI staining at ambient temperature for 1 h. The treated Caco-2 cells were visualized using a Zeiss LSM 780 confocal microscope (Carl Zeiss SAS, Jena, Germany), and the DP2-TWAN image-acquisition system (Olympus Corp., Tokyo, Japan).

2.11. Molecular Docking

The initial crystal structures of Keap1 and the Nrf2 (PDBID: 2FLU) were obtained from the RCSB PDB database (http://www.rcbs.org, accessed on 15 February 2022), as reported in previous literatures [28,29]. The 3D structure of the PAs was built by Chem3D Ultra 12.0, and energetically minimized with an MM2 force field. The interaction between Keap1, the Nrf2, and the PAs was investigated by using docking analysis. Schrodinger® docking suits were selected for the molecular docking studies, and a Glide® receptor grid generator was used to create the grid sites with default parameters. Both the protein and the ligand structures were refined using an OPLS3e forcefield, to get correct formal charges and protonation states. A receptor grid was then generated with the prepared structure, with correct formal charges and protonation states. Finally, the ligand was docked into the corresponding protein structure. After docking, the results were ranked according to a scoring function, combining GlideScore with Prime energies, and the complex features of the protein-ligand were visualized with Pymol (Delano Scientific LLC, San Carlos, CA, USA).

2.12. Western Blot

Western blot was conducted, following the method of Su, et al. [30]. In brief, the cytoplasmic protein was separated by NE-PER kit (n. 78833, Thermo Fisher Scientific, Rockford, IL, USA). The separated protein contained 1% phenylmethanesulfony fluoride (PMSF, ST506, Beyotime Biotech, China) for preventing the degradation of protein, and the concentrations were measured using the bicinchoninic acid method (BCA, P0012, Beyotime Biotech, China). The separated protein was produced by electrophoresis on SDS-polyacrylamide gels, then transferred to 0.45 μm polyvinylidene fluoride (PVDF) membranes (immobilon, Solarbio Science & Technology Co., Ltd., Beijing, China). After blocking with 5% non-fat dry milk, in PBS containing 0.1% Tween-20, the membrane was incubated with the primary antibody at 4 °C for 14 h. Then, the membrane was incubated with horseradish peroxidase-conjugated secondary antibodies (ab97205 and ab97215, Abcam, Cambridge, MA, USA) for 1–2 h after washing 3 times. Next, a chemiluminescent HRP substrate was used to visualize the immunoreactive protein bands (Millipore, WBKLS0100), after washing 3 times. Primary antibodies were used as follows: the nuclear factor erythroid 2-related factor 2 (Nrf2) for humans (ab62352), heme oxygenase-1 (HO-1) for humans (ab13243), and Kelch-like ECH-associated protein 1 (Keap1) for humans (ab139729) were procured from Abcam (Cambridge, MA, USA); superoxide dismutase 1 (SOD-1) for humans (sc-17767), quinone oxidoreductase 1 (NQO1) for humans (sc-271116), and B-cell lymphoma 2 (Bcl-2) for humans (sc-7382), were procured from Santa Cruz Biotechnology Inc. (California, MA, USA).

2.13. Statistical Analysis

Experiments were conducted in triplicate, and the results were presented as mean ± standard deviation. Data were further analyzed via one-way analysis of variance (ANOVA), using SPSS 19.0 (Chicago, IL, USA). Duncan and Least Significant Difference (LSD) analyses at $p < 0.05$ level were used to significantly differentiate the studied treatments. IC_{50} was calculated based on the regression equation.

3. Results and Discussion

3.1. Comparison of the Chemical Composition of PKLPs (FA and FB) by Reversed-Phase HPLC-QTOF-MS/MS

According to our previous study [7], FA contains small molecule flavonoids, including quercetin, isoquercetin, and polyphenols, such as procyanidins (Figure 1a). In the present study, we focused on analyzing the composition of PAs fractions without the acid hydrolysis post-purification process. In addition, we marked the retention time (Rt) of PAs with different degrees of polymerization. We also marked the main ionized fragments of PAs in QTOF-MS^2: molecular ions at m/z 447, with major fragment ions at m/z 301 ([M − H − 146]$^-$) and m/z 109 ([M − H − 338]$^-$), which were identified as quercetin based on their fragmentation patterns and retention time (Table 2). According to the results of QTOF-MS^2, the FA included one molecular ion peak [M − H]$^-$ at m/z 463, with ionic fragmentation MS^2 at m/z 301 and 463, which was consider as isoquercetin. Fraction A contained catechin, epicatechin, and procyanidins, with its molecular ion peak ([M − H]$^-$) at m/z 289 and 577, respectively, while tandem mass spectrometry yielded typical fragment ions at m/z 163, 137, 245, and 287, 289, 425, 451, respectively. The ionic fragmentations at m/z 163 and 451 ([M − H − 126]$^-$) resulted from the loss of a neutral molecule, A-ring (1,3,5-trihydroxybenzene, 126 Da), through heterocyclic ring fission (HRF). The typical fragment ions at m/z 137 and 425 derived from retro-diels-alder (RDA) reaction through the loss of a neutral fragment containing the B-ring (152 Da) [31].

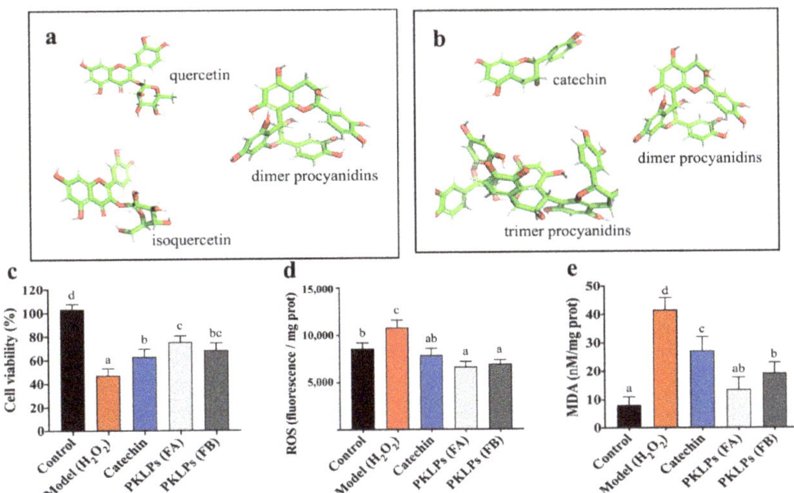

Figure 1. Chemical structure of eluted fractions: (**a**) FA; (**b**) FB. Enhancement of cell viability and protection of Caco-2 cells from H_2O_2-induced oxidant damage by pre-treatment with PKLPs: (**c**) cell viability; (**d**) intracellular generation of ROS; (**e**) malondialdehyde (MDA) level of Caco2 cells. The reported values are represented as mean $\pm$ SD (n = 3). Columns marked with different lowercase letters indicate the significant differences among treatments by using Duncan analysis.

Table 2. HPLC-QTOF-MS/MS of the different eluted fractions.

Fractions	$[M - H]^-$ (m/z)	Typical MS^2 Ions (m/z)	Molecular Formula	Compound Name	Rt (min)
		Fraction A (FA)			
FA	289	137; 163; 245	$C_{15}H_{14}O_6$	Catechin	8.7
FA	289	137; 163; 179; 289	$C_{15}H_{14}O_6$	Epicatechin	11.2
FA	447	109; 301; 447	$C_{21}H_{20}O_{11}$	Quercetin	13.3
FA	463	109; 301; 463	$C_{21}H_{20}O_{12}$	Isoquercetin	14.7
FA	577	287; 289; 425; 451; 577	$C_{30}H_{26}O_{12}$	Dimer procyanidins	18.2
FA	865	289; 577; 739; 711; 865	$C_{45}H_{38}O_{18}$	Trimer procyanidins	27.1
		Fraction B (FB)			
FB	289	137; 163; 245	$C_{15}H_{14}O_6$	Catechin	7.7
FB	289	137; 163; 179; 289	$C_{15}H_{14}O_6$	Epicatechin	9.7
FB	561	271; 289; 409; 561	$C_{30}H_{26}O_{11}$	Dimer propelargonidins	14.1
FB	577	287; 289; 451; 425; 577	$C_{30}H_{26}O_{12}$	Dimer procyanidins	16.7
FB	593	287; 305; 467; 593	$C_{30}H_{26}O_{13}$	Dimer prodelphindins	21.9
FB	865	289; 577; 739; 711; 865	$C_{45}H_{38}O_{18}$	Trimer procyanidins	27.5
FB	1153	289; 577; 865; 1153	$C_{60}H_{50}O_{24}$	Tetramer procyanidins	37.3

Fraction B contained monomer procyanidins and polymer PAs, such as catechin, epicatechin, procyanidins, prodelphindins, and propelargonidins (Figure 1b). For instance, FB showed the fragment ion ($[M - H - 288]^-$) at m/z 289 and 577, that derived from cleavage of the trimer procyanidins; it also contained one molecular ion peak $[M - H]^-$ at m/z 865, with MS^2 yielding typical fragment ions at m/z 577, 289, 739, and 713. The base ion $[M - H - 126]$ at m/z 739 came from HRF. The ion of m/z 713 was formed through an RDA fission, which characterizes hydroxvyinylbenzenediol elimination ($[M - H - 152]^-$). Compared with FA, the chemical composition of FB was purer. The concentration of procyanidins as extension units in FB was much higher than that in FA, as well as the quantity of monomers units in FB. On the other hand, there was a significant difference in the DP of the two fractions. Specifically, the mean degree of polymerization (mDP) values of the eluted fractions of FA and FB were 3.2 and 5.9, respectively, which implied that FA had a smaller molecular weight than FB.

3.2. PKLPs (FA and FB) Suppressed H_2O_2-Induced Oxidative Stress in Caco-2 Cells

As hydrogen peroxide (H_2O_2) was a product of the cellular oxygen metabolism that was a feature of the various metabolic and signaling cascades [12,32], the control of the physiological H_2O_2 intracellular concentration, as an antioxidant indicator, has a significant relationship to the cells' functionality and viability [33]. Therefore, the potential antioxidant activity of PKLPs on the cellular H_2O_2-induced oxidation model was established. In our previous study, we found that using 200 microM of H_2O_2 showed the most suitable impacts on Caco-2 cells, compared to 10, 25, 50, and 100 μM, which each caused too low an injury on the cells [6]. Therefore, 200 μM H_2O_2 was selected for the injury cell model.

In this study, a significant inhibition ($p = 0.01$; $47.25 \pm 5.72\%$) was observed after using 200 μM of H_2O_2 on treating Caco-2 cells for 4 h. On the other hand, pre-treatment with PKLPs observably increased the Caco-2 cell viability after using the same concentration of ROS. Meanwhile, FA cell viability ($73.16 \pm 7.27\%$) was higher than FB, with cell viability of $69.10 \pm 7.31\%$ (Figure 1c). This phenomenon was confirmed by the fluorescence microscope, which clearly defined the increase in the number of viable cells by FA pre-treatments compared to FB (Figure 2a).

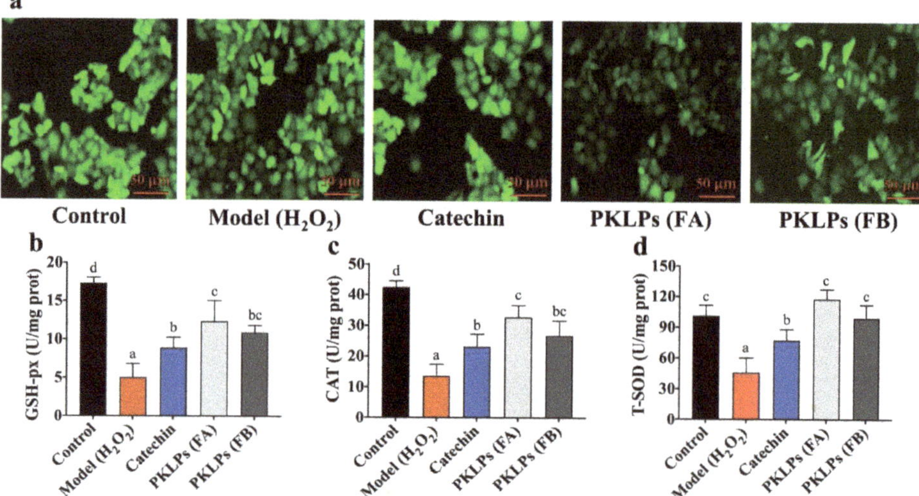

Figure 2. Effects of PKLPs on: (**a**) ROS level express mean fluorescence intensity; (**b**) GSH-px; (**c**) CAT; (**d**) T-SOD activities in H_2O_2-induced Caco-2 cells. The cells were pre-treated with catechin (75 mg/mL) and PKLPs (50 mg/mL) before being stimulated with H_2O_2 (200 μM/L) for 4 h. The reported values are represented as mean $\pm$ SD ($n = 3$). Columns marked with different lowercase letters indicate the significant differences among treatments by using Duncan analysis ($p < 0.05$).

The distinction between FA and FB came from the chemical structure of PKLPs with a higher DP of FB, which decreased its functionality [15]. Rauf, et al. [34], reported that extracted PAs revealed effective antioxidant activity as a result of their functional hydroxyl groups positions, compared to other phenolic compounds. Additionally, Koudoufio, Feldman, Ahmarani, Delvin, Spahis, Desjardins, and Levy [15] mentioned that pretreatment with PAs could significantly prevent the Caco-2/15 cell from oxidative damage to its biological macromolecules due to its exposure to the strong oxygen-radicals (Fe/Asc).

For the cytoprotective effect, the two fractions of PKLPs were evaluated for their intracellular free radicals scavenging ability on the ROS and MDA cellular levels (Figure 1d,e): the ROS levels reflected the antioxidant enzyme activity, and the MDA level reflected the cellular lipid oxidation degree that was damaged by the excessive exposure to free radicals [35,36].

This study shows that ROS levels significantly ($p = 0.03$) declined from $10.8 \pm 0.8 \times 10^3$ florescence/mg protein in the model group to $6.6 \pm 0.6 \times 10^3$ in the FA group (Figure 1d). Meanwhile, the Caco-2 treated with FB showed a significant decrease ($p = 0.04$) in the release of ROS compared to the normal cells (Figure 1d). Furthermore, the MDA of FB decreased ($p < 0.05$) to 18.58 ± 3.34 nM/mg protein compared to the model with 41.59 ± 4.23 nM/mg protein. Zhou, et al. [37], reported that PAs contain an abundance of hydroxyl groups, and can release H^+ to form cross-linkages with the radicals, which can significantly attenuate oxidative damage, by reducing the contents of MDA and ROS. Furthermore, Koudoufio, Feldman, Ahmarani, Delvin, Spahis, Desjardins, and Levy [15] reported that PAs can decrease lipid peroxidation through their ability to regulate the expression of TNFα, COX2, and NF-κB. On the other hand, Li, Liu, Li, McClements, Fu, and Liu [8] reported that reduction of the influence of PAs on Caco-2 cells viability is dependent on the specific mDP fractions that stimulate the apoptotic pathways of caspase-9, caspase-3, and caspase-8, which are generally increased by ROS generation. They also noted that the bioavailability of PAs can be increased through the hormesis effect, if it is under a lower mDP. Therefore, the study of PAs' specific fraction vital influence should be emphasized.

3.3. Protective Effect of PKLPs (FA and FB) in H_2O_2-Induced Caco-2 Cells

Glutathione peroxidase (GSH-Px) activity, catalase (CAT) activity, and total superoxide dismutase (T-SOD) activity are known to be a crucial enzyme-driven antioxidant defense system in organisms, which can scavenge free radicals to maintain the redox balance in cells [38]. As catechin has significant antioxidant activity, as reported previously, we selected it for the positive treatment group. In this study, there was a significant increase in the antioxidant enzyme biomarkers of the PKLPs compared to the oxidative stress model, in which regard, pre-treatment of H_2O_2-induced cells with PKLPs (FA or FB) significantly promoted GSH-Px activity when compared to the model group (Figure 2b). Meanwhile, FA exhibited a more noteworthy effect on enhancing antioxidant enzyme activity than FB, which significantly increased to 12.36 ± 2.70 U/mg protein in the FA group, and increased to 10.91 ± 0.93 U/mg protein in the FB group, respectively, when compared to the model group (4.39 ± 1.41 U/mg protein) ($p < 0.05$). Fujimaki, et al. [39], noted that PAs contain (+)-catechin as an upper unit that mainly increases its bioactivity in its unique structural combination with (−)-epicatechin-(4β→8)-(−)-epicatechin 3-O-gallate subunits. Moreover, Caco-2 treated with FB increased the release of CAT significantly ($p < 0.05$), to 23.21 ± 4.05 U/mg protein compared to the model group (Figure 2c).

As shown in Figure 2c,d, CAT and T-SOD activities in the PKLPs treatment group exhibited similar antioxidant activity to GSH-Px activity. From the results, we found that PKLPs pre-treatment could effectively improve the antioxidant status of cells exposed to free radicals damage by increasing the GSH-Px, CAT, and T-SOD activities, which indicated that PKLPs could activate an antioxidant signaling pathway [34]. Our findings are consistent with the previous study of Su, Li, Hu, Xie, Ke, Zheng, and Chen [30]: PAs can attenuate oxidative stress by scavenging excessive ROS, and protect the organism by increasing the activity of antioxidant enzymes.

3.4. Effect of PKLPs (FA and FB) on the Nrf2 and Its Downstream Target Genes Transcription

The employment of micro(m)RNA expression data as molecular markers could explain the antioxidant and anticancer activities potential pathway through the transcription factors [40]. For instance, the nuclear factor erythroid 2-related factor (Nrf2) is a transcription factor that plays a significant role in response to xenobiotics and oxidative stress, by binding to the antioxidant response element (ARE) [41].

In this study, the mRNA levels of the Nrf2 and Keap1, and the downstream target genes, were further investigated in Caco-2 cells for their potential response to various treatments. Quantitative real-time PCR (qRT-PCR) analysis of relative mRNA expression is shown in Figure 3. There were significant increases in the antioxidant mRNA biomarkers of the PKLPs pre-treated groups compared to the model group, in which regard, the Nrf2 levels were significantly increased to 1.86 ± 0.22 mRNA expression in the FA group compared to the catechins with 1.18 ± 0.11 mRNA expression (Figure 3a). Meanwhile, the Caco-2 treated with FA increased the release of HO-1 significantly ($p = 0.03$), to 1.29 ± 0.12 mRNA expression, compared to FB, with 0.97 ± 0.11 mRNA expression (Figure 3b), in which regard, the changes in FA and FB that impacted on the mRNA antioxidant expression could be ascribed to the higher molecular weight of FB, which may have blocked its accessibility into the Caco-2 cell, and therefore decreased its impacts [10] (Figure 3c). Ge, et al. [42], reported that PAs' impact on Caco-2 cell monolayers and cholesterol could explain their different dimers' impacts on the signaling pathways through the absorption affinity. The same trend was observed for the mRNA level of NQO1, SOD-1, CAT, and Bcl-2 (Figure 3c–e,g), while the mRNA expression of Keap1 and Bax showed different responses (Figure 3f,h). For instance, the Caco-2 pre-treated with FA significantly decreased the mRNA expression of Bax ($p = 0.02$), with 1.43 ± 0.13, compared to the model with 2.37 ± 0.18 (Figure 3h). Contrastingly, the PKLPs pre-treatment groups remarkably reversed the expression level of these genes compared to the model group, which indicates that PKLPs pre-treatment could significantly inhibit oxidative stress, by alleviating the alteration in Bcl-2 and Bax gene transcription (Figure 3g,h). Siddiqui, et al. [43], mentioned that Bax, as a pro-apoptotic protein, could promote the release of apoptotic molecules into the cytoplasm by competing with CAT, in which its expression was affected by the antioxidant molecules against ROS.

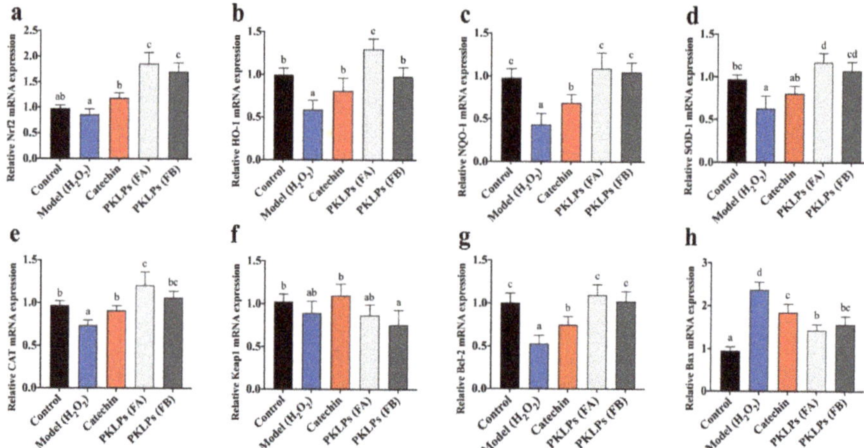

Figure 3. Effects of PKLPs on Nrf2-mediated antioxidants signaling transcripts were analyzed by qRT-PCR. The gene expression of the Nrf2 and its relative or downstream molecules: (**a**) Nrf2 mRNA; (**b**) HO-1 mRNA; (**c**) NQO1 mRNA; (**d**) SOD-1 mRNA; (**e**) CAT mRNA; (**f**) Keap1 mRNA; (**g**) Bcl-2 mRNA; (**h**) Bax mRNA. Data are shown as mean $\pm$ standard deviation ($n = 3$). Columns marked with different lowercase letters indicate the significant differences among treatments by using Duncan analysis ($p < 0.05$).

Our results are consistent with previous results, which confirmed that polyphenols extracted from plants or herb could increase antioxidative enzymes based on the upregulation of their antioxidative mRNA genes expression, such as NQO-1, HO-1, Nrf2, and SOD [44]. Hilary, et al. [45], reported that PAs' mode of action came from their hydroxyl groups, which could direct supporting antioxidant reactions or even mediate the occurrence of oxidant events. This phenomenon could be explained as, under normal cellular condition, the Nrf2 binds to Keap1 to form suitable complexes in cytoplasm; on the other hand, ROS cause the separation of the Nrf2 from Keap1 and, as a result, the Nrf2 transfers into the nucleus; meanwhile, Keap1 degrades in the cytoplasm [46] (Figure 4a). The reason for the dissociation of the Nrf2–Keap1 complex in the cytoplasm is the sulfhydryl modification and Nrf2 phosphorylation, which results in an uncoupled Nrf2 and Keap1 [47].

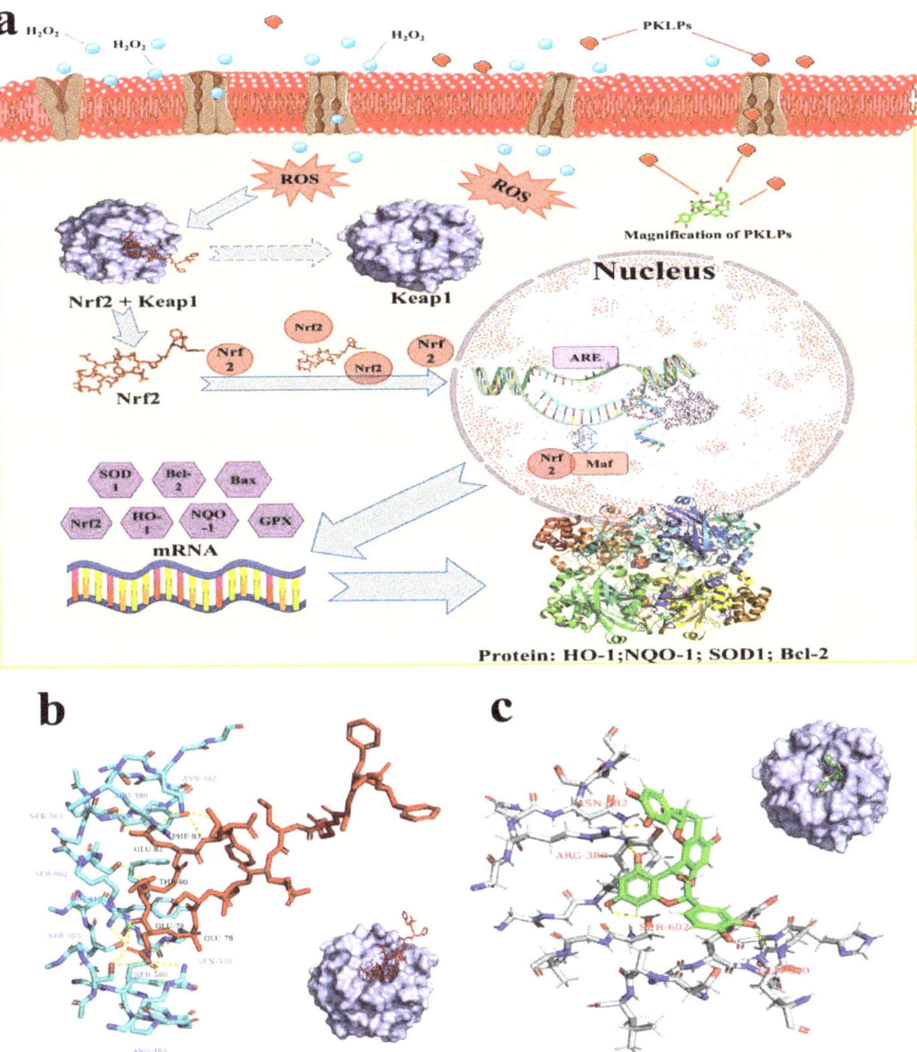

Figure 4. Comparison between the computational geometries of the binding of the Nrf2 and PKLPs with a pocket of Keap1 protein: (**a**) Proposed schematic diagram of the Nrf2 signaling pathway mechanism of the PKLPs in the H_2O_2-induced Caco-2 cells; (**b**) Magnification of the interaction site of the Nrf2–Keap1 complex motif, and its binding geometry; (**c**) Magnification of the interaction between the Keap1 pocket and the PKLPs. Hydrogen bonds are represented by yellow dashed lines, and the related amino acid residues and peptides are highlighted in the picture.

3.5. Effect of PKLPs (FA and FB) on the Nrf2 and Its Downstream Protein Expression

To further confirm the results of the Nrf2 and HO-1 expressions, the signaling pathway was further investigated through the molecular mechanism to underline the protective effect of PKLPs on H_2O_2-induced oxidative stress. The related antioxidant protein levels of the Nrf2, Keap1, SOD-1, HO-1, and NQO-1 were studied by Western blotting assay, and their changes were evaluated by confocal microscopy technology (Figures 5 and 6).

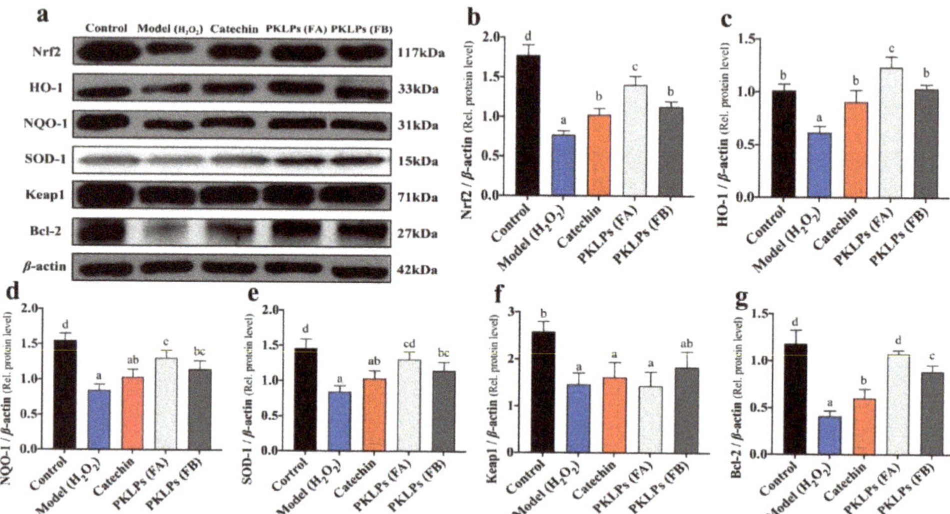

Figure 5. Effects of PKLPs on the protein expression of the Nrf2 signaling pathway: (**a**) Western blotting estimation of the Nrf2 signaling pathway protein expression in Caco-2 cells after pre-treatment with catechin (75 µg/mL) and PKLPs (50 µg/mL) for 24 h before stimulation with H_2O_2 (200 µM/L) for 2 h; (**b–g**) Quantification of the above-mentioned Western blots, using β-actin as a loading control. The results are presented as mean ± SD (*n* = 3). Columns marked with different lowercase letters indicate the significant differences among treatments by using Duncan analysis (*p* < 0.05).

As shown in Figure 5, compared to the normal control group, the antioxidant proteins were increased (*p* < 0.05) by pre-treatment of the PKLPs, when compared to the model that used the oxidative damage hydrogen peroxide as a negative control, in which regard, the Nrf2 levels were significantly increased to a 1.40 ± 0.10 Nrf2/β-actin relative level in the FA group, compared to the model with a 0.773 ± 0.05 Nrf2/β-actin relative level (Figure 5b). Additionally, the FA increased the release of HO-1 significantly (*p* < 0.05) to a 1.24 ± 0.10 HO-1/β-actin relative level compared to the catechins with a 0.91 ± 0.11 HO-1/β-actin relative level (Figure 5c). As shown in Figure 6, the confocal images showed that the distribution of the Nrf2 was increased inside the nucleus for the FA group compared to the model and catechin treated groups. It demonstrated that Nrf2 expression plays a critical role in oxidative stress response. This action is through the dissociative Nrf2 transfer from cytoplasm to nucleus, and binds to the antioxidant response element in the nucleus to exert function by activation of the gene transcription of the antioxidant enzyme [48].

Meanwhile, the obtained results revealed that the protein levels of the Nrf2 and its downstream protein were significantly increased in the FA pre-treated group, compared to those in other pre-treatment groups, such as the FB pre-treated group and the catechin pre-treated group. Based on our study, FA contained procyanidins, quercetin, and isoquercetin, while FB only contained polymer PAs. The relative antioxidative mRNA transcription and protein expression results implied that oligomer PAs and quercetin could work synergistically to regulate multiple and interactive molecular targets related to H_2O_2-induced oxidative damage and the involvement of Nrf2-associated pathways, which was consistent with previous reporting [30]. As for FB, the fraction had more chemical adducts that were causing bigger mDP, which might inhibit the bioavailability of PAs, and further inhibit their vital impact on enzyme activity and other bioactivities. Therefore, in this study, FA had overall activities higher than FB.

The mechanism of defense oxidative damage is clarified in Figure 4a. The insertion of FA inside the cells inhibited the ROS negative impacts, which significantly increased the excretion of Nrf2 and HO-1 antioxidant enzymes expressions. Similarly, the corresponding expression of the antioxidant enzymes (SOD-1, HO-1, NQO-1, and Bcl-2) were upregulated ($p < 0.05$) in Caco-2 cells against the H_2O_2 exposed group. The results of the molecular mechanisms, by qRT-PCR and the Western blot, indicated that both FA and FB of the PKLPs dramatically rescued previous effects caused by oxidative damage. H_2O_2 exposure inhibited the expression of the Nrf2 and its related or downstream genes such as SOD-1, HO-1, NQO-1, and Bcl-2 in organisms. Zhou, Chang, Gao, and Wang [3] reported that extracted plants' PAs protect the organisms against apoptosis caused by H_2O_2, through their significant impacts on the Nrf2-ARE pathway. Furthermore, the current results agree with the previous investigations, that H_2O_2 treatment suppressed the translocation of the Nrf2 signaling pathway [49].

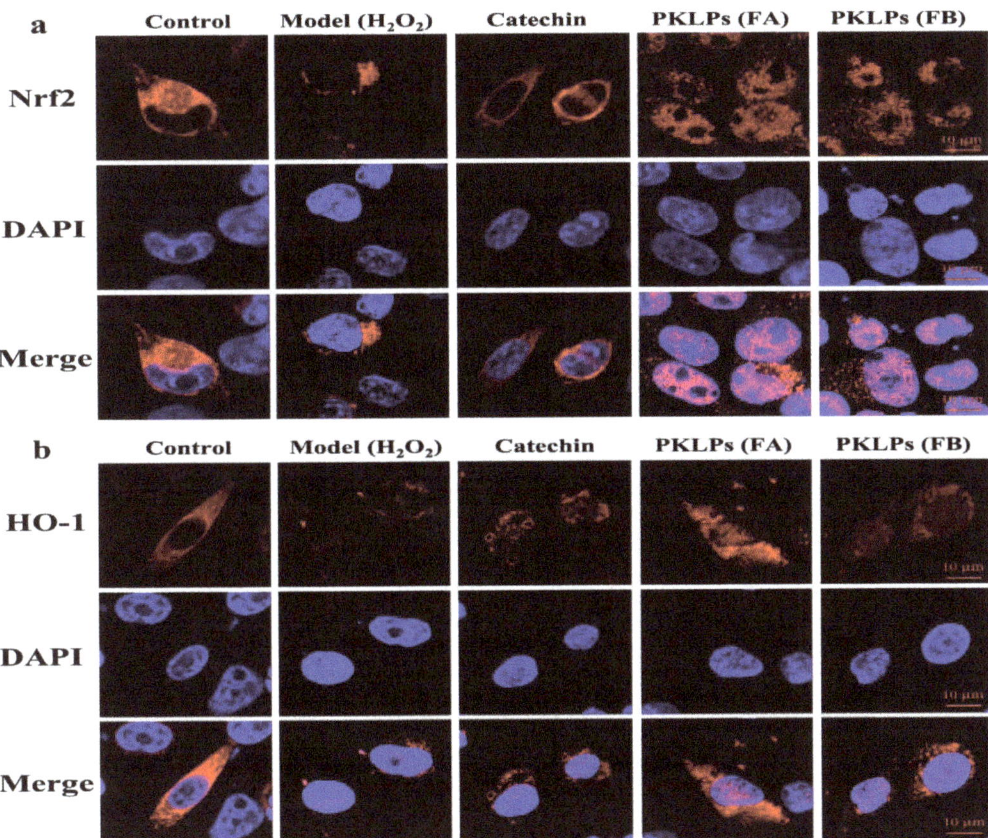

Figure 6. Confocal microscopy of single Caco-2 cells protein expression of (**a**) the Nrf2 and (**b**) HO-1 signaling pathways after pre-treatment with catechin (75 µg/mL) and PKLPs (50 µg/mL).

Molecular modelling of the interaction between Keap1, Nrf2, and PAs is shown in Figure 4b,c by using a molecular docking analysis with the binding mode of the Nrf2 to Keap1. The phenylalanine 83 (PHE-83) of the peptide was stabilized by a hydrogen bond with the asparaginate 382 (ASN-382) of the Keap1. Arginine 380 (ARG-380) and tyrosine 334 (TYR-334) from the Keap1 protein were simultaneously held together with glutamic acid 82 (GLU-82) on the Nrf2 protein by hydrogen bond. Moreover, serine 602 (SER-602) from the surface of the Keap1, was bonded to threonine 80 (THR-80) from the Nrf2 protein.

From the molecular docking simulation, it can be seen that PAs were simultaneously held together with the active cites amino acids, like aspartic acid 382 (ASN-382), arginine 380 (ARG-380), serine 602 (SER-602), and glutamine 530 (GLN-530), on Keap1 by hydrogen bonding [50]. It can be seen that PAs compete with the Nrf2 to promote the separation of the Nrf2 from the Nrf2–Keap1 complexes. The Nrf2 is then transported into the nucleus, where it binds to the antioxidant response element, and produces different antioxidant enzymes by activating relative antioxidant genes. Ma, et al. [51], noted that PAs bind with large molecular-weight proteins like β-casein through the interaction of their hydrophobic residues with the amino acid hydrophobic residues, including Ala-192, Pro-196, and Pro-201. Therefore, the increase in the affinity of carbonylic groups of the peptide backbone forming hydrogen bonds with other amino acids from Keap1 could be the best explanation of how the PAs stimulate the separation of the Nrf2 from Keap1 [47]. Under normal physiological conditions, Keap1 acts as a binder in the cytoplasm, which binds to the Nrf2 through Cul3 ubiquitin ligase, to regulate the ubiquitination of the Nrf2. As a result, the free Nrf2 is maintained at a low level in the cytoplasm.

4. Conclusions

Two PAs fractions (FA and FB) were considered as byproducts that were extracted from the leaves waste of kiwi fruit production. These compounds showed significant functional and vital antioxidant activity for green and sustainable production. The antioxidative mechanism of both fractions (FA and FB) was studied by using ROS measurement, oxidative stress indices, quantitative real-time PCR, Western blot analysis, confocal microscopy, and molecular docking in Caco-2 cell lines. The present study indicates that PLKPs have a potential key impact on promoting cells signaling pathways to construct the linkage between the Nrf2 and ARE in Caco-2 cells. In comparison with the models, the FA group had a more positive and significant effect compared to FB and catechin. Therefore, this study proposes the use of PAs for promoting the antioxidant capacity of life cells through enhancing antioxidant-related enzyme genes expressions; in which regard, their potential application as a medicinal bioactive ingredient in food cleaner production may enhance their lifestyle-related protective impact against oxidant-related diseases like cancers.

Author Contributions: Conceptualization, J.-M.L., M.G. and J.-C.C.; methodology, J.-M.L., M.G., X.-Q.Y. and J.-C.C.; software, J.-M.L. and X.-Q.Y.; validation, J.-M.L., M.G. and J.-C.C.; formal analysis, J.-M.L., M.G., X.-Q.Y., Z.-P.S. and J.-C.C.; investigation, J.-M.L., M.G., X.-Q.Y., Z.-P.S. and J.-C.C.; resources, X.-Q.Y., Z.-P.S. and J.-C.C.; data curation, J.-M.L. and M.G.; writing—original draft preparation, J.-M.L., M.G. and J.-C.C.; writing—review and editing, J.-M.L., M.G., X.-Q.Y., Z.-P.S. and J.-C.C.; visualization, X.-Q.Y., Z.-P.S. and J.-C.C.; supervision, J.-C.C.; project administration, X.-Q.Y. and Z.-P.S.; funding acquisition, X.-Q.Y. and J.-C.C. All authors have read and agreed to the published version of the manuscript.

Funding: This research and the APC were funded by the National Key Research and Development Program of China (2017YFD0400704).

Institutional Review Board Statement: Not applicable.

Informed Consent Statement: Not applicable.

Data Availability Statement: The data presented in this study are available in the article.

Acknowledgments: The authors acknowledge the support of Zhejiang University and the National Key Research and Development Program of China throughout the study.

Conflicts of Interest: The authors declare no conflict of interest.

Abbreviations

Proanthocyanidins: PAs, purified kiwifruit leaves PAs: PKLPs, fetal bovine serum: FBS, Dulbecco's Modified Eagle Medium: DMEM, penicillin/streptomycin: P/S, phosphate-buffered solution: PBS, cell counting kit-8: CCK-8, fraction A: FA, fraction B: FB, weight basis: WD, hydrogen peroxide: H_2O_2, reactive oxygen species: ROS, malondialdehyde: MDA, glutathione peroxidase: GSH-px, nuclear factor erythroid 2-related factor: Nrf2, Quantitative real-time PCR: qRT-PCR, dichlorofluorescein diacetate: DCF-DA, Kelch-like ECH-associated protein 1: Keap1, heme oxygenase-1: HO-1, antioxidant response element: ARE, quinone oxidoreductase 1: NQO-1, superoxide dismutase 1: SOD-1, catalase: CAT, B-cell lymphoma 2: Bcl-2, and Bcl-2-associated X protein: Bax, HRF: heterocyclic ring fission, RDA: Retro-diels-alder, QM: quinone methide.

References

1. Hou, K.; Bao, M.; Wang, L.; Zhang, H.; Yang, L.; Zhao, H.; Wang, Z. Aqueous enzymatic pretreatment ionic liquid–lithium salt based microwave–assisted extraction of essential oil and procyanidins from pinecones of pinus koraiensis. *J. Clean. Prod.* **2019**, *236*, 117581.
2. Li, Q.; Zhang, N.; Ni, L.; Wei, Z.; Quan, H.; Zhou, Y. One-pot high efficiency low temperature ultrasonic-assisted strategy for fully bio-based coloristic, anti-pilling, antistatic, bioactive and reinforced cashmere using grape seed proanthocyanidins. *J. Clean. Prod.* **2021**, *315*, 128148. [CrossRef]
3. Zhou, L.; Chang, J.; Gao, Y.; Wang, C. Procyanidin b2 protects neurons from cypermethrin-induced oxidative stress through the p13k/akt/nrf2 signaling pathway. *Nan Fang Yi Ke Da Xue Xue Bao J. South. Med. Univ.* **2021**, *41*, 1158–1164.
4. Gouda, M.; El-Din Bekhit, A.; Tang, Y.; Huang, Y.; Huang, L.; He, Y.; Li, X. Recent innovations of ultrasound green technology in herbal phytochemistry: A review. *Ultrason. Sonochem.* **2021**, *73*, 105538. [PubMed]
5. Gouda, M.; Huang, Z.; Liu, Y.; He, Y.; Li, X. Physicochemical impact of bioactive terpenes on the microalgae biomass structural characteristics. *Bioresour. Technol.* **2021**, *334*, 125232. [CrossRef] [PubMed]
6. Lv, J.M.; Gouda, M.; Zhu, Y.Y.; Ye, X.Q.; Chen, J.C. Ultrasound-assisted extraction optimization of proanthocyanidins from kiwi (*Actinidia chinensis*) leaves and evaluation of its antioxidant activity. *Antioxidants* **2021**, *10*, 1317. [CrossRef] [PubMed]
7. Lv, J.-M.; Gouda, M.; El-Din Bekhit, A.; He, Y.-K.; Ye, X.-Q.; Chen, J.-C. Identification of novel bioactive proanthocyanidins with potent antioxidant and anti-proliferative activities from kiwifruit leaves. *Food Biosci.* **2022**, *46*, 101554. [CrossRef]
8. Li, Q.; Liu, C.; Li, T.; McClements, D.J.; Fu, Y.; Liu, J. Comparison of phytochemical profiles and antiproliferative activities of different proanthocyanidins fractions from choerospondias axillaris fruit peels. *Food Res. Int.* **2018**, *113*, 298–308. [CrossRef]
9. Marangi, F.; Pinto, D.; de Francisco, L.; Alves, R.C.; Puga, H.; Sut, S.; Dall'Acqua, S.; Rodrigues, F.; Oliveira, M. Hardy kiwi leaves extracted by multi-frequency multimode modulated technology: A sustainable and promising by-product for industry. *Food Res. Int.* **2018**, *112*, 184–191. [CrossRef]
10. Ou, K.; Gu, L. Absorption and metabolism of proanthocyanidins. *J. Funct. Foods* **2014**, *7*, 43–53. [CrossRef]
11. Li, Q.; Chen, J.; Li, T.; Liu, C.; Liu, W.; Liu, J. Comparison of bioactivities and phenolic composition of choerospondias axillaris peels and fleshes. *J. Sci. Food Agric.* **2016**, *96*, 2462–2471. [PubMed]
12. Hernansanz-Agustín, P.; Enríquez, J.A. Generation of reactive oxygen species by mitochondria. *Antioxidants* **2021**, *10*, 415. [CrossRef] [PubMed]
13. Lian, P.; Braber, S.; Garssen, J.; Wichers, H.J.; Folkerts, G.; Fink-Gremmels, J.; Varasteh, S. Beyond heat stress: Intestinal integrity disruption and mechanism-based intervention strategies. *Nutrients* **2020**, *12*, 734. [CrossRef] [PubMed]
14. Herb, M.; Schramm, M. Functions of ros in macrophages and antimicrobial immunity. *Antioxidants* **2021**, *10*, 313. [CrossRef]
15. Koudoufio, M.; Feldman, F.; Ahmarani, L.; Delvin, E.; Spahis, S.; Desjardins, Y.; Levy, E. Intestinal protection by proanthocyanidins involves anti-oxidative and anti-inflammatory actions in association with an improvement of insulin sensitivity, lipid and glucose homeostasis. *Sci. Rep.* **2021**, *11*, 3878. [CrossRef]
16. Kellett, M.E.; Greenspan, P.; Pegg, R.B. Modification of the cellular antioxidant activity (caa) assay to study phenolic antioxidants in a caco-2 cell line. *Food Chem.* **2018**, *244*, 359–363. [CrossRef]
17. Zhao, H.; Eguchi, S.; Alam, A.; Ma, D. The role of nuclear factor-erythroid 2 related factor 2 (nrf-2) in the protection against lung injury. *Am. J. Physiol. Lung Cell. Mol. Physiol.* **2017**, *312*, L155–L162. [CrossRef]
18. Xu, L.; Zhao, Y.; Pan, F.; Zhu, M.; Yao, L.; Liu, Y.; Feng, J.; Xiong, J.; Chen, X.; Ren, F.; et al. Inhibition of the nrf2-trxr axis sensitizes the drug-resistant chronic myelogenous leukemia cell line k562/g01 to imatinib treatments. *BioMed Res. Int.* **2019**, *2019*, 6502793. [CrossRef]
19. Arroyave-Ospina, J.C.; Wu, Z.; Geng, Y.; Moshage, H. Role of oxidative stress in the pathogenesis of non-alcoholic fatty liver disease: Implications for prevention and therapy. *Antioxidants* **2021**, *10*, 174. [CrossRef]
20. Gulati, S.; Yadav, A.; Kumar, N.; Priya, K.; Aggarwal, N.K.; Gupta, R. Phenotypic and genotypic characterization of antioxidant enzyme system in human population exposed to radiation from mobile towers. *Mol. Cell. Biochem.* **2017**, *440*, 1–9. [CrossRef]
21. Pathak, R.K.; Kim, D.Y.; Lim, B.; Kim, J.M. Investigating multi-target antiviral compounds by screening of phytochemicals from neem (azadirachta indica) against prrsv: A vetinformatics approach. *Front. Vet. Sci.* **2022**, *9*, 854528. [CrossRef] [PubMed]

22. Gabr, S.K.; Bakr, R.O.; Mostafa, E.S.; El-Fishawy, A.M.; El-Alfy, T.S. Antioxidant activity and molecular docking study of erythrina × neillii polyphenolics. *S. Afr. J. Bot.* **2019**, *121*, 470–477. [CrossRef]

23. Chai, W.M.; Shi, Y.; Feng, H.L.; Xu, L.; Xiang, Z.H.; Gao, Y.S.; Chen, Q.X. Structure characterization and anti-tyrosinase mechanism of polymeric proanthocyanidins fractionated from kiwifruit pericarp. *J. Agric. Food Chem.* **2014**, *62*, 6382–6389. [CrossRef]

24. Tsolmon, B.; Fang, Y.; Yang, T.; Guo, L.; He, K.; Li, G.Y.; Zhao, H. Structural identification and uplc-esi-qtof-ms(2) analysis of flavonoids in the aquatic plant landoltia punctata and their in vitro and in vivo antioxidant activities. *Food Chem.* **2021**, *343*, 128392. [CrossRef]

25. Tang, W.; Shen, M.; Xie, J.; Liu, D.; Du, M.; Lin, L.; Gao, H.; Hamaker, B.R.; Xie, M. Physicochemical characterization, antioxidant activity of polysaccharides from mesona chinensis benth and their protective effect on injured nctc-1469 cells induced by h2o2. *Carbohydr. Polym.* **2017**, *175*, 538–546. [PubMed]

26. Rajput, S.A.; Shaukat, A.; Wu, K.; Rajput, I.R.; Baloch, D.M.; Akhtar, R.W.; Raza, M.A.; Najda, A.; Rafal, P.; Albrakati, A.; et al. Luteolin alleviates aflatoxinb1-induced apoptosis and oxidative stress in the liver of mice through activation of nrf2 signaling pathway. *Antioxidants* **2021**, *10*, 1268. [PubMed]

27. Gao, X.; Xiao, Z.; Li, C.; Zhang, J.; Zhu, L.; Sun, L.; Zhang, N.; Khalil, M.M.; Rajput, S.A.; Qi, D. Prenatal exposure to zearalenone disrupts reproductive potential and development via hormone-related genes in male rats. *Food Chem. Toxicol. Int. J. Publ. Br. Ind. Biol. Res. Assoc.* **2018**, *116*, 11–19. [CrossRef] [PubMed]

28. Tonolo, F.; Folda, A.; Cesaro, L.; Scalcon, V.; Marin, O.; Ferro, S.; Bindoli, A.; Rigobello, M.P. Milk-derived bioactive peptides exhibit antioxidant activity through the keap1-nrf2 signaling pathway. *J. Funct. Foods* **2020**, *64*, 103696. [CrossRef]

29. Jiang, C.S.; Zhuang, C.L.; Zhu, K.; Zhang, J.; Muehlmann, L.A.; Figueiro Longo, J.P.; Azevedo, R.B.; Zhang, W.; Meng, N.; Zhang, H. Identification of a novel small-molecule keap1-nrf2 ppi inhibitor with cytoprotective effects on lps-induced cardiomyopathy. *J. Enzym. Inhib. Med. Chem.* **2018**, *33*, 833–841. [CrossRef]

30. Su, H.; Li, Y.; Hu, D.; Xie, L.; Ke, H.; Zheng, X.; Chen, W. Procyanidin b2 ameliorates free fatty acids-induced hepatic steatosis through regulating tfeb-mediated lysosomal pathway and redox state. *Free Radic. Biol. Med.* **2018**, *126*, 269–286. [CrossRef]

31. Miranda-Hernandez, A.M.; Muniz-Marquez, D.B.; Wong-Paz, J.E.; Aguilar-Zarate, P.; de la Rosa-Hernandez, M.; Larios-Cruz, R.; Aguilar, C.N. Characterization by hplc-esi-ms(2) of native and oxidized procyanidins from litchi (litchi chinensis) pericarp. *Food Chem.* **2019**, *291*, 126–131. [CrossRef] [PubMed]

32. Zenin, V.; Ivanova, J.; Pugovkina, N.; Shatrova, A.; Aksenov, N.; Tyuryaeva, I.; Kirpichnikova, K.; Kuneev, I.; Zhuravlev, A.; Osyaeva, E.; et al. Resistance to h2o2-induced oxidative stress in human cells of different phenotypes. *Redox Biol.* **2022**, *50*, 102245. [CrossRef] [PubMed]

33. Sies, H.; Berndt, C.; Jones, D.P. Oxidative stress. *Annu. Rev. Biochem.* **2017**, *86*, 715–748. [CrossRef] [PubMed]

34. Rauf, A.; Imran, M.; Abu-Izneid, T.; Iahtisham Ul, H.; Patel, S.; Pan, X.; Naz, S.; Sanches Silva, A.; Saeed, F.; Rasul Suleria, H.A. Proanthocyanidins: A comprehensive review. *Biomed. Pharmacother. Biomed. Pharmacother.* **2019**, *116*, 108999. [CrossRef]

35. Wu, P.; Ma, G.; Li, N.; Deng, Q.; Yin, Y.; Huang, R. Investigation of in vitro and in vivo antioxidant activities of flavonoids rich extract from the berries of rhodomyrtus tomentosa(ait.) hassk. *Food Chem.* **2015**, *173*, 194–202. [CrossRef]

36. Zhao, Y.; Wang, T.; Li, P.; Chen, J.; Nepovimova, E.; Long, M.; Wu, W.; Kuca, K. Bacillus amyloliquefaciens b10 can alleviate aflatoxin b1-induced kidney oxidative stress and apoptosis in mice. *Ecotoxicol. Environ. Saf.* **2021**, *218*, 112286. [CrossRef]

37. Zhou, Q.; Han, X.; Li, R.; Zhao, W.; Bai, B.; Yan, C.; Dong, X. Anti-atherosclerosis of oligomeric proanthocyanidins from rhodiola rosea on rat model via hypolipemic, antioxidant, anti-inflammatory activities together with regulation of endothelial function. *Phytomedicine Int. J. Phytother. Phytopharm.* **2018**, *51*, 171–180. [CrossRef]

38. Wu, Y.; Wang, Y.; Nabi, X. Protective effect of ziziphora clinopodioides flavonoids against h2o2-induced oxidative stress in huvec cells. *Biomed. Pharmacother. Biomed. Pharmacother.* **2019**, *117*, 109156. [CrossRef]

39. Fujimaki, T.; Mori, S.; Horikawa, M.; Fukui, Y. Isolation of proanthocyanidins from red wine, and their inhibitory effects on melanin synthesis in vitro. *Food Chem.* **2018**, *248*, 61–69. [CrossRef]

40. Ahmed, F.E.; Gouda, M.M.; Hussein, L.A.; Ahmed, N.C.; Vos, P.W.; Mohammad, M.A. Role of melt curve analysis in interpretation of nutrigenomics' microrna expression data. *Cancer Genom. Proteom.* **2017**, *14*, 469–481.

41. Shaw, P.; Chattopadhyay, A. Nrf2-are signaling in cellular protection: Mechanism of action and the regulatory mechanisms. *J. Cell. Physiol.* **2020**, *235*, 3119–3130. [CrossRef] [PubMed]

42. Ge, Z.; Nie, R.; Maimaiti, T.; Yao, F.; Li, C. Comparison of the inhibition on cellular 22-nbd-cholesterol accumulation and transportation of monomeric catechins and their corresponding a-type dimers in caco-2 cell monolayers. *J. Funct. Foods* **2016**, *27*, 343–351. [CrossRef]

43. Siddiqui, W.A.; Ahad, A.; Ahsan, H. The mystery of bcl2 family: Bcl-2 proteins and apoptosis: An update. *Arch. Toxicol.* **2015**, *89*, 289–317. [CrossRef] [PubMed]

44. Zou, B.; Xiao, G.; Xu, Y.; Wu, J.; Yu, Y.; Fu, M. Persimmon vinegar polyphenols protect against hydrogen peroxide-induced cellular oxidative stress via nrf2 signalling pathway. *Food Chem.* **2018**, *255*, 23–30. [CrossRef]

45. Hilary, S.; Tomas-Barberan, F.A.; Martinez-Blazquez, J.A.; Kizhakkayil, J.; Souka, U.; Al-Hammadi, S.; Habib, H.; Ibrahim, W.; Platat, C. Polyphenol characterisation of phoenix dactylifera l. (date) seeds using hplc-mass spectrometry and its bioaccessibility using simulated in-vitro digestion/caco-2 culture model. *Food Chem.* **2020**, *311*, 125969. [CrossRef]

46. Song, M.Y.; Lee, D.Y.; Chun, K.S.; Kim, E.H. The role of nrf2/keap1 signaling pathway in cancer metabolism. *Int. J. Mol. Sci.* **2021**, *22*, 4376. [CrossRef]

47. Wang, M.; Chen, J.; Ye, X.; Liu, D. In vitro inhibitory effects of chinese bayberry (myrica rubra sieb. Et zucc.) leaves proanthocyanidins on pancreatic alpha-amylase and their interaction. *Bioorganic Chem.* **2020**, *101*, 104029. [CrossRef]
48. Zhu, J.; Wang, H.; Chen, F.; Fu, J.; Xu, Y.; Hou, Y.; Kou, H.H.; Zhai, C.; Nelson, M.B.; Zhang, Q.; et al. An overview of chemical inhibitors of the nrf2-are signaling pathway and their potential applications in cancer therapy. *Free Radic. Biol. Med.* **2016**, *99*, 544–556. [CrossRef]
49. Zhao, Y.; Li, M.Z.; Shen, Y.; Lin, J.; Wang, H.R.; Talukder, M.; Li, J.L. Lycopene prevents dehp-induced leydig cell damage with the nrf2 antioxidant signaling pathway in mice. *J. Agric. Food Chem.* **2020**, *68*, 2031–2040. [CrossRef]
50. Uddin, M.J.; Faraone, I.; Haque, M.A.; Rahman, M.M.; Halim, M.A.; Sonnichsen, F.D.; Cicek, S.S.; Milella, L.; Zidorn, C. Insights into the leaves of ceriscoides campanulata: Natural proanthocyanidins alleviate diabetes, inflammation, and esophageal squamous cell cancer via in vitro and in silico models. *Fitoterapia* **2022**, *158*, 105164. [CrossRef]
51. Ma, G.; Tang, C.; Sun, X.; Zhang, J. The interaction mechanism of β-casein with oligomeric proanthocyanidins and its effect on proanthocyanidin bioaccessibility. *Food Hydrocoll.* **2021**, *113*, 106485. [CrossRef]

Article

Investigation of Antioxidant Synergisms and Antagonisms among Phenolic Acids in the Model Matrices Using FRAP and ORAC Methods

Danijela Skroza [1,*], Vida Šimat [2], Lucija Vrdoljak [1], Nina Jolić [1], Anica Skelin [1], Martina Čagalj [2], Roberta Frleta [3] and Ivana Generalić Mekinić [1]

1　Department of Food Technology and Biotechnology, Faculty of Chemistry and Technology, University of Split, R. Boškovića 35, HR-21000 Split, Croatia
2　University Department of Marine Studies, University of Split, R. Boškovića 37, HR-21000 Split, Croatia
3　Center of Excellence for Science and Technology-Integration of Mediterranean Region (STIM), Faculty of Science, University of Split, HR-21000 Split, Croatia
*　Correspondence: danci@ktf-split.hr; Tel.: +385-21-329-458

Citation: Skroza, D.; Šimat, V.; Vrdoljak, L.; Jolić, N.; Skelin, A.; Čagalj, M.; Frleta, R.; Generalić Mekinić, I. Investigation of Antioxidant Synergisms and Antagonisms among Phenolic Acids in the Model Matrices Using FRAP and ORAC Methods. *Antioxidants* **2022**, *11*, 1784. https://doi.org/10.3390/antiox11091784

Academic Editors: Antonella D'Anneo and Marianna Lauricella

Received: 8 August 2022
Accepted: 7 September 2022
Published: 9 September 2022

Publisher's Note: MDPI stays neutral with regard to jurisdictional claims in published maps and institutional affiliations.

Abstract: The total antioxidant potential of a sample cannot be predicted from the antioxidant activity of its compounds; thus, scientists usually explain the overall activity through their combined effects (synergistic, antagonistic, or additive). Phenolic compounds are one of the most powerful and widely investigated antioxidants, but there is a lack of information about their molecular interactions. This study aimed to investigate the individual and combined antioxidant activity of equimolar mixtures (binary, ternary, quaternary, and quinary) of 10 phenolic acids (protocatechuic, gentisic, gallic, vanillic, syringic, *p*-coumaric, caffeic, ferulic, sinapic, and rosmarinic acid) at different concentrations using ferric reducing antioxidant power (FRAP) and oxygen radical absorbance capacity (ORAC) assays. Gallic acid showed the highest antioxidant activity, determined using the FRAP assay (494–5033 µM Fe^{2+}) and rosmarinic acid with the ORAC assay (50–92 µM Trolox Equivalents (TE)), while the lowest antioxidant potential was observed for *p*-coumaric acid (FRAP 24–113 µM Fe^{2+} and ORAC 20–33 µM TE). The synergistic effect (by FRAP) in the equimolar mixtures of hydroxybenzoic acids was confirmed for a large number of tested mixtures, especially at low concentrations. All mixtures containing gentisic acid showed a synergistic effect (28–89% difference). Using the ORAC method, only two mixtures of hydroxybenzoic acids showed an antagonistic effect, namely a mixture of gentisic + syringic acids (−24% difference) and gallic + vanillic acids (−30% difference), while all other mixtures showed a synergistic effect in a range of 26–236% difference. Among mixtures of hydroxycinnamic acids, the highest synergistic effect was observed for the mixtures of *p*-coumaric + ferulic acids and caffeic + sinapic acids with differences of 311% and 211%, respectively. The overall antioxidant activity of phenolic acids could be explained by the number or position of hydroxyl and/or methoxy functional groups as well as the compound concentration, but the influence of other parameters such as dissociation, intramolecular hydrogen bonds, and electron donating or withdrawing effect should not be neglected.

Keywords: phenolic acids; phenolic mixtures; interaction effect; antioxidant activity; FRAP; ORAC

1. Introduction

The research effort concerning the antioxidant behavior of phenolic compounds has significantly increased in recent decades, but the knowledge about their interaction in model mixtures is still scarce. Among the diverse and complex groups of phenolics that include simple phenols, flavonoids, stilbenes, tannins, and others, phenolic acids are the most distributed in nature. They have been found in various plants, fruits, vegetables, beverages, and agro-food by-products where they contribute to organoleptic attributes such as color, flavor, and odor but their true merits are numerous positive biological activities such as antibacterial, anti-inflammatory, antiallergenic, anticancer, cytotoxic, antitumor, cardioprotective, and antioxidant, which is among the most investigated [1–10].

Phenolic acids are represented by two main classes: hydroxybenzoic and hydroxycinnamic acids, containing seven (C1–C6) and nine (C3–C6) carbon atoms, respectively. Each phenolic acid is composed of an aromatic ring with hydroxyl (–OH) and carboxyl (–COOH) groups, and the main difference in the structure of these groups is the presence of one additional double bond between the –COOH group and the aromatic ring [3,11–13]. The phenolic acids also differ in type, number, and position of the attached functional groups on the aromatic ring (–OH, methoxy (–OCH$_3$)), and the research on their distribution is commonly used to find a relationship between structural features and compound activity, known as quantitative structure–activity relationship (QSAR) [14]. However, the knowledge of the mechanisms by which these molecules and their parts act in different reactions is limited. Scientific research indicated several factors with a possible impact on the mechanisms behind the compound's activity. Among them, the number and position of hydroxyl groups and their methylation, the distance between phenyl and carboxylic groups, and the concentration of the compound are suggested [15–21].

Like other phenolics, phenolic acids demonstrate different mechanisms of antioxidant action such as reduction of agents by hydrogen donation, quenching of singlet oxygen, or acting as chelators and trappers of free radicals, so usually, methods used to analyze their antioxidant activity are based on different mechanisms [3,14,22,23]. These methods may be generally classified as electron transfer (ET) and hydrogen atom transfer (HAT)-based assays [24]. The most accepted and widely used assays for the determination of antioxidant activity are Folin–Ciocalteu, FRAP (ferric reducing antioxidant power), ABTS/TEAC (2,2′-azinobis (3-ethylbenzothiazoline 6-sulfonate radical scavenging activity/Trolox equivalent antioxidant capacity), DPPH (2,2-diphenyl-1-picrylhydrazyl radical scavenging activity), and ORAC (oxygen radical absorbance capacity) [23–26]. Among these, Folin–Ciocalteu, FRAP, ABTS/TEAC, CUPRAC (cupric reducing antioxidant capacity), and DPPH methods are ET-based assays that provide information about reducing the capacity of an antioxidant, while the ORAC method is based on the HAT reaction mechanism. The FRAP method is often used to measure the reducing power of different samples and is considered one of the fastest, simplest, and less expensive methods, with reproducible results in a wide range of concentrations. On the other hand, the ORAC method uses a biologically relevant radical source (peroxyl radical), thus, the obtained activity could be used for interpreting activity in various biological systems [23,24,27]. For these reasons, the results obtained using different methods must be interpreted carefully, as due to differences in their mechanisms, the correlations between the obtained result often fail [23,26,28].

Although the antioxidant activity of phenolic acids is well studied using both in vitro and in vivo methods, the mechanisms of their action remain unclear and/or undefined [17–19]. An important factor that should be considered is their mutual interactions which can be synergistic, antagonistic, or additive (no interaction). Several studies aimed to investigate these interactions among phenolic acids using different antioxidant assays and confirmed both the occurrence of synergistic as well as antagonistic interactions [13,14,26,27,29–34]. The authors emphasized the influence of chemical structure and used concentrations on the overall activity of the tested mixtures. The efficiency of these interactions is also widely used to explain the activities of phenolic-rich extracts, where the dominant components cannot be identified as carriers of the total antioxidant activity [10,31,35].

In this regard, this study aimed to evaluate the antioxidant potential (reducing and free scavenging activity) of individual phenolic acids (protocatechuic, gentisic, gallic, vanillic, syringic, *p*-coumaric, caffeic, ferulic, sinapic, and rosmarinic) and their interactions in binary, ternary, quaternary, and quinary equimolar mixtures at different concentrations using FRAP and ORAC assays.

2. Materials and Methods

2.1. Preparation of Standard Solutions and Model Mixture

All used reagents and solvents were analytical or higher grade and purchased from Sigma (Sigma-Aldrich GmbH, Steinheim, Germany), Alkaloid AD (Skopje, North Macedonia), Merck (Darmstadt, Germany), Fluka (Buch, Switzerland), and Kemika (Zagreb, Croatia). The solutions of hydroxybenzoic acids (protocatechuic, gentisic, gallic, vanillic, and syringic) and hydroxycinnamic acids (*p*-coumaric, caffeic, ferulic, sinapic, and rosmarinic) (Sigma-Aldrich GmbH, Steinheim, Germany) shown in Table 1 were dissolved in an ethanol/water mixture (80:20, by volume) to the final concentration of 1000 µM. The experiment was divided in two parts. Firstly, all phenolic acids were individually tested for antioxidant activity at the concentrations of 2.5 and 5 µM in the ORAC assay, and 100, 500, and 1000 µM in the FRAP assay. Thereafter, the phenolic acids were mixed in binary, ternary, quaternary, and quinary equimolar combinations to reach the concentrations of 5 µM for ORAC and 100, 500, and 1000 µM for the FRAP assay.

Table 1. List and structural features of the investigated phenolic acids.

	Common Name	IUPAC Name	R1	R2	R3
	Hydroxybenzoic acids				
	Protocatechuic acid	3,4-dihydroxybenzoic acid	OH	OH	H
	Gentisic acid	2,5-dihydroxybenzoic acid	OH	H	OH
	Gallic acid	3,4,5-trihydroxybenzoic acid	OH	OH	OH
	Vanillic acid	4-hydroxy-3-methoxybenzoic acid	OCH$_3$	OH	H
	Syringic acid	4-hydroxy-3,5-dimethoxybenzoic acid	OCH$_3$	OH	OCH$_3$
	Hydroxycinnamic acids				
	p-coumaric acid	4-hydroxycinnamic acid	H	OH	H
	Caffeic acid	3,4-dihydroxycinnamic acid	OH	OH	H
	Ferulic acid	4-hydroxy-3-methoxycinnamic acid	OCH$_3$	OH	H
	Sinapic acid	4-hydroxy-3,5-dimethoxycinnamic acid	OCH$_3$	OH	OCH$_3$
	Rosmarinic acid	3,4-Dihydroxycinnamic acid (R)-1-carboxy-2-(3,4-dihydroxyphenyl)ethyl ester	Caffeic acid and 3,4-dihydroxyphenyllactic acid ester, with four OH groups		

2.2. Evaluation of the Antioxidant Activity

2.2.1. FRAP (Ferric Reducing Antioxidant Power) Assay

The reducing power of the samples detected with the FRAP method was measured according to the procedure described by Skroza et al. [36] and measurements were performed on a Tecan MicroPlate Reader, model Sunrise (Tecan Group Ltd., Männedorf, Switzerland). Analyses were completed in triplicates and results are expressed as µM of Fe^{2+}.

2.2.2. ORAC (Oxygen Radical Absorbance Capacity) Assay

Fluorimetric measurements in the ORAC assay were recorded on a microplate reader (Synergy HTX Multi-Mode Reader, BioTek Instruments, Inc., Winooski, VT, USA) following the procedure of Čagalj et al. [37]. The reaction was observed for 80 min and the results of three replicates are expressed in µM of Trolox Equivalents (µM TE).

2.3. Interaction and Statistical Analysis

The obtained results were analyzed using GraphPad Prism Version 4.03 for Windows (GraphPad Software, San Diego, CA, USA).

The interactions between phenolic acids were described as the difference in antioxidant activity between experimental and theoretical (calculated) values using the equation (Equation (1)) [36,38]:

$$Difference~(\%) = ((Combination~ab \times 100)/(Individual~a + Individual~b)) - 100 \qquad (1)$$

where combination *ab* is an experimentally obtained result for the binary mixture, while each *a/b* value was calculated individually for each compound. The theoretical values for each compound were calculated by dividing the experimental values by the number of compounds in the mixtures. Likewise, for ternary, quaternary, and quinary mixtures, the difference was calculated by subtracting the average of the individual three, four, or five compounds from the combination (Equations (2)–(4)):

$$Difference~(\%) = ((Combination~abc \times 100)/(a + b + c)) - 100 \qquad (2)$$

$$Difference~(\%) = ((Combination~abcd \times 100)/(a + b + c + d)) - 100 \qquad (3)$$

$$Difference~(\%) = ((Combination~abcde \times 100)/(a + b + c + d + e)) - 100 \qquad (4)$$

The obtained interactions (% difference) were used to determine potential synergistic (positive values, difference (%) > 0) or antagonistic (negative values, difference (%) < 0) effects. The additive effect was considered for the difference (%) $\cong 0 \pm 5\%$ when it can be considered that there was no interaction.

3. Results and Discussion

3.1. Antioxidant Activity of Individual Phenolic Acids

Previous studies confirmed that the number and arrangement of –OH and –OCH$_3$ groups (their mutual position) affect the antioxidant activity of the phenolic acids [14,27,31,39], but also other parameters, such as ionization, dissociation, and the rate constants of radical scavenging, resonance, solvent solvation effects, intramolecular hydrogen bonds, bond dissociation enthalpy, etc., should also be taken in consideration as reported by Sroka [16], Hanscha at al. [15], Foti et al. [17] Lucarini and Pedulli [18], Hang et al. [21], and Biela et al. [20].

The antioxidant activities of individual phenolic acids at different concentrations are provided in Table 2. Among the hydroxybenzoic phenolic acids, gallic acid showed the highest FRAP value, while the lowest activity was detected for vanillic acid. This was observed for all tested concentrations and linear dependence was confirmed, as expected. The highest reducing power of gallic acid could be related to its chemical structure and three –OH groups located at positions 3, 4, and 5. The gentisic acid, with two –OH groups in *para*-position to each other (at positions 2 and 5), and protocatechuic acid with two –OH groups at positions 3 and 4 (catechol structure) showed similar FRAP values at low concentrations, while at higher concentrations gentisic acid was superior. Considering the molecular structure, the activity of gentisic acid, an active metabolite of salicylic acid degradation, was investigated by Mardani-Ghahfarokhi and Farhoosh [40], and the authors reported its higher activity in comparison to α-resorcylic acid with the same type and number of substituents, but in *meta*-position. Sroka and Cisowski [16] reported that acids with two –OH groups exhibited higher antioxidant activity than those possessing only one. Cuvelier et al. [41] concluded that the introduction of –OH in *para*- or *ortho*-position enhances compound activity which increases also with the –OCH$_3$ substitution in *ortho*-positions to –OH group. The –COOH group, as one of the most common electron-withdrawing groups (containing an atom with a positive charge directly attached to a benzene ring) with a relatively strong effect, can affect other substituents and it has been known that this effect is the strongest on –OH groups in the *para*-position, and weaker on those in the *meta*-position [20,42]. Biela et al. [20] reported that phenolic acids in aqueous solutions are mostly dissociated (fully deprotonated) to carboxylate anions so the key role in the activity is carried by their phenolic –OH groups. It has been reported that the carboxylic substituent is a weak donor in the *meta*-position and has no effect in the *para*-position in comparison to the –COOH. Furthermore, the presence of other substituents, such as carboxylate anions, –OH and –OCH$_3$ can also form intramolecular hydrogen bonds that can affect the activity. The authors concluded also that –OH and –OCH$_3$ show an electron-donating effect in the *para*-position and an electron-withdrawing effect in the *meta*-position. Syringic acid, with two –OCH$_3$ groups at positions 3 and 5 (*meta*-position), and one –OH at the *ortho*-position, showed an increase in activity at higher concentrations reaching 3207 μM Fe^{2+} at 1000 μM, while vanillic acid, with an –OH group in the *para*-position, and –OCH$_3$ group in the *meta*-position had the lowest FRAP value. Spiegel et al. [27] investigated 22 phenolic acids using an FRAP assay and reported that the main structural feature of good antioxidants are two or more –OH groups in *ortho*- and *para*-positions, but the importance of the inductive effect of the carboxylic group should not be neglected. Protocatechuic acid has two –OH groups, the same as gentisic acid, but it showed a lower reducing effect so it can be concluded that again, besides the number and arrangement of attached functional groups (in this case, *ortho*- or *para*-position), the intramolecular hydrogen bonds between two –OH groups also have a notable impact. Additionally, intramolecular hydrogen bonds between the –OH and –COOH groups may affect antioxidant activity. The obtained results showed that hydroxylation at positions 2 and 5 and one intramolecular hydrogen bond between them contributes to the reducing ability of the compound [17,18,27]. Rice-Evans et al. [43] reported that the insertion of an additional –OH group at position 2 of hydroxybenzoic acids decreases the overall antioxidant capacity, while Sroka and Cisowski [16] showed that the antioxidant activity of phenolic acids correlates with the number of –OH groups in *ortho*- and *para*-positions, but also reported the importance of the position of –COOH and acetyl group near the –OH groups. Foti et al. [17], except for position and number of substituents, also indicate the importance of resonance stabilization and intramolecular hydrogen bonds between them, while Biskup [11] reported that the functional group binding site and the type of substitute affect the activity. Spiegel et al. [27] observed that the position of the second –OH group affected the reducing capacity and that two or more –OH groups located either in the vicinal position or in the opposite position to each other resulted in higher antioxidant activity. They also explained differences in antioxidant activity of phenolic acids using resonance stabilization of radicals by intermolecular hydrogen bonds between functional groups and a polar solvent. The influence of the hydrogen bonds is also discussed by Foti et al. [17] where the authors reported that only compounds that are non-hydrogen-bonded (free) possess activity (electron transfer mechanism) and that the rate of reaction depends on the strength of the hydrogen bond as well as on the used solvent (methanol or ethanol).

Table 2. Antioxidant activity of individual phenolic acids at different concentrations using the FRAP and ORAC methods.

	FRAP (μM Fe^{2+})			ORAC (μM TE)	
	100 μM	500 μM	1000 μM	2.5 μM	5 μM
Hydroxybenzoic acids					
Protocatechuic acid	282 ± 5	1014 ± 14	2341 ± 32	30 ± 0.9	57 ± 2
Gentisic acid	294 ± 4	1710 ± 81	3186 ± 104	29 ± 1	53 ± 0
Gallic acid	494 ± 3	2478 ± 17	5033 ± 106	23 ± 1	28 ± 1
Vanillic acid	193 ± 2	542 ± 4	850 ± 13	26 ± 1	56 ± 1
Syringic acid	245 ± 5	1514 ± 19	3207 ± 111	29 ± 2	40 ± 1
Hydroxycinnamic acids					
p-coumaric acid	23.9 ± 2	61.8 ± 1	113 ± 3	20 ± 2	33 ± 1
Caffeic acid	476 ± 17	1321 ± 2	2125 ± 18	31 ± 2	60 ± 1
Ferulic acid	260 ± 6	885 ± 2	1706 ± 38	26 ± 2	45 ± 0
Sinapic acid	235 ± 8	1201 ± 0	2186 ± 24	25 ± 1	44 ± 1
Rosmarinic acid	413 ± 14	1803 ± 50	3656 ± 148	50 ± 2	92 ± 2

The FRAP values of the hydroxycinnamic acids also showed a linear correlation with concentration (Table 2). Similarly, FRAP values increased with the introduction of –OH and –OCH$_3$ groups, which is in correlation with previous reports [14,27]. Among hydroxycinnamic acids, caffeic and rosmarinic acids had the highest FRAP values, while *p*-coumaric acid, with one –OH group, showed the lowest FRAP at all tested concentrations. Exceptionally good reducing power was also observed for sinapic and ferulic acids. In addition to one –OH group, these two acids also have –OCH$_3$ groups, sinapic acid two and ferulic only one. The rosmarinic acid, an ester of caffeic acid and 3,4-dihydroxyphenyllactic acid (diphenolic compound), also stands out, with FRAP values 2-fold higher than those obtained by caffeic acid alone at 1000 µM. This can be also connected with its structure (two phenolic rings with two –OH groups in the *ortho*-position and an unsaturated double bond and –COOH between them). Cao et al. [44] investigated the antioxidant activity of this compound given its molecular structure and reported a stronger electron-donating capability of the B ring. According to the authors, its activity is a result of the H-abstraction reactions on both rings. Therefore, the activity of rosmarinic acid can be related to two phenolic groups in both rings. The good reducing power of caffeic acid can be related to the catechol structure and distance between the –COOH group and functional groups [14,16,17,27]. It has also been reported that hydroxybenzoic acids have lower antioxidant activity in comparison to hydroxycinnamic acids when they have the same substituents at positions 2–6 [14,20,27,41,43,45]. These observations are also confirmed in the present study where syringic acid showed higher reducing activity than sinapic acid, and ferulic acid was superior to vanillic acid. In the case of the ORAC method, the values for ferulic and vanillic acids at 2.5 µM were similar, but at higher concentrations better activity was detected for vanillic acid.

The results of antioxidant activity tested using the ORAC method at two concentrations (2.5 and 5 µM) are shown in Table 2. Among hydroxybenzoic acids, protocatechuic acid showed the highest activity, while gallic acid had the lowest activity. At higher concentrations, 21–55% higher activity was detected for all acids, with exception of the gallic acid which had 19% higher activity. Protocatechuic acid, with two –OH groups at positions 2 and 3, had higher activity than other acids with the same number of –OH groups at other positions (gentisic acid) or even –OCH$_3$ groups (one in vanillic acid and two in syringic acid structure). Sroka and Cisowski [16] in their study also reported higher activity of protocatechuic acid against DPPH free radicals in comparison to the gentisic acid. Hang et al. [21] reported the influence of –OH and/or –OCH$_3$ groups at position 3 and/or 5 on the hydrogen transfer mechanism (characterized by bond-dissociation energy) using the hydroperoxyl radical scavenging assay. They conclude that the intramolecular hydrogen bond between the –OH group at the *ortho*-position and the –COOH group could be the main reason for the highest reduction of bond-dissociation energy, which indicated weaker antioxidant activity (radical scavenging activity followed the order: syringic acid > gentisic acid > gallic acid > vanillic acid > protocatechuic acid). For higher antioxidant activity measured using ORAC, the location of –OH groups in vicinal or in the opposite position side of the ring seems to be a more important factor than only the number of –OH groups.

The ORAC test for hydroxycinnamic acids (Table 2) showed the lowest results for *p*-coumaric acid, while rosmarinic acid had the highest antioxidant effect at tested concentrations. This could be again related to its chemical structure and four –OH groups. *p*-coumaric acid was previously described as a weak antioxidant [29], what is also confirmed in this research where the obtained results showed that caffeic, ferulic, and sinapic acids had 2-fold higher activity than this compound. While *p*-coumaric acid has only one –OH group, other acids have additional –OH and/or –OCH$_3$ groups. At the concentration of 5 µM, all acids showed approximately 45% higher values. Ferulic (one –OH and one –OCH$_3$) and sinapic (one –OH and two –OCH$_3$) acids showed almost the same activity, indicating that the additional –OCH$_3$ group does not have a significant impact on the activity. Interestingly, caffeic acid with a catechol group at the same position as protocatechuic acid showed similar antioxidant activity as this hydroxybenzoic acid, in accordance with the results of Sroka and Cisowski [16] who also confirmed a similar free radical scavenging activity. The authors pointed out that although these compounds have a similar model of substitution of the –OH and the incorporation of –CH$_2$ between –COOH and the phenyl group does not increase the antiradical activity of 3,- and 4- substituted acids. This is also in accordance with the conclusions that the catechol group enhances the radical scavenging activity of the compound [26,27,29]. However, ferulic acid was not effective in scavenging peroxyl radicals such as vanillic acid with the same structural features (–OH and –OCH$_3$ group), not in accordance with previous reports [26,27,29]. Additionally, greater distances between –COOH groups from the methoxylated ring do not enhance the antioxidant effect, as previous studies by Mathew et al. [19] and Spiegel et al. [27] suggested. Lucarini and Pedulli [18] reported the importance of the reaction medium. In their study on free radical scavenging activity of peroxyl radicals in autooxidation reactions, they reported the connection of bond dissociation enthalpies and rate constants with the antioxidant compound structure. The authors would like to point out once more the lack of systematic research on the antioxidant activity of phenolic acids using the ORAC method and investigations of their structure–activity relations.

3.2. Antioxidant Activity of Equimolar Mixtures of Phenolic Acids

The results of the antioxidant activity of the equimolar mixtures of two or more phenolic acids tested at different concentrations are shown in Tables 3–6. Based on the obtained data, the potential interaction (synergistic/additive/antagonistic) was determined and expressed as a percentage of the difference (%) between the experimental and theoretical (calculated) FRAP and ORAC values.

In the binary mixtures of hydroxybenzoic acids, at 100 µM, all mixtures containing gentisic acid showed a synergistic effect (28–89% difference). The mixture of protocatechuic and syringic acid showed an additive effect, while all others showed an antagonistic effect (up to −58% difference). At a concentration of 500 µM the synergic effect was observed only for the mixture of gentisic + syringic acids, while at 1000 µM the synergistic effect was not confirmed. Among ternary mixtures, it is interesting to highlight the mixture protocatechuic + gentisic + syringic acids, which showed the highest reducing power and the greatest synergistic effect (174% difference) at the lowest tested concentration. In comparison to the mixture of protocatechuic + gentisic acids and gentisic + syringic acids that also showed a high synergistic effect, and the mixture of these three phenolic acids showed a higher overall reducing capacity. When protocatechuic, gallic, and vanillic acids were combined they retained the antagonistic effect observed for mixtures of protocatechuic + gallic acid acids and protocatechuic + vanillic acids at all tested concentrations. Again, the mixtures containing either protocatechuic, gallic, and/or syringic acid along with gentisic acid show a synergistic effect leading to a conclusion that gentisic acid is a key component for the synergistic effect of these mixtures. However, the synergistic effect for protocatechuic + gentisic + vanillic acids and protocatechuic + gentisic + syringic acids observed at the concentration of 100 µM was confirmed also at 500 µM but not at a higher concentration where it was only detected for the protocatechuic + gentisic + gallic acids mixture. These results indicated that along with the compound ratio in the mixture, the compound concentration is also an important factor. The quaternary and quinary equimolar combinations showed a synergic effect only at the concentration of 100 µM, with the exception of the mixture of gallic + vanillic + syringic + protocatechuic acids which showed an antagonistic effect at all tested concentrations.

Table 3. Comparison of experimental and theoretical FRAP values and the interactions in equimolar phenolic mixtures of hydroxybenzoic acids at different concentrations.

	100 µM			500 µM			1000 µM		
	Experimental	Theoretical	Difference (%)	Experimental	Theoretical	Difference (%)	Experimental	Theoretical	Difference (%)
Combination of two acids (1:1)									
P + Ge	520 ± 2	288	**81**	1320 ± 15	1362	−3.1	2832 ± 22	2763	2.5
P + G	349 ± 7	388	−10	2181 ± 16	1746	**25**	3623 ± 7	3687	−1.7
P + V	101 ± 3	237	−58	559 ± 3	778	−28	1414 ± 8	1595	−11
P + Sy	274 ± 5	263	4.0	1280 ± 19	1264	**1.3**	2705 ± 10	2774	−2.5
Ge + G	542 ± 5	394	**38**	2034 ± 19	2094	−2.9	3937 ± 9	4109	−4.2
Ge + V	311 ± 3	243	**28**	825 ± 10	1126	−27	1440 ± 23	2018	−29
Ge + Sy	410 ± 3	270	**52**	2539 ± 4	1612	**58**	2974 ± 3	3197	−7.0
G + V	229 ± 5	343	−33	1237 ± 23	1510	−18	2861 ± 15	2942	−2.8
G + Sy	275 ± 2	370	−26	1814 ± 36	1996	−9.1	3599 ± 12	4120	−13
V + Sy	179 ± 0	219	−19	910 ± 5	1028	−12	1440 ± 23	2029	−29
Combination of three acids (1:1:1)									
P + Ge + G	166 ± 1	356	−54	2199 ± 28	1734	**27**	3811 ± 6	3520	**8.3**
P + Ge + V	317 ± 3	256	**24**	1338 ± 3	1089	**23**	1871 ± 5	2126	−12
P + Ge + Sy	749 ± 1	274	**174**	1509 ± 13	1413	**6.8**	2881 ± 6	2911	−1.1
P + G + V	248 ± 1	323	−23	1468 ± 15	1345	**9.2**	2591 ± 8	2741	−5.5
P + G + Sy	349 ± 6	340	2.6	1616 ± 8	1668	−3.2	2741 ± 50	3527	−22
P + V + Sy	210 ± 2	240	−13	947 ± 3	1023	−7.5	1786 ± 41	2133	−16
Ge + G + V	409 ± 3	327	**25**	1611.3 ± 8	1577	2.2	2430 ± 8	3023	−20
Ge + G + Sy	519 ± 3	344	**51**	1822.4 ± 12	1901	−4.1	3294 ± 47	3809	−14
G + V + Sy	245 ± 4	311	−21	1176.8 ± 6	1511	−22	2265 ± 21	3030	−25
V + Sy + Ge	335 ± 4	244	**37**	1023.5 ± 6	1255	−19	2014 ± 7	2414	−17
Combination of four acids (1:1:1:1)									
P + Ge + G + V	414 ± 3	316	**31**	1390 ± 2	1436	−3.2	2695 ± 20	2852	−5.5
P + Ge + V + Sy	367 ± 1	253	**45**	956 ± 4	1195	−20	1958 ± 4	2396	−18
Ge + G + V + Sy	412 ± 5	307	**34**	1236 ± 2	1561	−21	2470 ± 27	3069	−20
G + V + Sy + P	236 ± 0	303	−21	1291 ± 31	1387	−6.9	2789 ± 46	2858	−2.4
Sy + P + Ge + G	512 ± 5	329	**57**	1594 ± 15	1679	−5.1	3207 ± 43	3442	−6.8
Combination of five acids (1:1:1:1:1)									
P + Ge + G + V + Sy	439 ± 1	302	**46**	1257 ± 10	1452	−**13**	2876 ± 5	2923	−1.6

P—protocatechuic acid; Ge—gentisic acid; G—gallic acid; V—vanillic acid; Sy—syringic acid; *p*C—*p*-coumaric acid; C—caffeic acid; F—ferulic acid; Si—sinapic acid; R—rosmarinic acid; a difference (%) > 0 indicates a potential synergistic effect; a difference (%) < 0 shows an antagonistic and a difference (%) $\cong$ 0 or ± 5% shows an additive effect (no interaction).

Table 4. Comparison of experimental and theoretical FRAP values and the interactions in equimolar phenolic mixtures of hydroxycinnamic acids at different concentrations.

	100 μM			500 μM			1000 μM		
	Experimental	Theoretical	Difference (%)	Experimental	Theoretical	Difference (%)	Experimental	Theoretical	Difference (%)
Combination of two acids (1:1)									
pC + C	161 ± 10	250	−36	779 ± 18	691	**13**	1479 ± 42	1119	**32**
pC + F	95 ± 2	142	−33	537 ± 31	4474	−88	1046 ± 9	909	**15**
pC + Si	223 ± 10	130	**72**	655 ± 32	631	3.8	1321 ± 39	1150	**15**
pC + R	197 ± 10	219	−10	1357 ± 26	932	**46**	2635 ± 404	1885	**40**
C + F	314 ± 2	368	−15	1227 ± 28	5103	−76	2049 ± 28	1915	**7.0**
C + Si	342 ± 20	3560	−4.0	1362 ± 14	1261	**8.0**	2542 ± 87	2155	**18**
C + R	428 ± 44	445	−3.8	1751 ± 155	1562	**12**	3347 ± 122	2890	**16**
F + Si	501 ± 5	248	**102**	1234 ± 25	5043	−76	2303 ± 62	1946	**18**
F + R	322 ± 7	336	−4.2	1467 ± 19	5344	−73	2812 ± 57	2681	4.9
Si + R	307 ± 3	324	−5.2	1674 ± 56	1502	**11**	3394 ± 55	2921	**16**
Combination of three acids (1:1:1)									
pC + C + F	181 ± 2	253	−28	716 ± 18	3423	−79	1472 ± 26	1315	**12**
pC + C + Si	190 ± 9	245	−23	849 ± 19	861	−1.4	1559 ± 21	1475	**5.7**
pC + C + R	691 ± 39	305	**127**	1567 ± 32	1062	**48**	2178 ± 19	1965	**11**
pC + F + Si	171 ± 3	173	−0.9	737 ± 60	3383	−78	1436 ± 58	1335	**7.6**
pC + F + R	63 ± 7	232	−73	954 ± 8	3584	−73	1780 ± 54	1825	−2.5
pC + Si + R	216 ± 6	224	−3.6	947 ± 6	1022	−7.3	1925 ± 51	1985	−3.0
C + Si + R	328 ± 10	375	−13	1617 ± 22	1442	**12**	3214 ± 197	2656	**21**
C + F + Si	257 ± 5	324	−21	1044 ± 31	3803	−73	2241 ± 30	2005	**12**
C + F + R	337 ± 9	383	−12	1102 ± 23	4003	−73	2429 ± 54	2495	−2.7
F + Si + R	314 ± 3	303	3.9	1331 ± 71	3963	−66	2810 ± 43	2516	**12**
Combination of four acids (1:1:1:1)									
pC + C + F + Si	187 ± 8	249	−25	741 ± 22	2867	−74	1596 ± 44	1532	4.2
pC + C + F + R	224 ± 2	293	−24	967 ± 42	3018	−68	1917 ± 10	1900	0.9
C + F + Si + R	248 ± 11	346	−28	1164 ± 12	3303	−65	2576 ± 113	2418	**6.5**
C + Si + R + pC	245 ± 4	287	−15	1120 ± 12	1097	2.2	2176 ± 24	2020	**7.7**
R + pC + F + Si	201 ± 11	233	−14	1064 ± 21	2988	−64	1947 ± 36	1915	1.7
Combination of five acids (1:1:1:1:1)									
pC + C + F + Si + R	247 ± 1	282	−12	1107 ± 39	2655	−58	2077 ± 35	1957	**6.1**

P—protocatechuic acid; Ge—gentisic acid; G—gallic acid; V—vanillic acid; Sy—syringic acid; pC—*p*-coumaric acid; C—caffeic acid; F—ferulic acid; Si—sinapic acid; R—rosmarinic acid; a difference (%) > 0 indicates a potential synergistic effect; a difference (%) < 0 shows an antagonistic and a difference (%) $\cong$ 0 or ± 5% shows an additive effect (no interaction).

Table 5. Comparison of theoretical and experimental ORAC values and the interaction of equimolar phenolic mixtures (% difference) of hydroxybenzoic acids at a concentration of 5 µM.

	Experimental	Theoretical	Difference (%)
Combination of two acids (1:1)			
P + Ge	150 ± 0.4	55	172
P + G	55 ± 2	43	28
P + V	148 ± 1	56	162
P + Sy	51 ± 0.1	49	4.1
Ge + G	77 ± 5	41	89
Ge + V	150 ± 1	55	174
Ge + Sy	158 ± 2	47	236
G + V	29 ± 0.7	42	−30
G + Sy	26 ± 6	34	−24
V + Sy	149 ± 0.7	48	210
Combination of three acids (1:1:1)			
P + Ge + G	56 ± 0.5	46	22
P + Ge + V	82 ± 2	55	48
P + Ge + Sy	69 ± 2	50	38
P + G + V	61 ± 2	47	29
P + G + Sy	43 ± 1	42	1.9
P + V + Sy	65 ± 4	51	27
Ge + G + V	66 ± 4	46	45
Ge + G + Sy	50 ± 2	41	22
G + V + Sy	53 ± 1	41	28
V + Sy + Ge	65 ± 1	50	29
Combination of four acids (1:1:1:1)			
P + Ge + G + V	77 ± 2	49	58
P + Ge + V + Sy	86 ± 4	52	67
Ge + G + V + Sy	67 ± 2	45	50
G + V + Sy + P	65 ± 4	45	44
Sy + P + Ge + G	62 ± 9	45	39
Combination of five acids (1:1:1:1:1)			
P + Ge + G + V + Sy	60 ± 2	47	27

P—protocatechuic acid; Ge—gentisic acid; G—gallic acid; V—vanillic acid; Sy—syringic acid; *p*C—*p*-coumaric acid; C—caffeic acid; F—ferulic acid; Si—sinapic acid; R—rosmarinic acid; a difference (%) > 0 indicates a potential synergistic effect; a difference (%) < 0 shows an antagonistic and a difference (%) $\cong$ 0 or ± 5% shows an additive effect (no interaction).

The results of the reducing capacity for the mixtures of hydroxycinnamic acids are shown in Table 4. In contrast to the results for interactions obtained for hydroxybenzoic acids, a large number of tested mixtures of cinnamic acids showed lower antioxidant activity compared to the expected theoretical values, which indicated an antagonistic effect between these compounds. Among binary combinations, ferulic + sinapic acids showed the highest reducing power at 100 µM (501 µM Fe^{2+}), while at other concentrations the mixture of *p*-coumaric + ferulic acids exhibited the lowest activity. At the concentration of 100 µM, the greatest positive difference between theoretical and expected FRAP values, indicating the higher synergistic effect, was observed for the following mixtures: *p*-coumaric + sinapic acids (72% difference) and ferulic + sinapic acids (102% difference), while other mixtures showed an antagonistic effect. With the increase in concentration, the number of mixtures showing a synergistic effect rose to five at 500 µM and nine at 1000 µM. At the first higher concentration, only the combination of ferulic + rosmarinic acids had an antagonistic effect while at 1000 µM they showed only an additive interaction. The addition of *p*-coumaric acid to the mixture of ferulic and rosmarinic acid, at both concentrations of 500 and 1000 µM, resulted in a synergistic effect of the mixture (127 and 48% difference, respectively) which was strange due to the low FRAP value of this compound and only one –OH group in its structure. In ternary mixtures, only *p*-coumaric + caffeic + rosmarinic acids showed a synergistic effect at all tested concentrations, but the decrease in the difference with the increase in the concentration was recorded (127% at 100 μM > 48% at 500 μM > 11% at 1000 μM). The mixture of caffeic + sinapic + rosmarinic acids, with the highest number of –OH (seven) and –OCH$_3$ (two) groups showed an antagonistic effect at 100 μM and a synergistic effect at 500 and 1000 μM. In mixtures of *p*-coumaric + caffeic + sinapic acids and caffeic + sinapic + rosmarinic acids, an antagonistic effect passed to synergistic at 1000 μM. Among quaternary and quinary mixtures, at 1000 μM only mixtures of caffeic + ferulic + sinapic + rosmarinic acids, mixtures of caffeic + sinapic + rosmarinic + *p*-coumaric acids and *p*-coumaric + caffeic + ferulic + sinapic + rosmarinic acids showed a weak synergistic effect.

Table 6. Comparison of theoretical and experimental ORAC values and the interaction of equimolar phenolic mixtures (% difference) of hydroxycinnamic acids at a concentration of 5 μM.

	Experimental	Theoretical	Difference (%)
Combination of two acids (1:1)			
*p*C + C	52 ± 2	46	**13**
*p*C + F	160 ± 4	39	**311**
*p*C + Si	52 ± 3	39	**34**
*p*C + R	61 ± 7	63	−2.1
C + F	59 ± 3	49	**21**
C + Si	162 ± 4	52	**211**
C + R	163 ± 9	76	**115**
F + Si	51 ± 4	45	**13**
F + R	156 ± 1	69	**127**
Si + R	74 ± 4	68	**9**
Combination of three acids (1:1:1)			
*p*C + C + F	53 ± 1	46	**16**
*p*C + C + Si	47 ± 3	46	2.1
*p*C + C + R	66 ± 3	62	**7.3**
*p*C + F + Si	39 ± 1	41	−4.2
*p*C + F + R	55 ± 2	57	−3.6
*p*C + Si + R	54 ± 1	56	−3.7
C + Si + R	56 ± 2	65	−14
C + F + Si	43 ± 3	50	−13
C + F + R	60 ± 2	66	−8.4
F + Si + R	47 ± 2	61	−22
Combination of four acids (1:1:1:1)			
*p*C + C + F + Si	45 ± 2	45	1.7
*p*C + C + F + R	56 ± 2	57	−2.9
C + F + Si + R	52 ± 1	60	−14
C + Si + R + *p*C	52 ± 1	57	−9.2
R + *p*C + F + Si	77 ± 4	54	**44**
Combination of five acids (1:1:1:1:1)			
*p*C + C + F + Si + R	56 ± 3	55	2.7

P—protocatechuic acid; Ge—gentisic acid; G—gallic acid; V—vanillic acid; Sy—syringic acid; *p*C—*p*-coumaric acid; C—caffeic acid; F—ferulic acid; Si—sinapic acid; R—rosmarinic acid; a difference (%) > 0 indicates a potential synergistic effect; a difference (%) < 0 shows an antagonistic and a difference (%) ≅ 0 or ± 5% shows an additive effect (no interaction).

Hajimehdipoor et al. [32] confirmed the synergistic effect of the caffeic + rosmarinic acids mixture (38% difference) in different binary combinations. The authors mixed binary combinations of hydroxybenzoic and hydroxycinnamic acids (gallic, rosmarinic, caffeic, and chlorogenic), alone and with flavonoids (quercetin and rutine) and reported that binary mixtures show stronger synergistic effects than their ternary combinations. Olszowy-Tomczyk [23] also reviewed the available information in the literature about interactions among compounds in the binary mixtures of phenolic acids with other phenolics (flavonoids, catechins, stilbenes, etc.). Differences between the experimental and theoretical values for antioxidant activity among phenolic acids were observed also in binary mixtures of gallic + protocatechuic acids, gallic + vanillic acids [31], rosmarinic + caffeic acids [29], and gallic + caffeic acids [23].

The results of the theoretical and experimental ORAC values of phenolic mixtures, as well as their interactions, are presented in Tables 5 and 6. Only two mixtures of hydroxybenzoic acids showed an antagonistic effect, namely a mixture of gentisic + syringic acids (−24% difference) and gallic + vanillic acids (−30% difference), while all other mixtures had a synergistic effect. The antioxidant activities and synergistic effects of binary mixtures were higher in comparison to mixtures of three or more acids. The highest synergistic effect was observed for the mixture of gentisic + syringic acids with a total of two –OCH$_3$ and three –OH groups while the mixture of gallic + syringic acids with the highest number of substituents, four –OH and two –OCH$_3$ groups, showed an antagonistic effect. In the ORAC method, the presence of gentisic acid and protocatechuic acid resulted in a synergistic effect of the mixtures that contain these substances which could indicate that the presence of two –OH groups on the benzene ring (in *ortho*- or *para*-positions) is most likely responsible for this effect. In ternary, quaternary, and quinary mixtures, all combinations showed a synergistic effect except the mixture of protocatechuic + gallic + syringic acids which showed an additive effect. When gentisic acid was added to this mixture the effect was again synergistic.

The ORAC results obtained for the mixtures of hydroxycinnamic acids were lower and only a few mixtures showed the synergistic effect. The best antioxidant potential was confirmed for the mixture of *p*-coumaric + ferulic acids (160 μM TE) with a difference of 311% and a mixture of caffeic + sinapic acids (162 μM TE) with a difference of 211%. A positive interaction was observed also for the mixture of ferulic + rosmarinic acids (127%) and caffeic + rosmarinic acids (115%). Peyrat-Maillard et al. [29] also confirmed synergistic interaction between caffeic and rosmarinic acid at concentrations up to 5 μM by ORAC assay but in their study the concentration showed no effect on the interaction. In ternary, quaternary, and quinary mixtures of hydroxycinnamic acids only mixtures of *p*-coumaric + caffeic + ferulic acids, *p*-coumaric + caffeic + rosmarinic acids, and rosmarinic + *p*-coumaric + ferulic + sinapic acids showed a synergistic effect, while others showed a slight antagonistic or additive effect. In contrast to hydroxybenzoic acids, the number of the −OCH$_3$ group in the structure of hydroxycinnamic acids (e.g., in ternary, quaternary, and quinary mixtures with ferulic or sinapic acids) cannot be related to their higher antioxidant activity.

Palafox Carlos et al. [31] suggested that gallic, protocatechuic, and vanillic acids interact in a synergic way. Using a DPPH assay, the authors also confirmed the synergistic effect of the gallic and protocatechuic acid mixture relating this effect to the chemical structure of the compounds and the presence of the hydroxyl group. On the other hand, in their study, the mixture of protocatechuic and vanillic acid showed an antagonistic effect. Some authors suggested that interactions are concentration-related, rather than structure-related [39] or that presence or absence of the catechol group in the chemical structure of the compounds from the mixtures contributes to their synergic effect [13,46]. They investigated the interaction effect between caffeic, ferulic, and rosmarinic acid at different concentrations (50, 100, 200, and 250 μM) using the Briggs–Rauscher assay and reported the synergistic effect of the mixtures at concentrations ranging from 50 to 200 μM, and strong antagonism at 250 μM. The authors concluded that the antioxidant activity depends on compound structure (number and distribution of substituents) and concentration, which was opposite to some of the results obtained in the present study.

4. Conclusions

The results indicated that differences in antioxidant activity of the tested phenolic acids depend on their structure, as expected, regarding not only the type, number, and arrangement of substituents but also the compound concentration. The additional number of the –OCH$_3$ groups in the same positions in the phenolic ring in the hydroxybenzoic acids resulted in higher activity in comparison to the hydroxycinnamic acids with the same structural features. Among individual hydroxybenzoic acids, gallic acid showed the highest reducing activity, while the lowest activity was recorded with the ORAC assay. Among hydroxycinnamic acids the *p*-coumaric acid showed the lowest activity, using both methods at all tested concentrations. In the mixtures, synergistic effects were detected in several combinations, but special attention should be devoted to hydroxybenzoic acid mixtures containing gentisic acid, especially at lower concentrations where in all cases the positive differences were calculated. Similarly, in the ORAC method the presence of gentisic acid resulted in a synergistic effect of the mixtures, while low activity of the gallic acid obviously influenced the overall mixture activity as lower antioxidant or antagonistic effects are detected. Furthermore, it is obvious that other parameters such as the applied antioxidant method and solvent medium, the position of functional groups in relation to the –COOH group and other groups attached to the ring, ionization and bond dissociation enthalpies, intramolecular hydrogen bonding, etc., that were discussed but not investigated in this study, should be taken in consideration in further studies since they might have an impact on the overall antioxidant activity of the compounds and their mixtures.

Author Contributions: Conceptualization: D.S., V.Š. and I.G.M.; methodology and formal analysis: L.V., N.J., A.S., M.Č. and R.F.; data curation: D.S. and I.G.M.; supervision: D.S. and I.G.M.; writing—original draft: D.S., V.Š. and I.G.M.; writing—review and editing: D.S., V.Š., I.G.M., M.Č. and R.F. All authors have read and agreed to the published version of the manuscript.

Funding: This research is supported by Croatian Science Foundation (grant number IP-2014-09-6897) and the PRIMA program under project BioProMedFood (Project ID 1467). The PRIMA program is supported by the European Union.

Institutional Review Board Statement: Not applicable.

Informed Consent Statement: Not applicable.

Data Availability Statement: The data presented in this study are available in the article.

Conflicts of Interest: The authors declare no conflict of interest.

References

1. Miguel, M.G.; Neves, M.A.; Antunes, M.D. Pomegranate (Punica Granatum L.): A Medicinal Plant with Myriad Biological Properties-A Short Review. *J. Med. Plants Res.* **2010**, *4*, 2836–2847.
2. Ghasemzadeh, A.; Ghasemzadeh, N. Flavonoids and Phenolic Acids: Role and Biochemical Activity in Plants and Human. *J. Med. Plants Res.* **2011**, *5*, 6697–6703. [CrossRef]
3. De Ancos, B.; Colina-Coca, C.; González-Peña, D.; Sánchez-Moreno, C. Bioactive Compounds from Vegetable and Fruit By-Products. In *Biotechnology of Bioactive Compounds: Sources and Applications*; Gupta, V.K., Tuohy, M.G., Eds.; Wiely Online Library; John Wiley & Sons, Ltd.: Hoboken, NJ, USA, 2015; pp. 3–36. [CrossRef]
4. Gutiérrez-Grijalva, E.P.; Picos-Salas, M.A.; Leyva-López, N.; Criollo-Mendoza, M.S.; Vazquez-Olivo, G.; Heredia, J.B. Flavonoids and Phenolic Acids from Oregano: Occurrence, Biological Activity and Health Benefits. *Plants* **2018**, *7*, 2. [CrossRef] [PubMed]
5. Chaouch, M.A.; Benvenuti, S. The Role of Fruit By-Products as Bioactive Compounds for Intestinal Health. *Foods* **2020**, *9*, 1716. [CrossRef] [PubMed]
6. Dabulici, C.M.; Srbu, I.; Vamanu, E. The Bioactive Potential of Functional Products and Bioavailability of Phenolic Compounds. *Foods* **2020**, *9*, 953. [CrossRef]
7. Fierascu, R.C.; Sieniawska, E.; Ortan, A.; Fierascu, I.; Xiao, J. Fruits By-Products–A Source of Valuable Active Principles. A Short Review. *Front. Bioeng. Biotechnol.* **2020**, *8*, 319. [CrossRef]
8. Kiokias, S.; Proestos, C.; Oreopoulou, V. Phenolic Acids of Plant Origin-a Review on Their Antioxidant Activity in Vitro (O/W Emulsion Systems) along with Their in Vivo Health Biochemical Properties. *Foods* **2020**, *9*, 534. [CrossRef]
9. Gentile, C. Biological Activities of Plant Food Components: Implications in Human Health. *Foods* **2021**, *10*, 456. [CrossRef]
10. Zhang, Y.; Cai, P.; Cheng, G.; Zhang, Y. A Brief Review of Phenolic Compounds Identified from Plants: Their Extraction, Analysis, and Biological Activity. *Nat. Prod. Commun.* **2022**, *17*, 1–14. [CrossRef]
11. Biskup, I.; Golonka, I.; Zbigniew, S. Antioxidant Activity of Selected Phenols Estimated by ABTS and FRAP Methods. *Postep. Hig. Med. Dosw.* **2013**, *67*, 958–963. [CrossRef]
12. Stagos, D. Antioxidant Activity of Polyphenolic Plant Extracts. *Antioxidants* **2020**, *9*, 19. [CrossRef]
13. Gojak-Salimović, S.; Ramić, S. Investigation of the Antioxidant Synergisms and Antagonisms among Caffeic, Ferulic and Rosmarinic Acids Using the Briggs-Rauscher Reaction Method. *Bull. Chem. Technol. Bosnia Herzeg.* **2020**, *54*, 27–30.

14. Chen, J.; Yang, J.; Ma, L.; Li, J.; Shahzad, N.; Kim, C.K. Structure-Antioxidant Activity Relationship of Methoxy, Phenolic Hydroxyl, and Carboxylic Acid Groups of Phenolic Acids. *Sci. Rep.* **2020**, *10*, 5666. [CrossRef]

15. Corwin, H.; Leo, A.; Taft, R.W. A Survey of Hammett Substituent Constants and Resonance and Field Parameters. *Am. Chem. Soc.* **1991**, *91*, 165–195. [CrossRef]

16. Sroka, Z.; Cisowski, W. Hydrogen Peroxide Scavenging, Antioxidant and Anti-Radical Activity of Some Phenolic Acids. *Food Chem. Toxicol.* **2003**, *41*, 753–758. [CrossRef]

17. Foti, M.; Daquino, C.; Geraci, C. Esters with the DPPH ● Radical in Alcoholic Solutions. *J. Org. Chem* **2004**, *14*, 2309–2314. [CrossRef]

18. Lucarini, M.; Pedulli, G.F. Free Radical Intermediates in the Inhibition of the Autoxidation Reaction. *Chem. Soc. Rev.* **2010**, *39*, 2106–2119. [CrossRef]

19. Mathew, S.; Abraham, T.E.; Zakaria, Z.A. Reactivity of Phenolic Compounds towards Free Radicals under in Vitro Conditions. *J. Food Sci. Technol.* **2015**, *52*, 5790–5798. [CrossRef]

20. Biela, M.; Kleinová, A.; Klein, E. Phenolic Acids and Their Carboxylate Anions: Thermodynamics of Primary Antioxidant Action. *Phytochemistry* **2022**, *200*, 113254. [CrossRef]

21. Hang, D.T.N.; Hoa, N.T.; Bich, H.N.; Mechler, A.; Vo, Q.V. The Hydroperoxyl Radical Scavenging Activity of Natural Hydroxybenzoic Acids in Oil and Aqueous Environments: Insights into the Mechanism and Kinetics. *Phytochemistry* **2022**, *201*, 113281. [CrossRef]

22. Fernandez-Panchon, M.S.; Villano, D.; Troncoso, A.M.; Garcia-Parrilla, M.C. Antioxidant Activity of Phenolic Compounds: From in Vitro Results to in Vivo Evidence. *Crit. Rev. Food Sci. Nutr.* **2008**, *48*, 649–671. [CrossRef] [PubMed]

23. Olszowy-Tomczyk, M. *Synergistic, Antagonistic and Additive Antioxidant Effects in the Binary Mixtures*; Springer: Dordrecht, The Netherlands, 2020; Volume 19. [CrossRef]

24. Apak, R.; Özyürek, M.; Güçlü, K.; Çapanoğlu, E. Antioxidant Activity/Capacity Measurement. 1. Classification, Physicochemical Principles, Mechanisms, and Electron Transfer (ET)-Based Assays. *J. Agric. Food Chem.* **2016**, *64*, 997–1027. [CrossRef] [PubMed]

25. Dudonné, S.; Vitrac, X.; Coutière, P.; Woillez, M.; Mérillon, J.-M. Comparative Study of Antioxidant Properties and Total Phenolic Content of 30 Plant Extracts of Industrial Interest Using DPPH, ABTS, FRAP, SOD, and ORAC Assays-Journal of Agricultural and Food Chemistry (ACS Publications). *J. Agric. Food Chem.* **2009**, *57*, 1768–1774. [CrossRef] [PubMed]

26. Capitani, C.D.; Carvalho, A.C.L.; Botelho, P.B.; Carrapeiro, M.M.; Castro, I.A. Synergism on Antioxidant Activity between Natural Compounds Optimized by Response Surface Methodology. *Eur. J. Lipid Sci. Technol.* **2009**, *111*, 1100–1110. [CrossRef]

27. Spiegel, M.; Kapusta, K.; Kołodziejczyk, W.; Saloni, J.; Zbikowska, B.; Hill, G.A.; Sroka, Z. Antioxidant Activity of Selected Phenolic Acids–Ferric Reducing Antioxidant Power Assay and QSAR Analysis of the Structural Features. *Molecules* **2020**, *25*, 3088. [CrossRef]

28. Badarinath, A.V.; Rao, K.M.; Madhu, C.; Chetty, S.; Ramkanth, S.; Rajan, T.V.S.; Gnanaprakash, K. A Review On In-Vitro Antioxidant Methods: Comparisions, Correlations and Considerations. *Int. J. PharmTech Res.* **2010**, *2*, 1276–1285.

29. Peyrat-Maillard, M.N.; Cuvelier, M.E.; Berset, C. Antioxidant Activity of Phenolic Compounds in 2,2′-Azobis (2-Amidinopropane) Dihydrochloride (AAPH)-Induced Oxidation: Synergistic and Antagonistic Effects. *JAOCS J. Am. Oil Chem. Soc.* **2003**, *80*, 1007–1012. [CrossRef]

30. Yeh, C.T.; Shih, P.H.; Yen, G.C. Synergistic Effect of Antioxidant Phenolic Acids on Human Phenolsulfotransferase Activity. *J. Agric. Food Chem.* **2004**, *52*, 4139–4143. [CrossRef]

31. Palafox Carlos, H.; Gil-Chávez, J.; Sotelo-Mundo, R.R.; Namiesnik, J.; Gorinstein, S.; González-Aguilar, G.A. Antioxidant Interactions between Major Phenolic Compounds Found in "Ataulfo" Mango Pulp: Chlorogenic, Gallic, Protocatechuic and Vanillic Acids. *Molecules* **2012**, *17*, 12657–12664. [CrossRef]

32. Hajimehdipoor, H.; Shahrestani, R.; Shekarchi, M. Investigating the Synergistic Antioxidant Effects of Some Flavonoid and Phenolic Compounds. *Res. J. Pharmacogn.* **2014**, *1*, 35–40.

33. Olszowy, M.; Dawidowicz, A.L.; Jóźwik-Dolęba, M. Are Mutual Interactions between Antioxidants the Only Factors Responsible for Antagonistic Antioxidant Effect of Their Mixtures? Additive and Antagonistic Antioxidant Effects in Mixtures of Gallic, Ferulic and Caffeic Acids. *Eur. Food Res. Technol.* **2019**, *245*, 1473–1485. [CrossRef]

34. Yang, J.; Chen, J.; Hao, Y.; Liu, Y. Identification of the DPPH Radical Scavenging Reaction Adducts of Ferulic Acid and Sinapic Acid and Their Structure-Antioxidant Activity Relationship. *LWT* **2021**, *146*, 111411. [CrossRef]

35. Isaiah, O.O. The Synergistic Interaction of Phenolic Compounds in Pearl Millets with Respect to Antioxidant and Antimicrobial Properties. *Am. J. Food Sci. Health* **2020**, *6*, 80–88.

36. Skroza, D.; Generalić Mekinić, I.; Svilović, S.; Šimat, V.; Katalinić, V. Investigation of the Potential Synergistic Effect of Resveratrol with Other Phenolic Compounds: A Case of Binary Phenolic Mixtures. *J. Food Compos. Anal.* **2015**, *38*, 13–18. [CrossRef]

37. Čagalj, M.; Skroza, D.; Razola-Díaz, M.d.C.; Verardo, V.; Bassi, D.; Frleta, R.; Generalić Mekinić, I.; Tabanelli, G.; Šimat, V. Variations in the Composition, Antioxidant and Antimicrobial Activities of Cystoseira Compressa during Seasonal Growth. *Mar. Drugs* **2022**, *20*, 64. [CrossRef]

38. Hidalgo, M.; Sánchez-Moreno, C.; de Pascual-Teresa, S. Flavonoid-Flavonoid Interaction and Its Effect on Their Antioxidant Activity. *Food Chem.* **2010**, *121*, 691–696. [CrossRef]

39. Sonam, K.S.; Guleria, S. Synergistic Antioxidant Activity of Natural Products. *Ann. Pharmacol. Pharm.* **2017**, *2*, 1–6.

40. Mardani-Ghahfarokhi, A.; Farhoosh, R. Antioxidant Activity and Mechanism of Inhibitory Action of Gentisic and α-Resorcylic Acids. *Sci. Rep.* **2020**, *10*, 19487. [CrossRef]

41. Cuvelier, M.E.; Richard, H.; Berset, C. Comparison of the Antioxidative Activity of Some Acid-Phenols: Structure-Activity Relationship. *Biosci. Biotechnol. Biochem.* **1992**, *56*, 324–325. [CrossRef]

42. Collado, S.; Laca, A.; Diaz, M. Effect of the Carboxylic Substituent on the Reactivity of the Aromatic Ring during the Wet Oxidation of Phenolic Acids. *Chem. Eng. J.* **2011**, *166*, 940–946. [CrossRef]
43. Rice-Evans, C.; Miller, N.J.; Paganga, G. Structure-Antioxidant ACTIVITY Relationships of Flavonoids and Phenolic Acids. *Free Radic. Biol. Med.* **1995**, *20*, 933–956. [CrossRef]
44. Cao, H.; Cheng, W.X.; Li, C.; Pan, X.L.; Xie, X.G.; Li, T.H. DFT Study on the Antioxidant Activity of Rosmarinic Acid. *J. Mol. Struct. THEOCHEM* **2005**, *719*, 177–183. [CrossRef]
45. Piazzon, A.; Vrhovsek, U.; Masuero, D.; Mattivi, F.; Mandoj, F.; Nardini, M. Antioxidant Activity of Phenolic Acids and Their Metabolites: Synthesis and Antioxidant Properties of the Sulfate Derivatives of Ferulic and Caffeic Acids and of the Acyl Glucuronide of Ferulic Acid. *J. Agric. Food Chem.* **2012**, *60*, 12312–12323. [CrossRef] [PubMed]
46. Freeman, B.L.; Eggett, D.L.; Parker, T.L. Synergistic and Antagonistic Interactions of Phenolic Compounds Found in Navel Oranges. *J. Food Sci.* **2010**, *75*, 570–576. [CrossRef]

MDPI

St. Alban-Anlage 66

4052 Basel

Switzerland

www.mdpi.com

Antioxidants Editorial Office

E-mail: antioxidants@mdpi.com

www.mdpi.com/journal/antioxidants